# Environmental Mineralogy: Microbial Interactions, Anthropogenic Influences, Contaminated Land and Waste Management

Edited by

J. D. Cotter-Howells
*University of Aberdeen*

L. S. Campbell
*University of Salford*

E. Valsami-Jones
*Natural History Museum, London*

and

M. Batchelder
*Natural History Museum, London*

2000
Published by the Mineralogical Society of Great Britain & Ireland, London

## The Mineralogical Society

The Mineralogical Society of Great Britain & Ireland was instituted in 1876. The prime aim of the Society is to advance the knowledge not only of Mineralogy but also of Crystallography, Geochemistry and Petrology, together with kindred subjects. This is done principally through the publication of scientific journals, books and monographs, and through arranging or supporting scientific meetings. The Society speaks for Mineralogy in Great Britain, linking with British science in general through the Royal Society and cooperating closely with the Geological Society. It maintains liaison with European mineralogists as a member society of the European Mineralogical Union, and is the body that nominates British representatives to the International Mineralogical Association. There are some 900 members of the Society, of whom approximately 200 reside abroad.

The Society's contact details are:
Tel: +44 (0)20 7584 7516; Fax +44 (0)20 7823 8021;
E-mail: *info@minersoc.org*; *online: www.minersoc.org*

The Mineralogical Society Series, of which this book is number **9**, was originally published in conjunction with Chapman & Hall (London). Numbers 1–8 in the series are now marketed by Kluwer Academic Publishers at P.O. Box 322, 3300 AH Dordrecht, The Netherlands.

E-mail: *orderdept@wkap.nl*; Tel: +31 (0)78-6392392; Fax: +31 (0)78-6546474.

This book is available through Turpin Distribution Services Ltd, Blackhorse Road, Letchworth, Herts SG6 1HN, UK. Tel: +44 (0)1462 672555 A considerable discount is available to members of the Society.

First edition 2000.

Typeset in 10/12 pt Times by Westfield Typesetting, North Walsham, Norfolk, UK

Printed and bound by Alden Press Ltd, Oxford, UK

ISBN 0 903056 20 8



A catalogue record for this book is available from the British Library.

# Mineralogical Society Series

Series Editor
Dr P.J. Treloar

Numbers 1–8 in the Mineralogical Society Series were published in conjunction with Chapman & Hall. The Society has undertaken to publish this, the ninth volume in the series, independently.

## TITLES AVAILABLE

**1 Deformation Processes in Minerals, Ceramics and Rocks**
Edited by D.J. Barber and P.G. Meredith

**2 High-temperature Metamorphism and Crustal Anatexis**
Edited by J.R. Ashworth and M. Brown

**3 The Stability of Minerals**
Edited by G.D. Price and N.L. Ross

**4 Geochemistry of Cay-Pore Fluid Interactions**
Edited by D.A.C. Manning, P.L. Hall and C.R. Hughes

**5 Mineral Surfaces**
Edited by D.J. Vaughan, and R.A.D. Pattrick

**6 Microprobe Techniques in the Earth Sciences**
Edited by P.J. Potts, J.F.W. Bowles, S.J.B. Reed and M.R. Cave

**7 Rare Earth Minerals: Chemistry, Origin and Ore Deposits**
Edited by A.P. Jones, F. Wall and C.T. Williams

**8 Deformation-enhanced Fluid Transport in the Earth's Crust and Mantle**
Edited by M.B. Holness

# Preface

The past 10 years or so have seen the emergence of a discipline known as 'Environmental Mineralogy'. This should be regarded not as a new discipline *per se*, but as a new application of traditional mineralogy. Mineralogists have always sought to understand the chemical and physical environment under which a particular mineral forms and to determine the arrangement of atoms within that mineral. The field of Environmental Mineralogy asks the same questions in a different context. For example, can minerals assist in the remediation of contaminated soils and waters? Which minerals can potentially be deleterious to, *inter alia*, buildings, ecology and human health? Which minerals are suitable as containment for waste? How does the biota interact with minerals? Environmental Mineralogy is emerging as a field that seeks to define the roles of minerals in all environmental systems, and to work towards the preservation and restoration of such systems.

Environmental Mineralogy is achieving prominence because of increasing concern regarding the environments in which we live. Mineralogists have perceived a gap in our understanding of how minerals behave in the surface environment and a need for innovative, 'green' solutions to the problems of contamination and waste. However, the emergence of Environmental Mineralogy also owes much to modern analytical technology. Many minerals in the surface environment fall within the clay-grade range and therefore, demand high-resolution systems for analysis. Similarly, trace elements are now detectable at exceptionally low concentrations in a wide variety of matrices. Further, many mineral-environment interactions need to be examined at the atomic scale for a greater understanding of the interactive processes involved. This requires the application of the latest technologies such as X-ray photoelectron spectroscopy, X-ray absorption spectroscopy and atomic force microscopy to name but a few.

The aim of this monograph is to provide an up-to-date account of the state of this diverse subject area. With chapters containing a strong review element, it is hoped that this volume will appeal to both researchers and students alike. The volume is arranged in four sections: (1) mineral-microbe interactions; (2) anthropogenic influences on mineral interactions; (3) minerals in contaminated environments; and (4) minerals and waste management. These four sections by no means give exhaustive coverage of the subject area, but communicate some of the most important developments taking place at the present time.

The editors are indebted to the referees who all put in a considerable amount of time and effort to produce thoughtful, critical reviews of the chapters. International peer review, though very time-consuming and unrewarding for, often very busy, referees, continues to underpin scientific quality assurance for any research publication of this nature.

The basis of the volume is the group of papers which were originally presented at the Mineralogical Society's Winter Meeting 'Environmental Mineralogy', held in Aberdeen in January 1999. The meeting and hence this volume benefited from donations by English China Clays International, British Nuclear Fuels Ltd, WBB Technology Ltd and Rio Tinto plc. The editors acknowledge the contributions of the Geochemistry Group and of the Mineralogical Society officers and staff who assisted with the organisation of the meeting. Finally, the editors wish to express their gratitude to Kevin Murphy, the Mineralogical Society's Production Editor. Without Kevin's constant encouragement throughout the project, we might have given up at such a daunting task! We are grateful for all his hard work.

J.D. Cotter-Howells,
L.S. Campbell,
E. Valsami-Jones,
M. Batchelder

# Contents


*Preface* v

1 Section 1: Mineral-microbe interactions *J. D. Cotter-Howells* 1

1.1 Introduction 1
1.2 Biological processes and minerals 1
1.3 Future directions 3
1.4 The papers 3
References 4

2 Illustrations of the occurrence and diversity of mineral-microbe interactions involved in weathering of minerals *J. Berthelin, C. Leyval and C. Mustin* 7

2.1 Introduction 8
2.2 Main microbial dissolution and precipitation processes of inorganic elements 10
2.3 Weathering and transformation of a mica under the influence of roots and of their associated microorganisms in the plant rhizosphere 13
2.4 Interactions between *Thiobacillus ferrooxidans* and sulphide minerals (pyrite) during oxidation and dissolution (bioleaching) processes 16
2.5 Influence of Al and Mn substitution on the dissolution of ferric oxides (goethite) by Fe-reducing bacteria 20
2.6 Conclusions 21
References 23

3 Mineral dissolution by heterotrophic bacteria: principles and methodologies *E. Valsami-Jones and S. McEldowney* 27

3.1 Introduction 27
3.2 The principles of mineral dissolution 28
3.3 Bacteria as dissolution agents 29
3.4 Methods of studying heterotrophic bacteria 37
3.5 Field evidence of bacterial influence on mineral dissolution 44
3.6 Laboratory studies: mechanisms 45
3.7 Conclusions and future work 49
Acknowledgements 52
References 52

4 Heterotrophic solubilization of metal-bearing minerals by fungi *G. M. Gadd* 57

4.1 Introduction 57
4.2 Organic acid biosynthesis 58
4.3 Metal chemistry of citric and oxalic acid 59
4.4 Fungal organic acids and metal biogeochemistry 64
4.5 Fungal organic acid production and metal biotechnology 68
4.6 Conclusions 69
Acknowledgements 70
References 70

5 Weathering of rocks by lichens: fragmentation, dissolution and precipitation of minerals in a microbial microcosm *M. R. Lee* 77

5.1 Introduction 77
5.2 Background 79
5.3 Abiotic weathering and erosion of rock surfaces 83
5.4 A model of the lichen-rock interface 84
5.5 Photosynthetic zone 85
5.6 Direct biochemilithic zone 86
5.7 Indirect biochemilithic zone 90
5.8 Physicochemilitic zone 96
5.9 Rates of lichen-mediated weathering 96
5.10 Summary: the impact of lichen colonization on weathering rates 103
Acknowledgements 103
References 104

6 Section 2: Anthropogenic influences on mineral interactions *L. S. Campbell* 109

6.1 Introduction 109
6.2 Anthropogenic influences 110
6.3 Influences associated with the biosphere 112
6.4 Conclusion 114
References 114

7 Mechanisms and rates of sulphide oxidation in relation to the problems of acid rock (mine) drainage *C. N. Keith and D. J. Vaughan* 117

7.1 Introduction 117
7.2 Acid rock drainage: background to the problem 118
7.3 Sulphide minerals: chemistry and reactivity, oxidation mechanisms, rates and controls 123
7.4 Concluding remarks 135
Acknowledgements 136
References 136

8 The relationship of mineralogy to acid- and neutralization-potential values in ARD *J. L. Jambor* 141

8.1 Introduction 141
8.2 Static and kinetic tests 142
8.3 Acid-producing potential 143
8.4 Sulphide resistance and persistence 144
8.5 Neutralization potential 145
8.6 Weathering rates of minerals 146
8.7 Relationships to ARD 150
8.8 Conclusions 154
Acknowledgements 155
References 156

9 Mynydd Parys Cu-Pb-Zn Mines: mineralogy, microbiology and acid mine drainage *D. A. Jenkins, D. B. Johnson and C. Freeman* 161

9.1 Introduction 161
9.2 Geology and mining history 163
9.3 Surface mineralogy and geochemistry 165
9.4 Biology 170
9.5 Acid mine drainage 174
9.6 Conclusions 177
Acknowledgements 177
References 177

10 Decay effects associated with soluble salts on granite buildings of Braga (NW Portugal) *C. A. S. Alves and M. A. Sequeira Braga* 181

10.1 Introduction 181
10.2 Salt weathering on buildings: a review 182
10.3 Granite buildings of Braga: case studies 185
10.4 Characterization of soluble salts 186
10.5 Decay effects of soluble salts in granite buildings of Braga: a discussion 194
10.6 Conclusions 197
Acknowledgements 198
References 198

11 Section 3: Minerals in contaminated environments *E. Valsami-Jones* 201

11.1 Introduction 201
11.2 Mineral-pollutant interactions 201
11.3 Environmental minerals 202
11.4 Environmental Mineralogy and contaminated environments 203
References 204

12 Heavy metal-bearing Mn oxides in river channel and floodplain sediments *K. A. Hudson-Edwards* 207

12.1 Introduction 207
12.2 Mineralogy and heavy metal affinity of Mn oxides 208
12.3 Mn oxides in river sediments 209
12.4 Mn oxides in river channel and floodplain sediment in northeast England 211
12.5 Conclusions and environmental significance 221
Acknowledgements 222
References 222

13 Intercalation of organic and inorganic contaminants by expanding layer silicates *W. E. Dubbin* 227

13.1 Introduction 227
13.2 Structure and properties of expandable layer silicates 228
13.3 Intercalation of inorganic contaminants 230
13.4 Intercalation of organic contaminants 239
13.5 Summary 241
References 242

14 Uranium behaviour in natural environments *K. V. Ragnarsdottir and L. Charlet* 245

14.1 Introduction 245
14.2 Physical and chemical properties of uranium 246
14.3 Aqueous uranium chemistry 250
14.4 Transport of uranium 257
14.5 Sinks of uranium 266
14.6 Uranium mines and tailings 269
14.7 Conclusion – the uranium cycle 277
Acknowledgements 279
References 279

15 Metal phosphates and remediation of contaminated land *M. E. Hodson, E. Valsami-Jones and J. D. Cotter-Howells* 291

15.1 Introduction 291
15.2 Metal phosphates in the natural environment 292
15.3 Metal phosphate solubility 294
15.4 Metal phosphate formation in experimental systems using aqueous metal ions 298
15.5 Metal phosphate formation in batch experiments using metal-bearing solids 302
15.6 Metal phosphate formation in soil column and field experiments 306
15.7 Concluding remarks 308
Acknowledgements 308
References 308

16 Section 4: Minerals and waste management *L. S. Campbell* 313

16.1 Introduction 313
16.2 Interactive characteristics of minerals 313
16.3 Zeolites 314
16.4 Clay minerals 314
16.5 Carbonates 315
16.6 Other mineral groups 315
16.7 Conclusions 316
Acknowledgements 316
References 317

17 Applications of natural zeolites in the treatment of nuclear wastes and fall-out *A. Dyer* 319

17.1 Introduction 319
17.2 Historical perspective 321
17.3 Zeolites as specific ion exchangers for radioisotopes 322
17.4 Scavenging of radioisotopes from aqueous solution 327
17.5 Treatment of radioactive wastes from nuclear facilities 338
17.6 Clean-up of radioactive waste from nuclear accidents and fall-out 343
17.7 Waste containment uses of natural zeolites 348
17.8 Treatment of gaseous emissions from nuclear facilities 352
17.9 Summary and recommendations for future research 352
Acknowledgements 353
References 353

18 Gas entry into unconfined clay pastes at water contents between the liquid and plastic limits *A. T. Donohew, S. T. Horseman and J. F. Harrington* 369

18.1 Introduction 369
18.2 Gas entry and soil suction 374
18.3 Experiments 377
18.4 Results 380
18.5 Mechanisms 384
18.6 Discussion 386
18.7 Conclusions 389
Acknowledgements 391
References 391

19 Geosynthetic Clay Liners (GCLs) for municipal solid waste landfills *R. K. Rowe and C. B. Lake* 395

19.1 Introduction 395
19.2 Hydraulic conductivity and leachate compatibility 397
19.3 Diffusion research 399

19.4 GCL interaction with other components of the landfill system 401
19.5 Conclusion 404
References 404

INDEX 407

CHAPTER ONE

# Section 1: Mineral-microbe interactions

J. D. COTTER-HOWELLS*

*Department of Plant and Soil Science, University of Aberdeen, St. Machar Drive, Aberdeen AB24 3UU, UK*

## 1.1 Introduction

Mineral-microbe interaction is a broad, rather poorly understood field of science. Some aspects of mineral-microbe interactions are relatively well understood, e.g. the oxidation of pyrite and the industrial bioleaching of metals. However, other aspects of mineral-microbe interactions are less well understood, e.g. the significance of microbial weathering of silicate minerals. The field encompasses many interactions in addition to those considered in this short collection of papers. For example, the field encompasses bacterial control on biologically-induced mineralization of magnetic iron minerals (Bazylinski and Moskowitz, 1997), effects of clay colloids as a physical substrate for growth and adhesion of microbes (Stotzky, 1986) and the orientation of clay platelets surrounding bacteria. The papers presented here are largely restricted to one area of this broad field, that of mineral transformations mediated by microbes. This area is where much of the current research effort is being directed and advances in our understanding of mineral-microbe interactions are being made.

## 1.2 Biological processes and minerals

We live on (or, more properly, are a part of) a biologically-mediated planet. It is now well known that the actions of biological organisms are largely responsible for the precise balance of gases in the Earth's atmosphere that makes life possible. However, there is a tendency for geologists and mineralogists to think within the paradigm of physical and chemical factors

---

* Present address: Greenpeace Research Laboratories, Department of Biological Sciences, University of Exeter, Prince of Wales Road, Exeter EX4 4PS, UK
(E-mail: j.cotter-howells@exeter.ac.uk)

Cotter-Howells, J.D. (2000) Section 1: Mineral-microbe interactions. Pp. 1–5 in: *Environmental Mineralogy: Microbial Interactions, Anthropogenic Influences, Contaminated Land and Waste Management* (J.D. Cotter-Howells, L.S. Campbell, E. Valsami-Jones and M. Batchelder, editors). Mineralogical Society Series, **9**. Mineralogical Society, London. ISBN 0 903056 20 8.

only and to neglect biological factors. Whilst this may be justified for processes at very elevated *T* and/or *P* it is not likely to be justified for processes at surface or near-surface *PT* coditions.

When describing the weathering of igneous minerals to form either sedimentary rocks or soil, many textbooks list chemical and physical processes as primary factors and biological processes as an additional, relatively minor factor. This is not because biological processes are unimportant but because of the scale at which the majority of biological-mineralogical interactions occur. Much of the interaction between silicate rock-forming minerals and biological organisms is conducted by microorganisms, or microbes, such as bacteria and fungi. The scale of this interaction is microscopic (0.5–5 μm) and cannot be seen with the naked eye and this has hindered our observations of mineral-microbe interactions. Whilst there is an implicit understanding that microbes are an essential part of ecological cycling, e.g. by degrading large organic molecules, the same is not true for mineral weathering. The effects of mineral weathering are observed but attributed largely to chemical and physical processes. It is not known to what extent the action of microorganisms influences weathering.

### *1.2.1 Rock-forming minerals in ecosystems*

We are only just beginning to appreciate the enormous extent of the habitats for bacteria and fungi, from oil reservoirs beneath Earth's surface to clouds in the atmosphere. Bacteria have existed for at least 2,000–3,000 Ma, and fungi for 1,000–1,500 Ma. Fossilized plant roots in the Rhynie chert demonstrate that there was symbiotic interaction of fungi (mycorrhizas) and land plants by 400 Ma. Therefore, microbes have been evolving in juxtaposition with minerals for hundreds, or even thousands, of millions of years.

Rock-forming minerals are important in ecosystems as they are the primary pools of inorganic (sometimes called 'mineral') nutrients such as K, Ca, Mg, Fe and P as well as essential trace elements, e.g. Cu, Zn. The availability of nutrients (including C, N, O) and energy is important to all organisms and the same is true for microorganisms. Therefore, the ability to extract the inorganic nutrients directly from minerals would be an evolutionary advantage. Given the long period of microbe existence, it is likely that many species have developed efficient nutrient extraction techniques over the millennia. This is especially true of bacteria, which can evolve at very fast rates compared to more complex biological organisms, e.g. strains of bacteria can become resistant to an antibiotic within decades. Indeed, we now know that plant roots, bacteria and fungi exude organic acids which directly solubilize minerals and many of these acids then chelate the ion of interest, e.g. citric acid to form an Fe-citrate complex (Marschner, 1995; Gadd, 2000).

Once a nutrient ion has passed from a mineral to a microorganism, it has passed into the food chain. Microbes are the lowest trophic level in any ecosystem. They are predated upon (e.g. bacteria by amoeba) and form symbiotic relationships with other organisms, including higher organisms. For example, fungi and bacteria symbiotically associated with plant roots pass these inorganic nutrients to their host, possibly in exchange for carbon compounds (Marschner, 1995). The plant may then use that nutrient to form a leaf, which in turn may be eaten by a grazing animal. Thus, the actions of microbes could be instrumental in the biogeochemical cycling of inorganic nutrients, playing a pivotal role in transporting nutrients such as Fe, Mg etc. from the regosphere into the biosphere.

Lee (2000) considers the effects of lichens (a lichen is a symbiotic association of an alga with a fungus) on rock weathering. He discusses the idea that increased weathering of Mg- and Ca-rich silicate minerals due to colonization of rock surfaces by lichens during the late Precambrian and lower Palaeozoic would have increased carbonate precipitation, leading to enhanced removal of $CO_2$ from the atmosphere. Although only an idea, it demonstrates the possible importance of microbial interaction with minerals on a global scale.

## 1.3 Future directions

The ecological importance of the role of microbes in providing nutrients from minerals to biogeochemical cycles, both local and global is not known. The question then arises as to how we can quantify the effects that these organisms have on nutrient extraction from minerals? How important are these processes in the weathering rates of mineral material? Is it only in weathering that these organisms are important? Could microbes be important in the formation of neoformed clay minerals in sediments? Could microorganisms play a role in the formation of oil from the degradation of plant remains? It is estimated that only 5% of the total number of species of fungi and only 0.4% of bacteria have so far been identified (WRI, 1998). With the advent of DNA probes, and as this technology develops, more microbes will be identified and their roles elucidated. Therefore, the scientific field of microbe-mineral interactions is only just beginning. There are already two good books detailing microbe-mineral interactions, one by Huang and Schnitzer (1986) and the other by Banfield and Nealson (1997). I hope this short collection of papers provides the reader with background information and presents the state of our current knowledge on this exciting topic.

## 1.4 The papers

The session on mineral and microbe interactions at the Mineralogical Society's Winter Meeting in Aberdeen (January, 1999) was designed to introduce mineralogists to the types of interactions that occur between microbes and minerals. The emphasis of the session was on microbial action as mineral weathering/dissolution.

In their paper, **Berthelin, Leyval and Mustin (chapter 2)** introduce microorganisms and describe microbial dissolution of minerals (including pyrite) and the bioprecipitation of mineral deposits. They then describe an interesting experiment in which microorganisms associated with the roots of a pine tree were shown to increase the cation exchange capacity of phlogopite mica in the vicinity of the roots as a result of transformation of the mica to vermiculite. Finally, the role of solid substitution in controlling bacterial reduction of geothite is discussed.

**Valsami-Jones and McEldowney (chapter 3)** introduce bacteria in detail, including environmental controls on their distribution, and the fundamentals of their physiology. The chapter then describes the methods of culturing and imaging bacteria. Finally, laboratory and field experiments to observe effects of bacterial dissolution are discussed.

**Gadd (chapter 4)** introduces fungi and their synthesis of organic acids. The chemistry of these acids is discussed with respect to metal complex formation. The effects of these acid exudates on metal biogeochemistry are then described in detail with particular emphasis on the solubilization of non-silicate minerals containing toxic metals. The chapter concludes with a section on the biotechnology of fungal leaching of metals from waste and ore heaps.

**Lee (chapter 5)** provides a detailed discussion on the influence of lichens on the weathering rates of silicate minerals. The biology of lichens and techniques for observing the effects of lichen encrustation are described. The potential mineral interactions with both the photosynthetic zone and the biochemilithic zone are discussed and typical minerals resulting from lichen interaction with rocks described. Finally, there is a stimulating discussion as to whether lichen encrustation actually does affect the weathering rates of minerals contained in the rock.

## References

Banfield, J.F. and Nealson, K.H. (editors) (1997) *Geomicrobiology: Interactions between Microbes and Minerals*. Reviews in Mineralogy, **35**. Mineralogical Society of America, Washington D.C.

Bazylinski, D.A. and Moskowitz, B.M. (1997) Microbial biomineralisation of magnetic iron minerals: microbiology, magnetism and environmental significance. Pp. 181–224: *Geomicrobiology: Interactions between Microbes and Minerals* (J.F. Banfield and K.H. Nealson, editors). Reviews in Mineralogy, **35**. Mineralogical Society of America, Washington D.C.

Berthelin, J., Leyval, C. and Mustin, C. (2000) Illustrations of the occurrence and diversity of mineral–microbe interactions involved in weathering of minerals. Pp. 7–25 in: *Environmental Mineralogy: Microbial Interactions, Anthropogenic Influences, Contaminated Land and Waste Management* (J.D. Cotter-Howells, L.S. Campbell, E. Valsami-Jones and M. Batchelder, editors). Mineralogical Society Series, **9**. Mineralogical Society, London.

Gadd, G. (2000) Heterotrophic solubilization of metal-bearing minerals by fungi. Pp. 57–75 in: *Environmental Mineralogy: Microbial Interactions, Anthropogenic Influences, Contaminated Land and Waste Management* (J.D. Cotter-Howells, L.S. Campbell, E. Valsami-Jones and M.

Batchelder, editors). Mineralogical Society Series, **9**. Mineralogical Society, London.

Huang, P.M. and Schnitzer, M. (editors) (1986) *Interactions of Soil Minerals with Natural Organics and Microbes*. SSSA Special Publ., **17**. Soil Science Society of America, Madison, WI.

Lee, M.R. (2000) Weathering of rocks by lichens: fragmentation, dissolution and precipitation of minerals in a microbial microcosm. Pp. 77–107 in: *Environmental Mineralogy: Microbial Interactions, Anthropogenic Influences, Contaminated Land and Waste Management* (J.D. Cotter-Howells, L.S. Campbell, E. Valsami-Jones and M. Batchelder, editors). Mineralogical Society Series, **9**. Mineralogical Society, London.

Marschner, H. (1995) *Mineral Nutrition of Higher Plants* (2nd edition). Academic Press, London.

Stotzky, G. (1986) Influence of soil minerals colloids on metabolic processes, growth, adhesion, and ecology of microbes and viruses. Pp. 305–328 in: *Interactions of Soil Minerals with Natural Organics and Microbes* (P.M. Huang and M. Schnitzer, editors). SSSA Special Publ., **17**. Soil Science Society of America, Madison, WI.

Valsami-Jones, E. and McEldowney, S. (2000) Mineral dissolution by heterotrophic bacteria: principles and methodologies. Pp. 27–55 in: *Environmental Mineralogy: Microbial Interactions, Anthropogenic Influences, Contaminated Land and Waste Management* (J.D. Cotter-Howells, L.S. Campbell, E. Valsami-Jones and M. Batchelder, editors). Mineralogical Society Series, **9**. Mineralogical Society, London.

WRI World Resources Institute (1998) *World Resources 1998–1999: a Guide to the Global Environment*. Oxford University Press, Oxford.

Microbial weathering of mica (drawing courtesy of G.M. Gadd).

CHAPTER TWO

# Illustrations of the occurrence and diversity of mineral-microbe interactions involved in weathering of minerals

J. BERTHELIN, C. LEYVAL AND C. MUSTIN

*Centre de Pédologie Biologique, UPR 6831 du CNRS associée à l'Université Henri-Poincaré Nancy I, B.P. 5, 54501 Vandœuvre-lès-Nancy Cedex, France (E-mail: bertelin@cpb.cnrs-nancy.fr)*

## ABSTRACT

Microorganisms are widespread in all natural environments where, in order to generate energy to form new cell structures, they oxidize and reduce organic and inorganic materials, and form gaseous, liquid and solid metabolic compounds which they excrete into their environment.

The energetic and chemical activities of microorganisms are involved in the solubilization or fixing or precipitation of inorganic elements, in the weathering of minerals (silicates, phosphates, carbonates, sulphides, oxides) and in the formation of mineral deposits.

Both autotrophic (chemolithotrophic) and heterotrophic (chemo-organotrophic) microorganisms are involved and participate directly (mainly by oxidation-reduction processes) or indirectly (by metabolic products) in the weathering, transformation and evolution of minerals in soils and sediments.

Some examples are provided by: (1) the weathering of layer silicates in the rhizosphere of plants (root environment) as influenced by microorganisms (bacteria and mycorrhizal fungi) associated with the roots; (2) the autotrophic bacteria (*Thiobacilli*) which solubilize sulphides by oxidation of Fe and S that depend on the contact between bacteria and mineral and of mineral surface properties and electrochemical parameters; and (3) heterotrophic bacteria (*Bacilli*, *Clostridia*, etc.) which are able, using the available soil organic matter as source of C and energy, to dissolve ferric oxides (hematite, goethite) by reduction of insoluble ferric iron in soluble ferrous iron with rates depending on mineral element substitution in the oxide structure.

Such examples illustrate the importance, the interest and the diversity of 'microorganism–mineral interactions' and allow us to underline different incidences, applications and perspectives of research and development.

Berthelin, J., Leyval, C. and Mustin, C. (2000) Illustrations of the occurrence and diversity of mineral-microbe interactions involved in weathering of minerals. Pp. 7–25 in: *Environmental Mineralogy: Microbial Interactions, Anthropogenic Influences, Contaminated Land and Waste Management* (J.D. Cotter-Howells, L.S. Campbell, E. Valsami-Jones and M. Batchelder, editors). Mineralogical Society Series, **9**. Mineralogical Society, London., ISBN 0 903056 20 8.

## 2.1 Introduction

Microorganisms or protists (lower and higher protists) are characterized by the simple level of their biological organization. The lower protists, which are procaryotic organisms, include all the types of bacteria (eubacteria, filamentous and coryneform bacteria, cyanobacteria). They are the oldest living organisms on Earth and have, since at least 3500 million years ago developed a large diversity of energetic and nutritional strategies (Fig. 2.1). The higher protists including fungi, protozoa and algae are more recent (900 to 1400 million years) and, as eucaryotic organisms, have cell structures similar to those of plants and animals. They present a smaller energetic and nutritional diversity than the bacteria (lower protists). Both, and in particular bacteria, are widespread in all the natural earth environments (soils, sediments, fresh waters, oceans, etc.) where they contribute to the functioning of the biogeochemical cycles of major and trace elements (e.g. C, N, S, Fe, P, Hg, etc.) (Dommergues and Mangenot, 1970; Alexander, 1977; Berthelin, 1988; Ehrlich, 1996; Madigan *et al.*, 1996).

In fact, for their growth and reproduction, microorganisms (protists) can use, depending on their nutritional and energy requirements, inorganic or organic electron donors, inorganic or organic C sources and electromagnetic (photons) or chemical energy. So, it is possible to distinguish, on one hand photolithotrophic and photoorganotrophic microorganisms using photons as a source of energy, but inorganic or organic compounds as electron donors and C sources respectively. Conversely, using chemical energy, two main groups of

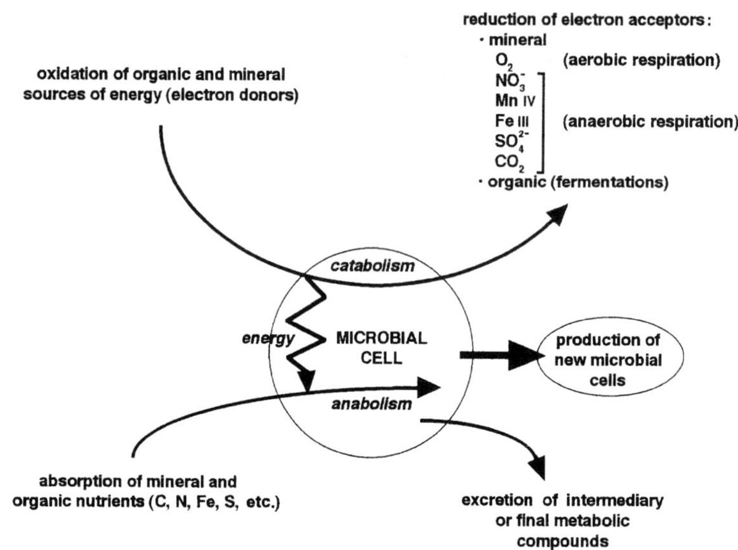


FIG. 2.1 Simplified diagram of microbial activities (from Berthelin, 1988).

microorganisms, chemolithotrophs and chemoorganotrophs, can be distinguished. Chemolithotrophic or autotrophic bacteria utilize the oxidation of inorganic components ($Fe^{2+}$, $S^{2-}$, $S^0$, $NH_4^+$, $Mn^{2+}$, etc.) to produce energy and $CO_2$ as the C source. The chemo-organotrophic or heterotrophic microorganisms (bacteria and fungi) use organic compounds as a source of C and energy and are the most numerous and largest group of microorganisms in the terrestrial ecosystems.

These two groups, chemolithotrophs and chemo-organotrophs, represent the majority of soil and sediment microorganisms that can also use different electron acceptors for their respiration purposes: inorganic, i.e. $O_2$ (in aerobic respiration), $NO_3^-$, $Mn^{4+}$, $Fe^{3+}$, $SO_4^{2-}$, $CO_2$ (in anaerobic respiration), or different organic compounds (in fermentation processes). They are also the largest microbial communities in soils (e.g. $10^6$ to $10^9$ bacteria $g^{-1}$ dry soil) and are responsible for the main biochemical and biogeochemical processes (degradation of the different organic compounds, sulphur oxidation, sulphate reduction, nitrification, denitrification, iron oxidation, phosphate dissolution, iron reduction, etc.) (e.g. Figs 2.1, 2.2).

During such metabolic processes, they also form gaseous, liquid, solid, alkaline, acid, complexing and reducing compounds that are excreted from the cell as metabolic end products or intermediary metabolites.

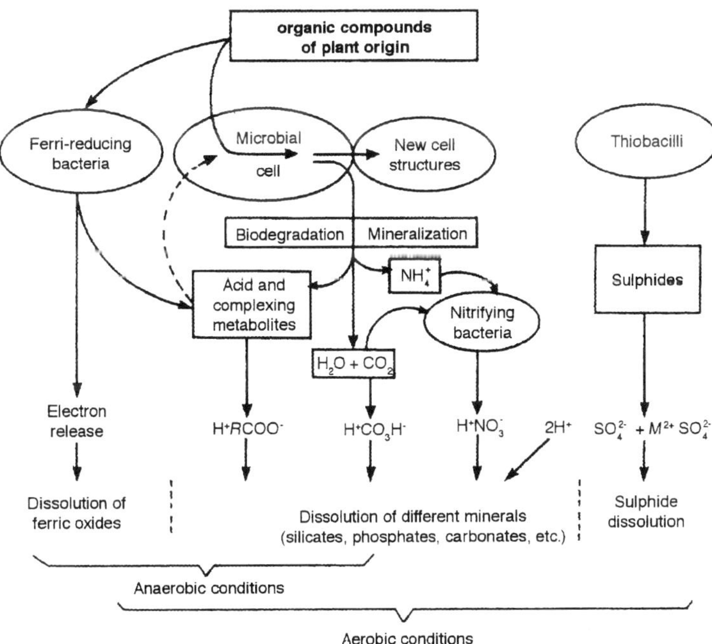


FIG. 2.2 The involvement of soil microorganisms in dissolution and weathering of minerals (from Berthelin, 1998).

Such rich diversity in energetic, nutritional and environmental strategies allow the different microbial (bacterial and fungal) communities to be directly or indirectly involved in the transformation, degradation, evolution and formation of mineral and organo-mineral compounds in their environment (Berthelin, 1983, 1988, 1998; Robert and Berthelin, 1986; McKeague *et al.*, 1986; Berthelin *et al.*, 1994; Ehrlich, 1996; Madigan *et al.*, 1996).

This chapter is not an exhaustive survey of the occurrence of microbial weathering processes but will provide some illustrations of these phenomena. After a general overview of the main microbial processes involved in dissolution and precipitation of elements (particularly metals), it will provide three examples from recent results to illustrate: (1) the weathering and transformation of a mica (phlogopite) under the influence of roots and of their associated heterotrophic microorganisms in the rhizosphere of plants; (2) the interactions between sulphide mineral (pyrite) surface properties with an autotrophic bacterium (*Thiobacillus*) during the oxidation of Fe and S and the consequent bioleaching of sulphide minerals; and (3) the effect of element substitution on iron oxide (goethite) dissolution and weathering by Fe-reducing bacteria.

## 2.2 Main microbial dissolution and precipitation processes of inorganic elements

Different mechanisms of microbial origin are involved in the dissolution and precipitation of elements (e.g. Berthelin, 1988; Berthelin *et al.*, 1994; Ehrlich, 1996). Some of them occur in the weathering and in the formation of minerals (Berthelin, 1983, 1988; Berthelin *et al.*, 1994; Ehrlich, 1996). They are presented below.

### *2.2.1 Microbial processes of dissolution and weathering of minerals*

Three main types of mechanisms can be distinguished: (1) dissolution and release of metal ions by acid and complexing compounds produced by heterotrophic and autotrophic microorganisms (fungi and bacteria); (2) oxidation and reduction of mineral elements (mainly Fe, Mn, S) promoting dissolution of sulphide minerals by oxidation or of oxyhydroxides by reduction; and (3) change in pH and Eh by uptake or release of different compounds or ions (e.g. $NH_4^+$, $Na^+$, etc.).

The first and second types of processes have been studied in greater depth than the third, due to their more widespread occurrence.

During degradation of organic matter, heterotrophic microorganisms (bacteria and fungi) can produce, according to the simplified equation 2.1, carbonic, aliphatic or aromatic acids which can be involved in the solubilization of cations and the breakdown of rocks and minerals according to the simplified equation 2.2:

$$\text{Organic matter } (CH_2O)_n \begin{cases} \rightarrow CO_2 \rightarrow H_2CO_3 \\ \rightarrow \text{organic acids } (RCOOH) \end{cases} \tag{2.1}$$

$$M^+ \text{ (Mineral)}^- + (H^+R^-) \rightarrow H^+ \text{ (Mineral)}^- + M^+R^- \tag{2.2}$$

where $M^+$ (Mineral)$^-$ = silicates, phosphates, carbonates, etc. from rocks; $M^+$ = metal ions and $R^-$ = $NO_3^-$, $SO_4^{2-}$, $HCO_3^-$, $RCOO^-$, etc.

Mineral acids ($HNO_3$, $H_2SO_4$) are produced by autotrophic microorganisms such as nitrifying bacteria and thiobacilli.

Different organic compounds such as oxalic and citric acids are much more efficient in the solubilization of minerals by formation of soluble and stable organo-metallic complexes (Robert and Berthelin, 1986; Berthelin, 1988; Berthelin *et al.*, 1994; McKeague *et al.*, 1986). The complexation dissolution processes are summarized by equations 2.3 and 2.4:

$$M^+\text{(Mineral)}^- + HL \rightarrow H^+ \text{ (Mineral)}^- + ML \tag{2.3}$$

$$HL + LM \rightarrow L_2M + H^+ \tag{2.4}$$

where $L$ = organic ligands and $M^+$ (Mineral)$^-$ are silicates, carbonates, phosphates, etc.

Different types of complexes can be formed depending on the nature of the ligands and of the metals, etc. Citric acid and ferric ions can form the $M_2(OH)_2L_2$ complex where $M$ = $Fe^{3+}$ and $L$ = citrate but an $ML$ complex can form at a lower pH (<4); the $M_2(OH)_2L_2$ complex exists in the largest range of pH (from acid conditions (pH 2.0) to elevated alkaline conditions) (Berthelin, 1988; Munier-Lamy *et al.*, 1991).

Some organic ligands, such as siderophores, produced by bacteria and fungi are very efficient and can have a specific affinity to metal ion complexation and transport (Berthelin, 1988; Berthelin *et al.*, 1994). Compounds such as citrate and oxalate, even if they have different affinity for different metal ions, can be considered as non-specific chelating agents. Siderophores (i.e. Fe chelating substances) on the other hand, such as trihydroxamate or tridiorthophenolate (Neilands, 1974; Reid *et al.*, 1984), are specific agents of ferric Fe chelation and transport. It has been shown that they dissolve ferric oxides (e.g. goethite) more specifically and much more efficiently than, for instance, dicarboxylic or tricarboxylic acids (Watteau and Berthelin, 1994).

Some bacteria are able to develop dissolution processes of elements which are oxidation-reduction reactions where the elements (S, Fe, etc.) are used as electron donors (energy sources) or as electron acceptors (in anaerobic respiration, or simultaneously in fermentation processes).

Thiobacilli (e.g. *Thiobacillus ferrooxidans* and *Thiobacillus thiooxidans*) that are acidophilic, chemolithotrophic bacteria, oxidize reduced S compounds (sulphide, elemental sulphur, thiosulphate, etc.). Hence, these bacteria are directly involved in the dissolution of sulphide minerals (pyrite, chalcopyrite, arsenopyrite, etc.) to form sulphates according to the simplified equation 2.5:

$$Fe(II)S_2 + 7/2O_2 + H_2O \rightarrow Fe(II)SO_4 + H_2SO_4 \quad (2.5)$$

Bacteria such as *T. ferrooxidans* can also oxidize Fe (II) to Fe (III). Thus the bacteria form a strong oxidizer, ferric sulphate (equation 2.6).

$$2Fe(II)SO_4 + H_2SO_4 + 1/2O_2 \rightarrow Fe_2(III)(SO_4)_3 + H_2O \quad (2.6)$$

Therefore, an indirect chemical oxidation can occur to dissolve sulphide and other minerals (e.g. uranium oxide) as a result of bacterial oxidation and sulphuric acid production (equations 2.7 and 2.8):

$$Fe(II)S_2 + Fe_2(III)(SO_4)_3 \rightarrow 3Fe(II)SO_4 + 2S \quad (2.7)$$

$$U(IV)O_2(\text{insoluble}) + 2Fe^{3+} + SO_4^{2-} \rightarrow U(VI)O_2SO_4(\text{soluble}) + 2Fe^{2+} \quad (2.8)$$

Conversely, in anoxic environments, aero-anaerobic or anaerobic chemo-organotrophic bacteria are able to reduce Mn or Fe (Mn(IV) or Mn(III) to Mn(II) and iron (Fe(III) to Fe(II)) that are, in such environmental conditions, electron acceptors either for anaerobic respiration or simultaneously for fermentation processes. The reactions occur according to the simplified equations 2.9 and 2.10:

$$\text{Organic source of C and energy (fermentable or not)} \rightarrow ne^- + H^+ \pm \text{fermentation products:} \quad (2.9)$$

$$Fe(OH)_3(s) + 3H^+(aq) + e^- \rightarrow Fe^{2+}(aq) + 3H_2O(aq) \quad (2.10)$$

(s) and (aq) represent solid and aqueous physical states, respectively.

Such reactions are significant in the dissolution of ferric oxides.

Figure 2.2 gives a representation of these different microbial weathering processes occurring in soils.

### *2.2.2 Microbial processes of bio-precipitation of elements and formation of mineral deposits*

Separate from or concurrently with solubilization processes, heterotrophic and autotrophic microorganisms (essentially bacteria and fungi) are involved in precipitation and accumulation processes. They are related mainly either to organo-mineral associations or to oxidation and reduction mechanisms. These processes are presented below in simplified forms. More details of the reactions are available in Berthelin (1988), Berthelin *et al.* (1994) and Ehrlich (1996).

The processes of precipitation involving microbial oxidation or reduction mechanisms involve mainly, on one hand Fe and Mn oxidation, which results in their precipitation as hydroxides and, on the other hand, sulphate reduction producing sulphides that can precipitate metals as insoluble sulphide minerals.

In acidic aerated conditions, *T. ferrooxidans* oxidizes Fe so that it is very soluble in strong acid conditions (pH 1.5–2.0) but will precipitate in different forms when the pH increases (Ivarson and Sojak, 1978; Ehrlich, 1996). Other bacteria such as the mixotrophic sheath bacteria *Leptothrix ochracea* or *Sphaerotilus natans,* or the autotrophic stalk bacteria *Gallionella ferruginea* (Houot and Berthelin, 1992; Ehrlich, 1996) can, in natural or slightly acid conditions and in micro-aerobic or aerobic environments, oxidize Fe and Mn as represented in the simplified equations 2.11 and 2.12 that then precipitate as the hydroxides:

$$Fe^{2+}(aq) + \tfrac{1}{4}O_2 + H^+ \rightarrow Fe^{3+} + \tfrac{1}{2}H_2O \qquad (2.11)$$

$$Fe^{3+} + 3OH^- \rightarrow Fe(OH)_3(s) \qquad (2.12)$$

In strictly anaerobic conditions, sulphate can be reduced by sulphate-reducing bacteria using fermentation products (lactic acid, acetic acid, ethanol, hydrogen, etc.) as energy sources and sulphate as electron acceptor in anaerobic respiration as summarized in equation 2.13:

$$2CH_3CHOH\ COOH + SO_4^{2-} \rightarrow 2CH_3COOH + 2CO_2 + S^{2-} + 2H_2O \qquad (2.13)$$

This type of reaction (2.13) plays a major role in sulphide deposits.

Three main types of precipitation or accumulation processes involving organo-mineral associations can be distinguished: the biodegradation of soluble organo-metallic complexes, the biosorption of metals and the bioaccumulation of metals (Hutchins *et al.*, 1986; Berthelin, 1988; Berthelin *et al.*, 1994). In biodegradation processes, the organic ligands (or anionic part of an organo-metallic complex) are used as sources of C and energy by the microorganisms. The metal is then released to the solution and, depending on the environmental conditions (Eh, pH, ionic strength), will either precipitate or remain soluble and hence contribute to accumulation and deposits.

Microorganisms can be involved in metal accumulation and concentration processes by adsorption-complexation on cell wall constituents or by absorption into the cell in excess of the nutritional requirements. The occurrence of either living or non-living cells allows distinction between bioaccumulation (active processes) from biosorption (non active) processes (Volesky, 1990, 1999; Hutchins *et al.*, 1986). Bioaccumulation phenomena can correspond to mechanisms of resistance or tolerance to metal, e.g. the production of proteins that specifically fix Cd or Cu, or the production of exopolysaccharides that immobilize metals by complexation, or the modification of cell wall constituents, such as greater formation of diaminoacids in the cell wall structure with increasing heavy metals accumulation (Hutchins *et al.*, 1986; Berthelin, 1988; Berthelin *et al.*, 1994).

## 2.3 Weathering and transformation of a mica under the influence of roots and of their associated microorganisms in the plant rhizosphere

The ability of roots and of soil microorganisms to weather minerals has been reported by different authors (Berthelin, 1983, 1988; Robert and Berthelin, 1986; Hinsinger *et al.*, 1993; Berthelin *et al.*, 1994) but few data are available on the interaction phenomena between microorganisms–roots and mineral soil constituents. In the rhizosphere, defined as the root environment, large amounts of energy are available to microorganisms living in symbiotic or non-symbiotic associations with the roots. In such environments, both microorganisms and roots can be involved in the weathering of soil minerals through the processes of uptake via ion-exchange of inorganic elements that can occur in excess of the nutritional requirements.

In greenhouse experiments, plants (pine and beech trees) have been cultivated in well defined conditions: special lysimeters were installed, and, as source of Mg, K and Fe for plant growth, phlogopite was mixed in the sand which filled the lysimeters (Leyval and Berthelin, 1991). The plant roots were inoculated (in some cases, not in others) with a bacterium (*Agrobacterium* sp.) and/or with an ectomycorrhizal fungus, *Laccaria laccata*, both selected from a pine rhizosphere for their ability to solubilize insoluble phosphates *in vitro*. Over one- and two-year periods of cultivation, analyses were performed on the soil and the solution of the lysimeters, the plants, the microorganisms and the minerals (Leyval and Berthelin, 1991, 1993). In the lysimeters inoculated either with bacteria or with the mycorrhizal fungi or with both, the bacterial communities have grown in the rhizosphere of pine roots where $10^6$ to $10^{10}$ CFU (colony forming units) were counted per g of rhizospheric dry matter. The biomass of the mycorrhiza (symbiotic association between *Laccaria laccata* and pine roots) was also well developed and comprised 2.6–8.6% of dry root matter. Significant weathering of a mica (phlogopite) was observed in the rhizosphere environment. Firstly the size frequency distribution analysis of the phlogopite flakes showed the breakdown of particles under the influence of rhizosphere microorganisms. Strong chemical and mineralogical changes were also noted.

The cation exchange capacity (CEC) of phlogopite increased significantly under the effect of roots (from <5 to 12 cmol $kg^{-1}$) and much more under the effect of roots inoculated with the bacterium and/or with the ectomycorrhizal fungus, and reached 18 and 25 cmol $kg^{-1}$, respectively.

In addition, mineralogical transformation was observed. This comprised transformation of phlogopite into vermiculite, as characterized by an X-ray diffraction (XRD) peak at 14 Å. This mineralogical transformation is stronger when the pine roots were inoculated either with the bacteria and/or the ectomycorrhizal fungi (Fig. 2.3). It was also noted that bacteria have a greater effect because vermiculite formation is also observed in the non-rhizospheric soil of the pine lysimeters inoculated only with the bacteria (Fig. 2.3) (Leyval and Berthelin, 1991). This mineral transformation is related to the production of aliphatic carboxylic acids (e.g. lactic, fumaric, malic, citric) (Table 2.1) in the rhizosphere by the bacteria that use the root exudates or root excretions and root residues as sources of C and energy for their growth (Table 2.1). Such organic acids are efficient agents of weathering. Mycorrhizas, on the other hand, have improved the surface and volume of root prospection and root

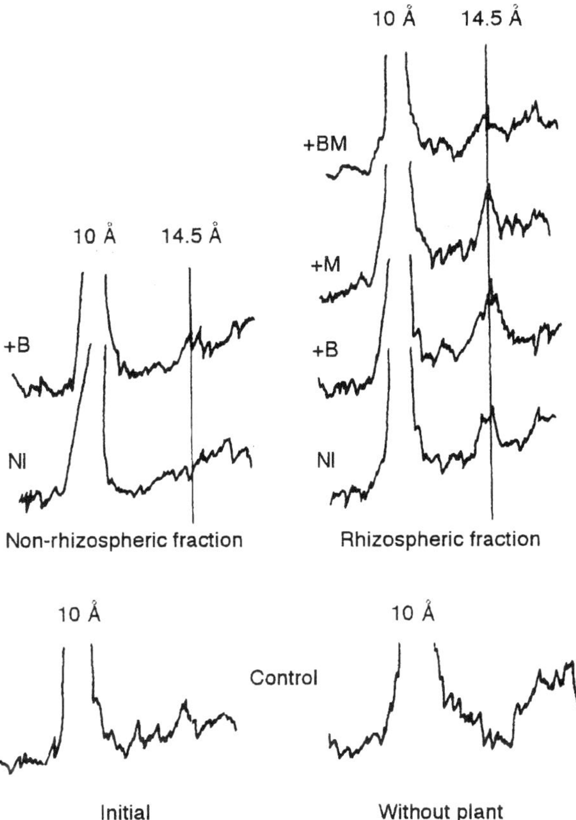


FIG. 2.3 XRD patterns of initial phlogopite and after 2 years of treatment without plant and in rhizospheric and non-rhizospheric soil fractions (NI. non-inoculated, plant alone; B, M and BM: inoculated with bacteria, mycorrhiza and both, respectively) (from Leyval and Berthelin, 1991).

contact with the mineral particles (Leyval and Berthelin, 1991, 1993) and, as a consequence, increase metal uptake through cation exchange ($H^+$ for $M^+$) on mineral surfaces.

Such results allow the fundamental role played by the roots of plants and their associated microorganisms in the weathering of minerals to be underlined. The results of such experiments show that symbiotic (mycorrhiza) and non-symbiotic (e.g. the bacteria living in the rhizosphere) microorganisms develop physical and chemical processes that have a significant effect on mineral transformation and evolution in the rhizosphere.

TABLE 2.1 Amounts of organic acids collected in the rhizospheric soil of pine after 2 years' growth in the lysimeters of a greenhouse experiment (from Leyval and Berthelin, 1991).

| Acids | Pine root treatment | | | |
|---|---|---|---|---|
| | | Inoculated | | |
| | Non-inoculated | Bacterium | Mycorrhiza | Dual |
| | | μmol plant$^{-1}$ | | |
| Citric | trace | 9.4 | 0.0 | 0.0 |
| Malic | 17.2 | 20.3 | 0.0 | 0.0 |
| Fumaric | 0.04 | 4.0 | 0.1 | 0.03 |
| Gluconic | 7.0 | 0.0 | 0.0 | 0.0 |
| Lactic | 14.7 | 26.5 | 0.0 | 0.0 |

## 2.4 Interactions between *Thiobacillus ferrooxidans* and sulphide minerals (pyrite) during oxidation and dissolution (bioleaching) processes

The acidophilic Fe- and S-oxidizing bacterium *Thiobacillus ferrooxidans* has been and is used extensively for recovery of metals (e.g. Cu, U) from low grade ores (Torma, 1986), and more recently for beneficiation of precious refractory ores to improve gold extraction (Hutchins *et al.*, 1986; Ehrlich and Brierley, 1990), and for Co extraction (Morin, 1998). Such bacteria are also involved in acidification, corrosion and weathering processes in different environments. A better knowledge of the bioleaching process requires fundamental studies not only of the physiology, biochemistry and genetics of this bacterium, but also of the nature of the bacterium–mineral interface and the reactions occurring there. Both the bacterial activity and the inorganic chemical reactions are involved in pyrite dissolution (Moses *et al.*, 1987), which can be controlled by semi-conducting properties (Lowson, 1982; Crundwell, 1988), the presence of impurities or pre-existing mineral defects (Southwood and Southwood, 1985) and the nature of the superficial pellicular phases (McKibben and Barnes, 1986; De Donato *et al.*, 1991). The nature of the sulphide surface can modify the physicochemical and electrochemical properties of the reactive surface and hence the kinetics of oxidation (Chander and Briceno, 1987). In this volume, two chapters deal specifically with acid mine drainage (Jenkins *et al.*, 2000; Keith and Vaughan, 2000), but the purpose of this chapter is to focus on the reactions at the interface between bacteria and minerals and mineral surface reactivity.

Other authors (Mustin *et al.*, 1993; Dziurla *et al.*, 1997; Toniazzo *et al.*, 1999*a,b*) have described the effect and the role of sulphide (pyrite) surface properties and the contact between bacteria and minerals on pyrite bio-oxidation, bioleaching and weathering. Some of their results and a discussion follow below.

One set of bioleaching experiments (Mustin *et al.*, 1993) was performed to determine the role of oxidized mineral species (elemental S, sulphates, oxides) occurring naturally at the pyrite surface. The bacterium *T. ferrooxidans* was cultivated in the presence of a pyrite with and without naturally occurring S at the surface. For the treatment without elemental S ($S_0$ or $S_8$) on the pyrite surface, S was removed in a liquid chromatographic device, which compared to a control, did not modify (except S removal) the pyrite surface and pyrite properties. Bioleaching experiments were performed in a pulp reactor equipped with a three electrode open circuit including a special ground pyrite-packed electrode (Mustin *et al.*, 1991) to follow the electrochemical behaviour of untreated (with S on the surface) and cleaned pyrites (without S) (Mustin *et al.*, 1993).

The results of the bioleaching experiments, as presented in Fig. 2.4, show that the oxidation rates of Fe, determined by solubilized ferric iron, were lower for 'cleaned' pyrite than for 'untreated' pyrite. Small experimental errors on the solubilization curves (<5%) (error bars are included in the square on the curves in Figs 2.4, 2.5 and 2.6) and a good reproducibility confirmed that the observed difference in the oxidation of both pyrites (with and without S) is significant. Similar results were observed for S oxidation (Mustin *et al.*, 1993). A significant delay in the lag phase and in the first step of the bacterial oxidation was observed: a delay of >24 h extra with the same inoculum of $10^7$ bacteria $ml^{-1}$ was necessary for the pyrite without S at the surface for the same amount of Fe to be released.

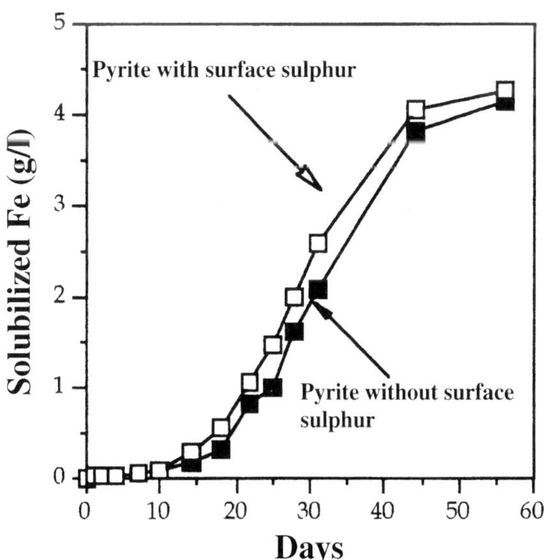


Fig. 2.4 Dissolution of total Fe by *T. ferrooxidans* from the 'cleaned' (after sulphur removal) and from the untreated (or 'uncleaned') pyrite (from Mustin *et al.*, 1993).

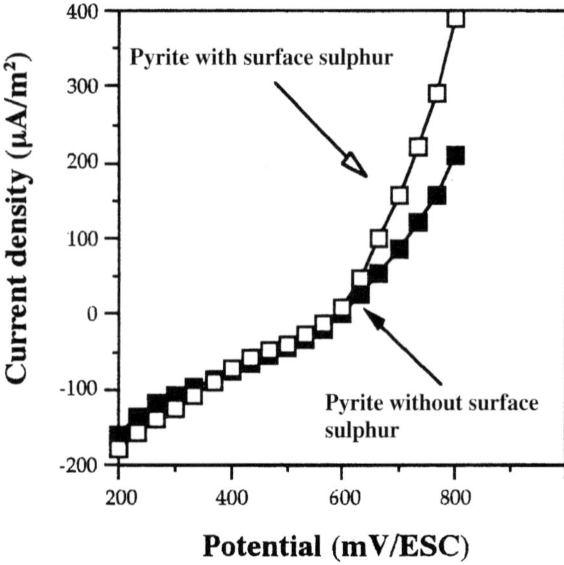


FIG. 2.5 Voltamperometric curves (current potential curves) of pyrite in the nutrient medium before and after removal of S at the pyrite surface (from Mustin *et al.*, 1993).

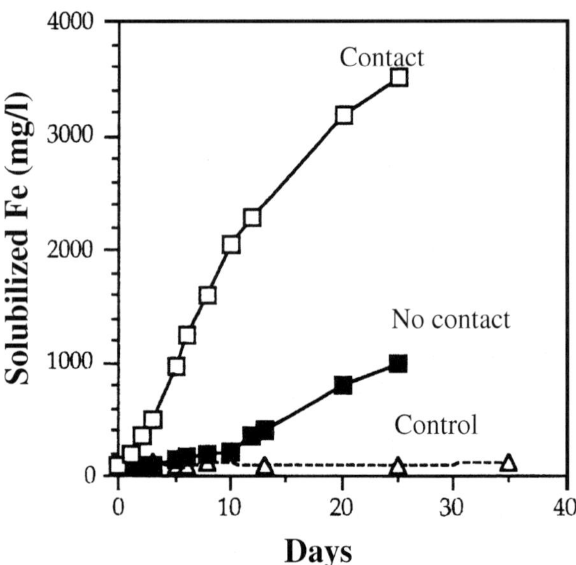


FIG. 2.6 Dissolution of total Fe (Fe mg $l^{-1}$) during pyrite leaching processes with and without contact between *T. ferrooxidans* and the mineral particles (from Dziurla *et al.*, 1997).

Bacterial growth, CFUs in the liquid medium, (results not shown, see Mustin *et al.*, 1993) showed a greater adsorption of bacteria in the presence of surface S at the beginning of the incubation. Sulphur can render the sulphide surface more hydrophobic but also more electroactive. Voltammetric studies (Fig. 2.5) highlighted the differences in the electrochemical properties of the two pyrites. The 'untreated' pyrite presents a rest potential of ~600 mV/SHE (Standard Hydrogen Electrode) in the nutrient medium and an anodic current greater than for the pyrite without S. The cathodic behaviour of pyrite was not affected by the removal of elemental S. However anodic behaviour was modified and a decrease of 200 $\mu A\ m^{-2}$ determined at 850 mV/SHE, showed a lower tendency to oxidize pyrite without S. So, the S at the surface enhanced the bacterial oxidation of the sulphide (pyrite) by providing a more accessible energetic substrate and a more readily available source of electrons for *T. ferrooxidans*.

Such results emphasize that the composition and properties of the surface can lead to notable differences in the electrochemical behaviour of pyrites and as a consequence, in the bacterial oxidation rates of these minerals.

Another set of experiments was devoted to the study of interactions between *T. ferrooxidans* and sulphides to establish in greater detail the effect of contact between the bacteria and the mineral (Dziurla *et al.*, 1997). Bioleaching experiments have been performed using a two-compartment reactor that allowed separation of bacteria from pyrite particles. The compartments were separated by a semi-permeable 0.22 μm filter membrane to avoid bacterial contact with mineral particles but to allow ionic and molecular exchange between the compartments (Dziurla *et al.*, 1997). Different parameters were monitored (bacterial growth, pH and Eh evolution, solubilized Fe and S, mineral particle corrosion and weathering) but only the results relating to the kinetics of Fe solubilization are reported here (Fig. 2.6). Iron oxidation and solubilization were very low in abiotic control (Fig. 2.6). When the bacteria were in contact with the sulphide, the lag phase of Fe solubilization was very short (no more than 1 day) and the mineral oxidation rate was, in terms of solubilized iron, 212 mg Fe $l^{-1}$ $day^{-1}$ from 1 to 10 days and then 116 mg Fe $l^{-1}$ $day^{-1}$ thereafter (Fig. 2.6). In treatments where no contact was possible between the bacterium and pyrite particles (due to the semi-permeable membrane), the first step occurred after 5–12 days in which the rate of Fe solubilization was similar to the rate observed in the abiotic control. This is a very long lag phase compared to the treatment where *T. ferroxidans* was in contact with the mineral. In a second step, the rate of Fe solubilization ranged from 46 to 59 mg Fe $l^{-1}$ $day^{-1}$ occurred, and was followed by a third step where the oxidation rate reached a maximum, in terms of solubilized Fe, of 73 mg Fe $litre^{-1}$ $day^{-1}$ (Fig. 2.6).

Hence, it appears that with bacterial contact, the optimal oxidation rate was three times greater (Fig. 2.6) and that the level of corrosion, determined by the density of corrosion pores (results not presented), was greater than that without contact. Simultaneously, significant transformation and evolution of pyrite surface properties (presence of elemental S and of sulphate) were also observed when bacteria were in contact with the mineral (results not presented, see Dziurla *et al.*, 1997).

### 2.5 Influence of Al and Mn substitution on the dissolution of ferric oxides (goethite) by Fe-reducing bacteria

Iron oxides (goethite, hematite, lepidocrocite) are common minerals, formed under various weathering conditions on Earth. The solubility of iron oxides in pure water is generally extremely low (Schwertmann, 1991). However, dissolution is promoted by strong acids, and by reducing and complexing agents. Iron in Fe(III) oxides can also be reduced by bacteria which use organic energy sources to convert Fe(III) to Fe(II) in anaerobic or facultatively anaerobic conditions (Ghiorse, 1988; Lovley, 1991, 1993) (see equations 2.9 and 2.10). Natural iron oxides generally contain other cations, particularly Al, Mn, Cr and Ni which can modify the reactivity and stability of the mineral species and which can be dispersed in the environment during oxide dissolution. The presence of these metals can modify the chemical and bacterial dissolution processes and rates and also the activity of the Fe-reducing bacteria. Experiments have been performed to determine the influence of Al and Mn substitution on the reduction and dissolution rates of goethite by a fermentative Fe-reducing bacterium. A non-substituted synthetic goethite and two other goethites with Al and Mn substitutions were used and placed in batch devices that received nutrient media and a bacterial inoculum of *Clostridium butyricum* isolated from soils (Bousserrhine *et al.*, 1998, 1999). The metal substitution ratio was 0.051 and 0.050 for Al and Mn (mole/mole ratio), respectively.

The dissolution curves presented in Fig. 2.7 show that no dissolution occurred in the control (without bacteria). The bacteria reduced Fe and rapidly dissolved pure goethite (Fig. 2.7*a*). The dissolution of Mn-substituted goethite is a little slower than that of the pure goethite but the dissolution of Mn is strongly correlated with Fe (Fig. 2.7*a,c*). Manganese, which is present as Mn(III) within goethite, as shown by Cornell and Schwertmann (1996) can be reduced to the soluble oxidation state, Mn(II) in a similar fashion to Fe. Its presence, however, delayed the bacterial release of Fe. The dissolution rate of Al-substituted goethite is much lower than that of pure or Mn-substituted goethite and $<40\%$ of the iron is reduced and dissolved (Fig. 2.7*b*) compared to almost 100% obtained with pure goethite whilst Al is solubilized at a lower rate than Fe (Fig. 2.7*b,d*). The kinetics and the relatively high solubilization of Al observed in the second step of the bacterial reduction of Fe (curve with open circles Fig. 2.7*d*) indicate that Al is released as a result, firstly, of bacterial dissolution of goethite by Fe reduction of the Fe oxide, and secondly, of acidification and decrease of the pH by acid (butyric and acetic) production which improves the solubility of Al (Bousserrhine *et al.*, 1999).

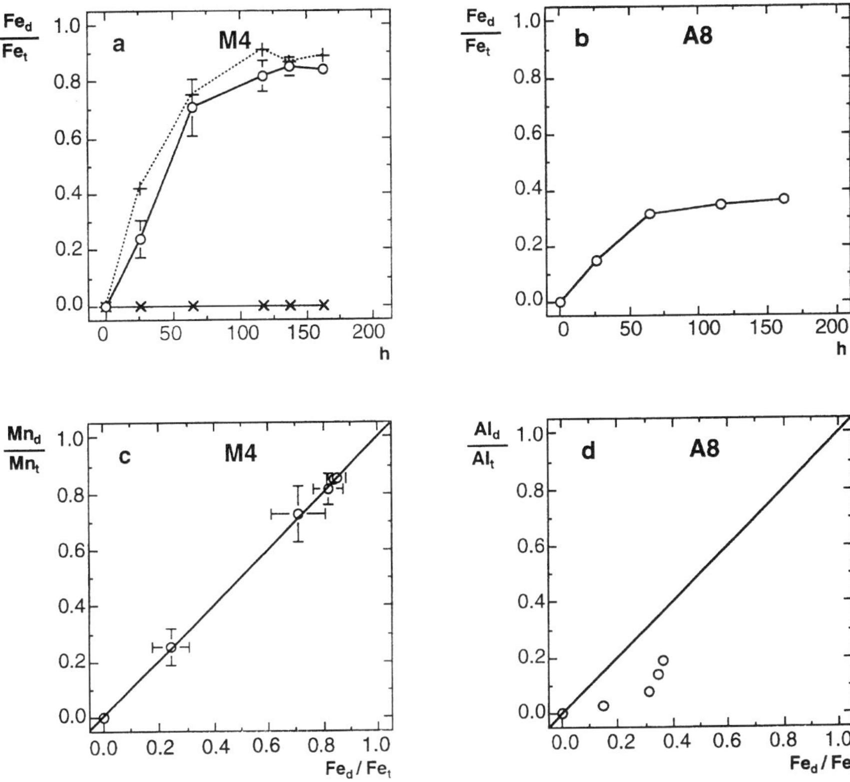


FIG. 2.7 Metal release from pure goethite (M4, dotted line), from Mn-substituted (Fe and Mn) goethite (M4, full line) and from Al-substituted goethite (Fe and Al) (A8) by an Fe-reducing bacterium: (*a,b*) rate of dissolution of Fe (dissolved Fe/total Fe:$Fe_d/Fe_t$) *vs.* time (h); (*c,d*) comparison between the dissolution rate of Mn and Al and the dissolution rate of Fe (from Bousserrhine *et al.*, 1999).

The presence of structural Al plays a greater role in controlling bacterial reduction of goethites than does Mn. Inhibition of Fe(III) reduction and Fe release may be related to Al accumulation on the goethite surface, blocking bacterial access to sites used in Fe reduction. Such inhibition can be prevented if Al remained in solution, e.g. as in the presence of an Al complexing agent.

## 2.6 Conclusions

Transformation of minerals and the behaviour of metals in soils and sediments are affected not only by physicochemical parameters but also by biological factors and particularly by microbial activity. Both chemolithotrophic and chemo-organotrophic microorganisms are involved in dissolution and precipitation-accumulation processes of major and trace elements by direct or indirect mechanisms. Such phenomena are a function of complex interactions between minerals–organic compounds and microorganisms. These interactions strongly influence the mobility and availability of major and trace elements and the evolution of soils and sediments (Berthelin, 1988; Berthelin *et al.*, 1994).

Microbial weathering processes in soils depend on several environmental factors (Berthelin, 1988, Berthelin *et al.*, 1994):

(1) the nature, amounts and availability of C and energy sources either organic or inorganic;

(2) the availability of mineral nutrients (e.g. mineral nutrient deficiencies can promote production of chelating agents);

(3) the antimicrobial compounds of plant or microbial origin that modify the activities of microbial communities;

(4) the aeration conditions and waterlogging which control respiration processes;

(5) the crystallochemical properties, particle size and surface properties of minerals that control their reactivity; and

(6) the microbial and plant species that control interaction between organisms (e.g. roots and microorganisms) and determine the amounts and quality of nutrients available.

All these microbial weathering processes exert a strong influence on: the functioning of terrestrial environments and ecosystems; the behaviour and cycling of elements including contaminants (e.g. heavy metals, radionuclides); plant nutrition; soil and water quality; and the treatment of raw materials, ores, industrial minerals, polluted soils and wastes.

This large and important field of fundamental and applied interest suffers from limited knowledge of:

(1) the microbial ecology and geomicrobiology to develop knowledge of the structure and function of microbial communities;

(2) the physiology, biochemistry and genetics of the microorganisms involved;

(3) the definition of limiting factors and main environmental parameters controlling the processes; and

(4) the development in modelling needed to advance monitoring and to determine useful confidence limits.

Finally, as previously noted for the advancement of knowledge of organo-mineral associations in soils (McKeague *et al.*, 1986) "a team approach is highly desirable". To obtain an adequate understanding of these mineral-microbe interactions, multiscale and multidisciplinary approaches where microbiologists, biologists, chemists, geochemists, soil scientists, etc. work together to develop complementary research in a common framework, are necessary.

## References


Alexander, M. (1977) *Introduction to Soil Microbiology*, 2nd edition. John Wiley and Sons, New York.

Berthelin, J. (1983) Microbial weathering processes. Pp. 223–62 in: *Microbial Geochemistry* (W.E. Krumbein, editor). Blackwell Science Publishers, London.

Berthelin, J. (1988) Microbial weathering processes in natural environments. Pp. 33–59 in: *Physical and Chemical Weathering in Geochemical Cycles* (A. Lerman and M. Meybeck, editors). Kluwer Academic Press, Dordrecht, The Netherlands.

Berthelin, J. (1998) Les microorganismes dans la transformation des minéraux: incidence sur la formation, le fonctionnement et l'évolution des sols. Pp. 83–91 in: *Sol: Interface Fragile* (P. Stengel and S. Gelin, coordinators). INRA, Paris.

Berthelin, J., Leyval C. and Toutain, F. (1994) Biologie des sols. Rôle des organismes dans l'altération et l'humification. Pp. 143–237 in: *Pédologie Tome 2. Constituants et Propriétés du Sol* (M. Bonneau and B. Souchier, editors). Masson, Paris.

Bousserrhine, N., Gasser, U.G., Jeanroy, E. and Berthelin, J. (1998) Effect of aluminium substitution on ferri-reducing bacterial activity and dissolution of goethites. *C.R. Acad. Sci. Paris,* série II, **326**, 617–24.

Bousserrhine, N., Gasser, U.G., Jeanroy, E. and Berthelin, J. (1999) Bacterial and chemical reductive dissolution of Mn, Co, Cr and Al substituted goethites. *Geomicrobiol. J.*, **16**, 245–58.

Chander, S. and Briceno, A. (1987) Kinetics of pyrite oxidation. *Min. Metal. Proc.*, 171–6.

Cornell, R.M. and Schwertmann, U. (1996) *The Iron Oxides. Structure, Properties, Reactions, Occurrence and Use* 2nd edition. VCH Publishers, Weinheim, Germany.

Crundwell, F.K. (1988) The influence of the electronic structure of solids on the anodic dissolution and leaching of semiconducting sulphide minerals. *Hydrometallurgy*, **22**, 141–57.

De Donato, Ph., Mustin, C., Berthelin, J. and Marion, Ph. (1991) An infrared investigation of pellicular phases observed on pyrite by scanning electron microscopy during its bacterial oxidation. *C.R. Acad. Sci. Paris*, série II, **312**, 241–8.

Dommergues, Y. and Mangenot, F. (1970) *Ecologie Microbienne du Sol.* Masson, Paris.

Dziurla, M.-A., Monroy, M.G., Dumestre, A., Barres, O. and Berthelin, J. (1997) Importance of contact between *Thiobacillus ferrooxidans* and pyrite particles, in oxidation and dissolution processes. Pp. 503–15 in: *Solid-Liquid Separation Hydro- and Biohydrometallurgy* (H. Hoberg and H. von Blottnitz, editors). Vol. **4**, XX Int. Mineral Proc. Cong. Aachen, Germany, GDMB Informations gesellschaft mbH.

Ehrlich, H.L. (1996) *Geomicrobiology* (3rd edition revised and expanded). Marcel Dekker Inc., New York.

Ehrlich, H.L. and Brierley, C.L. (1990) *Microbial Mineral Recovery*. McGraw-Hill, New York.

Ghiorse, W.C. (1988) Microbial reduction of manganese and iron. Pp. 305–27 in: *Biology of Anaerobic Microorganisms* (A.J.B. Zehnder, editor). John Wiley and Sons, New York.

Hinsinger, P., Elsass, F., Jaillard, B. and Robert, M. (1993) Root-induced irreversible transformation of a trioctahedral mica in the rhizosphere of rape. *J. Soil Sci.*, **44**, 535–45.

Houot, S. and Berthelin, J. (1992) Submicroscopic studies of iron deposits occurring in field drains: formation and evolution. *Geoderma*, **52**, 209–22.

Hutchins, S.R., Davidson, M.S., Brierley, J.A. and Brierley, C.L. (1986) Microorganisms in reclamation of metals. *Ann. Rev. Microbiol.*, **40**, 311–36.

Ivarson, K.C. and Sojak, M. (1978) Microorganisms and ochre deposits in field drains of Ontario. *Canad. J. Soil Sci.*, **58**, 1–17.

Jenkins, D.A., Johnson, D.B. and Freeman C. (2000) Mynydd Parys Cu-Pb-Zn mines: mineralogy, microbiology and acid mine drainage. Pp. 169–79 in: *Environmental Mineralogy: Microbial Interactions, Anthropogenic Influences, Contaminated Land and Waste Management* (J.D. Cotter-Howells, L.S. Campbell, E. Valsami-Jones and

M. Batchelder, editors). Mineralogical Society Series, **9**. Mineralogical Society, London.

Keith, C.N. and Vaughan, D.J. (2000) Mechanisms and rates of sulphide oxidation in relation to acid rock (mine) drainage. Pp. 117–39 in: *Environmental Mineralogy: Microbial Interactions, Anthropogenic Influences, Contaminated Land and Waste Management* (J.D. Cotter-Howells, L.S. Campbell, E. Valsami-Jones and M. Batchelder, editors). Mineralogical Society Series, **9**. Mineralogical Society, London.

Leyval, C. and Berthelin, J. (1991) Weathering of a mica by roots and rhizospheric microorganisms of Pine. *Soil Sci. Soc. Amer. J.,* **55**,.1009–16.

Leyval, C. and Berthelin, J. (1993) Rhizodeposition and net release of soluble organic compounds by pine and beech seedlings inoculated with rhizobacteria and ectomycorrhizal fungi. *Biol. Fertil. Soils*, **15**, 259–67.

Lovley, D.R. (1991) Dissimilatory Fe(III) and Mn(IV) reduction. *Microbiol. Rev.*, **55**, 259–87.

Lovley, D.R. (1993) Microbial reduction of iron, manganese and other metals. *Adv. Agron.*, **12**, 175–229.

Lowson, R.T. (1982) Aqueous oxidation of pyrite by molecular oxygen. *Chem. Rev.*, **82**, 461–97.

Madigan, M.T., Martinko, J.M. and Parker, J. (1996) *Brock Biology of Microorganisms*, 8th edition. Prentice Hall, New Jersey.

McKeague, J.A., Cheshire, M.V., Andreux, F. and Berthelin, J. (1986) Organo-mineral complexes in relation to pedogenesis. Pp. 549–92 in: *Interactions of Soil Minerals with Natural Organics and Microbes*. SSSA Spec. Publ. **17**, Soil Science Society of America, Madison, WI.

McKibben, M.A. and Barnes, H.L. (1986) Oxidation of pyrite in low temperature acidic solutions, rate laws and surface textures. *Geochim. Cosmochim. Acta*, **50**, 1509–20.

Morin, D. (1998) Des bactéries vont extraire du cobalt. Les plus gros réacteurs de biolixiviation jamais conçus. *La Recherche,* **312**, 38–40.

Moses, C.O., Nordstrom, D.K., Herman, J.S. and Mills, A.L. (1987) Aqueous pyrite oxidation by dissolved oxygen and by ferric iron. *Geochim. Cosmochim. Acta,* **51**, 1561–71.

Munier-Lamy, C., Adrian, Ph. and Berthelin, J. (1991) Fate of organo-heavy metal complexes of sludges from domestic wastes in soils: a simplified modelization. *Toxic. Environ. Chem.,* **31–32**, 527–38.

Mustin, C., Marion, Ph., Berthelin, J., De Donato, Ph. and Monroy, M. (1991) An original technique to study bacterial oxidation of pyrite: packed electrodes. *C.R. Acad. Sci. Paris*, série II, **312**, 1197–203.

Mustin, C., De Donato, P., Berthelin, J. and Marion, Ph. (1993) Surface sulphur as promoting agent of pyrite leaching by *Thiobacillus ferrooxidans. FEMS Microbiol. Rev.,* **11**, 71–8.

Neilands, J.B. (1974) Iron and its role in microbial physiology. Pp. 3–34 in: *Microbial Iron Metabolism* (J.B. Neilands, editor). Academic Press, New York.

Reid, R.K., Reid, C.P.P., Powell, P.E. and Szaniszlo, P.J. (1984) Comparison of siderophores concentrations in aqueous extracts of rhizosphere and adjacent bulk soil. *Pedobiologia*, **26**, 263–6.

Robert, J. and Berthelin, J. (1986) Role of biological and biochemical factors in soil minerals weathering. Pp. 453–95 in: *Interactions of Soil Minerals with Natural Organics and Microbes*. SSSA Spec. Publ., **17**, Soil Science Society of America, Madison, WI.

Schwertmann, U. (1991) Solubility and dissolution of iron oxides. *Plant and Soil,* **130**, 1–25.

Southwood, M.J. and Southwood, A.J. (1985) Mineralogical observations on the bacterial leaching of auriferous pyrite: a new mathematical model and implications for the release of gold. Pp. 98–113 in: *Fundamental and Applied Biohydrometallurgy* (R.W. Lawrence, R.M. Branion and H.G. Ebner, editors). Elsevier, Amsterdam.

Toniazzo, V., Mustin, C., Portal, J.M., Humbert, B., Benoit, R. and Erre, R. (1999*a*) Elemental sulphur at the pyrite surfaces: speciation and quantification. *Appl. Surf. Sci.*, **143**, 229–37.

Toniazzo, V., Lazaro, L., Humbert, B. and Mustin, C. (1999*b*) Bioleaching of pyrite by *Thiobacillus ferrooxidans*: fixed grains electrode to study superficial oxidized compounds. *C.R. Acad. Sci, Paris*, **328**, 535–40.

Torma, A.F. (1986) Biohydrometallurgy as an emerging technology. Pp. 23–33 in: *Biotechnology for the Mining, Metal Refining and Fossil Fuel Processing Industries* (H.L. Ehrlich and D.S. Holmes, editors). John Wiley and Sons, New York.

Volesky, B. (1990) *Biosorption of Heavy Metals.* CRC Press, Boca Raton, FL, USA.

Volesky, B. (1999) Biosorption for the next Century. Pp. 161–73 in: *Biohydrometallurgy and the Environment Toward the Mining of the 21st Century* (R. Amils and A. Ballester, editors). Elsevier, Amsterdam.

Watteau, F. and Berthelin, J. (1994) Microbial dissolution of iron from minerals: efficiency and specificity of hydroxamate siderophore in the dissolution of ferric oxides. *Eur. J. Soil Biol.*, **30**, 1–9.

CHAPTER THREE

# Mineral dissolution by heterotrophic bacteria: principles and methodologies

E. VALSAMI-JONES[1] AND S. MCELDOWNEY[2]

[1] *Department of Mineralogy, The Natural History Museum, Cromwell Road, London SW7 5BD, UK (E-mail: evj@nhm.ac.uk)*

[2] *School of Biosciences, University of Westminster, 115 New Cavendish Street, London W1M 8JS, UK (E-mail: mceldows@westminster.ac.uk)*

**ABSTRACT**

Mineral dissolution, the primary mechanism of release of inorganic species/nutrients to the Earth's surface, can be influenced by heterotrophic microbial activity. Heterotrophic (also known as organotrophic) bacteria may affect mineral dissolution directly or indirectly via a number of mechanisms, which can be grouped in two main categories: (1) dissolution from bacterially-formed organic and/or inorganic products; and (2) chelation or uptake by cells, either in suspension or attached to mineral surfaces. In this chapter, the principles of mineral dissolution and of heterotrophic biology in relation to dissolution are discussed. A review of recent evidence of the role of heterotrophic bacteria in mineral dissolution, with particular emphasis on the mechanisms involved, is presented.

## 3.1 Introduction

Mineral dissolution on the Earth's surface represents the primary source of inorganic species into the geo- and biosphere, and as such, controls element cycling in soil, ground and surface water. It also plays a role in global climate control, via the primary release of non-carbonate Ca and Mg, which ultimately fixes $CO_2$ through precipitation of Ca and Mg carbonates from the sea and in sediments.

Microorganisms are also prime agents of element cycling; some of the mineralogical transformation resulting from microbial interactions are discussed in other chapters in this volume (Berthelin *et al.*, 2000; Gadd, 2000). The variety of ways in which microorganisms contribute to the biogeochemical cycles and the formation and dissolution of minerals on this planet has been well reviewed in the recent past. Ehrlich (1996) provided a comprehensive review of geomicrobiology, and covered some of the very early work on the

---

Valsami-Jones, E. and McEldowney, S. (2000) Mineral dissolution by heterotrophic bacteria: principles and methodologies. Pp. 27–55 in: *Environmental Mineralogy: Microbial Interactions, Anthropogenic Influences, Contaminated Land and Waste Management* (J.D. Cotter-Howells, L.S. Campbell, E. Valsami-Jones and M. Batchelder, editors). Mineralogical Society Series, **9**. Mineralogical Society, London. ISBN 0 903056 20 8.

subject. Douglas and Beveridge (1998) and Fortin *et al.* (1997) focused on mineral formation by bacteria, whereas Silver (1997) reviewed bacterial interactions with aqueous ions and explained strategies and justification for the uptake or exclusion for most elements in the periodic table. Nordstrom and Southam (1997) considered the oxidative dissolution of sulphides by autotrophic bacteria. An earlier, comprehensive review of microbe-mineral interactions has been given by Robert and Berthelin (1986).

The focus for this chapter is to consider the influence of heterotrophic bacteria on mineral (mainly silicate) stability in soils and aquifers, including consideration of: (1) the principles of mineral dissolution; (2) some basic features of bacterial biology, and how these may influence bacteria/mineral interactions; (3) methods of culturing and imaging bacteria, applicable to geo-microbial interactions; (4) field evidence of mineral dissolution controlled by heterotrophic bacteria; and (5) laboratory evidence of effects on mineral dissolution by heterotrophic bacteria, with emphasis on the potential mechanisms involved.

## 3.2 The principles of mineral dissolution

Mineral dissolution, the breakdown of a mineral framework to release its components, is believed to be a process controlled by the mineral surface that takes place in two steps. The first step involves the attachment of the reactant ($H^+$, $OH^-$, or other ligands) on the mineral surface. The second step is the detachment of the newly-formed surface species (Stumm and Wollast, 1990), according to the reactions shown below in equations 3.1 and 3.2 (Stumm and Morgan, 1996):

$$\text{surface sites} + \text{reactants } (H^+, OH^-, \text{ligand}) \xrightarrow{\text{fast}} \text{surface species} \tag{3.1}$$

$$\text{surface species} \xrightarrow{\text{slow}} \text{aqueous species} \tag{3.2}$$

Reactant attachment (3.1) occurs rapidly. It is the slower detachment reaction (3.2) that controls the overall rate of dissolution. The above reaction (3.1) indicates that the dissolution of minerals is dependent upon the concentration of attached protons on the mineral surface, and is therefore a function of pH. Dissolution rates show a minimum at the point of zero charge ($pH_{pzc}$), where the net mineral surface charge is zero. This has been shown experimentally for some silicates and oxides (e.g. plagioclase, Brady and Walther, 1989; quartz, Berger *et al.*, 1994).

Mineral reactivity is not only a function of pH, but also depends on mineral structure and particle size. Poorly crystallized or amorphous minerals and minerals with structural defects tend to be more reactive. Small particle size (<0.2 μm) also enhances reactivity. Furthermore, within the same mineral, some surfaces are more reactive than others, depending on their crystallographic orientation and their position with respect to the mineral external topography, i.e. steps are more reactive than flat surfaces (e.g. Bosbach *et al.*, 1998).

A final control on mineral tendency to dissolve is exerted by other ions in solution. Such ions may adsorb onto the mineral surface and interfere with dissolution, either by displacing the reactants and blocking the formation of the surface species, thus hindering dissolution, or by forming alternative surface species and thus enhancing the dissolution rate.

Bacteria may potentially affect dissolution either through direct contact phenomena with the mineral surface, or through their role in influencing solution chemistry when not attached. Heterotrophic bacteria are well known to accumulate cations from solution (e.g. Beveridge and Fyfe, 1985), and therefore effectively act as organic ligands and promote dissolution following the same mechanism as shown in equations 3.1 and 3.2. In addition, heterotrophic bacteria produce a range of organic compounds e.g. polysaccharide exopolymers and organic acid waste products, outside the cell; these may also act as ligands for cations, thus enhancing mineral dissolution. If attached cells accumulate cations directly from the mineral surface and subsequently detach (McEldowney and Fletcher, 1988*a*), dissolution again may be enhanced. Alternatively, attached bacterial cells and/or their exoproducts may bind to active mineral sites and act as dissolution inhibitors.

## 3.3 Bacteria as dissolution agents

### *3.3.1 The basics of bacterial biology*

Bacteria are morphologically simple unicellular organisms, but their cell structure is so different from that of any other organism that they represent a unique evolutionary branch (together with blue-green algae, the cyanobacteria), the *procaryotes*. To put this into context, there are two further kingdoms that form the living world. The first is the *eucaryote* kingdom, which includes other microorganisms such as fungi and protists as well as plants and animals. The second is the *archaebacteria*. Archaebacteria were thought to be procaryotes until the 1970s, when Carl Woese demonstrated that they were as different from procaryotes as from eucaryotes and so formed a distinct kingdom (Woese, 1987). The archaebacteria include the methanogens (methane producing bacteria), the extreme halophiles (capable of living in concentrated salt brines) and thermoacidophiles (capable of living at high temperatures in highly acidic conditions) (Barns and Nieizwicki-Bauer, 1997).

The procaryotic bacterial cell consists of a nuclear region, rather than the true nucleus of eucaryotic cells. The cytoplasm surrounds the nuclear region and contains ribosomes but none of the organelles present in eucaryotes. The ribosomes are the site of structural and catalytic protein production within the cell. The cytoplasm is enclosed within a semi-permeable membrane (e.g.

Brock, 1970). This cytoplasmic membrane has associated membrane-bound enzymes which are important in a range of cell physiological functions, e.g. substrate transport into the cell and respiration. This cytoplasmic membrane acts as a boundary between the environment and the inside of the cell allowing or facilitating transfer of some molecules and ions into and out of the cell while excluding others (Neidhart *et al.*, 1990; Ferguson, 1992).

In the majority of bacteria a more rigid layer, the cell wall, surrounds the cell membrane. This structure gives the cell strength and protects it from osmotic shock. There are two main types of cell wall structure, Gram-positive and Gram-negative. Both types of cell wall contain a complex polymer unique to procaryotes called peptidoglycan, although the amount varies between Gram-positive and Gram-negative walls (Fig. 3.1). It is the peptidoglycan that confers much of the cell wall strength. The Gram-positive cell wall consists of a relatively thick layer of peptidoglycan overlaying the cell membrane. Gram-negative cell walls have a periplasmic space outside the cytoplasmic membrane, over which is a layer of peptidoglycan (smaller than that in Gram-positive cells). Beyond this layer is the so-called outer membrane which consists of phospholipids and lipoproteins. Bacteria often have a coating external to their cell walls. This is called a glycocalyx, and consists either of proteins or polysaccharides. If the glycocalyx forms a discrete layer around the cell it is often called a capsule, if it is more diffuse and less rigid it may be

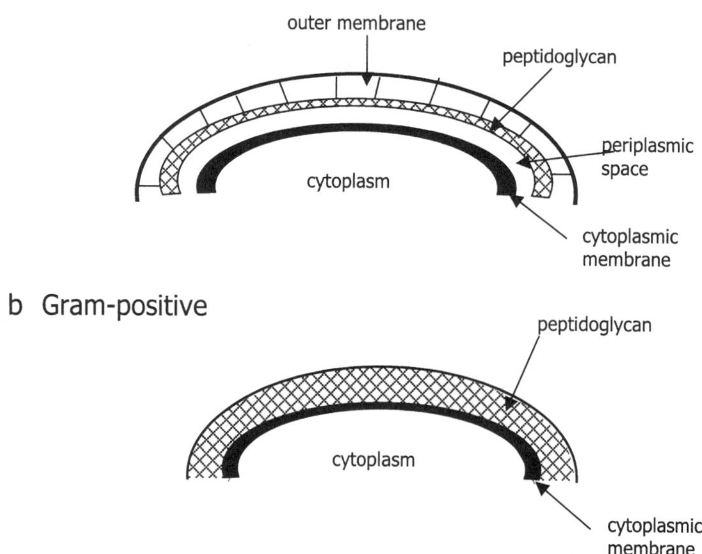


FIG. 3.1. Diagrammatic representation of Gram-positive and Gram-negative cell walls.

referred to as a 'slime layer' or simply diffuse exopolymer (Hammond *et al.* 1984; Neidhart *et al.*, 1990).

Bacterial cells have a limited range of shapes: spherical (coccus), rod-shaped (bacillus), helical-rod shaped (spirillum) and V-shaped (vibrio). The cell wall provides the structural strength underpinning the cell shape. Some bacterial species have thin proteinaceous appendages protruding through the cell wall. These flagella confer motility on the cell. The number and distribution of flagella round the cell varies with species (Stanier *et al.*, 1987; Neidhardt *et al.*, 1990).

Bacteria multiply by binary fission (individual cells splitting into two daughter cells). Sometimes, however, the cells do not separate after division, remaining attached in characteristic patterns. Individual cells vary in size from 0.1 μm to a few microns. The volume of a medium-sized bacterium is 1 $\mu m^3$. This size confers a high surface to volume ratio on bacteria which plays a major role in their chemical reactivity (Schlegel, 1993).

Bacteria have traditionally been identified and classified on the basis of a range of characteristics from cell morphology and staining reaction to biochemical, nutritional and physiological criteria (Stanier *et al.*, 1987). Recently, molecular techniques have provided many insights into bacterial diversity and evolution and have been applied to identification and classification (Barns and Nierzwicki-Bauer, 1997). In the present review, we focus on classification by the physiological and nutritional characteristics of bacteria.

A major physiological distinction within the procaryotic bacteria is the source of carbon used to produce cellular components. Heterotrophs (also known as organotrophs) use organic carbon as their carbon source (or substrate), while autotrophs (or lithotrophs) obtain their carbon from $CO_2$. Cells must also obtain energy in order to grow and survive. Some heterotrophs and autotrophs obtain their energy from sunlight through photosynthesis and are termed photoheterotrophs and photoautotrophs respectively. A number of different autotrophic bacteria obtain their energy by the oxidation of a reduced inorganic compound, e.g. ammonia oxidation by *Nitrobacter* sp. These organisms are called chemoautotrophs. Many heterotrophic bacteria obtain energy from organic C and are termed chemoheterotrophs (Stanier *et al.*, 1987; Nealson and Stahl, 1997); it is this physiological type of bacterium that is the main focus of this chapter. For the sake of simplicity, chemoheterotrophs will be referred to as heterotrophs in this chapter. The characteristics of heterotrophs are considered in more detail below in terms of their interactions with the environment and potential significance for mineral dissolution.

### *3.3.2 Bacterial distribution and numbers in the environment*

Bacteria are ubiquitous, surviving and growing in all the diverse habitats that constitute the biosphere including some particularly extreme habitats, e.g. deep subsurface, submarine hydrothermal vents, polar ice. Their ubiquity is ample evidence for their physiological and metabolic diversity and adaptability, key in their significance as agents of geological change.

In soils, average bacterial populations range between $10^6$–$10^9$ cells $cm^{-3}$ (Atlas and Bartha, 1998), compared with $10^4$–$10^6$ cells $cm^{-3}$ for fungi, $10^3$–$10^5$ cells $cm^{-3}$ for protozoa and $10^2$–$10^4$ cells $cm^{-3}$ for algae/cyanobacteria (Alexander, 1977). Both autotrophs and heterotrophs are common.

Bacteria are prevalent in the oceans, particularly the euphotic zone, because of the relatively high proportion of organic and inorganic nutrients. They are also present in nutrient deficient marine environments, e.g. at abyssal depths, although their metabolic activity and growth rates are reduced in such habitats (Ehrlich, 1996). A large proportion of marine bacteria are heterotrophs, although chemoautotrophs become dominant in the extreme deep-sea hydrothermal vent environments. Microbial populations in rivers and lakes are mainly heterotrophic whereas groundwater harbours a mixture of heterotrophic and autotrophic bacteria. In terms of numbers, Ehrlich (1996) reports $10^6$ cells $ml^{-1}$ in shallow aquifers; even in deep subsurface environments similar numbers have been reported (Sinclair and Ghiorse, 1989; Stevens and McKinley, 1995). This compares with $10^2$–$10^5$ cells $ml^{-1}$ in lake waters, $10^9$ cells $ml^{-1}$ in river waters and $10^6$ cells $g^{-1}$ in lake sediment.

### *3.3.3 Energy generation by bacteria*

Bacteria use a variety of electron donors, energy sources, and carbon sources during growth, which ultimately determine their impact on the inanimate world. Metabolic energy is obtained by transferring electrons from one compound (the electron donor), in the case of heterotrophs an organic compound, to another, the electron acceptor (Neidhart *et al.*, 1990; Chapelle, 1993). Heterotrophs utilize organic compounds as carbon sources and their energy is provided through oxidation-reduction reactions of these organic compounds (Stanier *et al.*, 1987). Aerobic bacteria use oxygen as the oxidized compound, which acts as the terminal electron acceptor in this process called respiration. In the absence of oxygen, both inorganic and organic compounds can be used as electron acceptors. Anaerobic respiration may involve inorganic electron acceptors (e.g. nitrate, sulphate and ferric ions) or organic electron acceptors; the latter is called fermentation (Chapelle, 1993; Stanier *et al.*, 1987). The use of different electron acceptors has considerable implications for the chemical environment and potential mineral dissolution. Three examples serve to illustrate this point. Firstly, sulphate-reducing bacteria transfer the electron produced during respiration to sulphate and produce hydrogen sulphide as the highly reactive product (Nealson and Stahl, 1997). Secondly, fermentation results in the production of a range of organic by-products which are either acids or capable of acting as ligands for metal ion species (Gadd, 1992; Hardman *et al.*, 1993). Finally, during respiration of organic carbon under both aerobic and anaerobic conditions $CO_2$ is produced making the aqueous phase more acidic. Some heterotrophs are obligate anaerobes, only able to survive and grow through anaerobic metabolism in the absence of oxygen. Others are facultative anaerobes preferring oxygen as the terminal electron acceptor in respiration, but capable of switching metabolism in the absence of oxygen to use another inorganic acceptor (Stanier *et al.*, 1987; Atlas and Bartha, 1998).

Heterotrophic growth is the dominant form of metabolism in the bacteria successfully isolated and grown in laboratory culture and may represent the dominant form of bacterial metabolism in the biosphere. Certainly heterotrophs are common in soils, groundwater and even the deep subsurface and undoubtedly affect the ecosystems they inhabit (Chapelle, 1993; Nealson and Stahl, 1997).

There is a great deal of physiological diversity within the heterotrophic bacteria (Stanier *et al.*, 1987) that has implications for considering their role in mineral dissolution. They vary in their carbon substrate specificities, some using only a few types of organic carbon sources, others capable of using many. As a group, however, heterotrophs are capable of using a diverse range of organic compounds from simple sugars and amino acids, to more complex molecules such as polysaccharides, proteins, nucleic acids, cellulose, and even highly complex polymers such as lignin (Stanier *et al.*, 1987). Polymer degradation requires the production of exoenzymes, e.g. cellulases and lignases. Even single bacterial strains are commonly able to utilize a variety of carbon sources (indeed the type of organic compounds metabolized by heterotrophs is used as one criterion in identification and classification) and switch between them depending on their concentration and availability in the environment (Neidhart *et al.*, 1990). Thus, there is potential for heterotrophic activity in any habitat in which organic compounds are present. Even in the deep subsurface (Sinclair and Ghiorse, 1989), deep sediments in the Pacific Ocean (Parkes *et al.*, 1994) and in rock strata (Krumholz *et al.*, 1997) heterotrophs are present. Indeed, it is possible that in some of these habitats the heterotrophic utilization of organic compounds supports the food-web as a whole (Krumholz, 1998). This broad distribution of heterotrophic bacteria suggests it may be possible for them to contribute to mineral dissolution in many different habitats throughout the biosphere.

### *3.3.4 Heterotrophic requirements for growth*

The growth of heterotrophs is not only controlled by the availability of organic carbon and electron acceptors but also by the availability of a range of inorganic elements essential for growth. These mainly are $K^+$, $Mg^{2+}$, $Co^{2+}$, $Cu^{2+}$, $Fe^{3+}$, $Mn^{2+}$, $Ni^{2+}$, $Zn^{2+}$, $PO_4^{3-}$, $NO_3^-$, $NH_4^+$, and $SO_4^{2-}$(Silver, 1997).

These essential elements can be divided into major nutrients, such as S, P and N, minor nutrients, e.g. Mg and K, and trace elements such as Zn (Stanier *et al.*, 1987; Neidhardt *et al.*, 1990). The concentrations of these nutrients in a habitat can be limiting and reduce the growth rates and density of heterotrophs. It is interesting to speculate if such limitation may drive the development of specific bacterial mechanisms for mineral dissolution. Certainly there is evidence that heterotrophs increase K, Fe and Mg release from minerals to solution (e.g. micas, Barker *et al.*, 1998) and that these inorganic nutrients are subsequently accumulated by the organisms. As for all forms of life on this planet, water is also essential for bacteria and its availability will significantly influence their growth rates and the survival. Some heterotrophs also require certain organic molecules, e.g. vitamins, as growth factors because they are unable to synthesize these compounds themselves.

The growth rate, physiological status, density and types of heterotrophs in any habitat will be determined by the availability of carbon source, electron acceptor and nutrients. In addition to these factors, other physical and chemical characteristics of the abiotic environment will influence heterotrophic activity and density, e.g. temperature, pH, osmotic pressure, solution composition, redox potential. The degree and rate at which mineral dissolution may be influenced by heterotrophic activity is, then, likely to vary with the physicochemical environment because of differences in bacterial activity, numbers and types. It is important to note that heterotrophs are inherently metabolically flexible and have the facility to be active and grow under a range of environmental conditions. They have a significant ability to adjust their metabolism to changing conditions, provided the changes are not too extreme. For example, they can survive surprisingly long periods of desiccation, even grow under some conditions albeit at slower rates (McEldowney and Fletcher, 1988*b*).

In many soil and subsurface environments, heterotrophs may live almost a feast-famine existence depending on the availability of these necessities of life. This has consequences for the manner of their growth and the production of exo-metabolites that may contribute to mineral dissolution. Certain heterotrophs will use a period of surfeit for rapid growth and reproduction. These organisms can be referred to as ‘zymogenous’ (opportunistic) bacteria; they have high maximum growth rates and low substrate affinity (i.e. they are poor at taking up substrates) (Hardman *et al.*, 1993; Atlas and Bartha, 1998). During their growth they produce a range of primary metabolites (small molecules required during growth), secondary metabolites (small molecules not contributing to growth), exopolymers (often polysaccharide), and metabolic waste products which are excreted outside the cell. Any of these may contribute to mineral dissolution through acting as metal ligands or binding to mineral surfaces. The characteristics of the exoproducts produced will vary with the bacterial species involved, the environmental conditions and the stage in any feast-famine cycle to which the organisms are exposed. For example, there are major differences in exo-products between aerobic and anaerobic metabolism, the latter is often far less efficient with more organic waste products formed (Stanier *et al.*, 1987). If the organisms are growing in a high carbon environment, there is a tendency for more exopolysaccharide to be produced (McEldowney and Fletcher, 1988*c*). If there is a decrease in the availability of ferric ion, a key inorganic nutrient, then some heterotrophs will secrete siderophores into the environment. Iron is likely to be most commonly limiting in aerobic systems. Siderophores are small molecules which have a high affinity for the ferric ion facilitating bacterial capture of the inorganic nutrient. They can also bind other divalent or trivalent metal ions, e.g. $Al^{3+}$, $Pb^{2+}$, though at lower affinity (Gadd, 1992; Hardman *et al.*, 1993, Watteau and Berthelin, 1994). Zymogenous organisms often cause substantial environmental change by their growth and activity. They include members of genera such as *Pseudomonas* and *Bacillus*.

If nutrient concentrations are too low to support the growth of zymogenous heterotrophs or if the organic material has a complex composition, e.g. humus, then physiologically specialized organisms will dominate the microbial community, e.g. soil streptomycetes, which can degrade complex polymers, via the production of exo-enzymes. Those heterotrophs, physiologically specialized for growth under low nutrient conditions, are called 'oligotrophs'; examples include *Caulobacter* and marine vibrios. They have slow maximum growth rates but very high affinity uptake systems, i.e. take up nutrients at very low concentrations with great efficiency (Atlas and Bartha, 1998; Stanier *et al.*, 1987). It is interesting to speculate on the possible contribution of these different types of heterotrophs in mineral dissolution under different nutrient conditions.

### *3.3.5 Environmental controls*

Soils and subsurface environments are in essence particulate environments. In such habitats, organic compounds will not only be found in the liquid phase surrounding these solid substrates but the polymeric and hydrophobic organic compounds will become immobilized on the surfaces as conditioning films (Wimpenny *et al.*, 1993; Savage and Fletcher, 1985). Bacteria attach to solid surfaces and are capable of growing and dividing on the surface using macromolecules in the conditioning film and those that diffuse from the surrounding liquid phase (Wimpenny *et al.*, 1993; Savage and Fletcher, 1985; Denyer *et al.*, 1993). Some bacteria may not become permanently attached, but are more loosely associated with the surface. The close proximity of attached and surface-associated heterotrophs to mineral surfaces might impact on the process of mineral dissolution in two ways: firstly, by possibly increasing dissolution through their metabolic activity, e.g. via inorganic nutrient uptake or the production of exo-metabolites; and secondly, through producing exopolymers, which adsorb to the mineral surface potentially inhibiting dissolution (see below). Many of the heterotrophs are motile and show positive chemotaxis (swimming in response to a chemical stimulus) to organic nutrients (Neidhart *et al.*, 1990). In conditions where nutrients are lacking in the aqueous phase, the bacteria would tend to swim towards the surfaces where there is a concentration of organic nutrients in the conditioning film. Again this may have consequences for any mineral dissolution process catalysed by heterotrophs, since it is possible that in nutrient deficient conditions, surfaces, such as mineral surfaces, may hold the bulk of the bacterial numbers in a given habitat.

The soil and subsurface environment are highly heterogeneous environments for heterotrophs. There may be major differences in the abiotic part of the environment within a few μm, certainly within cm and it is on this scale that differences can be significant for bacteria which are colloid-sized (~1 μm long). For example, there can be anaerobic and aerobic physiological reactions occurring within a few μm of each other, depending on the availability of oxygen. In addition, those bacteria attached to surfaces are exposed to very different environmental conditions than those free in the liquid phase (Savage and Fletcher, 1985). It is possible to visualize local scale variations in the impact of heterotrophs on mineral dissolution as a result of these differences.

The consideration of heterotrophs and their interactions provided above illustrates an important point. Heterotrophs are highly plastic in their physiological responses to the environment (Hardman *et al.*, 1993). Individual cells are able to adapt their metabolism to accommodate changes in environmental conditions, provided the changes are not too extreme. They are dynamic living organisms continually responding to their environment, e.g. moving from aerobic to anaerobic metabolism. In natural environments, responses to environmental changes also occur at community level with different species dominating in any given set of conditions, e.g. obligate anaerobes selected for in a newly oxygen-deficient environment. Species that are less suited to the new conditions with their growth rate less than optimum will be out-competed by the organisms better suited to the new environment. Such phenotypic shifts in metabolic characteristics and community structure are likely to influence the role heterotrophs play and mechanisms by which they influence mineral dissolution. It is also important to remember that heterotrophs often co-operate within consortia (different species acting together). For example, this is well known in the degradation of some synthetic organic pollutants with each organism in a consortium catalysing a different stage in the degradation (Hardman *et al.*, 1993; Atlas and Bartha, 1998). It is possible that consortia are important in mineral dissolution.

If an environmental change becomes a permanent feature of the environment then there will be evolutionary pressures exerted on the bacteria. Heterotrophs are capable of rapid evolution, they are genetically highly flexible (Hardman *et al.*, 1993). In part, this is due to their rapid reproductive rate, but they also have some unique features in the organization of genetic material, e.g. the presence of extrachromosomal elements such as plasmids, which facilitate the rapid transfer of genetic traits within and indeed between species (Neidhart *et al.*, 1990; Veal *et al.*, 1992). If a genetic change in a heterotroph provides a competitive advantage in 'new' environmental conditions, these bacteria will be selected for and succeed within the community. In terms of mineral dissolution it is interesting to speculate on the possible significance of such rapid evolutionary capability. What if heterotrophic enhancement of mineral dissolution provides a selective advantage in a mineral nutrient deficient habitat? This scenario has been noted in the context of selective feldspar dissolution (Rogers *et al.*, 1998).

In summary, hetrotrophs are physiologically diverse showing a range of adaptations for life in a variety of habitats. They are inherently responsive to the environment modulating their metabolism to achieve the highest growth rate in any particular condition. They are capable of rapid evolution and show great facility for developing adaptations that provide a competitive advantage in the environment. They have a range of characteristics allowing them to act as direct or indirect agents in dissolution of minerals.

Possible mechanisms of bacterial dissolution include:

(1) the production of a range of chemical species; these can be grouped in five categories: (a) inorganic acids, including carbonic (from the production of $CO_2$ via respiration), sulphuric and nitric, and inorganic bases, e.g. ammonia; (b) organic acids (such as acetic, citric and oxalic) produced by catabolism and bases (such as ethanol, propanol, ketones); (c) siderophores, produced as a means of acquiring necessary iron in deficient environments (see above); (d) exopolysaccharides, secreted externally around bacterial cells (see above), as a mechanism of cell protection; and (e) exoenzymes.

(2) The accumulation of cations by cells; this can be divided in two categories: (a) accumulation of free cations within the aqueous phase by suspended cells, either internally or at the cell surface (McEldowney, 1990; Gadd, 1992), thus driving dissolution by maintaining an under-saturated solution; (b) accumulation through direct attachment of the bacteria cells to the surface. Attachment occurs through a physico-chemical reaction between the cell and mineral surfaces. The resulting close proximity of cells to mineral may have implications for dissolution, if attached cells act as ligands for cations at the surface of the mineral lattice, binding surface species with higher affinity than the mineral.

These mechanisms may also have the reverse effect, i.e. dissolution inhibition, by inactivation of dissolution sites. They will be considered in greater detail later.

## 3.4 Methods of studying heterotrophic bacteria

### *3.4.1 Methods of culturing heterotrophs*

In order to undertake laboratory-based experiments to establish the possible role of heterotrophs in mineral dissolution, it is necessary to first culture and grow the bacteria in laboratory media. The aim of culturing is to provide all the nutritional and growth requirements needed by bacteria and at the same time ensure environmental conditions favourable for growth. The types of media used for culturing heterotrophs can vary from rich media consisting of a wide variety of organic nutrients, growth factors and inorganic nutrients, e.g. nutrient agar/broth, or peptone yeast extract, to highly controlled, minimal media. A minimal medium contains only a carbon source, electron acceptor, and essential inorganic salts, a few may include essential growth factors, e.g. vitamins (Chapelle, 1993; Stanier *et al.*, 1987). These media can either be in liquid form, or solidified with agar. In both cases pH, availability of oxygen or other electron acceptor, redox potential and incubation temperature will be controlled to some extent.

It is important to note when isolating heterotrophs from the environment, e.g. from soil or the surface of minerals, that the medium and growth conditions chosen will tend to select for certain types of organisms. For example, if a glucose minimal medium is used then only heterotrophs capable of growing on glucose will be isolated. If the medium is adjusted to a slightly acid pH, then the selection is for species that have the highest growth rates at that pH. In reality no matter how many different types of media and conditions are used, the heterotrophs isolated will probably only represent a small fraction of those present in the original sample. It is estimated that <10% of soil bacteria and between 0.001 and 0.1% of marine bacteria have been cultured successfully (Hardman *et al.*, 1993). This sets something of a dilemma in that it may be those heterotrophs which we cannot isolate and grow in the laboratory that have the most significant impact on mineral dissolution. Field experiments, of course, do not have this limitation, but they may not provide information on the mechanisms of microbially-enhanced dissolution. It must be advantageous for field-based experiments to be conducted in tandem with *in vitro* experiments.

A number of different culture techniques are likely to be useful in studies on the impact of heterotrophs on mineral dissolution. The batch or closed culture involves the growth of heterotrophs in a finite volume of liquid medium of choice, incubated under defined conditions, very similar in fact to batch systems used for mineral dissolution experiments. These closed systems have a high initial concentration of all nutrients, no throughput of material, and a continuously changing environment. This variable environment is caused by an increase in exo-metabolites and bacterial biomass with time and a depletion of nutrients (Hardman *et al.*, 1993).

In a closed system, heterotrophs will grow in a characteristic cycle (Fig. 3.2). When the heterotrophs are inoculated into the medium there is a period during which they do not grow called the lag phase. The length of the lag phase varies with the past history of the inoculum and growth conditions. After the lag phase the bacteria enter a period of exponential growth, the exponential or log phase. The cells are at their most active during this phase

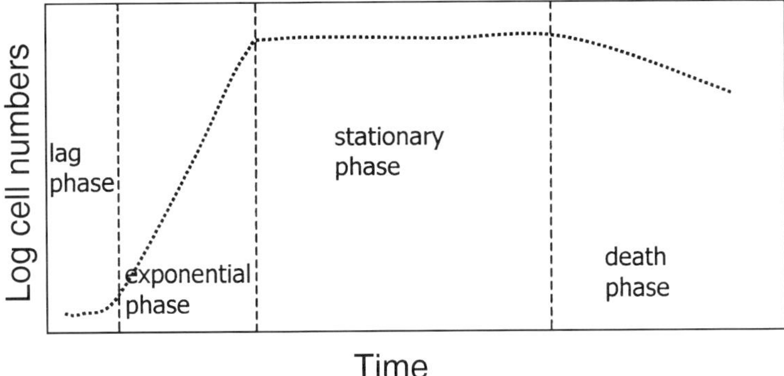


FIG. 3.2 Bacterial growth phases in batch culture experiments, expressed by the logarithm of cell numbers as a function of time.

and their maximum growth rate under the imposed growth conditions can be determined. During stationary phase, nutrients within the culture become limiting and the build-up of waste metabolites inhibitory, and there is no net increase or decrease in cell numbers. Cell growth does not normally continue in the stationary stage but cells remain active in terms of energy metabolism and even some biosynthesis of cell components. Also, it is possible for slow growth to occur during the stationary phase. This is normally balanced out by the death of other cells, so that no change in cell numbers occurs. During stationary stage many secondary metabolites are formed. The next stage in the growth cycle is the death phase. During this stage the cells die and "lyse" (disintegrate; Neidhardt *et al.*, 1990; Stanier *et al.*, 1987). Although the onset of death phase is difficult to predict, cultures normally remain in stationary phase for periods of the order of days to a few weeks.

Batch culture, or closed systems, may be particularly useful to model environments where interaction with minerals occurs in high nutrient and substrate conditions, perhaps in soils or some groundwater systems. It is possible, of course, to reduce the concentration of any of the nutrients supplied in batch flasks so applying a stress on the organisms growing in the culture. This would tend to elicit different physiological responses potentially altering any impacts on mineral dissolution. For example, if low concentrations of ferric ion are provided, the bacteria may produce siderophores to overcome the limitation with possible implications for mineral dissolution. Ultimately though, even if some nutrients are reduced in concentration, it is generally the fast-growing zymogenous heterotrophs that will do well in this type of culture. It is not possible to establish a stable bacterial community in batch culture because of the continually changing nature of the culture, making this an unsuitable technique for investigating the role of consortia in dissolution. Thus batch culture is particularly suited to study mineral dissolution by pure cultures of zymogenous heterotrophs in short-term experiments.

Flow-through, or open systems, also called continuous culture, provide very different conditions from closed cultures. In open systems, nutrients are continuously added, while used medium, exo-metabolites and biomass are continuously removed. In terms of experimental set-up, this is very similar to flow-through reactors used for assessment of mineral dissolution rates (e.g. Stumm and Morgan, 1996), although in the latter systems, minerals are retained in the reactor. Continuous culture allows the establishment and maintenance of steady-state conditions, i.e. defined conditions can be maintained for extended periods of time. In continuous culture, the bacteria do not go through the whole growth cycle, as they do in batch culture, but are maintained in an actively growing state; a proportion of the cells are washed-out in the outflow, but a relatively constant number is retained. Their density within the culture remains constant provided there is no fluctuation in conditions (Hardman *et al.*, 1993). The commonest form of open system is called a chemostat (Fig. 3.3). Normally a single nutrient is provided in limiting concentrations, e.g. the carbon source. The growth rate of the organism(s) in the chemostat is controlled by the rate at which medium flows into the culture vessel. This specific growth rate equals the dilution rate (the rate of input of growth medium divided by the culture volume) (Atlas and Bartha, 1998). An open system allows an active population of bacteria to be maintained under carefully defined conditions for long experimental periods. Since there are steady-state conditions, it is possible to maintain a stable community or consortium of bacteria, and would be particularly useful in mineral dissolution

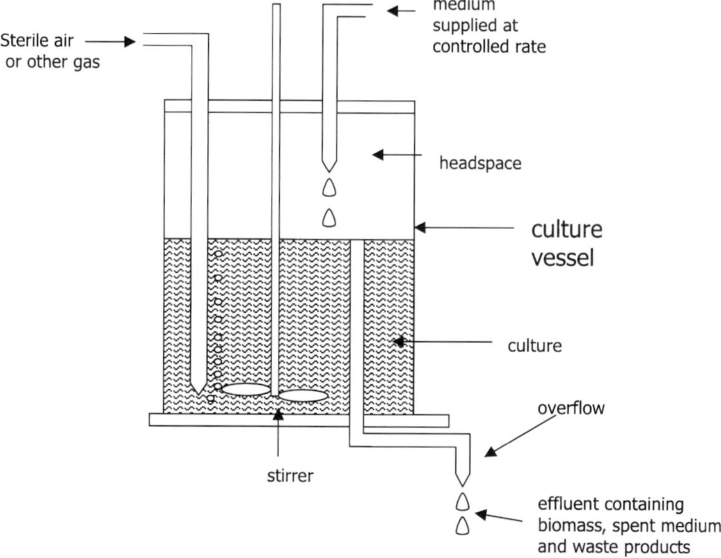


FIG. 3.3 Diagramatic representation of a chemostat.

experiments where community-based mechanisms are to be studied. Continuous culture also allows for the establishment of stable cultures of oligotrophic bacteria. This is because the whole of the cell cycle is not supported in continuous culture allowing exceptionally low nutrient levels to be used (Hardman *et al.*, 1993). Oligotrophs may have impacts on mineral dissolution in low nutrient environments and could be studied using this culture technique. Indeed the impact of other specialized heterotrophs growing slowly on relatively recalcitrant compounds could be studied using continuous culture.

One possible disadvantage of continuous culture studies is the fact that cells do not move through the whole cell cycle. During stationary phase in particular, a range of exo-metabolites may be produced that might influence dissolution. It is possible, of course, for the flow of nutrients into the continuous culture system to be stopped for periods causing the bacteria to move through the cell cycle. If this is done then steady state conditions are no longer maintained, growth rates will vary and any community of heterotrophs will be disrupted. Another possible disadvantage of continuous culture is the capacity of bacteria for rapid evolutionary change. This results in the selection of cells adapted to the conditions provided in the reactor. It is important to remember that this problem occurs in any long-term laboratory culture of bacteria.

There are a number of other culture systems used to model the natural environment worthy of mention here. The aim of using these model systems is to attempt to incorporate some element of the heterogeneity encountered by bacteria in the environment. There are multi-stage systems that can be simply described as a series of chemostats joined in a cascade with nurients supplied at one end of the systems only. There is unidirectional flow through this type of system (Wimpenny *et al.*, 1983; Wimpenny, 1992). This type of apparatus has been used to model sediment systems in order to investigate the sequential use of electron acceptors by the bacterial community (Wimpenny *et al.*, 1983).

Gradostats, and other similar systems, have been developed in order to model bi-directional flow through a habitat. They are comparable to the multi-stage chemostat described above except that there is input of material at the opposite end of the cascade. In these models, the introduced heterogeneity in the system allows for the whole cell cycle to be represented in different parts of the reactor. There are also interactions between the different components in the model (Wimpenny, 1992).

One further culture system used in the modelling of natural environments may be particularly interesting in terms of mineral dissolution mechanisms: the percolating column. Here, particles, e.g. of soil, are packed into a column, and medium fed to the top of the column is allowed to percolate down. Alternatively, medium can be fed continuously upwards through the porous medium. These systems do not achieve steady state since nutrients entering the column will be utilized and their concentrations reduced by the first 'layer' of microorganisms encountering them. The next layer of bacteria will therefore be exposed to lower concentrations of nutrients, and they will reach a lower cell density as a result. In addition, waste products of bacterial metabolism from each layer will flow to the next. The amount and type of these products may vary with time. Thus gradation will be established throughout the column and substrate utilization and microbial numbers will be dependent on spatial factors and on time. It is possible to build sampling ports into these columns along the length of the apparatus, allowing some determination of differences with depth (Wimpenny, 1992). Percolating columns allow the study of bacterial impacts on mineral dissolution in highly heterogeneous particulate systems under varying nutrient conditions. They may not, however, be appropriate for determining mechanisms underlying any dissolution processes.

It is also possible to design *in situ* bacterial growth systems. The design of these is highly varied and often specifically developed for the habitat in question, for example aquatic (e.g. Williams *et al.*, 1992) or groundwater (e.g. Rogers *et al.*, 1998). They are placed in the natural environments and can yield data on field-based events.

### *3.4.2 Methods of imaging bacteria*

Imaging bacteria is essential in order to assess their interactions with geological substrates. For example, imaging allows the evaluation of the degree of bacterial attachment, whether it is random or preferential, whether bacterial exoproducts are involved, etc. Epifluorescence microscopy, involving staining of the specimen with fluorescent dyes and subsequent viewing/imaging with a standard light microscope, is commonly used for low-resolution observations. It can be used to differentiate physiologically active cells from inactive cells, and can provide a method for assessing the density and distribution of each population. One problem associated with epifluorescence is high background fluorescence, which can be encountered in samples from natural ecosystems.

A variety of more complex imaging techniques is available. Sample preparation for more detailed *ex situ* imaging techniques (resolution of ~1 μm or better) is, however, far from simple. Air-drying causes dehydration of the bacterial mass. The loss of water, 70–86% of the biomass (Schlegel, 1993), means that only semi-transparent outlines of the bacterial cells are visible and that exoproducts (99% water; Sutherland, 1972) are lost altogether.

A substantial improvement on preservation comes with methods such as critical point drying. This involves chemical fixation of the sample using, for example, a glutaraldehyde solution, followed by dehydration, e.g. in a graded series of ethanol solutions and finally a critical point drying stage. This entails exchanging the alcohol for $CO_2$ and is followed by either drying and coating (e.g. Au) for Scanning Electron Microscopy (SEM), or by embedding in resin for Transmitted Electron Microscopy (TEM). It has been reported, however, that the harsh chemical treatment involved in this process can damage delicate cell structures (Barker *et al.*, 1998); there is also a substantial loss of shape of the extracellular material (Fig. 3.4). Knowledge of the structure and complexity of the extracellular material comes from observations of untreated and uncoated samples of bacterial biofilms, using environmental SEM (e.g. Little *et al.*, 1997), although a disadvantage of this method is a relatively poor resolution.

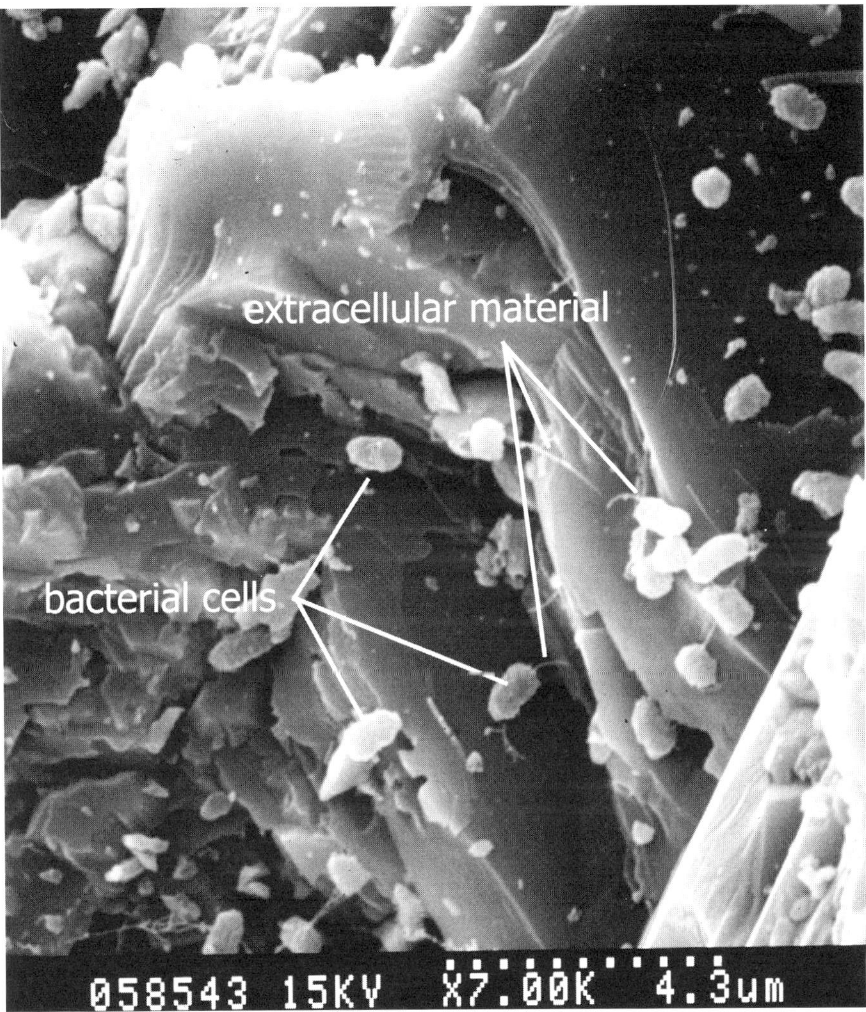


FIG. 3.4 Bacterial cells attached on feldspar surfaces, demonstrating imaging problems. The photo was taken 48 h after the beginning of the experiment, and there is only a small proportion of extracellular material. However the distortion is apparent, as the extracellular material should appear surrounding the cells, rather than in the form of strings stretching away from the cells.

Cryofixation, followed by cryomicroscopy appears to offer the best potential for preservation of the delicate bacterial structures for *ex situ* imaging. This method involves rapid freezing of the specimen using liquid nitrogen, so that the water forms vitreous ice, rather than ice crystals; the specimen is then cryo-coated and examined under cryo-SEM at −95°C (Barker *et al.*, 1998). Detailed images of cells and associated extracellular polymers attached to feldspar surfaces were produced by Barker *et al.* (1998) using cryofixation.

Recently developed high-resolution *in situ* methods for bacterial observations include atomic force microscopy (AFM) and confocal scanning laser microscopy (CSLM). The AFM using Fluid Tapping Mode has allowed direct observations of living bacteria attached to mineral surfaces (e.g. Grantham and Dove, 1996). The method allows cross-sectioning of the specimen surface to record relief variations and time series data recording of experiments in progress. The CSLM also allows *in vivo* examination of untreated samples. Vertical profiling, via multiple scans of a fluorescing specimen surface, provides a 3-dimentional digital image (Caldwell *et al.*, 1992). Ratiometric pH-sensitive fluorescent probes have been developed to measure pH gradient in the studied specimens (Barker *et al.*, 1998). There are difficulties with this technique, however, because it does not easily permit observations of both mineral and bacteria simultaneously, to relate surface morphology to bacterial attachment.

### 3.5 Field evidence of bacterial influence on mineral dissolution

Numerous studies have provided evidence, mostly circumstantial, for microbial influences in dissolution (e.g. Robert and Berthelin, 1986, and references therein). Microbial leaching of building stone, glass and other substrates have been reported (e.g. Krumbein *et al.*, 1991; Palmer *et al.*, 1991) and again offer circumstantial support for a microbial role in dissolution. On a larger scale, a microbially-mediated process was implicated in the generation of secondary porosity in sedimentary formations (McMahon and Chapelle, 1991).

In most field studies, however, it is difficult to differentiate bacterial effects from those of other microorganisms, especially fungi. Another shortfall of field studies is that although they often demonstrate close association between microbes and mineral surfaces, e.g. bacterial attachment, this is not necessarily evidence for enhanced dissolution by bacteria. A few studies where bacteria were specifically implicated in mineral dissolution are discussed below.

Enhanced silicate dissolution was shown to occur in an environment of high microbial activity, at a shallow petroleum-contaminated aquifer colonized by indigenous bacteria. Dissolution rates at this location were faster than theoretically predicted and mineral morphologies had apparently been influenced by bacterial activity (Hiebert and Bennett, 1992). Field experiments carried out on the same site, using microcosms introduced into the acquifer for several months, indicated that bacterial colonization and concomitant weathering can be selective (Rogers *et al.*, 1998). The minerals (feldspars) preferentially colonized were found to contain phosphorus mineral inclusions, a nutrient that is often limiting in such environments. This is an interesting observation, but it is too early to draw conclusions as to whether preferential attachment to certain minerals is driven by nutrient limitation and the potential release of nutrient elements within the mineral. Bacterial attachment is a non-specific physico-chemical process involving the solid surface, bacterial surface and the liquid phase and will vary with changes in any of these components (McEldowney and Fletcher, 1986; McEldowney, 1994). It is possible that the selective nature of colonization is simply a reflection of variations in the attachability of bacteria to different minerals in *in situ* conditions rather than the bacteria gaining a growth advantage on the minerals with phosphate inclusions due to solubilization of phosphorus. It is interesting to note, however, that bacterial dissolution of phosphate minerals in soils is well documented (e.g. Babenko *et al.*, 1984) (see below) and is known as an important mechanism of releasing previously unavailable phosphate for plant uptake (Atlas and Bartha, 1998).

A study of altered volcanic glasses from Iceland indicated that dissolution features identified by SEM could have been caused by bacterial activity, based on morphological observations of dissolution pits and evidence of bacterial attachment (Thorseth *et al.*, 1992).

The dissolution of pyrite and other sulphide minerals, and the generation of acid mine drainage has been unequivocally associated with the acid generation activity of lithoautotrophic bacteria, such as *Thiobacillus ferrooxidans*. There has been extensive research into this type of oxidative dissolution, mainly due to its industrial and environmental applications (metal recovery and acid mine drainage). The reader is referred to recent reviews elsewhere, e.g. Ehrlich (1996) and Nordstrom and Southam (1997). The dissolution of Mn compounds as a result of Mn-reducing bacteria is an example of a microbial weathering mechanism developed specifically to release this nutrient and the reader is referred to Ehrlich (1996) for a detailed discussion.

### 3.6 Laboratory studies: mechanisms

Several laboratory studies have demonstrated the ability of soil bacterial strains to influence (mostly increase) silicate and phosphate mineral dissolution (Babenko *et al.*, 1984; Duff *et al.*, 1963; Kutuzova, 1969). Most studies to date suggest that this process is ligand promoted, as a result of the bacterially produced organic acids, such as 2-ketogluconate, gluconate, lactate and acetate (Duff *et al.*, 1963; Vandevivere *et al.*, 1994; Webley *et al.*, 1960). However, the overall mechanism(s) by which bacteria influence dissolution is still unclear. As discussed above, a number of factors are likely to operate in tandem, grouped in two main categories: (1) bacterial exoproduct action; and (2) chelation and/or accumulation directly by the cell.

### *3.6.1 Exoproducts*

The extent of mineral dissolution by bacterial exoproducts will be a function of the type of exoproduct and the environment. For example, $CO_2$, which is a product of all bacterial metabolism, is only a weak acid. Carbonic acid ($H_2CO_3$) will have a minor effect in mineral dissolution as a ligand, but it may play a more important role by influencing pH. A typical example is the biodeterioration of concrete, where $CO_2$ production reduces the pH in the surface of concrete from 12.5 to 8.5. This not only makes concrete more susceptible to weathering, but also a better habitat for the growth of microorganisms that cannot tolerate high alkalinity (Sand, 1997). In most soils or aquifers, however, production of $CO_2$ is unlikely to have a major effect outside its pH-adjusting capacity. Even so it is possible to visualize small-scale local effects induced by attached bacteria respiring oxygen.

Sulphuric and nitric are strong acids, but are produced only by specialized autotrophic bacteria. For the production of sulphuric acid, the presence of a source of reduced sulphur, e.g. sulphide minerals, is a prerequisite, and thus this mechanism is specific to certain environments, e.g. acid mine drainage, sulphide mining heaps (Nordstrom and Southam, 1997). Nitrifying bacteria require the presence of ammonia, which is oxidized to nitric acid. Lebedeva *et al.* (1979) reported the dissolution of ultrabasic rocks by nitrifying bacteria. In terms of weathering capacity, both sulphur oxidizing bacteria and nitrifying bacteria are limited by the fact that they require aerobic conditions for growth. Besides, autotrophic dissolution is outside the scope of this review and will not be discussed further.

Bacterially-derived ammonia and other organic bases (ketones, ethanol, propanol), produced to the greatest extent in anaerobic conditions, may also influence pH and hence mineral dissolution. No studies appear to have been undertaken as to the effect of such bases in mineral dissolution, probably because these are chemically less potent weathering agents, as they would form weaker bonds to mineral surfaces than acids (Welch and Vandevivere, 1994).

There is a range of bacterially produced low molecular weight organic acids formed in both aerobic and anaerobic conditions that can influence mineral dissolution in the environment. A list of acids produced by organisms is presented, for example, in Robert and Berthelin (1986; see also Berthelin *et al.*, 2000). Organic acids tend to be weak, and thus their dissociation is moderate. As a result, organic acids have a smaller weathering effect due to proton-promoted dissolution compared to inorganic acids. They do, however, tend to be good ligands, associating with cations in solution, and generating apparent undersaturation and conditions for more intense mineral dissolution. This is further supported by evidence suggesting that organic acids enhance mineral dissolution rates by a factor of up to 10 compared to inorganic acids; their effect being most prominent at near-neutral pH (Welch and Ullman,

1993, 1996). Vandevivere *et al.* (1994) also investigated the effect of organic acids at neutral pH. They found enhanced mineral dissolution under these conditions, providing evidence that the mode of action of organic acids is predominantly through ligand binding rather through acidity effects.

All bacterially-produced organic acids may be important locally, but the most abundant types, such as acetic or oxalic, can have large-scale effects. Oxalic acid has been reported to release $Al^{3+}$ from plagioclase feldspar at higher rates than $Si^{4+}$ at neutral pH, an effect opposite to that observed in the presence of inorganic acids under the same conditions (Welch and Ullman, 1993). Such observations are very important, as they suggest that it may be possible to differentiate between microbial and inorganic dissolution in natural systems.

Cation binding by siderophores may potentially enhance dissolution of common rock and soil minerals that do not contain Fe, such as feldspars, due to their affinity for other cations, e.g. $Al^{3+}$, as discussed earlier. This hypothesis has not yet been tested.

Large organic molecules, such as polysaccharides, have been shown to both enhance and hinder dissolution, as discussed above. These studies have considered the action of polysaccharides *in vitro*. Ullman *et al.* (1996) discussed the potential of polysaccharides to inhibit dissolution *in vitro*, and this was demonstrated experimentally by Welch and Vandevivere (1994) and Welch *et al.* (1999), who showed that polysaccharides such as alginate, pectin, gum xanthan and polyaspartate play a complex role in feldspar dissolution. They play a predominantly inhibitory role, although their behaviour may vary as a function of pH and the type of polysaccharide. Avakyan *et al.* (1986) and Malinovskaya *et al.* (1990) on the other hand, reported only enhancement of dissolution of silicates and release of Si in the presence of polysaccharides excreted by bacteria.

An experimental study of microbial colonization of metal surfaces indicated that the development of a biofilm increased the overall susceptibility of the surface to corrosion (Little *et al.*, 1997). It is likely that in nature the effect of such exoproducts, excreted by the bacteria to 'condition' the environment around them, might have an effect difficult to predict from laboratory experiments. Certainly polysaccharides, as macromolecules, will be irreversibly adsorbed at surfaces (Savage and Fletcher, 1985). Polysaccharides are highly hydrated and anionic polymers, which bind cations with high efficiency (Ferris *et al.*, 1989). Whether this might act to enhance or reduce *in situ* dissolution requires further investigation. Indeed, it is possible that the impact will vary with mineral type and environmental conditions. Bacteria attached to solid surfaces often produce exopolysaccharides that become permanently adsorbed to the surface. Bacterial growth after attachment to a surface ultimately results in the formation of a biofilm that consists of cells and exopolymer (Caldwell *et al.*, 1992). The proportions of cells to polymer vary with growth conditions and with time, thus any impact on dissolution caused by polysaccharide associated with attached cells may well vary in time and with conditions. In addition, mineral surfaces vary in their physicochemical characteristics so that bacterial 'attachability' to one part of a mineral may be greater than to another (see below). Thus, amounts of cell-associated exopolymer are likely to vary across the mineral surface as will dissolution impacts caused by polysaccharides.

Microorganisms produce exoenzymes to degrade complex and large organic substrates such as cellulose and lignin to smaller units, dimers and monomers, that can be taken up by the bacteria (Neidhart *et al.*, 1990). There is, as yet, no evidence to suggest that bacterial exoenzymes contribute to mineral dissolution processes (Sand, 1997).

### *3.6.2 Bacterial cells*

*3.6.2.1 Cells in suspension.* There are many studies on the potential of bacterial cells to act as chelating agents, accumulating cations from solution internally or externally at the cell surface. Accumulation at the cell surface is a physicochemical process, which is rapid. Internal accumulation is often a slower process, and may involve cell physiological activity and specialized cell membrane transport processes. Even toxic cations with no known biological function are accumulated. Cation accumulation can be to exceptionally high levels (McEldowney, 1990; Hardman *et al.*; 1993; Gadd, 1992). Such accumulation by bacteria may significantly influence cation partitioning in natural environments. It is not yet known how significant such direct chelation by the cells may be in mineral dissolution. It has, however, been shown that metal adsorption onto bacteria can be studied using equilibrium thermodynamics for the interactions occurring at the cell-solution interface. It is, therefore, possible to make theoretical predictions of the stability constants for 'bacteria-metal' species in solution, as shown by Fein *et al.* (1997). A further comparative study of two different species of bacteria by the same research group, however, showed that the two species had different relative and absolute concentrations of binding surface sites and slightly different adsorption constants (Daughney *et al.*, 1998). The basis for these differences is probably found in variations in cell wall characteristics.

Metal binding by bacteria is likely to vary *in situ* in both time and space. This is due to the variability of bacterial cell surfaces (McEldowney and Fletcher, 1986; McEldowney, 1994), and therefore surface accumulation, with growth condition and species (McEldowney, 1990), and the impact of environmental conditions on cell activity, and therefore internal accumulation. As with all bacterial-based mechanisms for mineral dissolution, there is likely to be variability in impact because of the inherent flexibility in the physiological response of bacteria to their environment. The relative importance of cation chelation by bacterial cells to mineral dissolution, compared to other mechanisms, cannot be assessed at the present moment.

*3.6.2.2 Cells attached.* Mineral dissolution via direct bacterial attachment may cause preferential dissolution of certain surfaces. Preferential dissolution may either be due to the close proximity of bacteria to surface, increasing the impact of exoproducts (well documented in corrosion studies, e.g. Palmer *et al.*, 1991) or possibly through the introduction of cell-mineral interactions that do not occur when bacteria are free in the aqueous phase.

Bacterial surface characteristics are complex, with a range of negatively and positively charged groups, the ionization of which varies with pH, and hydrophobic sites exposed to the surrounding environment. The overall surface charge, however, is negative, regardless of whether the bacterium is Gram-negative or Gram-positive (Beveridge and Fyfe, 1985). Minerals also have a surface charge, which is mainly a function of pH and solution composition, as discussed earlier. It is interactions between the charged cell and mineral sites that determine the bacterial attachment to surfaces. Bacterial attachment will also be affected by environmental factors such as temperature and variations in growth conditions, which affect the cell surface (Savage and Fletcher, 1985; McEldowney and Fletcher, 1986). Surface roughness also affects attachment, possibly through the exposure of more or different attachment sites at the mineral surface (Ehrlich, 1996) or through an impact on surface energy. Bacterial attachment can be described in terms of free energy interactions (Savage and Fletcher, 1985; Denyer *et al.*, 1993). It is a common observation that bacteria tend to colonize rough surfaces (Fig. 3.5), suggesting that minerals with a high proportion of rough surfaces (e.g. phylosilicates) might be colonized more than minerals occurring as smoother crystals. It has also been shown that bacteria can attach preferentially to certain substrates, i.e. Fe-oxide coated rather than uncoated quartz surfaces, particularly under nutrient-limited conditions (e.g. Grantham and Dove, 1996). In nutrient-deficient conditions, bacterial attachment is often increased, possibly as a result of cell surfaces becoming more hydrophobic. Grantham and Dove (1996) showed that the imprint of a bacterial cell was apparent after the cell was removed, indicating direct damage to the surface underneath.

Once attached to a mineral surface, all the mechanisms of dissolution discussed above may be operative, but in closer proximity to the mineral. It may also be possible that further novel dissolution mechanisms do occur through cell surface contact with the mineral. Attachment to surfaces results in considerable changes in macromolecular conformation of the cell wall and in changes in the charge density and distribution across the cell wall and membrane (Savage and Fletcher, 1985). Might such changes have a further impact on mineral dissolution?

## 3.7 Conclusions and future work

The focus of this chapter has been a consideration of the interactions between heterotrophic bacteria and minerals, both in physical and chemical terms.

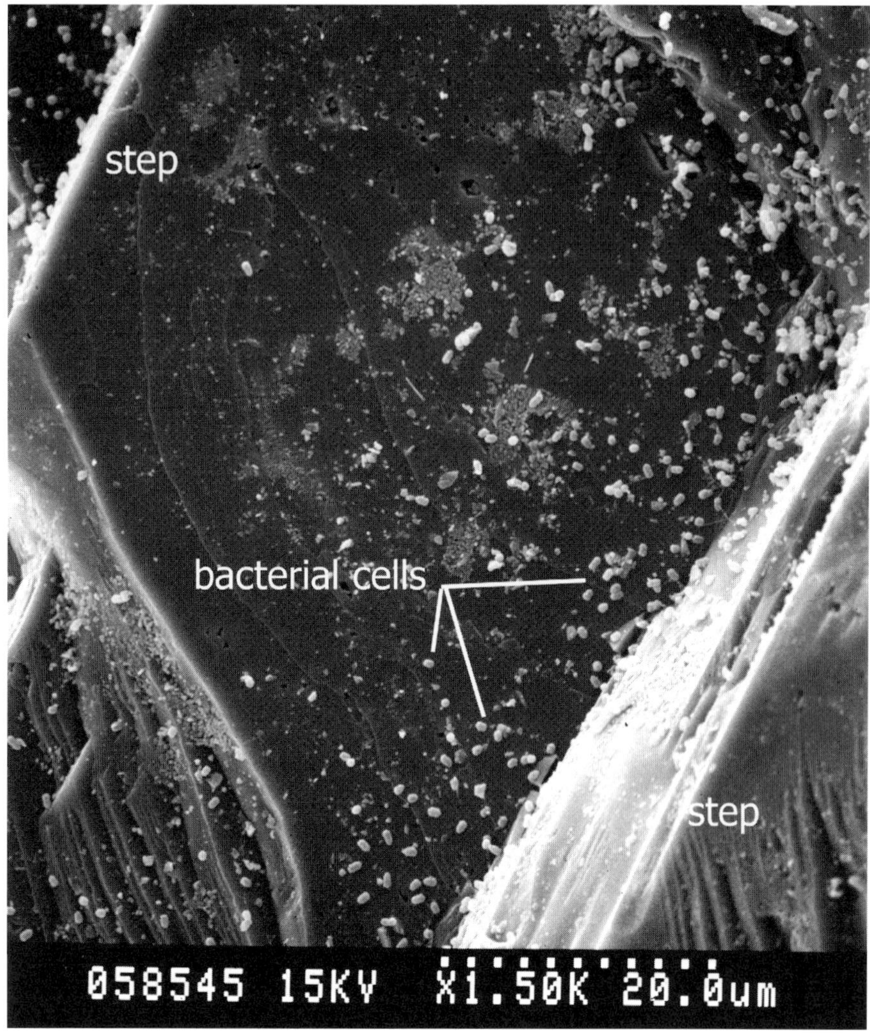


FIG. 3.5 Bacterial attachement patterns on a feldspar, 48 h after the beginning of an experiment. Cells show higher affinity for the rough surfaces near and along the steps, rather than the flat surface in between.

Bacterial biology was discussed with emphasis on the inherent adaptability and flexibility of heterotrophs, as well as their dynamic reactions to and interactions with the environment. Their abundance in ordinary, as well as the most extreme environments was explained, as well as field evidence on their role in mineral dissolution. Experimental work using bacteria, or their exoproducts, with emphasis on potential dissolution mechanisms was discussed.

Controls of the experimental circumstances (apparatus, conditions, nutrients) on the results were put into context.

Laboratory conditions represent a point worthy of further discussion, particularly as experimental work to date has reached a critical mass. Although it has been demonstrated in several studies that heterotrophs are capable of influencing mineral dissolution, it is still unclear to what extent this happens in nature. This is partly because field and laboratory studies are not carried out in tandem, but also because of the inherent difficulties of assessing mineral reactivity in the field. It is also argued here that some of the existing work is difficult to interpret due to limitations of the chosen experimental conditions. One of the main objectives of the above discussion was to emphasize the complexity of work involving the two systems, bacteria and minerals, combined, so as to encourage more comprehensive and rigorous experiments in the future.

For example, bacterial strains used in experimental work can be 'off the shelf' laboratory cultures. However, our experience is that after prolonged growth on high nutrient medium the behaviour of bacteria may change significantly (Valsami-Jones and McEldowney, unpublished data). Such results may therefore not reflect the real behaviour of the bacterial species in the environment. The recommended approach is to collect fresh samples from appropriate sites and use them to isolate a heterotrophic population for laboratory work.

Most experiments performed to date have been carried out in one type of medium only. These, however, do not reflect the conditions encountered by heterotrophs in nature; the scenario of feast-famine conditions experienced by heterotrophs was discussed above. Future work should assess bacterially induced mineral dissolution in a range of media, ideally including media that would introduce the type of stress heterotrophs may encounter naturally, e.g. low nutrients, limited carbon source or lack of iron. The type of exoproducts that may be excreted under such conditions could give us a better understanding of what exactly are the main contributing factors to dissolution. Another important issue when conducting or interpreting laboratory studies is that of the life cycle (Fig. 3.2 and discussion above) of bacteria in batch cultures. Bacteria excrete different products during active growth compared to stationary phase. Also, the biochemistry of lysing cells, during death phase, is distinct and may have a different effect to mineral dissolution. It is important that the cell population is recorded in parallel to other variables (chemistry, pH) during experiments, for the results to be meaningful.

Studies of oligotrophs are also needed, and of bacteria acting as consortia, which is common in natural systems: there is a likelihood that a synergistic effect may be observed. In fact, a very small number of heterotrophs have been studied to date, and few have been assessed in a range of conditions. It is likely that we have only observed a very small proportion of the range of effects heterotrophs may have on mineral dissolution. It is, therefore, fair to say that we are only beginning to understand the effects of heterotrophs in environmental mineralogy and chemistry; further diversity and significance of heterotrophic mineral dissolution is likely to be uncovered in future work.

## Acknowledgements

The authors acknowledge a grant by The Natural History Museum Research Fund. William Gaze, Judith McLean, Heiko Hinrichs and Anna Poli have worked with the authors to develop laboratory protocols for bacteria-mineral research. Thoughtful reviews by Janet Cotter-Howells and an anonymous reviewer substantially improved the clarity of the manuscript.

## References

Alexander, M. (1977) *Introduction to Soil Microbiology.* John Wiley & Sons, New York.

Atlas, R.M. and Bartha, R. (1998) *Microbial Ecology. Fundamentals and Applications* 4th edition. Benjamin/Cummings, CA, USA.

Avakyan, Z.A., Pivovarova, T.A. and Karavaiko, G.I. (1986) Properties of a new species *Bacillus mucilaginosus*. *Mikrobiologiya*, **55**, 369–74.

Babenko, Yu.S., Tyrygina, G.I., Grigor'ev, E.F., Dolgikh, L.M. and Borisova, T.I. (1984) Biological activity and physiological-biochemical properties of phosphate dissolving bacteria. *Mikrobiologiya*, **53**, 427–33.

Barker, W.W., Welch, S.A., Chu, S. and Banfield, J.F. (1998) Experimental observations of the effects of bacteria on aluminosilicate weathering. *Amer. Mineral.*, **83**, 1551–63.

Barns, S.M. and Nierzwicki-Bauer, S. (1997) Microbial diversity in ocean, surface and subsurface environments. Pp. 35–79 in: *Geomicrobiology: Interactions between Microbes and Minerals* (J.F. Banfield and K.H. Nealson, editors). Reviews in Mineralogy, **35**. Mineralogical Society of America, Washington D.C.

Berger, G., Cadore, E., Schott, J. and Dove, M. (1994) Dissolution of quartz in lead and sodium electrolyte solutions between 25 and 300°C: Effect of the nature of surface complexes and reaction affinity. *Geochim. Cosmochim. Acta*, **58**, 541–51.

Berthelin, J., Leyval, C. and Mustin, C. (2000) Illustrations of the occurrence and diversity of mineral-microbe interactions involved in the weathering of minerals. Pp. 7–25 in: *Environmental Mineralogy: Microbial Interactions, Anthropogenic Influences, Contaminated Land and Waste Management* (J.D. Cotter-Howells, L.S. Campbell, E. Valsami-Jones and M. Batchelder, editors). Mineralogical Society Series, **9**. Mineralogical Society, London.

Beveridge, T.J. and Fyfe, W.S. (1985) Metal fixation by bacteria cell walls. *Canad. J. Earth Sci.*, **22**, 1893–8.

Brady, P.V. and Walther, J.V. (1989) Controls on silicate dissolution rates in neutral and basic pH solutions at 25°C. *Geochim. Cosmochim. Acta*, **53**, 2823–30.

Bosbach, D., Hall, C. and Putnis, A. (1998) Mineral precipitation and dissolution in aqueous solution: in-situ microscopic observations on barite (001) with atomic force microscopy. *Chem. Geol.*, **151**, 143–60.

Brock, T.D. (1970) *Biology of Microorganisms.* Prentice-Hall Inc., New Jersey.

Caldwell, D.E., Korber, D.R. and Lawrence, J.R. (1992) Confocal laser microscopy and digital image analysis in microbial ecology. *Adv. Microbial Ecol.*, **12**, 1–67.

Chapelle, F.H. (1993) *Ground-water Microbiology and Geochemistry.* John Wiley & Sons, New York.

Daughney C.J., Fein, J.B. and Yee N. (1998) A comparison of the thermodynamics of metal adsorption onto two common bacteria. *Chem. Geol.*, **144**, 161–76.

Denyer, S.P., Hanlon, G.W. and Davies, M.C. (1993) Mechanisms of microbial adherence. Pp. 13–28 in: *Microbial Biofilms: Formation and Control* (S.P. Denyer, S.P. Gorman and M. Sussman, editors). Society for Applied Bacteriology, Technical Series, **30**. Blackwell Scientific Publications, Oxford.

Douglas, S. and Beveridge, T.J. (1998) Mineral formation by bacteria in natural microbial communities. *FEMS Microbiol. Ecol.*, **26**, 79–88.

Duff, R.B., Webley, D.M. and Scott, R.O. (1963) Solubilization of minerals and related materials by 2-ketogluconic acid-producing bacteria. *Soil. Sci.*, **95**, 105–14.

Ehrlich, H.L. (1996) *Geomicrobiology*. Marcel Dekker, Inc., New York.

Fein, J.B., Daughney, C.J., Yee, N. and Davis, T. (1997) A chemical equilibrium model of metal adsorption onto bacterial surfaces. *Geochim. Cosmochim. Acta*, **61**, 3319–28.

Ferguson, S. (1992) The periplasm. Pp. 311–40 in: *Prokaryotic Structure and Function: a New Perspective* (S. Mohan, C. Dow and J.A. Cole, editors). Society for General Microbiology Symposium **47**. Cambridge University Press, Cambridge, UK.

Ferris, F.G., Schultze, S., Witten, T.C., Fyfe, W.S. and Beveridge, T.J. (1989) Metal interactions with microbial biofilms in acidic and neutral pH environments. *Appl. Environ. Microbiol.*, **55**, 1249–57.

Fortin, D., Ferris, F.G. and Beveridge, T.J. (1997) Surface-mediated mineral development by bacteria. Pp. 161–80 in: *Geomicrobiology: Interactions between Microbes and Minerals* (J.F. Banfield and K.H. Nealson, editors). Reviews in Mineralogy, **35**. Mineralogical Society of America, Washington D.C.

Gadd, G.M. (1992) Microbial control of heavy metal pollution. Pp. 59–88 in: *Microbial Control of Pollution* (J.C. Fry, G.M. Gadd, R.A. Herbert, C.W. Jones and I.A. Watson-Craik, editors). Society for General Microbiology Symposium **48**. Cambridge University Press, Cambridge, UK.

Gadd, G. (2000) Heterotrophic solubilization of metal-bearing minerals by fungi. Pp. 57–75 in: *Environmental Mineralogy: Microbial Interactions, Anthropogenic Influences, Contaminated Land and Waste Management* (J.D. Cotter-Howells, L.S. Campbell, E. Valsami-Jones and M. Batchelder, editors). Mineralogical Society Series, **9**. Mineralogical Society, London.

Grantham, M.C. and Dove, P.M. (1996) Investigation of bacterial-mineral interactions using Fluid Tapping Mode Atomic Force Microscopy. *Geochim. Cosmochim. Acta*, **60**, 2473–80.

Hammond, S.M., Lambart, P.A. and Rycroft, A.N. (1984) *The Bacterial Cell Surface*. Croom Helm, London.

Hardman, D., McEldowney, S. and Waite, S. (1993) *Pollution: Ecology and Biotreatment*. Longman, Essex, UK.

Hiebert, F.K. and Bennett, P.C. (1992) Microbial control of silicate weathering in organic-rich ground water. *Science*, **258**, 278–81.

Krumbein, W.E., Urzi, C.E. and Gehrmann, C. (1991) Biocorrosion and biodeterioration of antique and medieval glass. *Geomicrobiol. J.*, **9**, 139–60.

Krumholz, L.R. (1998) Microbial ecosystems in the earth's subsurface. *ASM News*, **64**, 197–202.

Krumholz, L.R., McKinley, J.P., Ulrich, F.A. and Suflita, J.M. (1997) Confined subsurface microbial communities in Cretaceous rock. *Nature*, **386**, 64–6.

Kutuzova, R.S. (1969) Release of silica from minerals as a result of microbial activity. *Mikrobiologiya*, **38**, 596–602.

Lebedeva, E.V., Lyalikova, N.N. and Bugel'skii, Y.Y. (1979) Participation of nitrifying bacteria in the weathering of serpentinized ultrabasic rocks. *Mikrobiologiya*, **47**, 898–904.

Little, B.J., Wagner, P.A. and Lewandowski, Z. (1997) Spatial relationships between bacteria and mineral surfaces. Pp. 123–59 in: *Geomicrobiology: Interactions between Microbes and Minerals* (J.F. Banfield and K.H. Nealson, editors). Reviews in Mineralogy, **35**. Mineralogical Society of America, Washington D.C.

Malinovskaya, I.M., Kosenko, L.V., Votselko, S.K. and Podgorskii, V.S. (1990) Role of *Bacillus mucilaginosus* polysaccharide in degradation of silicate minerals. *Mikrobiologiya*, **59**, 49–55.

McEldowney, S. (1990) Microbial bioaccumulation of radionuclides in waste stream treatment. *Appl. Biochem. Biotechnol.*, **26**, 159–79.

McEldowney, S. (1994) The effect of cadmium and zinc on the attachment and detachment interactions of *Pseudomonas fluorescens H2* with glass. *Appl. Environ. Microbiol.*, **60**, 2759–65.

McEldowney, S. and Fletcher, M. (1986) Effect of growth conditions and surface characteristics of aquatic bacteria on their attachment to solid surfaces. *J. Gen. Microbiol.*, **132**, 513–23.

McEldowney, S. and Fletcher, M. (1988*a*) Bacterial desorption from food container and food processing surfaces. *Microbial Ecol.*, **15**, 229–37.

McEldowney, S. and Fletcher, M. (1988*b*) The effect of temperature and relative-humidity on the survival of bacteria attached to dry solid surfaces. *Lett. Appl. Microbiol.*, **7**, 83–6.

McEldowney, S. and Fletcher, M. (1988*c*) Effect of pH, temperature, and growth conditions on the adhesion of a gliding bacterium and three nongliding bacteria to polystyrene. *Microbial Ecol.*, **16**, 183–95.

McMahon, P.B. and Chapelle, F.H. (1991) Microbial production of organic acids in aquitard sediments and its role in aquifer geochemistry. *Nature*, **349**, 233–5.

Nealson, K.H. and Stahl, D.A. (1997) Microorganisms and biogeochemical cycles: what can we learn from layered microbial communities. Pp. 5–34 in: *Geomicrobiology: Interactions Between Microbes and Minerals* (J.F. Banfield and K.H. Nealson, editors). Reviews in Mineralogy, **35**. Mineralogical Society of America, Washington D.C.

Neidhardt, F.C., Ingraham, J.L. and Schaechter, M. (1990) *Physiology of the Bacterial Cell. A Molecular Approach.* Sinaues Ass. Inc., Massachusetts.

Nordstrom, D.K. and Southam, G. (1997) Geomicrobiology of sulfide mineral oxidation. Pp. 361–90 in: *Geomicrobiology: Interactions Between Microbes and Minerals* (J.F. Banfield and K.H. Nealson, editors). Reviews in Mineralogy, **35**. Mineralogical Society of America, Washington D.C.

Palmer, R.J., Jr., Siebert, J. and Hirsch, P. (1991) Biomass and organic acids in sandstone of a weathered building: Production by bacterial and fungal isolates. *Microbial. Ecol.*, **21**, 253–66.

Parkes, R. J., Cragg, B.A., Bale, S.J., Getliff, J.M., Goodman, K., Rochelle, P.A., Fry, J.C., Weightman, A.J. and Harvey, S.M. (1994) Deep bacterial biosphere in Pacific Ocean sediments. *Nature*, **371**, 410–3.

Robert, M. and Berthelin J. (1986) Role of biological and biochemical factors in soil mineral weathering. Pp. 453–95 in: *Interactions of Soil Minerals with Natural Organics and Microbes* (P.M. Huang and M. Schnitzer, editors). SSSA special publication, **17**. Soil Science Society of America, Wisconsin, USA.

Rogers, J.R., Bennett, P.C. and Choi, W.J. (1998) Feldspars as a source of nutrients for microorganisms. *Amer. Mineral.*, **83**, 1532–40.

Sand, W. (1997) Microbial mechanisms of deterioration of inorganic substrates – a general mechanistic overview. *Int. Biodet. Biodeg.*, **40**, 183–90.

Savage, D.C. and Fletcher M. (1985) *Bacterial Adhesion: Mechanisms and Physiological Significance*. Plenum Press, New York.

Schlegel, H.G. (1993) *General Microbiology* (7th Edition). Cambridge University Press, Cambridge, UK.

Silver, S. (1997) The bacterial view of the periodic table: specific functions for all elements. Pp. 345–60 in: *Geomicrobiology: Interactions between Microbes and Minerals* (J.F. Banfield and K.H. Nealson, editors). Reviews in Mineralogy, **35**. Mineralogical Society of America, Washington D.C.

Sinclair, J.L. and Ghiorse, W.C. (1989) Distribution of aerobic-bacteria, protozoa, algae, and fungi in deep subsurface sediments. *Geomicrobiol. J.*, **7**, 15–31.

Stanier, R.Y., Ingraham, J.L., Wheelis, M.L. and Painter, P.R. (1987) *General Microbiology* (5th edition). Macmillan Education, Hong Kong.

Stevens, T.O. and McKinley, J.P. (1995) Lithoautotrophic microbial ecosystems in deep basalt aquifers. *Science*, **270**, 450–4.

Stumm, W. and Morgan, J.J. (1996) *Aquatic Chemistry. Chemical Equilibria and Rates in Natural Waters* (3rd edition). John Wiley & Sons, New York.

Stumm, W. and Wollast, R. (1990) Coordination chemistry of weathering: kinetics of the surface-controlled dissolution of oxide minerals. *Rev. Geophys.*, **28**, 53–69.

Sutherland, I.W. (1972) Bacterial exopolysaccharides. *Adv. Microbial Physiol.*, **6**, 142–213.

Thorseth, I.H., Furnes, H. and Heldal, M. (1992) The importance of microbiological activity in the alteration of natural basaltic glass. *Geochim. Cosmochim. Acta*, **56**, 845–50.

Ullman, W.J., Kirchman, D.L., Welch, S.A. and Vandevivere, P. (1996) Laboratory evidence of microbially mediated silicate mineral dissolution in nature. *Chem. Geol.*, **132**, 11–7.

Vandevivere, P., Welch, S.A., Ullman, W.J. and Kirchman, D.L. (1994) Enhanced dissolution of silicate minerals by bacteria at near-neutral pH. *Microbial. Ecol.*, **27**, 241–51.

Veal, D.A., Stokes, H.W. and Daggard, G. (1992) Genetic exchange in natural microbial communities. *Adv. Microbial Ecol.*, **12**, 383–430.

Watteau, F. and Berthelin, J. (1994) Microbial dissolution of iron and aluminium from soil minerals: efficiency and specificity of hydroxamate siderophores compared to aliphatic acids. *Eur. J. Soil Biol.*, **30**, 1–9.

Webley, D.M., Duff, R.B. and Mitchell, W.A. (1960) A plate method for studying the breakdown of synthetic and natural silicates by soil bacteria. *Nature*, **188**, 766–7.

Welch, S.A. and Ullman, W.J. (1993) The effect of organic acids on plagioclase dissolution rates and stoichiometry. *Geochim. Cosmochim. Acta*, **57**, 2725–36.

Welch, S.A. and Ullman, W.J. (1996) Feldspar dissolution in acidic and organic solutions: Compositional and pH dependence of dissolution rate. *Geochim. Cosmochim. Acta*, **60**, 2939–48.

Welch, S.A. and Vandevivere, P. (1994) Effect of microbial and other naturally occurring polymers on mineral dissolution. *Geomicrobiol. J.*, **12**, 227–38.

Welch, S.A., Barker, W.W. and Banfield, J.F. (1999) Microbial extracellular polysaccharides and plagioclase dissolution. *Geochim. Cosmochim. Acta*, **63**, 1405–19.

Williams, H.G., Day, M.J. and Fry, J.C. (1992) Detecting natural transformation of *Acinetobacter calcoaceticus, in situ*, within natural epilithosis of the River Taff. Pp. 253–6 in: *The Release of Genetically Modified Microorganisms* (D.E.S. Stewart-Tull and M. Sussman, editors). REGEM 2. Federation of European Microbiological Societies (FEMS) symposium no **63**.

Wimpenny, J.W.T. (1992) Microbial systems: patterns in time and space. *Adv. Microbial Ecol.*, **12**, 469–522.

Wimpenny, J.W.T., Lovitt, R.W. and Coombs, J.P. (1983) Laboratory model systems for the investigation of spatially and temporally organised microbial ecosystems. Pp 67–118 in: *Microbes in the Natural Environments* (J.H. Slate, R. Whillenbery and J.W.T. Wimpenny, editors). The Society for General Microbiology, Symposium **34**.

Wimpenny, J.W.T., Kinniment, S.L. and Stonefield, M.A. (1993) The physiology and biochemistry of biofilm. Pp. 51–94 in: *Microbial Biofilms: Formation and Control* (S.P. Denyer, S.P. Gorman and M. Sussman, editors). Society for Applied Bacteriology Technical Series, **30**. Blackwell Scientific Publications, Oxford.

Woese, C.R. (1987) Bacterial evolution. *Microbiol. Rev.*, **51**, 221–71.

CHAPTER FOUR

# Heterotrophic solubilization of metal-bearing minerals by fungi

G. M. GADD

*Department of Biological Sciences, University of Dundee, Dundee DD1 4HN, UK (E-mail: g.m.gadd@dundee.ac.uk)*

**ABSTRACT**

The production of organic acids by fungi has profound implications for metal speciation and biogeochemical cycles. Metal-complexing properties of organic acids, e.g. citric and oxalic acid, assists essential metal and anionic (e.g. phosphate) nutrition of fungi, other microorganisms and plants, and affects metal speciation and mobility in the environment, including transfer between terrestrial and aquatic habitats, biocorrosion and weathering. Metal solubilization processes also have potential for metal recovery from contaminated solid wastes, soils and low-grade ores. Such 'heterotrophic leaching' can occur by several mechanisms but organic acids occupy a central position in the overall process supplying both protons and metal-complexing organic acid anions, e.g. citrate. Most simple metal oxalates (except those of alkali metals, Fe(III) and Al) are sparingly soluble and precipitate as crystalline or amorphous solids. Calcium oxalate is the most important oxalate in the environment and is ubiquitously associated with free-living, symbiotic and pathogenic fungi. The main forms are the monohydrate (whewellite) and the dihydrate (weddelite) and their formation affects nutritional heterogeneity in soil, especially that of Ca, P, K and Al, while in semi-arid environments, calcium oxalate formation is important in the development of terrestrial subsurface limestones. The formation of insoluble toxic metal oxalates, e.g. Cu, may ensure fungal survival in the presence of elevated metal concentrations.

## 4.1 Introduction

Fungi are eukaryotic heterotrophic organisms, requiring organic substances for their nutrition, and are ubiquitous and important in natural environments and industrial processes. While several important fungi exist in unicellular forms, e.g. yeasts, the majority exhibit a filamentous mode of growth which enables efficient colonization of substrates. Each filament is called a hypha, which exhibits apical growth and branching, and all the hyphae constitute the mycelium. This growth form is functionally efficient and underpins many of the key roles played by fungi in the environment (see Gow and Gadd, 1995; Frankland *et al.*, 1996; Gow *et al.*, 1999). In the terrestrial environment, fungi are of fundamental importance as decomposer organisms, plant pathogens and symbionts (mycorrhizas) and in soil can comprise the largest pool of biomass

Gadd, G.M. (2000) Heterotrophic solubilization of metal-bearing minerals by fungi. Pp. 57–75 in: *Environmental Mineralogy: Microbial Interactions, Anthropogenic Influences, Contaminated Land and Waste Management* (J.D. Cotter-Howells, L.S. Campbell, E. Valsami-Jones and M. Batchelder, editors). Mineralogical Society Series, **9**. Mineralogical Society, London. ISBN xxxx-xxxx-xxxx.

including other microorganisms and invertebrates (Metting, 1992). Fungi are therefore biologically distinct from bacteria and have some different ecological roles. It should be noted, however, that both these groups of organisms are intimately involved in the biogeochemical cycling of elements in the biosphere (Frankland *et al.*, 1996) (see Valsami-Jones and McEldowney, 2000, for bacteria). Metals and their compounds can interact with fungi in various ways depending on the metal species, organism and environment, while fungal metabolism also influences metal speciation and mobility (Gadd, 1993; Wainwright and Gadd, 1997). Some mechanisms mobilize metals into forms available for cellular uptake and/or leaching from the system, e.g. organic acid complexation, siderophore excretion (Francis, 1994), while immobilization may result from sorption onto cell components, and intra- and extracellular sequestration or precipitation (Morley and Gadd, 1995; Gadd, 1996; Sayer and Gadd, 1997; White *et al.*, 1997). These opposing processes of solubilization and immobilization are components of biogeochemical cycles for indigenous or introduced metals, as well as associated elements, e.g. S and P, and important determinants of fungal growth and morphogenesis (Morley *et al.*, 1996; Ramsay *et al.*, 1999).

Organic acids have important roles in fungal nutrition and physiology but are also important in metal biogeochemistry (Gadd, 1999). The production of metal-complexing organic acids assists essential metal and anionic nutrition of fungi and plants by means of the solubilization of, e.g. phosphate and sulphate, from insoluble metal-containing substances and is also important in biodeterioration and weathering. The formation of organic acid complexes is also important in metal transfer between terrestrial and aquatic habitats, and metal recovery from wastes and low grade ores ('heterotrophic leaching') (Francis *et al.*, 1992; Burgstaller and Schinner, 1993; Francis and Dodge, 1994; Gadd, 1999). Metal immobilization by insoluble metal oxalate formation is of environmental significance in the contexts of fungal survival, biodeterioration, soil weathering, mineral formation and metal detoxification. Oxalate-containing or derived minerals include humboldtine (ferrous oxalate dihydrate, $FeC_2O_4{\cdot}2H_2O$), whewellite (calcium oxalate monohydrate, $CaC_2O_4{\cdot}H_2O$) and weddelite (calcium oxalate dihydrate, $CaC_2O_4{\cdot}2H_2O$), with significant fungal involvement in their production. This chapter will outline the physiology and chemistry of citric and oxalic acid production in fungi and their importance in metal biogeochemistry and bioremediation.

## 4.2 Organic acid biosynthesis

### *4.2.1 Fungal biosynthesis of citric and oxalic acid*

Citric acid is an intermediate in the tricarboxylic acid cycle (see Kubicek, 1998; Wolschek and Kubicek, 1999). Growth-limiting concentrations of Mn, Fe and Zn are essential to achieve high yields of citric acid, e.g. Mn concentrations as low as 40 nM can markedly reduce acid production (Mattey,

1992). Oxalic acid is a by-product of citric acid biosynthesis, and its synthesis depends on whether glucose or citric acid is used as the carbon source (Wolschek and Kubicek, 1999). Biosynthesis on glucose occurs by hydrolysis of oxaloacetate to oxalate and acetate catalysed by cytosolic oxaloacetase (oxaloacetate (acetyl)hydrolase); this process competes with citrate production (Wolschek and Kubicek, 1999). Where citric acid is used, oxalic acid production occurs by means of the glyoxylate cycle (Hodgkinson, 1977; Dutton and Evans, 1996; Wolschek and Kubicek, 1999). Oxalic acid/oxalate production is widespread in fungi and this can be affected by the type of carbon and nitrogen source(s) and the pH of the environment or culture medium (Micales, 1994, 1995; Dutton and Evans, 1996; Shimada *et al.*, 1997; Gharieb and Gadd, 1999).

## 4.3 Metal chemistry of citric and oxalic acid

### *4.3.1 Metal-complex formation*

Organic acids can form coordination compounds or complexes with metals. If the organic acid, e.g. citric and oxalic acid, contains two or more electron donor groups so that ring-like structures are formed, then the resulting complexes are metal chelates. Citric acid (Fig. 4.1*a*) can form mononuclear, binuclear or polynuclear complexes depending on the metal (Fig. 4.1*b,c,d*). $Ca^{2+}$, $Fe^{3+}$ and $Ni^{2+}$ form bidentate, mononuclear complexes with two carboxylic acid groups (Fig. 4.1*b*) while $Cu^{2+}$, $Fe^{2+}$, $Cd^{2+}$ and $Pb^{2+}$ form tridentate mononuclear complexes with two carboxylic acid groups and the hydroxyl group (Fig. 4.1*c*). Such complex formation affects metal mobility, toxicity and biodegradation: the recalcitrance of metal citrate complexes may be important in the migration of hazardous metals from metal and nuclear disposal sites (Francis *et al.*, 1992). In the biosphere, oxalic acid (Fig. 4.1*e*) is widely found and can be produced by plants, animals and microorganisms including fungi, occurring as the free acid but more commonly as the K or Ca salt (Hodgkinson, 1977). The oxalate ion, $C_2O_4^{2-}$ (Fig. 4.1*f*), is a bidentate ligand, and forms complexes when the metal is coordinated by, e.g. 1, 2 or 3 oxalate anions (Fig. 4.1*g,h,i*). Most simple oxalates are sparingly soluble in water except those of the alkali metals (Li, Na, K), ammonium and Fe(III). Divalent metal oxalates are of similar solubilities with the most soluble being magnesium oxalate and the least soluble being calcium and lead oxalate (Table 4.1). Precipitation occurs according to the following equation:

$$M^{2n+}(aq) + nC_2O_4^{2-}(aq) + xH_2O \rightarrow M(C_2O_4)_n{\cdot}xH_2O(s) \qquad (4.1)$$

### *4.3.2 Calcium oxalate*

Calcium oxalate is the most common metal oxalate in the terrestrial environment. The ubiquitous association of calcium oxalate with fungi from

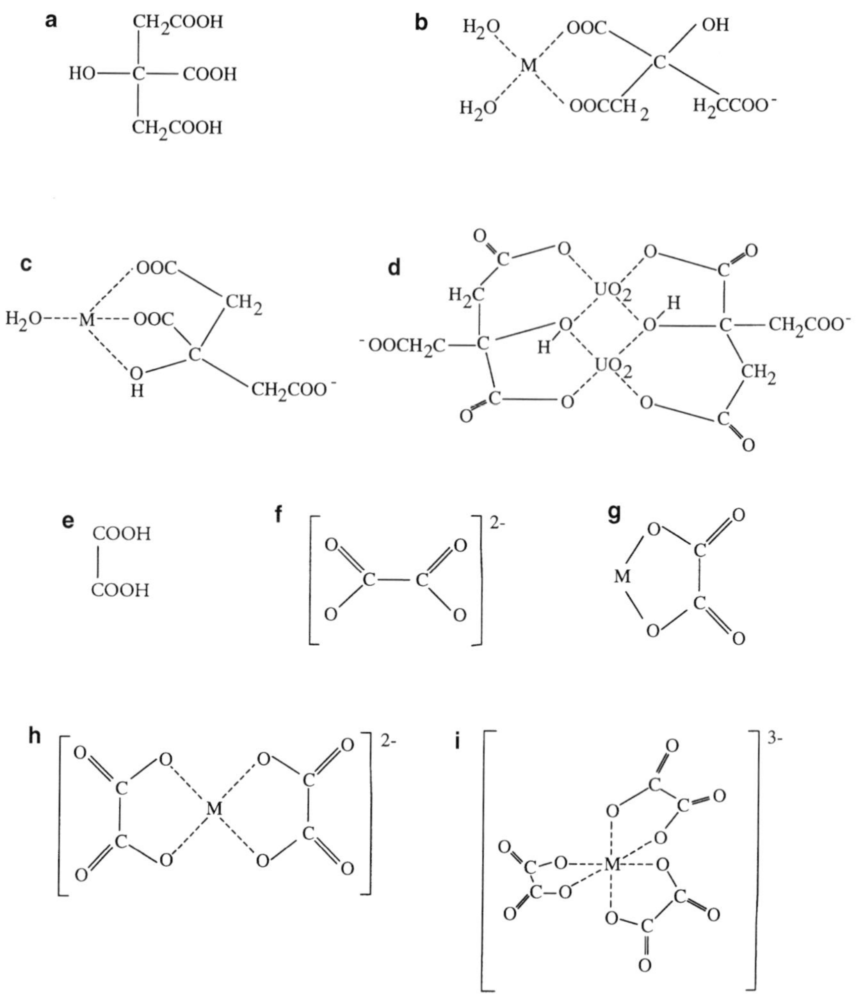

FIG. 4.1 Metal complex (M) formation by citric and oxalic acid. (*a*) citric acid; (*b*) bidentate citric acid complex (e.g. $(\mathrm{Ca\ cit})^-$, $(\mathrm{Ni\ cit})^-$, $(\mathrm{Fe(OH)_2\ cit})^{2-}$); (*c*) tridentate citric acid complex (e.g. $(\mathrm{Cd\ cit})^-$, $(\mathrm{Fe\ cit})^-$, $(\mathrm{FeOH\ cit})^{2-}$), $(\mathrm{Pb\ cit})^-$, $(\mathrm{Cu\ cit})^{2-}$); (*d*) binuclear citric acid complex ($((\mathrm{UO_2})_2\ \mathrm{cit_2})^{2-}$); (*e*) oxalic acid; (*f*) oxalate; (*g*) bidentate metal oxalate complex formation; (*h*) complex anion formation of oxalate with metals which form square planar four-coordinate complexes (e.g. $Cu^{2+}$, $(\mathrm{Cu\ oxal})^{2-}$); (*i*) complex anion formation of oxalate with metals which form octahedral six-coordinate complexes (e.g. $Al^{3+}$, $Fe^{3+}$, $Cr^{3+}$, $(\mathrm{Al\ oxal})^{3-}$), $(\mathrm{Fe\ oxal})^{3-}$), $(\mathrm{Cr\ oxal})^{3-}$)) (adapted from Francis *et al.*, 1992; Gadd, 1999).

all major classes has long been evident (Arnott, 1982, 1995; Dutton and Evans, 1996) as well as in symbiotic fungal associations like mycorrhizas (Lapeyrie *et al.*, 1984; Arnott, 1995) and lichens (Purvis and Halls, 1996; Lee, 2000).

TABLE 4.1 Solubility products of some metal oxalates (Chang, 1993).

| Metal oxalate | Temperature (°C) | Solubility product |
|---|---|---|
| Cadmium (trihydrate) $CdC_2O_4{\cdot}3H_2O$ | 25 | $1.42 \times 10^{-8}$ |
| $CdC_2O_4{\cdot}3H_2O$ | | |
| Calcium (monohydrate) | 25 | $2.57 \times 10^{-9}$ |
| $CaC_2O_4{\cdot}H_2O$ | | |
| Copper (II) | 25 | $2.87 \times 10^{-8}$ |
| $CuC_2O_4$ | | |
| Lead | 18 | $2.74 \times 10^{-11}$ |
| $PbC_2O_4$ | 25 | $8.51 \times 10^{-10}$ |
| Magnesium (dihydrate) | 18 | $8.57 \times 10^{-5}$ |
| $MgC_2O_4{\cdot}2H_2O$ | 25 | $4.83 \times 10^{-6}$ |
| Manganese (II) (dihydrate) | 25 | $1.70 \times 10^{-7}$ |
| $MnC_2O_4{\cdot}2H_2O$ | | |
| Silver (I) | 25 | $5.40 \times 10^{-12}$ |
| $Ag_2C_2O_4$ | | |
| Mercury (I) oxalate | 25 | $1.75 \times 10^{-13}$ |
| $Hg_2C_2O_4$ | | |
| Strontium (monohydrate) | 18 | $5.61 \times 10^{-8}$ |
| $SrC_2O_4{\cdot}H_2O$ | | |
| Zinc (dihydrate) | 25 | $1.37 \times 10^{-9}$ |
| $ZnC_2O_4{\cdot}2H_2O$ | | |

Calcium oxalate crystals can be associated with hyphae as well as fruiting bodies, with the main forms being the monohydrate (whewellite) and the dihydrate (weddelite), the latter being the more ubiquitous. Sometimes the two hydration states can be distinguished by their morphology as the monohydrate belongs to the monoclinic system while the dihydrate is tetragonal in crystallization (Fig. 4.2) (Arnott, 1995). It is possible that the monohydrate arises from dihydrated forms although there is much uncertainty in the literature (see Gadd, 1999). Calcium oxalate formation by free-living and symbiotic mycorrhizal fungi is frequently observed in soil , decomposing wood and in leaf litter (Graustein *et al.*, 1977; Cromack *et al.*, 1979; Lapeyrie *et al.*, 1984; Cairney and Clipson, 1991; Horner *et al.*, 1995; Tewari *et al.*, 1997), and is also associated with certain plant pathogenic fungi (Punja and Jenkins, 1984; Yang *et al.*, 1993; Arnott, 1995). A variety of calcium oxalate crystal formations have been described with many fitting into the four groups: tetragonal bipyramids, prisms, tablets and needles (Keller, 1985; Arnott, 1995; Whitney and Arnott, 1987, 1988) (Figs 4.2, 4.3). In basidiomycete rhizomorphs, some of the interwoven hyphae are encrusted with crystals, and 'vessel hyphae' within may also possess crystalline deposits (Cairney *et al.*, 1989; Cairney and Clipson, 1991). It is therefore thought that the ability of litter degraders to translocate and precipitate calcium as calcium oxalate could

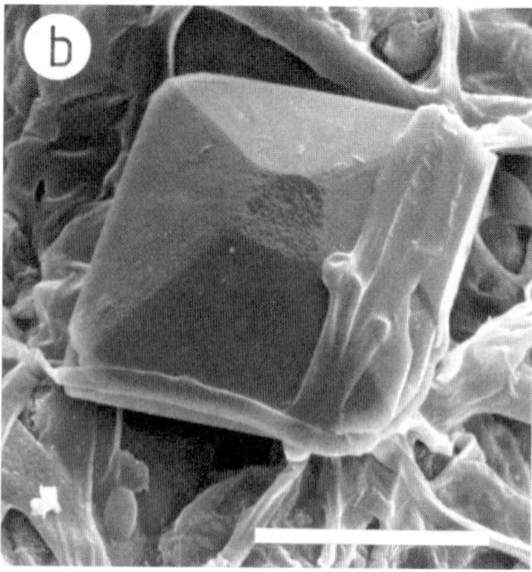


FIG. 4.2 Scanning electron micrographs of crystalline structures found in leaf litter microcosms. (*a*) needle-like crystals characteristic of calcium oxalate monohydrate, (*b*) prism-like crystals characteristic of calcium oxalate dihydrate (see Tait *et al.*, 1999). Scale bars = 10 μm.

contribute to nutritional heterogeneity in soil with concomitant influences on, e.g. P, Al and K distribution (Connolly and Jellison, 1995, 1997). As well as calcium oxalate, fungi can precipitate other metal oxalates (Fig. 4.4). Copper oxalate (moolooite) has been observed around hyphae growing on copper-

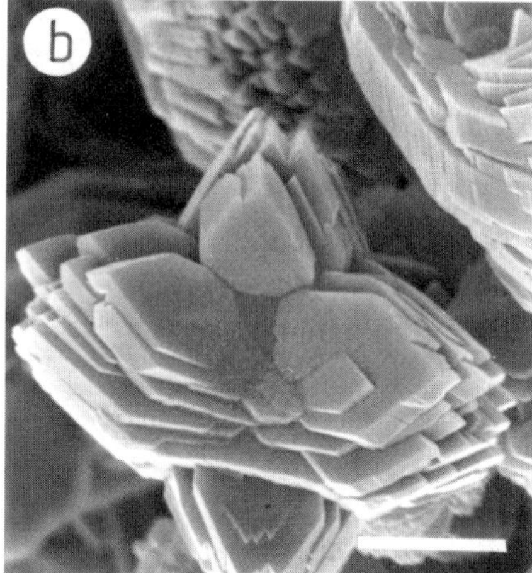


FIG. 4.3 Scanning electron micrographs of purified calcium oxalate crystals formed in malt extract agar (MEA) amended with 0.5%(w/v) gypsum under growing colonies of (*a*) *Serpula himantioides* and (*b*) *Aspergillus niger*. The crystals were identified using a combination of GC-MS, HPLC and X-ray microanalysis: the hydration state is not known but the morphologies suggest the dihydrate (see Gharieb *et al.*, 1998; Gharieb and Gadd, 1999). Scale bars = 10 μm.

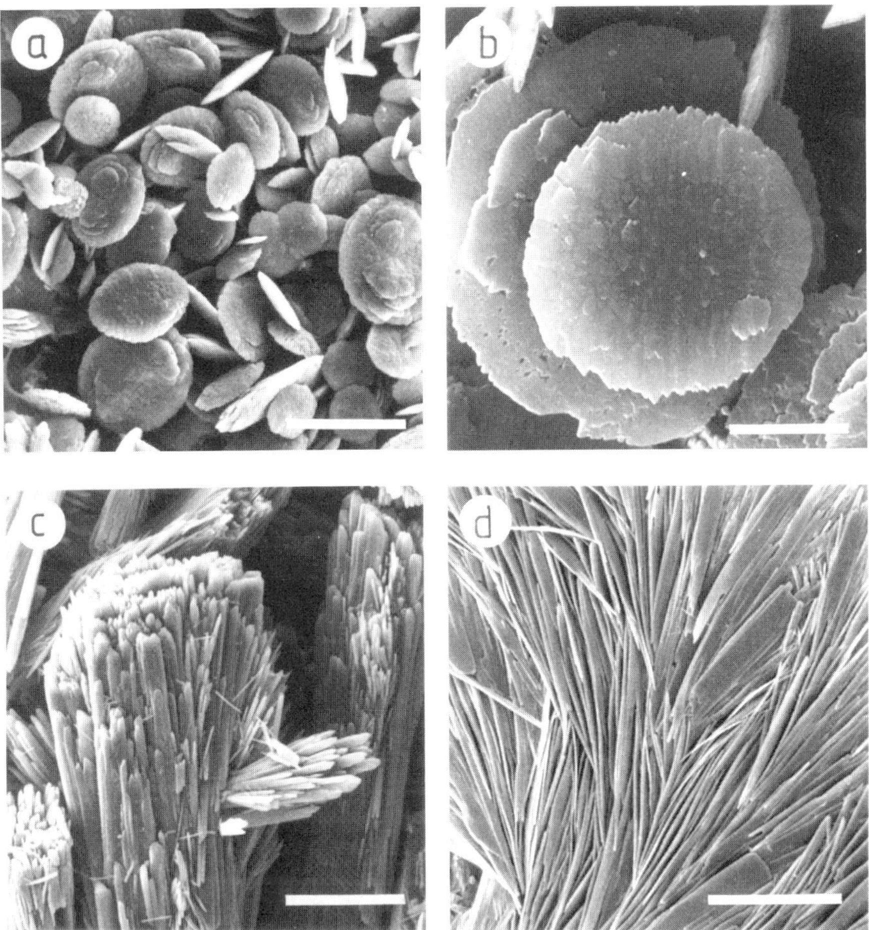


FIG. 4.4 Scanning electron micrographs of crystals purified from under growing colonies of *Aspergillus niger* on MEA supplemented with 0.5%(w/v) (*a,b*) $Zn_3(PO_4)_2$, (*c*) $Co_3(PO_4)_2$, and (*d*) rhodochrosite ($Mn(CO_3)_x$). The crystals formed were identified as the corresponding metal oxalates using a combination of GC-MS, HPLC and X-ray microanalysis: precise mineralogy is not established because of the sparse data on hydrated metal oxalates in the ICDD Powder Diffraction File (see Sayer and Gadd, 1997; Sayer *et al.*, 1997). Scale bars = (*a,c*, 50 μm), (*b*, 10 μm), (*d*, 100 μm).

treated wood (Sutter *et al.*, 1984; Murphy and Levy, 1983). *A. niger* can form metal oxalate crystals after 1–2 days when grown on culture medium amended with a range of metal compounds including insoluble metal phosphates (Fig. 4.4) (Sayer and Gadd, 1997) and powdered metal-bearing minerals (Fig. 4.4) (Sayer *et al.*, 1997, 1999; Gharieb *et al.*, 1998). *A. niger* can produce metal oxalates with many different metals, e.g. Ca, Cd, Co, Cu, Mn, Sr and Zn

(Fig. 4.4) (Sayer and Gadd, 1997). The formation of oxalates containing potentially toxic metals may provide a mechanism whereby oxalate-producing fungi can tolerate environments containing high concentrations of toxic metals. Copper oxalate has also been observed in lichens growing on Cu-rich rocks where the precipitation of copper oxalate could also be a detoxification mechanism (Purvis and Halls, 1996).

## 4.4 Fungal organic acids and metal biogeochemistry

### *4.4.1 Metal and anion solubilization*

Solubilization of insoluble metal compounds is an important aspect of fungal physiology for the release of anions, such as phosphate, and essential metal cations into forms available for intracellular uptake but also into biogeochemical cycles. Fungal solubilization of insoluble metal compounds, including certain oxides, phosphates, sulphides and mineral ores, and leaching of metals from soil and other solids, can occur by several mechanisms with organic acids occupying a central position in the overall process. Solubilization can occur by protonation of the anion of the metal compound, decreasing its availability to the cation. Nutrient-proton antiport, the plasma membrane proton-translocating ATPase and the production of organic acids are all potential sources of protons for such 'acidolysis' (Burgstaller and Schinner, 1993; Gadd, 1993; Morley *et al.*, 1996; Sayer *et al.*, 1995; Karamushka *et al.*, 1996; Dixon-Hardy *et al.*, 1998). In addition, organic acid anions are frequently capable of soluble metal complex formation ('complexolysis') (Burgstaller and Schinner, 1993; White *et al.*, 1997). Another example of organic acid mediated solubilization is 'redoxolysis' where oxalic acid can reduce Fe(III) to Fe(II) thus increasing iron solubility (Ghiorse, 1988). A rather different means of metal solubilization is the production of iron-chelating siderophores which are coordination compounds which specifically solubilize Fe(III) (Crichton, 1991). The incidence of metal-solubilizing ability among natural soil fungal communities appears to be high; in one study approximately one third of the isolates tested were able to solubilize at least one of $Co_3(PO_4)_2$, ZnO or $Zn_3(PO_4)_2$, and approximately one tenth were able to solubilize all three compounds (Sayer *et al.*, 1995). Using this method, *A. niger* was found to be capable of solubilizing a wide range of insoluble metal compounds, including CdS, $Cu_3(PO_4)_2$, $Ni_3(PO_4)_2$, MnS and metal-bearing mineral ores including cuprite (CuS), rhodochrosite ($Mn(CO_3)_x$) (Sayer *et al.*, 1997), gypsum ($CaSO_4.2H_2O$) (Gharieb *et al.*, 1998) and pyromorphite ($Pb_5(PO_4)_3Cl$) (Sayer *et al.*, 1999).

As well as such influences on metal solubility, organic acids concomitantly increase the availability of, e.g. phosphate and sulphate (Evans and Anderson, 1990; Drever and Vance, 1994; Cajuste *et al.*, 1996; Gharieb *et al.*, 1998). Most phosphate fertilizers are applied in a solid form (e.g. calcium phosphate), and need to be solubilized before becoming available to plants and other organisms (Asea *et al.*, 1988; Leyval *et al.*, 1993; Bojinova *et al.*, 1997). In fact, increased phosphate uptake by mycorrhizal plants is believed to be due to the phosphate-solubilizing ability of the mycorrhizas (Lapeyrie *et al.*, 1991). In Canada, a formulation containing spores of *Penicillium bilaii* has been registered for application to crops because of its phosphate-solubilizing ability (Asea *et al.*, 1988; Cunningham and Kuiack, 1992). Soil fungi, including mycorrhizas, may increase the mobility of other anions, e.g. sulphate (Gharieb *et al.*, 1998).

In the soil, oxalate is a common low molecular weight organic anion (Graustein *et al.*, 1977) and this increases the solubility of both P and Al, thus influencing nutrient availability and soil weathering (Fox and Comerford, 1992). Oxalate concentrations in forest soil solution can range from 25 to 1000 μM, and may be the only low molecular weight organic acid found in non-rhizosphere soil (Fox and Comerford, 1992). Ligand-exchange reactions at oxide surfaces directly release P while surface complexation also enhances dissolution of Al-oxide surfaces, increasing both Al and P solubility. Such solubilization increases as the organic acid anion concentration increases (Fox *et al.*, 1990). Oxalic acid may also solubilize P bound by $Ca^{2+}$ and $Fe^{3+}$ (Jurinak *et al.*, 1986; Cajuste *et al.*, 1996), e.g.:

$$Ca_5(PO_4)_3OH(s) + 5H_2C_2O_4\ (aq) \rightarrow 5CaC_2O_4(s) + 3H_3PO_4(aq) + H_2O \quad (4.2)$$

Similar reactions occur with Al and Fe phosphate minerals. The P release is generally greater with acids that form stable metal complexes (Fox *et al.*, 1990; Bolan *et al.*, 1994). Oxalate concentrations are generally higher in the rhizosphere which has obvious implications for P availability to the plant (Fox and Comerford, 1992). It has been found that the amounts of Al and P released are proportional to the cumulative oxalate loading rate suggesting that the continuous release of even small amounts of organic anions could solubilize large amounts of P and Al on an annual basis (Fox and Comerford, 1992). Oxalic acid is produced by plant symbiotic fungi including ectomycorrhizas and this may be significant in increasing P availability to host plants (Graustein *et al.*, 1977). Oxalate is produced by *Paxillus involutus* and *Pisolinthus tinctorius*, and both fungi are able to weather a phlogopite mica, displacing interlayer $K^+$, $Al^{3+}$ and/or $Mg^{2+}$ from within the mineral (Paris *et al.*, 1995*a*). The role of $H^+$ in $K^+$ replacement was invoked with the formation of metal oxalates acting on the multivalent cations (Paris *et al.*, 1996). Oxalic acid from *P. involutus* is also involved in vermiculite weathering (Paris *et al.*, 1995*b*). Citric acid is also able to increase soluble P (as are other organic acids not discussed in this article, e.g. acetic, formic, lactic, malic, tartaric (Bolan *et al.*, 1994; Kpomblekou and Tabatabai, 1994) and in fact tricarboxylic acids (e.g. citric acid) were more effective than dicarboxylic acids (e.g. oxalic acid) (Bolan *et al.*, 1994). In one study, 1 mM citric and oxalic acids were more effective than 1 mM $H_2SO_4$ in phosphate release from phosphate-containing rocks (Kpomblekou and Tabatabai, 1994).

The complexing ability of oxalate and effect on Al mobility is also a factor in secondary-porosity development in sandstones by increasing feldspar solubility and dissolution rate (Huang and Longo, 1992; Franklin *et al.*, 1994). In some studies, higher dissolution rates in the presence of organic acids were ascribed to $H^+$ activity rather than complexation (Manley and Evans, 1986). Although organic acids may accelerate dissolution by lowering pH, this may only be significant below ~pH 5 (Drever and Stillings, 1997). Quartz dissolution and iron crystallite formation was associated with fungal-derived oxalic acid (Feldmann *et al.*, 1997). Although gross concentrations of organic anions in the soil solution may appear insufficient to cause significant increases in dissolution rates of silicate minerals, higher concentrations are likely to be present in microenvironments surrounding fungal hyphae (Drever and Stillings, 1997). Individual organic acids in the soil solution can exceed millimolar concentrations with extremely high concentrations in the vicinity of certain plants and fungal hyphae (Cromack *et al.*, 1979). It is now known that weatherable minerals under many European coniferous forests contain a network of pores, believed to be formed by fungal organic acids (Jongmans *et al.*, 1997). The pores (~3–10 μm) sometimes contain hyphae, and are widespread in feldspars and hornblendes from a variety of locations. Organic acid concentrations in the soil ranged from μM to mM and included citrate and oxalate: it was calculated that Ca-rich feldspars could be dissolved to form pores at the rate of 0.3–30 μm $year^{-1}$ (Jongmans *et al.*, 1997).

Solubilization of insoluble toxic metal compounds in the environment can have adverse effects if potentially toxic metal ions are released into soil and/or water systems (Francis *et al.*, 1992). Metal-citrate complexes are highly mobile and are not readily degraded by microorganisms, degradation depending on the type of complex formed (Francis *et al.*, 1992). Some rock phosphate fertilizers contain Cd, and solubilization of these to release the phosphate could also release Cd (Leyval *et al.*, 1993).

### *4.4.2 Role of organic acids in stone and buildings corrosion*

Organic acid production can lead to corrosion of building materials (Little *et al.*, 1994) including stone (Hirsch *et al.*, 1995; Garcia-Valles *et al.*, 1996). Acidogenic fungi, as mentioned, can cause extensive deterioration of clays, micas and feldspars due to oxalic, citric and gluconic acid production (De la Torre *et al.*, 1993). Contributions from different groups of organisms appear to act synergistically in stone deterioration with, e.g. algal/cyanobacterial lysis providing nutrients for heterotrophic microorganisms (which derive energy from the oxidation of organic compounds) resulting in subsequent development of complex microbial communities (Gomez-Alarcon *et al.*, 1994; Hirsch *et al.*, 1995). Fungal-produced organic acids are also capable of cement degradation, a phenomenon not only relevant to building deterioration but also nuclear waste repositories (Perfettini *et al.*, 1991).

Oxalic acid is also involved in the corrosion of stonework by lichens (Lee, 2000). Calcium oxalates (whewellite and weddelite) can be significant chemical components in surface patinas, particularly on limestone (Edwards *et al.*, 1994; Russ *et al.*, 1996). Weddelite may serve as a water-absorbing substrate which transforms to whewellite when the humidity falls (Lamprecht *et al.*, 1997). At the thallus-substrate interface, oxalic and polyphenolic lichen acids are produced by the mycobiont which act to remove metals from the mineral substrate while carbonic acid arises from respiratory $CO_2$. Calcium oxalate can therefore be a major constituent of lichen thallus encrustation, as well as calcium carbonate (Edwards *et al.*, 1994, 1997; Prieto *et al.*, 1997) with the nature of the metal oxalate often depending on the chemical composition of the rock (Edwards and Lewis, 1994). For example, *Pertusaria corallina* produces calcium oxalate on a basalt substrate but magnesium oxalate forms during growth of *Lecanora atra* on Mg-rich serpentine (Edwards and Lewis, 1994). Hydrated copper oxalate, moolooite $CuC_2O_4{\cdot}nH_2O$ ($n \approx 0.4-0.7$), is found within certain lichen species growing on Cu-bearing rocks (Chisholm *et al.*, 1987). On granite, a mono- or multi-layered patina can arise from surface biofilm activity, mineral deposition and dust trapping which can be regarded as a microstromatolitic structure, containing calcite, calcium oxalate, gypsum and calcium phosphate (Blazquez *et al.*, 1997). Oxalate has been shown to be the only organic acid detectable in cryptoendolithic (hidden-in-rock) lichen-dominated microbial communities in Antarctica with a proposed significant role in affecting dissolution of Si, Fe, Al, P and K (Johnson and Vestal, 1993).

### *4.4.3 Role of fungal oxalate in limestone biomineralization*

Fungal activity, in influencing cation mobility and acidification, is an important factor in early lithification, alteration and diagenesis of terrestrial subsurface limestones (Verrecchia and Dumont, 1996). Weathered profiles which develop on chalky parent rocks in arid regions usually include a layer called a 'platy caliche/calcrete' or 'platy zone' (Verrecchia and Dumont, 1996). While physicochemical phenomena contribute to this process, fungal activity is of fundamental importance to the mineral dynamics of the pore/carbonate system. Fungal hyphae are present in the pores (Verrecchia *et al.*, 1990, 1993) and are encrusted with sharp crystalline spikes. Up to ~80% of the limestone micropores are inhabited by fungi, and in some places, pores are partly infilled with needle-fibre calcite (Verrecchia *et al.*, 1990; Verrecchia and Verrecchia, 1994) with the matrix around the pore being enriched in $CaCO_3$, sometimes being recrystallized and separated from the pore by a mat of calcite (Verrecchia and Dumont, 1996). The crystalline needles are calcium oxalate monohydrate (whewellite) and arise from oxalic acid production by the fungi and subsequent precipitation of calcium oxalate:

$$C_2O_4^{2-}(aq) + Ca^{2+}(aq) + H_2O \rightarrow CaC_2O_4{\cdot}H_2O(s) \qquad (4.3)$$

It is likely that polyhydrate oxalates form first which, after dehydration, evolve into the monohydrate (Verrecchia *et al.*, 1993). A biogeochemical model of this system proposes that organic acid excretion, and calcium oxalate formation, leads to dissolution of the internal pore walls so that the solution becomes enriched in dissolved carbonates. During passage of this solution through the pore walls, $CaCO_3$ recrystallizes due to changes in $pCO_2$ which contributes to hardening of the material. Subsequent biodegradation of the oxalate leads to transformation into carbonate which results in precipitation of calcite in the pore interior or periphery, the latter ultimately leading to closure of the pore system and hardening of the chalky parent material (Verrecchia *et al.*, 1990; Verrecchia and Dumont, 1996). Ultimately, the pores become completely filled and a hard limestone ('desert crust') results (Verrecchia, 1990; Verrecchia and Dumont, 1996).

## 4.5 Fungal organic acid production and metal biotechnology

### *4.5.1 Metal solubilization for recovery and bioremediation*

Fungal removal or leaching of metals ('heterotrophic leaching') from industrial wastes and by-products, low grade ores (Dave and Natarajan, 1981; Burgstaller and Schinner, 1993) and metal-bearing minerals (Tzeferis *et al.*, 1994; Drever and Stillings, 1997), is a process relevant to metal recovery and recycling and/or bioremediation of solid wastes (Groudev, 1987; Burgstaller and Schinner, 1993; Hahn *et al.*, 1993; White *et al.*, 1997; Krebs *et al.*, 1997). Heterotrophic leaching may also have application to phosphate (or other solubilized anions) recovery from, e.g. rock phosphate or other sources (Parks *et al.*, 1990; Vassilev *et al.*, 1997; Bojinova *et al.*, 1997). Most processes of biological leaching utilize chemoautotrophic bacteria (organisms that generate energy by the oxidation of inorganic compounds and fix $CO_2$), particularly *Thiobacillus* spp. (Ewart and Hughes, 1991). However, the pH of culture media may be increased by industrial metal-containing wastes such as filter dust/oxides from metal processing, and most Thiobacilli cannot solubilize metals effectively above pH 5.5, with an optimum of pH 2.4 (Burgstaller and Schinner, 1993). Although fungi need aeration and a carbon source, they can function at higher pH values and so could have potential where chemoautotrophic bacterial leaching is not possible. However, fungi may not readily be used *in situ*, and their use in specific bioreactors is generally envisaged. Leaching of metals with fungi can be very effective although a high level of organic acid production may need to be maintained. For example, laboratory-scale leaching of Ni, Co and Mn from low grade laterite ores has been carried out using strains of *Aspergillus* and *Penicillium* (Tzeferis *et al.*, 1994). 55–60% Ni was leached when the fungi were grown in the presence of the ore, and 70% was leached at high temperature (95°C) by application of metabolites produced on cultivation of the fungi at 30°C in a glucose and sucrose medium (Tzeferis, 1994). In most fungal strains, leaching occurs mainly by the production of organic acids which provide protons and complexing organic acid anions

(Burgstaller and Schinner, 1993; Müller *et al.*, 1995; Sayer and Gadd, 1997). The pH of *A. niger* cultures can fall to between 1.5 and 2.0 due to high citric acid production, the optimal pH being below 3.5, with manganese deficiency ($<10^{-8}$ M) leading to the production of large amounts of citric acid (up to ~600 mM) (Mattey, 1992). Oxalic acid production can also be manipulated for solubilization of metals like Al and Fe; up to ~430 mM oxalic acid was excreted in a stirred-tank reactor, controlled at pH 6 (Strasser *et al.*, 1994).

Fly ash from municipal waste incineration contains, e.g. Cd, Zn, Ni, Pb and Cu, with Al and Zn being present in amounts considered economical for recovery (Torma and Singh, 1993; Bosshard *et al.*, 1996). A two-step process, where citric acid first produced by control growth of *A. niger* was used as the leaching agent, provided a system only slightly less efficient than leaching with commercial citric acid of equal molarity (Bosshard *et al.*, 1996). A strain of *Penicillium simplicissimum* which was isolated from a metal-contaminated site produced citric acid (>100 mM) in the presence of metal contaminants, and was used to leach Zn from insoluble ZnO in industrial filter dust (Schinner and Burgstaller, 1989; Franz *et al.*, 1993; Burgstaller and Schinner, 1993). Acidolysis was found to be the main ZnO solubilization mechanism:

$$ZnO(s) + 2H^{+}(aq) \rightarrow Zn^{2+}(aq) + H_2O \tag{4.4}$$

with the protons from the citric acid produced being responsible for nearly all the solubilization, prior to possible formation of a zinc citrate complex which may prevent $Zn^{2+}$ toxicity (Burgstaller *et al.*, 1992). It can be noted that metal-citrate complexes could eventually be degraded for ultimate metal recovery (Francis, 1994). In fact, another bioremediation scheme uses commercial citric acid to remove metals and radionuclides from contaminated wastes by forming water-soluble metal citrate complexes. The resulting metal citrate complexes are then microbially and photochemically degraded and the metals recovered in concentrated form with the degrading bacterial biomass (Francis *et al.*, 1992; Francis and Dodge, 1994). Another possible application of fungal metal solubilization could be the removal of unwanted contaminants such as phosphates (Burgstaller and Schinner, 1993), while interaction of leaching technologies with biosorption is also a possibility (Gadd, 1993; Tobin *et al.*, 1994).

## 4.6 Conclusions

The production of organic acids by fungi has profound implications for metal speciation, physiology and biogeochemical cycles. The metal-complexing properties of organic acids like citric and oxalic acid assists both essential metal and anionic nutrition (e.g. phosphate, sulphate) of fungi, other microbes and plants, and can determine metal speciation and mobility in the environment. Metal solubilization processes are also of potential for metal recovery and reclamation from contaminated solid wastes, soils and low grade ores. Most simple metal oxalates (except those of alkali metals, Fe(III) and Al) are sparingly soluble and precipitate as crystalline or amorphous solids. Calcium oxalate is the most important manifestation of this in the environment and is ubiquitously associated with free-living, symbiotic and pathogenic fungi in a variety of crystalline forms. The main forms are the monohydrate (whewellite) and the dihydrate (weddelite) and their formation is of significance in biomineralization and affecting nutritional heterogeneity in soil, especially Ca, P, K and Al cycling, while the formation of insoluble toxic metal oxalates, e.g. Cu, may confer tolerance and ensure survival in contaminated environments. In semi-arid environments, calcium oxalate formation is important in the formation and alteration of terrestrial subsurface limestones. Further research will undoubtedly reveal other examples of 'fungal biogeochemistry' as well as accurately defining the examples detailed in this chapter and their applied and environmental significance.

## Acknowledgements

The author gratefully acknowledges research support from the Biotechnology and Biological Sciences Research Council (SPC 02812), the Royal Society (London), the Royal Society of Edinburgh, NATO (Envir. LG 950387) and the Nuffield Foundation. Martin Kierans, Margaret Gruber and Ewan Starke are also acknowledged for assistance with scanning electron microscopy and photography.

## References

Arnott, H.J. (1982) Calcium oxalate (weddelite) crystals in forest litter. *Scanning Electron Microsc.*, **P3**, 1141–9.

Arnott, H.J. (1995) Calcium oxalate in fungi. Pp. 73–111 in: *Calcium Oxalate in Biological Systems* (S.R. Khan, editor). CRC Press, Boca Raton, FL, USA.

Asea, P.E.A., Kucey, R.M.N. and Stewart, J.W.B. (1988) Inorganic phosphate solubilization by two *Penicillium* species in solution culture and soil. *Soil Biol. Biochem.,* **20**, 459–64.

Blazquez, F., Garcia-Valles, M., Krumbein, W.E., Sterflinger, K. and Vendrell-Saz, M. (1997) Microstromatolitic deposits on granitic monuments: development and decay. *Eur. J. Mineral.,* **9**, 889–901.

Bojinova, D., Velkova, R., Grancharov, I. and Zhelev, S. (1997) The bioconversion of Tunisian phosphorite using *Aspergillus niger*. *Nutrient Cycl. Agroecos.,* **47**, 227–32.

Bolan, N.S., Naidu, R., Mahimairaja, S. and Baskaran, S. (1994) Influence of low-molecular-weight organic acids on the solubilization of phosphates. *Biol. Fertil. Soil.,* **18**, 311–9.

Bosshard, P.P., Bachofen, R. and Brandl, H. (1996) Metal leaching of fly ash from municipal water incineration by *Aspergillus niger*. *Environ. Sci. Technol.,* **30**, 3066–70.

Burgstaller, W. and Schinner, F. (1993) Leaching of metals with fungi. *J. Biotechnol.,* **27**, 91–116.

Burgstaller, W., Strasser, H., Wöbking, H. and Schinner, F. (1992) Solubilization of zinc oxide from filter dust with *Pencicllium simplicissimum*: bioreactor leaching and stoichiometry. *Environ. Sci. Technol.,* **26**, 340–6.

Cairney, J.W.G. and Clipson, N.J.W. (1991) Internal structure of rhizomorphs of *Techispora vaga. Mycol. Res.*, **95**, 764–7.

Cairney, J.W.G., Jennings, D.H. and Veldtkamp, C.J. (1989) A scanning electron microscope study of the internal structure of mature linear mycelial organs of four basidiomycete species. *Canad. J. Bot.*, **67**, 2266–71.

Cajuste, L.J., Laird, R.J., Cajuste, B. and Cuevas, B.G. (1996) Citrate and oxalate influence on phosphate, aluminium and iron in tropical soils. *Comm. Soil Sci. Plant Anal.*, **27**, 1377–86.

Chang, J.C. (1993) Solubility product constants. P. 39 in: *CRC Handbook of Chemistry and Physics* (D.R. Lide, editor). CRC Press, Boca Raton, FL, USA.

Chisholm, J.E., Jones, G.C. and Purvis, O.W. (1987) Hydrated copper oxalate, moolooite, in lichens. *Mineral. Mag.*, **51**, 715–8.

Connolly, J.H. and Jellison, J. (1995) Calcium translocation, calcium oxalate accumulation and hyphal sheath morphology in the white-rot fungus *Resinicium bicolor*. *Canad. J. Bot.*, **73**, 927–36.

Connolly, J.H. and Jellison, J. (1997) Two-way translocation of cations by the brown rot fungus *Gloeophyllum trabeum*. *Int. Biodet. Biodeg.*, **39**, 181–8.

Crichton, R.R. (1991) *Inorganic Biochemistry of Iron Metabolism*. Ellis Horwood, Chichester, UK.

Cromack, K. Jr., Sollins, P., Graustein, W.C., Speidel, K., Todd, A.W., Spycher, G., Li, C.Y. and Todd, R.L. (1979) Calcium oxalate accumulation and soil weathering in mats of the hypogeous fungus *Hysterangium crassum*. *Soil Biol. Biochem.*, **11**, 463–8.

Cunningham, J.E. and Kuiack, C. (1992) Production of citric and oxalic acids and solubilization of calcium phosphate by *Penicillium bilaii*. *Appl. Environ. Microbiol.*, **58**, 1451–8.

Dave, S.R. and Natarajan, K.A. (1981) Leaching of copper and zinc from oxidized ores by fungi. *Hydrometallurgy*, **7**, 235–42.

De la Torre, M.A., Gomez-Alarcon, G., Vizcaino, C. and Garcia, M.T. (1993) Biochemical mechanisms of stone alteration carried out by filamentous fungi. *Biogeochem.*, **19**, 129–47.

Dixon-Hardy, J.E., Karamushka, V.I., Gruzina, T.G., Nikovska, G.N., Sayer, J.A. and Gadd, G.M. (1998) Influence of the carbon, nitrogen and phosphorus source on the solubilization of insoluble metal compounds by *Aspergillus niger*. *Mycol. Res.*, **102**, 1050–4.

Drever, J.I. and Stillings, L.L. (1997) The role of organic acids in mineral weathering. *Coll. Surf.*, **120**, 167–81.

Drever, J.I. and Vance, G.F. (1994) Role of soil organic acids in mineral weathering processes. Pp. 138–61 in: *Organic Acids in Geological Processes* (E.D. Pittman and M.D. Lewan, editors). Springer, New York.

Dutton, M.V. and Evans, C.S. (1996) Oxalate production by fungi: its role in pathogenicity and ecology in the soil environment. *Canad. J. Microbiol.*, **42**, 881–95.

Edwards, H.G.M. and Lewis, I.R. (1994) FT-Raman spectroscopic studies of metal oxalates and their mixtures. *Spectrochim. Acta*, **50**, 1891–8.

Edwards, H.G.M., Edwards, K.A.E., Farwell, D.W., Lewis, I.R. and Seaward, M.R.D. (1994) An approach to stone and fresco lichen biodeterioration through Fourier-transform Raman microscopic investigation of thallus substratum encrustations. *J. Raman Spectr.*, **25**, 99–103.

Edwards, H.G.M., Russell, N.C. and Seaward, M.R.D. (1997) Calcium oxalate in lichen biodeterioration studied using FT-Raman spectroscopy. *Spectrochim. Acta*, **53**, 99–105.

Evans, A. and Anderson, T.J. (1990) Aliphatic acids: influence on sulphate mobility in a forested cecil soil. *Soil Sci. Soc. Amer. J.*, **54**, 1136–9.

Ewart, D.K. and Hughes, M.N. (1991) The extraction of metals from ores using bacteria. *Adv. Inorg. Chem.*, **36**, 103–35.

Feldmann, M., Neher, J., Jung, W. and Graf, F. (1997) Fungal quartz weathering and iron crystallite formation in an Alpine environment, Piz Alv, Switzerland. *Eclogae Geol. Helv.*, **90**, 541–56.

Fox, T.R. and Comerford, N.B. (1992) Influence of oxalate loading on phosphorus and aluminium solubility in spodosols. *Soil Sci. Soc. Amer. J.*, **56**, 290–4.

Fox, T.R., Comerford, N.B. and McFee, W.W. (1990) Phosphorus and aluminium release from a spodic horizon mediated by organic acids. *Soil Sci. Soc. Amer. J.,* **54**, 1763–7.

Francis, A.J. (1994) Microbial transformations of radioactive wastes and environmental restoration through bioremediation. *J. Alloys Comp.,* **213/214**, 226–31.

Francis, A.J. and Dodge, C.J. (1994) Influence of complex structure on the biodegradation of iron-citrate complexes. *Appl. Environ. Microbiol.,* **59**, 109–13.

Francis, A.J., Dodge, C.J. and Gillow, J.B. (1992) Biodegradation of metal citrate complexes and implications for toxic metal mobility. *Nature,* **356**, 140–2.

Frankland, J.C., Magan, N. and Gadd, G.M. (editors) (1996) *Fungi and Environmental Change.* Cambridge University Press, Cambridge, UK.

Franklin, S.P., Hajash, A., Dewers, T.A. and Tieh, T.T. (1994) The role of carboxylic acids in albite and quartz dissolution: an experimental study under diagenetic conditions. *Geochim. Cosmochim. Acta,* **58**, 4259–79.

Franz, A., Burgstaller, W. Müller, B. and Schinner, F. (1993) Influence of medium components and metabolic inhibitors on citric acid production by *Penicillium simplicissimum. J. Gen. Microbiol.,* **139**, 2101–7.

Gadd, G. M. (1993) Interactions of fungi with toxic metals. *New Phytol.* **124,** 25–60.

Gadd, G.M. (1996) Roles of microorganisms in the environmental fate of radionuclides. *Endeavour,* **20**, 150–6.

Gadd, G.M. (1999) Fungal production of citric and oxalic acid; importance in metal speciation, physiology and biogeochemical proceses. *Adv. Microb. Physiol.,* **41**, 47–92.

Garcia-Valles, M., Blazquez, F., Molera, J. and Vendrell-Saz, M. (1996) Studies of patinas and decay mechanisms leading to the restoration of Sta. Maria de Montblanc (Catalonia, Spain). *Stud. Conserv.,* **41**, 1–8.

Gharieb, M.M. and Gadd, G.M. (1999) Influence of nitrogen source on the solubilization of natural gypsum ($CaSO_4.2H_2O$) and the formation of calcium oxalate by different different oxalic and citric acid-producing fungi. *Mycol. Res.,* **103**, 473–81.

Gharieb, M.M., Sayer, J.A. and Gadd, G.M. (1998) Solubilization of natural gypsum ($CaSO_4.2H_2O$) and the formation of calcium oxalate by *Aspergillus niger* and *Serpula himantiodes. Mycol. Res.,* ***102***, 825–30.

Ghiorse, W.C. (1988) Microbial reduction of manganese and iron. Pp. 178–210 in: *Biology of Anaerobic Microorganisms* (A.J.B. Zehnder, editor). Wiley, New York.

Gomez-Alarcon, G., Munoz, M.L. and Flores, M. (1994) Excretion of organic acids by fungal strains isolated from decayed sandstone. *Int. Biodet. Biodeg.,* **34**, 169–80.

Gow, N.A.R. and Gadd, G.M. (editors) (1995) *The Growing Fungus.* Chapman & Hall, London.

Gow, N.A.R., Robson, G.D. and Gadd, G.M. (editors) (1999) *The Fungal Colony.* Cambridge University Press, Cambridge, UK.

Graustein, W.C., Cromack, K. Jr. and Sollins, P. (1977) Calcium oxalate: occurrence in soils and effects on nutrient and geochemical cycles. *Science,* **198**, 1252–4.

Groudev, S.N. (1987) Use of heterotrophic microorganisms in mineral biotechnology. *Acta Biotechnol.,* **7**, 299–306.

Hahn, M., Willscher, S. and Straube, G. (1993) Copper leaching from industrial wastes by heterotrophic microorganisms. Pp. 99–108 in: *Biohydrometallurgical Technologies Vol. 1* (A.E. Torma, M.L. Apel and C.L. Brierley, editors). The Minerals, Metals and Materials Society, Warrendale.

Hirsch, P., Eckhardt, F.E.W. and Palmer, R.J. (1995) Fungi active in weathering of rock and stone monuments. *Canad. J. Bot.,* **73**, S1384–90.

Hodgkinson, A. (1977) *Oxalic Acid in Biology and Medicine.* Academic Press, New York.

Horner, H.T., Tiffany, L.H. and Knaphus, G. (1995) Oak leaf litter rhizomorphs from Iowa and Texas: calcium oxalate producers. *Mycologia,* **87**, 34–40.

Huang, W.L. and Longo, J.M. (1992) The effects of organics on feldspar dissolution and the development of secondary porosity. *Chem. Geol.,* **98**, 271–92.

Johnston, C.G. and Vestal, J.R. (1993) Biogeochemistry of oxalate in the Antarctic cryptoendolithic lichen-dominated community. *Microb. Ecol.*, **25**, 305–19.

Jongmans, A.G., Van Breeman, N., Lundstrom, U., Van Hees, R.A.W., Finlay, R.D., Srinivasan, M., Unestam, T., Giesler, R., Melkerud, P.-A. and Olsson, M. (1997) Rock-eating fungi. *Nature*, **389**, 682–3.

Jurinak, J.J., Dudley, L.M., Allen, M.F. and Knight, W.G. (1986) The role of calcium oxalate in the availability of phosphorus in soils of semi-arid regions: a thermodynamic study. *Soil Sci.*, **142**, 255–61.

Karamushka, V.I., Sayer, J.A. and Gadd, G.M. (1996) Inhibition of $H^+$ efflux from *Saccharomyces cerevisiae* by insoluble metal phosphates and protection by calcium and magnesium: inhibitory effects a result of soluble metal cations? *Mycol. Res.*, **100**, 707–13.

Keller, J. (1985) Les cystides crystalliferes des Aphyllophorales. *Mycol. Helvet.*, **1**, 277–305.

Kpomblekou, A.K. and Tabatabai, M.A. (1994) Effect of organic acids on release of phosphorus from phosphate rocks. *Soil Sci.*, **158**, 442–53.

Krebs, W., Brombacher, C., Bosshard, P.P., Bachofen, R. and Brandl, H. (1997) Microbial recovery of metals from solids. *FEMS Microbiol. Rev.*, **20**, 605–17.

Kubicek, C.P. (1998) Citric acid production. Pp. 236–57 in: *Commercial Fermentations and Nutritional Needs of their Microorganisms* (T. Nagodawithana and G. Reed, editors). Esteekay Associates Inc., Milwaukee, WI, USA.

Lamprecht, I., Reller, A., Riesen, R. and Wiedemann, H.G. (1997) Ca-oxalate films and microbiological investigations of the influence of ancient pigments on the growth of lichens – thermogravimetric/thermomicroscopic analyses. *J. Thermal Anal.*, **49**, 1601–7.

Lapeyrie, F., Perrin, M., Pepin, R. and Bruchet, G. (1984) Extracellular formation of weddelite cultured *in vitro* by *Paxillus involutus*: significance of this production for ectomycorrhizal symbiosis. *Canad. J. Bot.*, **62**, 1116–21.

Lapeyrie, F., Ranger, J. and Vairelles, D. (1991) Phosphate solubilizing activity of ectomycorrhizal fungi *in vitro*. *Canad. J. Bot.*, **69**, 342–6.

Lee, M.R. (2000) Weathering of rocks by lichens: fragmentation, dissolution and precipitation of minerals in a microbial microcosm. Pp. 77–107 in: *Environmental Mineralogy: Microbial Interactions, Anthropogenic Influences, Contaminated Land and Waste Management* (J.D. Cotter-Howells, L.S. Campbell, E. Valsami-Jones and M. Batchelder, editors). Mineralogical Society Series, **9**. Mineralogical Society, London.

Leyval, C., Surtiningsih, T. and Berthelin, J. (1993) Mobilization of P and Cd from rock phosphates by rhizosphere microorganisms (phosphate dissolving bacteria and ectomycorrhizal fungi). *Phosphorus, Sulfur, Silicon*, 77, 133–6.

Little, B., Ray, R., Hart, K. and Wagner, P. (1994) Fungal-induced corrosion of wire rope exposed in humid atmospheric conditions. *Corr. Prev. Control*, **41**, 89–93.

Manley, E.P. and Evans, L.J. (1986) Dissolution of feldspars by low-molecular-weight aliphatic and aromatic acids. *Soil Sci.*, **141**, 106–12.

Mattey, M. (1992) The production of organic acids. *CRC Crit. Rev. Biotechnol.*, **12**, 87–132.

Metting, F.B. (1992) Structure and physiological ecology of soil microbial communities. Pp. 3–25 in: *Soil Microbial Ecology, Applications and Environmental Management* (F.B. Metting, editor). Marcel Dekker, New York.

Micales, J.A. (1994) Induction of oxalic acid by carbohydrate and nitrogen sources in the brown-rot fungus *Postia placenta*. *Mater. Org.*, **28**, 197–207.

Micales, J.A. (1995) Oxalate decarboxylase in the brown-rot wood decay fungus *Postia placenta*. *Mater. Org.*, **29**, 177–86.

Morley, G.F. and Gadd, G.M. (1995) Sorption of toxic metals by fungi and clay minerals. *Mycol. Res.*, **99**, 1429–38.

Morley, G.F., Sayer, J.A., Wilkinson, S.J., Gharieb, M.M. and Gadd, G.M. (1996) Fungal sequestration, solubilization and transformation of toxic metals. Pp. 235–56 in: *Fungi and Environmental Change* (J.C. Frankland, N. Magan and G.M. Gadd, editors). Cambridge

University Press, Cambridge, UK.

Müller, B., Burgstaller, W., Strasser, H., Zanella, A. and Schinner, F. (1995) Leaching of zinc from industrial filter dust with *Penicillium*, *Pseudomonas* and *Corynebacterium* – citric acid is the leaching agent rather than amino acids. *J. Ind. Microbiol.*, **14**, 208–12.

Murphy, R.J. and Levy, J.F. (1983) Production of copper oxalate by some copper tolerant fungi. *Trans. Br. Mycol. Soc.*, **81**, 165–8.

Paris, F., Bonnaud, P., Ranger, J. and Lapeyrie, F. (1995*a*) *In vitro* weathering of phlogopite by ectomycorrhizal fungi. 1. Effect of K and Mg deficiency on phyllosilicate evolution. *Plant and Soil*, **177**, 191–201.

Paris, F., Bonnaud, P., Ranger, J., Robert, M. and Lapeyrie, F. (1995*b*) Weathering of ammonium or calcium saturated 2:1 phyllosilicates by ectomycorrhizal fungi *in vitro*. *Soil Biol. Biochem.*, **27**, 1237–44.

Paris, F., Botton, B. and Lapeyrie, F. (1996) In vitro weathering of phlogopite by ectomycorrhizal fungi. II. Effect of $K^+$ and $Mg^{2+}$ deficiency and N sources on accumulation of oxalate and $H^+$. *Plant and Soil,* **179**, 141–50.

Parks, E.J., Olson, G.J., Brinckman, F.E. and Baldi, F. (1990) Characterization by high performance liquid chromatography (HPLC) of the solubilization of phosphorus in iron ore by a fungus. *J. Ind. Microbiol.*, **5**, 183–90.

Perfettini, J.V., Revertegat, E. and Langomazino, N. (1991) Evaluation of cement degradation by the metabolic products of 2 fungal strains. *Experientia*, **47**, 527–33.

Prieto, B., Silva, B., Rivas, T., Wierzchos, J. and Asdaso, C. (1997) Mineralogical transformation and neoformation in granite caused by the lichens *Tephromela atra* and *Ochrolechia parella*. *Int. Biodet. Biodeg.*, **40**, 191–9.

Punja, Z.K. and Jenkins, S.F. (1984) Light and scanning electron microscopic observations of calcium oxalate crystals produced during growth of *Sclerotium rolfsii* in culture and in infected tissue. *Canad. J. Bot.*, **62**, 2028–32.

Purvis, O.W. and Halls, C. (1996) A review of lichens in metal-enriched environments. *Lichenol.*, **28**, 571–601.

Ramsay, L.M., Sayer, J.A. and Gadd, G.M. (1999) Stress responses of fungal colonies towards toxic metals. Pp. 179–201 in: *The Fungal Colony* (N.A.R. Gow, G. Robson and G.M. Gadd, editors). Cambridge University Press, Cambridge, UK.

Russ, J., Palma, R.L., Lloyd, D.H., Boutton, T.W. and Coy, M.A. (1996) Origin of the whewellite-rich rock crust in the Lower Pecos region of Southwest Texas and its significance to paleoclimate reconstructions. *Quat. Res.*, **46**, 27–36.

Sayer, J.A. and Gadd, G.M. (1997) Solubilization and transformation of insoluble inorganic metal compounds to insoluble metal oxalates by *Aspergillus niger*. *Mycol. Res.*, **101**, 653–61.

Sayer, J.A., Raggett, S.L. and Gadd, G.M. (1995) Solubilization of insoluble metal compounds by soil fungi: development of a screening method for solubilizing ability and metal tolerance. *Mycol. Res.*, **99**, 987–93.

Sayer, J.A., Kierans, M. and Gadd, G.M. (1997) Solubilization of some naturally-occurring metal-bearing minerals, limescale and lead phosphate by *Aspergillus niger*. *FEMS Microbiol. Lett.*, **154**, 29–35.

Sayer, J.A., Cotter-Howells, J.D., Watson, C., Hillier, S. and Gadd, G.M. (1999) Lead mineral transformation by fungi. *Curr. Biol.*, **9**, 691–4.

Schinner, F. and Burgstaller, W. (1989) Extraction of zinc from industrial waste by a *Penicillium* sp. *Appl. Environ. Microbiol.*, **55**, 1153–6.

Shimada, M., Akamatsu, Y., Tokimatsu, T., Mii, K. and Hattori, T. (1997) Possible biochemical roles of oxalic acid as a low molecular weight compound involved in brown-rot and white-rot wood decays. *J. Biotechnol.*, **53**, 103–13.

Strasser, H., Burgstaller, W. and Schinner, F. (1994) High-yield production of oxalic acid for metal leaching processes by *Aspergillus niger*. *FEMS Microbiol. Lett.*, **119**, 365–70.

Sutter, H.-P., Jones, G.E.B. and Wälchi, O. (1984) Occurrence of crystalline hyphal sheaths in

*Poria placenta* (Fr) Cke. *J. Inst. Wood Sci.*, **10**, 19–23.

Tait, K., Sayer, J.A., Gharieb, M.M. and Gadd, G.M. (1999) Fungal production of calcium oxalate in leaf litter microcosms. *Soil Biol. Biochem.*, **31**, 1189–92.

Tewari, J.P., Shinners, T.C. and Briggs, K.G. (1997) Production of calcium oxalate crystals by two species of *Cyathus* in culture and infested plant debris. *Z. Naturforsch.*, **52c**, 421–5.

Tobin, J.M., White, C. and Gadd, G.M. (1994) Metal accumulation by fungi: applications in environmental technology. *J. Indust. Microbiol.*, **13**, 126–30.

Torma, A.E. and Singh, A.K. (1993) Acidolysis of coal fly ash by *Aspergillus niger*. *Fuel*, **12**, 1625–30.

Tzeferis, P.G. (1994) Leaching of a low-grade hematitic laterite ore using fungi and biologically produced acid metabolites. *Int. J. Mineral Proc.*, **42**, 267–83.

Tzeferis, P.G., Agatzini, S. and Nerantzis, E.T. (1994) Mineral leaching of non-sulphide nickel ores using heterotrophic micro-organisms. *Lett. Appl. Microbiol.*, **18**, 209–13.

Valsami-Jones, E. and McEldowney, S. (2000) Mineral dissolution by heterotrophic bacteria: principles and methodologies. Pp. 27–55 in: *Environmental Mineralogy: Microbial Interactions, Anthropogenic Influences, Contaminated Land and Waste Management* (J.D. Cotter-Howells, L.S. Campbell, E. Valsami-Jones and M. Batchelder, editors). Mineralogical Society Series, **9**. Mineralogical Society, London.

Vassilev, N., Vassileva, M. and Azcon, R. (1997) Solubilization of rock phosphate by immobilized *Aspergillus niger*. *Biores. Technol.*, **59**, 1–4.

Verrecchia, E.P. (1990) Lithodiagenetic implications of the calcium oxalate-carbonate biogeochemical cycle in semi-arid calcretes, Nazareth, Israel. *Geomicrobiol. J.*, **8**, 89–101.

Verrecchia, E.P. and Dumont, J.-L. (1996) A biogeochemical model for chalk alteration by fungi in semiarid environments. *Biogeochem.*, **35**, 447–70.

Verrecchia, E.P. and Verrecchia, K.E. (1994) Needle-fiber calcite: a critical review and a proposed classification. *J. Sedim. Res.*, **A64**, 650–64.

Verrecchia, E.P., Dumont, J.L. and Rolko, K.E.. (1990) Do fungi building limestones exist in semi-arid regions? *Naturwissenschaften*, **77**, 584–6.

Verrecchia, E.P., Dumont, J.L. and Verrecchia, K.E. (1993) Role of calcium oxalate biomineralization by fungi in the formation of calcretes: a case study from Nazareth, Israel. *J. Sed. Petrol.*, **63**, 1000–6.

Wainwright, M. and Gadd, G. M. (1997) Fungi and industrial pollutants. Pp. 85–97 in: *The Mycota IV. Environmental and Microbial Relationships* (D.T. Wicklow and B.E. Soderstrom, editors). Springer-Verlag, Berlin.

White, C., Sayer, J.A. and Gadd, G.M. (1997) Microbial solubilization and immobilization of toxic metals: key biogeochemical processes for treatment of contamination. *FEMS Microbiol. Rev.*, **20**, 503–16.

Whitney, K.D. and Arnott, H.J. (1987) Calcium oxalate crystal morphology and development in *Agaricus bisporus*. *Mycologia*, **80**, 180–7.

Whitney, K.D. and Arnott, H.J. (1988) The effect of calcium on mycelial growth and calcium oxalate crystal formation in *Gilbertella persicaria* (Mucorales) *Mycologia*, **80**, 707–15.

Wolschek, M.F. and Kubicek, C.P. (1999) Biochemistry of citric acid accumulation by *Aspergillus niger*. Pp. 11–31 in: *Citric Acid Fermentation* (B. Kristiansen and M. Mattey, editors). Taylor & Francis, London.

Yang, J., Tewari, T.P. and Verma, P.R. (1993) Calcium oxalate crystal formation in *Rhizoctonia solani* AG 2-1 culture and infected crucifer tissue: relationship between host calcium and resistance. *Mycol. Res.*, **97**, 1516–22.

CHAPTER FIVE

# Weathering of rocks by lichens: fragmentation, dissolution and precipitation of minerals in a microbial microcosm

M. R. LEE

*Division of Earth Sciences, University of Glasgow, Gregory Building, Lilybank Gardens, Glasgow G12 8QQ, UK (E-mail: m.lee@geology.gla.ac.uk)*

**ABSTRACT**

Despite the abundant evidence that rock-encrusting lichens can weather their substrates, it is currently unclear whether rates of lichen-mediated weathering are faster or slower than rates of abiotic weathering of otherwise identical rock surfaces (i.e. are lichens biodestructive or bioprotective?). This question is of considerable academic and commercial importance. Lichens weather rocks by a combination of biophysical and biochemical mechanisms. Fungal hyphae can penetrate into rocks at $\leqslant$~0.1 mm $y^{-1}$ and sandstones and limestones are especially vulnerable. Biophysical weathering leads to the fragmentation of minerals, exposing grain interiors. These fresh surfaces may be attacked by a variety of compounds including extracellular polysaccharides, lichen acids and oxalic acid. Evidence for the effectiveness of these biochemical agents includes etched and leached mineral grains and reaction products such as oxalate salts, clay minerals and Fe-hydroxides. Few studies have quantified rates of lichen-mediated weathering and fewer still have compared these data with the weathering rate of unencrusted rock surfaces. The conclusion of this work is that lichens enhance the weathering rate of rock surfaces relative to identical but abiotic substrates. As weathered mineral grains and weathering products are bound within the lichen, these materials will not be eroded until the lichen dies after ~$10^1-10^3$ y. Thus, despite being active agents of weathering, lichens should stabilize and protect rock surfaces over the short term. Studies of dated surfaces of a variety of rock types colonized by diverse lichen populations are essential before the impact of lichen colonization on rates of rock weathering can be accurately quantified and predicted.

## 5.1 Introduction

The impact of rock-encrusting lichens on the mineralogical and chemical composition and structural integrity of their substrate is of considerable current interest to scientists from many fields including geologists and mineralogists, microbiologists, architects and archaeologists. The lichen-rock interface is

Lee, M.R. (2000) Weathering of rocks by lichens: fragmentation, dissolution and precipitation of minerals in a microbial microcosm. Pp. 77–107 in: *Environmental Mineralogy: Microbial Interactions, Anthropogenic Influences, Contaminated Land and Waste Management* (J.D. Cotter-Howells, L.S. Campbell, E. Valsami-Jones and M. Batchelder, editors). Mineralogical Society Series, **9**. Mineralogical Society, London. ISBN 0 903056 20 8.

especially important because it represents a biologically, biochemically and mineralogically simplified 'soil' that is also scaled down to a millimetric size, making it highly amenable to characterization by modern microanalytical techniques. This interface, the 'lichen-mineral microcosm' (Banfield *et al.*, 1999), is an excellent natural laboratory within which to explore interactions between mineral surfaces and a wide range of organic compounds.

One of the prime reasons that geologists and mineralogists study lichen-encrusted rocks is to determine whether growth of lichens on or beneath a rock surface enhances or slows its weathering rate relative to an identical but lichen-free surface: are lichens 'biodestructive' or 'bioprotective'? Much of the work on lichen-rock interactions in the geological literature has focused on attempting to quantify the impact of microorganisms on the weathering rate of Mg- and Ca-rich silicate minerals over the long term ($>10^5$ y). These minerals are of particular significance because the precipitation of Mg and Ca liberated by chemical weathering on the continents as carbonate minerals in the oceans will, on a geological timescale, lock up atmospheric $CO_2$, an important greenhouse gas. One possible path of the reactions is summarized below (after Berner, 1992). In equation (5.1) anorthite feldspar reacts with carbonic acid during weathering on land to yield kaolinite, calcium and bicarbonate. In equation (5.2) the calcium and bicarbonate combine in the oceans to form calcium carbonate (calcite or aragonite) that is subsequently buried and lithified, producing carbonate rocks:

$$2CO_2 + 3H_2O + CaAlSi_2O_8 \rightarrow Al_2Si_2O_5(OH)_4 + Ca^{2+} + 2HCO_3^- \quad (5.1)$$

$$2HCO_3^- + Ca^{2+} \rightarrow CaCO_3 + CO_2 + H_2O \quad (5.2)$$

*Via* these reactions therefore, mineral weathering has a direct impact on climate. In the long period of time between the evolution of terrestrial microbial communities in the Precambrian and development of vascular land plants in the Silurian (~400 Ma ago), the continents may have been colonized by algae, fungi and lichens. Would the weathering rate of this hypothetical microbially-covered landscape have been any different to that of a landscape with bare rock surfaces? If so, algae, bacteria, fungae and lichens may have played a major role in reducing the Earth's high initial atmospheric $CO_2$ concentrations and global temperatures, thus making land surfaces more hospitable for higher forms of life. Berner (1992, 1995) and Schwartzman and Volk (1989, 1991) provide models of palaeoclimate evolution incorporating contrasting microbiological impacts on rock weathering. Note that, although the oldest fossil lichens are early Devonian in age, as Seaward (1997) has pointed out, lichens probably developed soon after the appearance of algae and fungi because such organisms are better able to cope with extreme environments in a symbiotic association. Lichens are believed to have evolved independently a number of times. As they are the dominant biological species on 8% of the present-day land surface (Paradise, 1997), lichens may still be significant agents of rock weathering on a global scale.

Other issues that interest geologists, architects and archaeologists, relate to the effect of lichen encrustation on the durability of historic monuments and works of art. This field has expanded greatly over the last few years creating a large literature. The central question is similar to that faced by those interested in palaeoclimate modelling but on a more practical level: will more damage be done to a natural or synthetic surface by removing encrusting lichens than by leaving them *in situ*? The answer again focuses on whether the biodestructive effects of lichens outweighs their bioprotective function, but over timescales many orders of magnitude shorter than those of interest to the palaeoclimate modellers ($\sim 10^1 - 10^3$ y). Interestingly, with the recent decline in pollution in western Europe, lichens have just started to colonize monuments in urban areas and understanding their detrimental effects is becoming a more urgent issue (Seaward, 1988).

The purpose of this review is to discuss evidence for lichen-mediated weathering of rocks and allied materials from a mineralogical perspective. Much of the work that has been published in this field, especially in the building stone preservation literature, is highly descriptive. This chapter will focus on those studies that have provided models of lichen–rock interactions and that have attempted to quantify the impact of lichen colonization on the weathering rate of rock surfaces. The reader is also directed to a number of other recent reviews that focus on different aspects of the problem: Ascaso and Wierzchos (1995) (description of lichen–rock interactions and historical review of techniques), Banfield *et al.* (1999) (the lichen–mineral interface as a model for the rhizosphere), Barker *et al.* (1997) (biochemical weathering of silicate minerals), Easton (1995) (biology of lichens from a geological perspective and their use in mineral prospecting), Jones and Wilson (1986) (biomineralization in lichens), Piervittori *et al.* (1994) (publications related to lichen–rock interactions) and Wilson (1995) (crystallization of biochemical reaction products).

## 5.2 Background

### *5.2.1 Biology of lichens*

Lichens are a symbiotic association of two organisms, a photobiont and a mycobiont that together form the thallus, which is the main body of the lichen. Photobionts of most lichens are green algae, but cyanobacteria (blue-green algae) may also fulfil the photosynthetic role. The mycobiont is a fungus and is the dominant partner and of most importance with regard to rock weathering. The relationship between the two organisms is referred to by some as controlled parasitism, with the mycobiont obtaining more benefit than the photobiont. Most lichens, including the saxicolous (rock-encrusting) species, can be placed into one of three morphological groups: crustose, foliose and fruticose. Saxicolous lichens can also be divided into two by habitat: epilithic (the more abundant variety in which the thallus is mainly above the rock surface) and endolithic (the thallus is predominantly beneath the rock surface).

Crustose lichens are the most abundant of the three morphological groups and the entire thallus is in contact with the substratum *via* a mass of fungal hyphae. As they are very firmly attached to the substrate, many rock weathering studies have focused on the activity of these species. Epilithic crustose lichens typically have a three-layered structure in cross-section. A thin outer cortex of fungal hyphae overlies an algal layer and protects it from desiccation, grazing and overexposure to sunlight. Beneath the algal layer is a thick medulla of interwoven fungal hyphae that is in direct contact with the rock surface (Fig. 5.1). Foliose lichens are flat and leaf-like and are only partially attached to the rock *via* rhizines, root-like masses of fungal hyphae. Many fruticose lichens are attached to the substrate *via* a holdfast, from which hyphae may penetrate beneath the rock surface. Typical radial growth rates of crustose lichens are 0.5–2.0 mm $y^{-1}$ (Hale, 1974); rates of penetration into the substrate, which are of more interest here, are more difficult to quantify

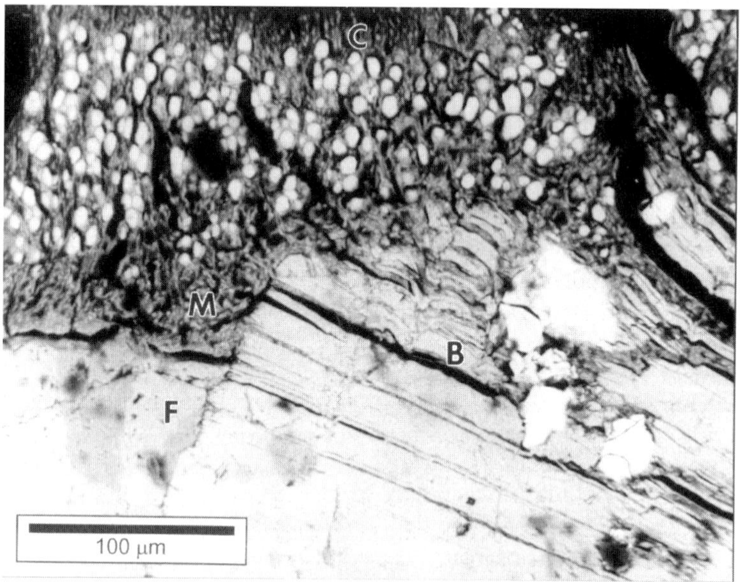


Fig. 5.1 Backscattered electron image (high vacuum SEM, 20 kV) of a polished cross-section of the crustose lichen *Rhizocarpon geographicum* on a $\leqslant$122 y old surface of the Lower Devonian Shap Granite, northwest England. The polished surface has been fixed with glutaraldehyde and osmium tetroxide ($OsO_4$). The thallus has a three-fold structure comprising a thin outer cortex (C), algal zone (the algal cells have preferentially taken the $OsO_4$ stain thus rendering them white in this image) and medulla (M), which is in direct contact with the granite. Fungal hyphae have not penetrated into alkali feldspar (F) but have extensively biophysically weathered a biotite grain (B) by exploiting its perfect cleavages.

(section 5.6.1). As lichens lack roots, they are very dependent for the nutrients that are essential for metabolic function on atmospheric sources including gases, precipitation and wind-blown dust. However, the intense substrate specificity exhibited by certain saxicolous lichens (sometimes called ornithocoprophilous species) indicates that they must also gain essential nutrients by weathering of rocks (Brodo, 1973).

### *5.2.2 Techniques for characterizing the mineralogy of the lichen–rock interface*

A thorough discussion of techniques for characterization of mechanisms and rates of lichen–rock interactions is beyond the scope of this review. The reader is directed to Ascaso and Wierzchos (1995). All that is possible here is to highlight those mineralogical techniques that are currently significantly advancing our understanding of the subject area.

Electron microscopy is one of the main tools used by mineralogists for characterization of mixed organic and inorganic materials. Work over the last ~10–15 y has concentrated on applications of scanning electron microscopy (SEM), in particular backscattered electron imaging of polished cross-sections of lichen-encrusted rock surfaces used in combination with qualitative and quantitative X-ray microanalysis. These techniques have been used very effectively by Dorn and co-workers (e.g. Dorn 1995; Brady *et al.* 1999) and by Ascaso and co-workers (Ascaso and Wierzchos, 1994, 1995; Wierzchos and Ascaso, 1994; Ascaso *et al.,* 1998). The latter workers pioneered fixing and staining techniques which enhance ultrastructural detail (Fig. 5.1). Recent work by Barker and Banfield (1996, 1998) has highlighted the applications of transmission electron microscopy (TEM) for characterizing the interfaces between organic polymers and mineral surfaces at very high (crystal lattice scale) resolutions. Barker and Banfield (1998) showed that new energy-filtering TEM techniques are especially well suited for imaging fine-scale mixtures of organic and inorganic materials.

Among the drawbacks of conventional electron microscopy are that parts of the lichen are destroyed during sample preparation, the samples need to be coated with a conductive layer of gold or carbon and the organic materials shrink under high vacuum conditions. One solution to the last two problems in particular has come from the recent development of SEMs that can operate at conditions much closer to those encountered at the Earth's surface. These instruments come in two varieties, the controlled pressure scanning electron microscope (CP-SEM) (see Lee, 1999, for applications to lichen–rock interactions) and more sophisticated environmental scanning electron microscope (ESEM). Figure 5.2*a,b* are images of a mass of fungal hyphae associated with the crustose lichen *Trapelia placodioides* taken using a CP-SEM and ESEM respectively. As the CP-SEM uses backscattered electrons to form the image whereas the ESEM is able to use secondary electrons, these two

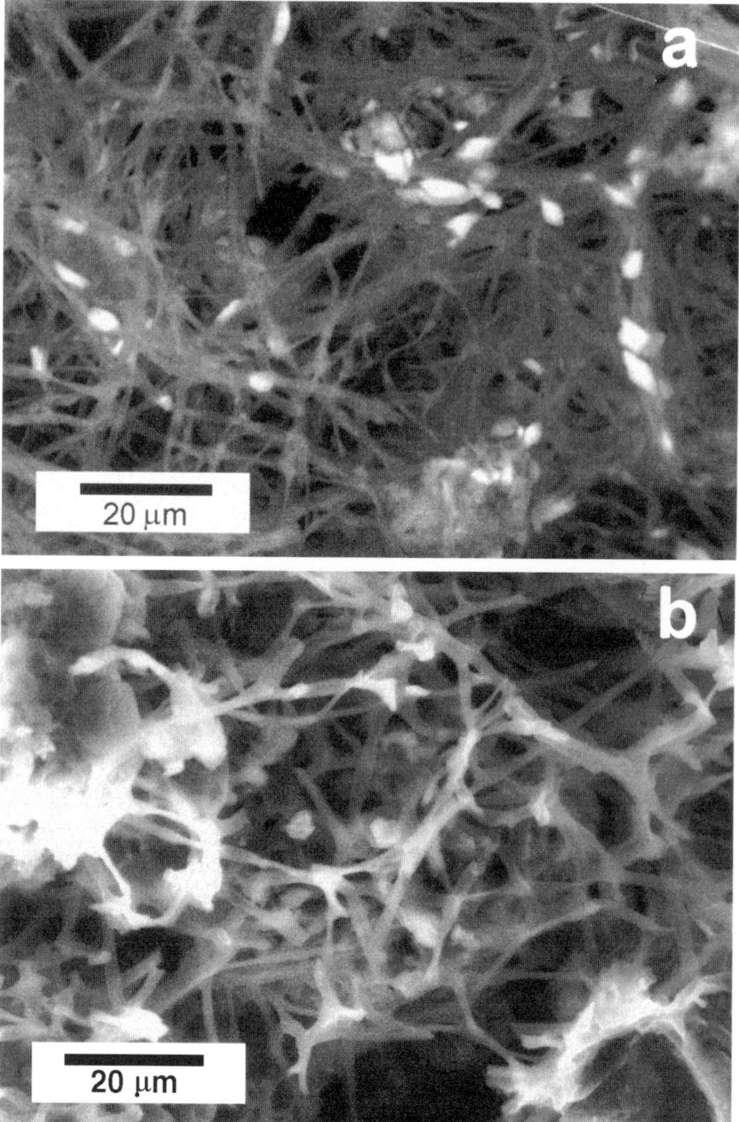


FIG. 5.2 Images of fungal hyphae within pores in a sandstone wall from the author's Edinburgh garden. This wall has been colonized by the crustose lichen *Trapelia placodioides* for <101 y. (*a*) Backscattered electron image (CP-SEM, 20 kV, 0.3 mbar) showing a mass of fungal hyphae that are encrusted with bipyramidal crystals of calcium oxalate (weddellite) a few μm in size. These oxalate crystals can be distinguished by their greater mean atomic number than the fungal hyphae. (*b*) Secondary electron image (ESEM, 20 kV, 7.9 mbar, 90% relative humidity) in which the resolution of fungal hyphae is far superior to the CP-SEM image. However oxalate minerals, if present, will be difficult to locate.

techniques yield complementary information. Conventional high-vacuum SEMs that use cold field emission electron guns offering high resolutions at very low accelerating voltages, thus obviating the need for a conductive coating, also have considerable potential (e.g. Barker *et al.*, 1997, Fig. 10). Lastly, Raman spectroscopy is a completely non-destructive technique that has been applied extensively to studying works of art encrusted by lichens and in particular to the identification of biochemical weathering products (see Edwards, 1997, for review).

## 5.3 Abiotic weathering and erosion of rock surfaces

Before considering the mechanisms by which lichens weather their substrates, it is necessary to summarize those weathering processes that are not directly related to organic activity (i.e. physical and chemical weathering as opposed to biophysical and biochemical weathering). Mechanisms of physical weathering include growth of ice crystals (frost wedging), growth and hydration of salt crystals, fracturing from thermal expansion and contraction and exfoliation resulting from pressure release. Physical weathering increases the surface area to volume ratio of a rock, making it more susceptible to chemical weathering by processes including dissolution and hydrolysis, in which carbonic acid plays a major role (equation 5.1), and oxidation. Ions liberated by chemical weathering may contribute to the formation of new minerals such as clays or may be removed in solution (equation 5.1). Note that as respiration and degradation of organic matter is a major source of carbonic acid for weathering in soils, the distinction between chemical and biochemical weathering by carbonic acid is indistinct. Physical and chemical weathering usually take place simultaneously and one process enhances the rate of the other by positive feedback mechanisms (e.g. Lee *et al.,* 1998). For a comprehensive recent review of chemical weathering of silicate minerals see White and Brantley (1995).

Where the gradients of rock surfaces are relatively low and rates of chemical and physical weathering are high, weathered minerals and weathering products accumulate to form soils or saprolites. This review is more concerned with the weathering of bare rock surfaces that are found where rates of chemical weathering are low (e.g. in hot and cold deserts) and/or where slopes are steep (e.g. the faces of buildings). In these environments, the removal of weathered mineral grains and weathering products by erosive agents including wind, water and ice is also an important process.

### *5.3.1 Potential impact of lichen colonization on abiotic weathering and erosion*

The microenvironment of a lichen-colonized rock surface will differ considerably from that of a lichen-free surface and these differences may have a significant impact on the mechanisms and rates of physical and chemical weathering and erosion listed above. Thalli of crustose species are likely to have a thermal insulating effect, which may be magnified by differences in albedo between the lichen and adjacent rock surface. For example, Wessels and Wessels (1995) demonstrated that the temperature of the thallus of the endolithic lichen *Lecidea* aff. *sarcogynoides* 2 mm below the rock surface was consistently greater than the surrounding air temperature in the South African spring, reaching 11 K greater at some times during the day. Such temperature differences will elevate rates of chemical weathering. Barker and Banfield (1998) stress that one of the most important effects of lichen colonization may simply be increased residence time of water in the vicinity of mineral surfaces, which is bound within extracellular polymers. Encrustation by lichens should also protect rock surfaces against direct ingress of precipitation and runoff. As the thallus binds fragmented mineral grains and weathering products and will protect a rock surface from the direct action of wind and running water, lichen colonization should significantly slow the rate of erosion of a rock surface. This protective effect will only last for as long as the lichen remains alive ($\sim10^1-10^3$ y), after which time the thallus together with mineral grains will be susceptible to erosion.

## 5.4 A model of the lichen–rock interface

A model that describes the range of mechanisms and products of biologically-mediated weathering at the lichen–rock interface has recently been proposed by Barker and Banfield (1998) (Fig. 5.3). This model is based on electron microscope studies of an amphibole syenite substrate that has been colonized by the crustose lichens *Rhizocarpon grande* and *Porpidia albocaerulescens* for $\leqslant$90 y. Here, this model will be used as a framework within which previous studies of lichen-mediated weathering can be reviewed.

Barker and Banfield (1998) recognized four distinct zones of organic–mineral interactions. Zone 1 (photosynthetic zone) is in the upper thallus and contains the photobiont. The significant processes in this zone include exchange of mineral grains and solutes with the atmosphere and hydrosphere and synthesis of lichen acids. Zone 2 (direct biochemilithic zone) is in the lower thallus and contains substrate-derived mineral fragments whose surfaces are coated with organic polymers. Weathering mechanisms include fragmentation of mineral grains by biophysical action, alteration of mineral surfaces by organic polymers, acids and ligands and crystallization of weathering products. Zone 3 (indirect biochemilithic zone) is in the intact rock substrate below zone 2 and also in the interior of mineral fragments within zone 2. The important weathering processes in this zone are etching and leaching of mineral grains and precipitation of biochemical reaction products due to soluble acids and ligands. This zone shows microbially-influenced weathering but inter- and intra-granular spaces are too small for organisms to penetrate. Zone 4 (physicochemilithic zone) has no direct biological weathering effects. It

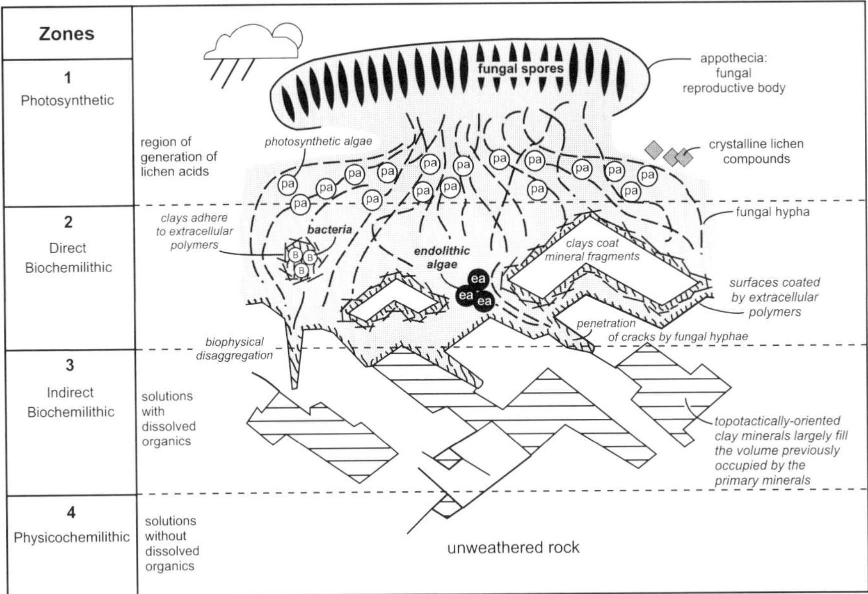


FIG. 5.3 Diagram showing the four weathering zones on a lichen-encrusted rock surface. From Barker and Banfield (1998). Reproduced with permission from Taylor & Francis.

contains unweathered rock and predominantly inorganically-formed weathering products. Below, evidence for the operation of these weathering mechanisms and the formation of weathering products is discussed. It must be emphasized that any one weathering process or product is likely to feature in more than one zone, but they are discussed in the context of individual zones for brevity.

## 5.5 Photosynthetic zone

### *5.5.1 Exchange of particles and solutes with the atmosphere and hydrosphere*

Barker and Banfield (1998) did not recognize organic–mineral interactions taking place in this mixed fungal and algal/cyanobacterial zone and state that substrate-derived mineral grains are absent. However, the outermost region of lichens is where solids can enter the lichen–rock system, principally as wind-borne dust, and ions in solution can be introduced *via* runoff from surrounding areas and from precipitation, especially 'occult' precipitation (i.e. fog and dew) (Nash, 1996*b*). The ability of lichens to trap and bind atmospheric particulates and also to tolerate high metal concentrations accounts for their use in monitoring past and present-day levels of environmental pollutants (Richardson, 1992). Lee (1999) provides CP-SEM images of fly ash derived from power stations embedded within a lichen. The degree to which wind-borne dust is utilized by lichens as a nutrient source is presently unclear and Nash (1996*b*) suggests that most such particles are probably unaffected by lichen-mediated weathering. The introduction of solutes from the atmosphere or hydrosphere may however be a more significant process than is generally recognized. This is because oxalate minerals, whose presence in the lichen's thallus is almost universally presented as *de facto* evidence for biochemical weathering of the lichen's substrate, may in fact form by combination of oxalic acid secreted by the mycobiont with solutes such as Ca or Cu derived from runoff or occult precipitation (see 5.7.5). Thus, care must be taken when interpreting oxalate salts as products of biochemical weathering of the rock substrate.

### *5.5.2 Synthesis of lichen acids*

In addition to being the locus for input of particulates and solutes, lichen acids, which are able to complex metal cations, are generated in the photosynthetic zone, probably by interactions between the mycobiont and photobiont (Barker *et al.*, 1997). Lichen acids are a group of water-soluble ($5-57$ mg $l^{-1}$) mainly polyphenolic compounds that are produced by all species of lichen. The best evidence for the involvement of lichen acids in biochemical weathering comes from Purvis' work on lichens encrusting copper-rich rocks. Crystalline copper-nostictic acid and copper-psoromic acid was found in these lichens (Purvis *et al.,* 1985, 1987, 1990), demonstrating conclusively the complexing ability of lichen acids. However, as other evidence for the involvement of lichen acids in biochemical weathering is scarce, most authors conclude that these compounds are of relatively minor importance in mineral weathering (Barker *et al.,* 1997).

## 5.6 Direct biochemilithic zone

### *5.6.1 Fragmentation of mineral grains by biophysical weathering*

The most obvious effect of saxicolous lichens on their substrate, which has been recognized for many years, is biophysical weathering (Fry, 1924, 1927). Syers and Iskandar (1973) proposed two mechanisms by which biophysical weathering may take place: (1) penetration of the rock by medullary hyphae (crustose lichens) and rhizines (foliose species); and (2) expansion and contraction of the thallus.

Early work reported in Syers and Iskandar (1973) demonstrated a strong lithological control on the depth of rhizine penetration. Minerals with a well developed cleavage, such as biotite, are especially susceptible whereas framework silicates including feldspar and quartz are less prone (Figs 5.4, 5.5). Carbonate rocks are most readily exploited, with maximum depths reported by Syers (1964) of 16 mm. As limestones are typically tightly cemented, it is likely that such depths of penetration reflect the biochemical weathering ability of hyphal tips more than their mechanical action. Data on penetration rates are summarized in Table 5.1.

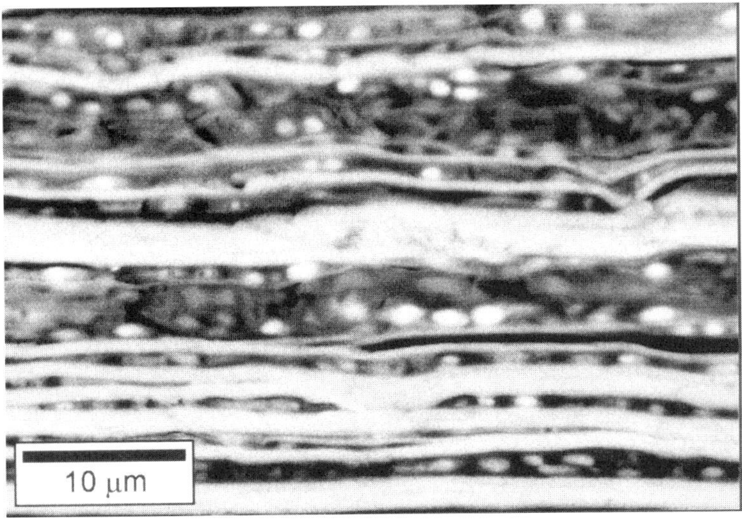


FIG. 5.4 Backscattered electron image (high vacuum SEM, 20 kV) of an $OsO_4$-stained polished cross-section of the crustose lichen *Rhizocarpon geographicum* on a $\leqslant$122 y old surface of the Lower Devonian Shap Granite, northwest England. The image shows a biotite grain that has been exfoliated by growth of algal and fungal cells between cleavage planes. The thinnest biotite sheets are ~0.5 µm thick. Reprinted from *Chemical Geology*, **161**, Lee, M. R. and Parsons, I., Biophysical and biochemical weathering of lichen-encrusted granite: textural controls on organic-mineral interactions and deposition of silica-rich layers, pp. 385–97. Copyright (1999), with permission from Elsevier Scientific Publishers.

Penetration of rhizines may be facilitated by expansion and contraction of the lichen's thallus following wetting and drying respectively. A lichen may alternate between a hydrated and desiccated state on a daily basis, the transition taking minutes to hours (Nash, 1996*a*). These cycles may have the effect of tearing away minerals in the lower part of the medullary layer. Syers and Iskandar (1973) suggested that foliose lichens may be more effective in biophysical weathering than crustose species because of their greater freedom of movement.

Data on rates of hyphal penetration into rock substrates are scarce and must be treated with caution owing to the large number of potential variables, especially rock type (mineralogy, texture, porosity and permeability) and the growth stage of the lichen in question. A study of the endolithic lichen *Lecidea* aff. *sarcogynoides* by Wessels and Wessels (1995) showed that early (vegetative) stages of growth of the thallus have a vertical:lateral component of 1:12 mm, whereas in later (reproductive) stages this changed to 1:30 mm; average vertical and lateral growth rates were 4.0 mm $y^{-1}$ and 0.1 mm $y^{-1}$ respectively. Data on penetration rates summarized in Table 5.1 suggest an upper limit of ~0.1 mm $y^{-1}$ for non-carbonate rocks.

FIG. 5.5 Backscattered electron image (high vacuum SEM, 20 kV) of an $OsO_4$-stained polished cross-section of the crustose lichen *Rhizocarpon geographicum* on a <10,000 y old surface of the Lower Devonian Shap Granite, northwest England. Plagioclase feldspar in the centre of the image has been fragmented into angular blocks by biophysical weathering. Surrounding grains have been penetrated by hyphae but remain intact. The three-fold cross-sectional structure of the lichen is well defined. Reprinted from *Chemical Geology*, **161**, Lee, M.R. and Parsons, I., Biophysical and biochemical weathering of lichen-encrusted granite: textural controls on organic-mineral interactions and deposition of silica-rich layers, pp. 385–97, Copyright (1999), with permission from Elsevier Scientific, Publishers.

One of the most significant impacts of biophysical weathering is to create fresh mineral surfaces. Fragmented mineral grains are then physically bound within an organic matrix forming a primitive (cryptogamic) soil (Fig. 5.5). Stabilization of mineral grains within this soil will render them more susceptible to biochemical weathering than if they had been eroded and deposited in a terrestrial or marine sediment. Schwartzman and Volk (1989) and Schwartzman (1993) suggested that stabilization of mineral grains and exposure to biogenic acids may have been the most significant weathering-related impact of the possible colonization of late Precambrian/lower Palaeozoic land surfaces by lichens.

### 5.6.2 *Alteration of mineral surfaces by organic polymers*

Most biochemical weathering within the direct biochemilithic zone is, by definition, mediated by extracellular organic polymers (polysaccharides) that coat the surfaces of biophysically fragmented mineral grains. These polymers, which are produced by the microorganisms and serve to attach cells to mineral

TABLE 5.1 Depths and rates of hyphal penetration into rock substrates.

| Authors | Lithology | Lichen species | Rate/depth |
|---|---|---|---|
| **Depths of penetration** | | | |
| Cooks and Otto (1990) | Quartzite | *Lecidea* aff. *sarcogynoides* | 1.12 mm |
| Smith (1921 in Syers and Iskandar, 1973) | Limestone | not stated | 15 mm |
| Syers (1964) | Limestone | not stated | 0.3–16 mm |
| Viles (1987) | Limestone | not stated | ⩽ 2 mm |
| **Rates of penetration** | | | |
| Barker and Banfield (1996) | Syenite | *Rhizocarpon grande* and *Porpidea albocaerulescens* | >0.11 mm $y^{-1}$ |
| Lee and Parsons (1999)[1] | Granite | *Rhizocarpon geographicum* | ⩾0.002 mm $y^{-1}$ |
| Lee and Parsons (1999)[2] | Granite | *Rhizocarpon geographicum* | >0.003 mm $y^{-1}$ |
| This study | Sandstone | *Trapelia placodioides* | 0.017 mm $y^{-1}$ |
| **Rates of penetration calculated from published data** | | | |
| Wessels and Wessels (1995)[3] | Sandstone | *Lecidea* aff. *sarcogynoides* | 0.1 mm $y^{-1}$ |
| Wessels and Schoeman (1988)[3] | Sandstone | *Lecidea* aff. *sarcogynoides* | 0.08 mm $y^{-1}$ |

[1] A ⩽122 y old granite substrate
[2] A ⩽10 ky granite substrate
[3] Data on rates of penetration of an endolithic lichen normal to the rock surface

surfaces, have been characterized in detail by Barker and Banfield (1996, 1998) using TEM techniques (Fig. 5.6). Barker *et al.* (1997) and Welch *et al.* (1999) have summarized a number of mechanisms by which extracellular polysaccharides participate in the weathering of mineral surfaces: (1) they serve to retain water at the mineral surface, thus facilitating hydrolysis and diffusion of ions; (2) they provide protons that participate directly in dissolution; (3) they bind with metal ions in solution, decreasing saturation states and enhancing dissolution rates; and (4) they provide an environment within which reaction products can precipitate and recrystallize. Barker and Banfield (1996, 1998) showed that reaction products on alkali feldspar, amphibole and quartz in a lichen-encrusted amphibole syenite may include sub-micrometre sized crystallites of clay and phyllosilicate that are contained within the extracellular polymers (Fig. 5.6). One aspect of the role of polysaccharides in mineral weathering has been quantified in laboratory studies by Welch *et al.* (1999). From batch dissolution experiments on bytownite feldspar, they found that mildly acidic (pH ≈ 4) solutions of acid polysaccharides accelerate rates of release of Al and Si from the feldspar relative to an inorganic control acid, in some cases by two orders of magnitude. These results are highly significant because such acidities may be expected in natural environments.

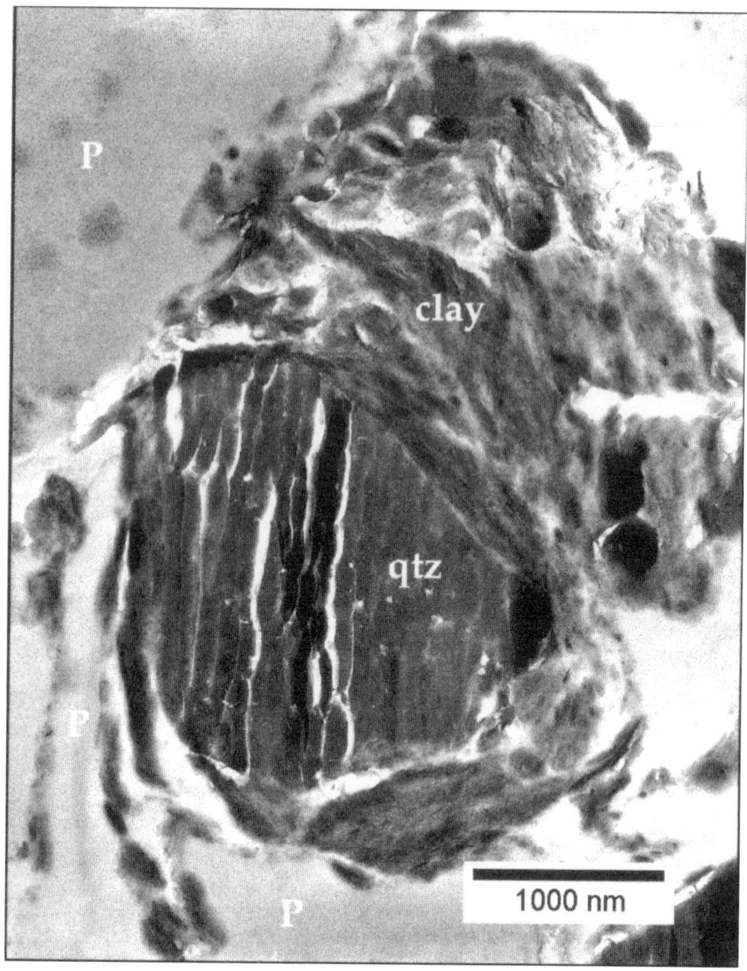


FIG. 5.6 Zero-loss energy-filtered TEM image of a quartz grain (qtz) surrounded by clays and extracellular polymers (P). The quartz has been fractured during sample preparation by ultramicrotome. Previously unpublished image, courtesy of Bill Barker.

## 5.7 Indirect biochemilithic zone

Much of the previous work on lichen–rock interactions has focused on characterizing weathering within Barker and Banfield's indirect biochemilithic zone. The acids and ligands that may mediate biochemical weathering are described first, followed by evidence for their interaction with rock surfaces, which includes etched and leached mineral grains and biochemical reaction products, principally crystalline oxalates, iron-rich minerals, clays and phyllosilicates.

### 5.7.1 *Carbonic acid*

$CO_2$ produced by microbial respiration and degradation of organic matter may combine with water within the thallus to form carbonic acid ($H_2CO_3$), a common weathering agent. However, as carbonic acid is relatively weak, most authors have concluded that it will not have a significant impact on biochemical weathering rates (Syers and Iskandar, 1973; Barker *et al.* 1997). Wilson and Jones (1983) and Jones and Wilson (1985) noted that weathering of silicate rocks in environments where carbonic acid dominates leads to the accumulation of clay minerals such as kaolinite (equation 5.1). As such, weathering products are rare in lichens (see below); Jones and Wilson (1985) used this as further evidence that carbonic acid does not play a major role in weathering within the lichen–rock system.

### 5.7.2 *Oxalic acid*

There is now a large body of evidence demonstrating that oxalic acid, $(COOH)_2{\cdot}2H_2O$, excreted by the mycobiont, is one of, if not the most, important agents of biochemical weathering. The clearest illustration of the role of oxalic acid is provided by the occurrence of oxalate salts in many species of lichen. Wilson and colleagues showed that the cation composition of the salts within any one saxicolous lichen correlates closely with the cation composition of its rock substrate (e.g. calcium oxalate is found within lichens encrusting Ca-rich rocks, etc.). More details on oxalate mineralogy are given in section 5.7.5 below.

Oxalic acid mediates biochemical weathering by a number of mechanisms: (1) it provides protons that directly attack metal-oxygen bonds at the mineral surface; (2) oxalate ions form soluble complexes with polyvalent cations at the mineral surface (ligand-promoted dissolution); and (3) in forming metal oxalate complexes the saturation state of the solution is reduced (Welch and Ullman, 1993, 1996; Barker *et al.*, 1997). Experiments by Welch and Ullman (1993, 1996) demonstrated that dissolution rates of framework silicates, mainly plagioclase feldspars, in organic acids varies with Al concentration in the mineral; those with higher concentrations of Al dissolved faster than those with lower concentrations. These results are consistent with dissolution in organic acids, including oxalic acid, being promoted by adsorption of organic ligands such as the oxalate ion at Al sites on the mineral surface, in addition to attack of Al sites by protons. Welch and Ullman (1996) also documented a smaller effect of ligand adsorption at Si sites on the mineral surface. Stillings *et al.* (1996) suggested that in addition to the oxalate anion ($C_2O_4^{2-}$), the bioxalate anion ($HC_2O_4^-$) also plays a role in dissolution, although details of its interaction with mineral surfaces remain unclear. The impact of ligand-promoted dissolution is most apparent at circumneutral pH, under which conditions the effects of proton-promoted dissolution are smallest (Welch and Ullman, 1993, 1996). A number of other experimental studies on aluminosilicate minerals have shown that oxalate forms soluble complexes with ions including $Al^{3+}$ and $Fe^{3+}$, leaving a silica-rich residual solid. Silica-rich gels and siliceous pseudomorphs of aluminosilicate minerals are commonly encountered at the lichen–rock interface, again supporting the involvement of oxalic acid in lichen-mediated weathering (see section 5.7.4).

### 5.7.3 *Etched mineral surfaces*

Some of the clearest evidence for biochemical weathering in the indirect biochemilithic zone is the occurrence of etched mineral grains. The most obvious expression of biological involvement is the presence of tubes within grains that contain fungal hyphae or are the same size and shape as hyphae. Carbonate minerals and rocks are especially susceptible and commonly contain fungal borings (Fig. 5.7). It is possible that similar features may be found in silicate minerals. Jongmans *et al.* (1997) described alkali and plagioclase feldspar and hornblende from European soils that contain 3–10 μm wide tubular pores, some of which were filled by fungal hyphae. They suggested that excretion of organic

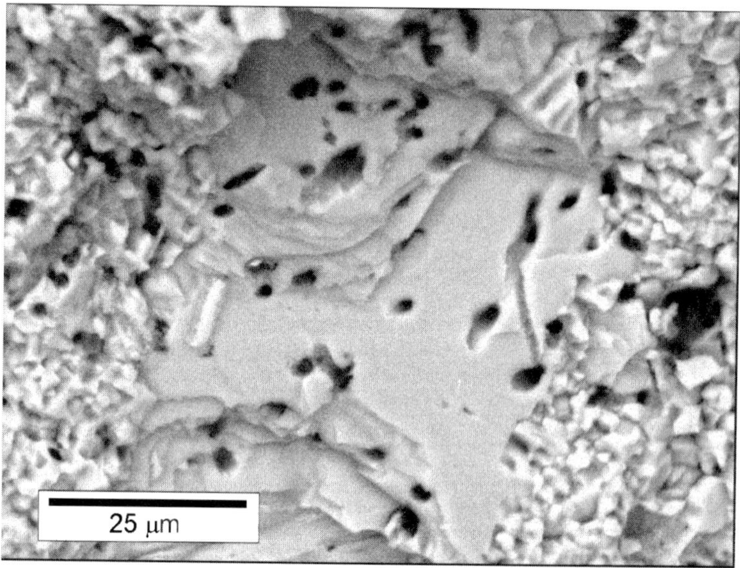


FIG. 5.7 Backscattered electron image (CP-SEM, 20 kV, 0.4 mbar) of a cross-sectional fracture surface of Carboniferous limestone from the Mendip Hills, southern England, which is colonized by a crustose lichen. Fungal hyphae have penetrated $\geqslant 0.8$ mm into the interior of this tightly-cemented rock. In the centre of the image is a relatively large single crystal of calcite ($CaCO_3$) surrounded by smaller crystals that have broken to leave an irregular surface. Hyphae have penetrated through the calcite producing tubes, various sections of which are seen in the image. Those tubes containing hyphae are especially easy to recognize due to significant differences in mean atomic number between calcite (light grey) and organic material (dark grey).

acids enables the hyphae tips to extend into Ca-rich plagioclase, a relatively soluble silicate mineral, at a rate of 0.3–30 μm $y^{-1}$. Although the tubes illustrated by Jongmans *et al.* (1997) were formed by soil fungi, the potential for lichenized fungi to have the same impact is clear especially since lichens are microbial communities and may contain several species of fungi.

The external surfaces of mineral grains also show good evidence for etching by acidic solutions; carbonates and ferromagnesian silicate minerals are especially susceptible (Wilson, 1995) (Fig. 5.8). Using SEM, Jones *et al.* (1981) and Wilson and Jones (1983) described the etched surfaces of feldspars (albite, labradorite and perthitic alkali feldspar) and ferromagnesian minerals (augite and olivine) beneath lichens. Etching produces distinctive features including sub-micrometre to μm-sized pits, which form at sites where dislocations intersect the grain surface, and troughs, which develop by preferential etching of one mineral relative to another in lamellar intergrowths. Jones *et al.* (1981) were able to reproduce troughs on the surface of plagioclase feldspar (labradorite) from a lichen-encrusted basalt by incubating fresh labradorite with *Aspergillus niger*, a free-living fungus known to excrete oxalic acid, for 3 months; comparable textures were formed by treatment of

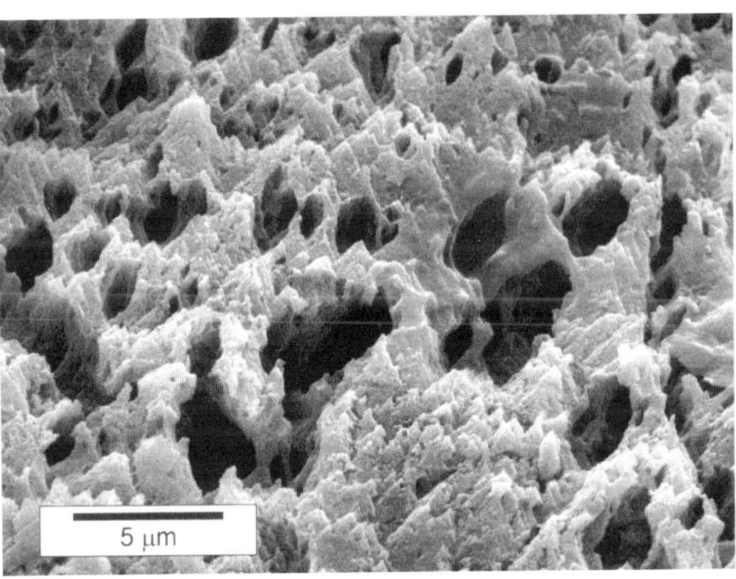


FIG. 5.8 Secondary electron image (high vacuum SEM, 20 kV) of a dolomite crystal from within sandstone from the author's Edinburgh garden that has been colonized by a *Trapelia placodioides*. Dissolution of the dolomite has produced deep prismatic etch pits. Note the apparent absence of organic materials and reaction products in direct contact with the dolomite crystal, indicating that dissolution was probably mediated by a fluid and that solutes were removed from the reaction site.

fresh labradorite with 0.5 M oxalic acid for 18 days. If the size and abundance of etch pits on a grain's surface can be quantified, for example by atomic force microscopy (AFM), minimum biochemical weathering rates can be calculated. An example of this technique to determine alkali feldspar weathering rates in soils is given by Lee *et al.* (1998).

### *5.7.4 Amorphous silica*

Using X-ray microanalysis and X-ray diffraction (XRD) techniques, a number of authors have noted that aluminosilicate minerals in the vicinity of the lichen–rock interface have commonly been leached heavily to leave siliceous relicts. For example, Wilson *et al.* (1981) described an X-ray amorphous silica gel from the surface of a serpentine whose encrusting lichen, *Lecanora atra*, had been removed by treatment with hydrogen peroxide. The gel, which had a fibrous morphology, was inferred to have formed by leaching of Mg from fibrous chrysotile by oxalic acid excreted by the mycobiont. Magnesium subsequently reprecipitated within the thallus as the magnesium oxalate, glushinskite ($MgC_2O_4{\cdot}2H_2O$). Wilson *et al.* (1981) were able to form fibrous gels and Mg oxalates by experimentally reacting fresh chrysotile with oxalic acid, thus confirming their model. Wilson and Jones (1983) also reported biotite from beneath a crustose lichen whose major cations, apart from Si, had been leached, leaving a siliceous relic; they noted the same effects for anorthite. Jones *et al.* (1981) experimentally leached biotite flakes to siliceous relics by incorporating the biotite into a culture containing the oxalic acid-producing fungus *A. niger*.

Despite this evidence, Barker and Banfield (1996) suggested that the amorphous silica found in natural samples may be an artefact resulting from laboratory oxidation of rock-encrusting organics by $H_2O_2$. This suggestion was prompted by their observation that mineral grains surrounded by organic materials in a syenite, whose surface has been encrusted by *Rhizocarpon grande* and *Porpidea albocaerulescens* for $\leqslant$90 y, were crystalline to within 1 nm of their surface; amorphous silica layers would have been readily identifiable with the high-resolution TEM techniques they used. However, Lee and Parsons (1999) have described Si-rich layers coating grain surfaces from a granite encrusted by *Rhizocarpon geographicum*. As these layers were observed *in situ* in cross-sections of the interface by backscattered electron SEM imaging, their primary origin is unequivocal, although Lee and Parsons (1999) were unable to determine if the layer was amorphous. As the Si-rich material also cements fractures and pores within minerals beneath the interface, Lee and Parsons (1999) suggested that the layer was precipitated from silica in solution, probably ultimately derived from biochemically leached biotite. What is clear from differences between the various studies described above is that mechanisms and products of biochemical weathering are likely to differ considerably between species of lichens (possibly itself reflecting the 'cocktail' of biogenic compounds, both oxalic acid and lichen acids, secreted by the mycobiont), the age of the lichen, the mineralogy of the substrate and especially the abundance of sheet silicates including biotite.

### 5.7.5 *Oxalate minerals*

Crystalline oxalates are the most commonly encountered biochemical reaction products in lichens. As the cation composition of oxalates in any saxicolous lichen is, in general, closely comparable to that of its rock substrate, these minerals are interpreted to form by combination of oxalic acid secreted by the mycobiont with cations derived from the host rock. Most authors suggest that formation of oxalates is one of the ways in which lichens deal with intolerably high metal concentrations; they thus represent a "detoxification" mechanism (Wilson *et al.,* 1981; Purvis, 1984). See Gadd (2000) for descriptions of oxalates produced by fungi.

The most frequently reported oxalates are the monohydrate and dihydrate of calcium oxalate whewellite ($CaC_2O_4{\cdot}H_2O$) and weddellite ($CaC_2O_4{\cdot}2H_2O$), which can comprise up to 60% of the dry weight of lichens on calcareous substrates (Syers *et al.,* 1967). Other oxalates include glushinskite ($MgC_2O_4{\cdot}2H_2O$, Wilson *et al.,* 1980; Wilson and Bayliss, 1987), moolooite ($CuC_2O_2{\cdot}nH_2O$, Purvis, 1984; Chisholm *et al.*, 1987) and an unnamed dihydrate of manganese oxalate ($MnC_2O_4{\cdot}2H_2O$, Wilson and Jones, 1984). Although most research on oxalates has employed X-ray microanalysis in the SEM and XRD techniques, recent work by Edwards and co-workers (e.g. Edwards, 1997) has shown that Fourier transform Raman spectroscopy is a powerful tool for identifying oxalates and has the advantage of being completely non-destructive.

The occurrence of oxalates is, in most cases, very good evidence for biochemical weathering of a rock substrate, but in some situations the appropriate cations may be introduced *via* runoff or even occult precipitation. For example, oxalates may occur in lichens on wood substrates. In a study of the copper oxalate hydrate moolooite, Purvis (1984) and Chisholm *et al.* (1987) concluded that much of this oxalate had formed by reaction of dissolved copper within surface runoff with oxalic acid, although direct interaction of oxalic acid within the lichen with the Cu-rich substrate was also likely. The location of oxalates within the thallus may be a good guide to the origin of divalent cations (atmosphere *vs.* hydrosphere *vs.* substrate).

### 5.7.6 *Iron oxides and hydroxides*

Significantly, the iron oxalate dihydrate, humboldtine ($Fe^{2+}C_2O_4{\cdot}2H_2O$), has yet to be found in lichens. Instead, iron oxides and hydroxides including hematite, ferrihydrite and goethite occur at the interface between lichens and especially basic igneous rock substrates (Jackson and Keller, 1970*a,b*; Jones *et al.,* 1980; Adamo *et al.,* 1997), but they also occur in association with more acid igneous and metamorphic rocks (Ascaso *et al.,* 1976; Galvan *et al.,* 1981; Jones *et al.,*

1981; Barker and Banfield, 1996, 1998). Schwertmann and Taylor (1989) suggest that the iron oxides and hydroxides associated with lichens are formed by oxidation of organic molecules that have complexed substrate-derived iron.

### *5.7.7 Clays and phyllosilicates*

The status of clays as biochemical weathering products has been ambiguous because of the potential for contamination by wind-borne dust and inorganically-formed clays within the substrate itself (Wilson, 1995). However, high-resolution TEM work has now demonstrated conclusively that clays can form as a direct result of biochemical weathering in the indirect biochemilithic zone. Barker and Banfield (1996, 1998) described alteration of Fe-rich amphibole (ferrohastingite) to nontronitic smectite associated with aluminous Fe-oxyhydroxides.

There is good evidence that some clays, specifically vermiculite, can also form by biologically-mediated replacement of biotite. Using TEM, Barker and Banfield (1996) illustrated partial replacement of biotite (ferriannite) by vermiculite following oxidation of $Fe^{2+}$ and loss of interlayer K, although they could not unambiguously state that replacement was wholly biologically promoted. However in a subsequent TEM study of granite that had been colonized by *Parmelia conspersa* for ~15 y, Wierzchos and Ascaso (1998) also illustrated the interstratification of vermiculite with biotite and concluded that transformation of biotite to vermiculite was the result of biophysical followed by biochemical weathering. These TEM studies confirm results from earlier work using X-ray chemical analysis which have suggested that sheet silicates in general, and biotite in particular, are especially susceptible to biologically-mediated weathering (Wilson *et al.,* 1981; Wilson and Jones, 1983; Wierzchos and Ascaso, 1996; Lee and Parsons, 1999). Their susceptibility is probably due to a perfect cleavage, facilitating biophysical weathering (Lee and Parsons, 1999) and for biotite in particular, a product of the presence of loosely held interlayer K, an important nutrient.

## 5.8 Physicochemilitic zone

This zone is characterized by unweathered rock and physical and chemical weathering. Prior to colonization by a lichen, the physicochemilitic zone will constitute the whole rock surface, but will then migrate into its interior as the lichen grows. Banfield *et al.* (1999) stressed the importance of physical and chemical weathering of a surface to create sufficient porosity and permeability for the establishment of microbial communities.

## 5.9 Rates of lichen-mediated weathering

Work described in previous sections provides unambiguous evidence for biophysical and biochemical weathering of lichen-encrusted rock surfaces. In this section two significant questions will be addressed: (1) can the rate of weathering beneath lichens be quantified? and (2) is this rate greater or slower than the weathering rate of identical but lichen-free rock surfaces?

### *5.9.1 The debate on lichen-mediated weathering of Hawaiian lavas*

An excellent example of the difficulties inherent in attempting to quantify the impact of lichens on rock weathering is provided by the debate during the 1990s between Berner, Jackson and others. Jackson and Keller (1970*a*) studied the weathering of Recent Hawaiian basalts by the fruticose lichen *Stereocaulon vulcani*. They found that weathering crusts, a form of incipient lateritic soil composed of ferrihydrite-rich gels (Jackson, 1993), occur on the lavas. These crusts were consistently thicker on lichen-encrusted than on lichen-free parts of the same lava flows. For example, the crust on a 1907 flow had a mean thickness of 0.142 mm where lichen-encrusted, but <0.002 mm where lichen-free. An average of the mean crust thickness for five flows of different ages gives a crust growth rate of $32 \times 10^{-4}$ mm $y^{-1}$ where lichen colonized and $<1 \times 10^{-4}$ mm $y^{-1}$ where lichen free. Jackson and Keller (1970*a*) concluded that encrustation by *S. vulcani* greatly accelerates weathering rates.

From the perspective of palaeoclimate modelling, Berner (1992, 1995) and Cochran and Berner (1993*a,b*, 1996) re-examined the Hawaiian lavas and concluded that the ferrihydrite-rich crusts are predominantly accumulations of wind-borne dust, albeit biochemically modified, and not weathering products. Berner (1992) concluded that "On either a decadal or thousand-year time scale, lichens on Hawaii do not bring about much greater weathering than that achieved by simple water–rock interaction". These conclusions were strongly refuted by Jackson (1993) and questioned by Schwartzman (1993). In their most recent contribution to this debate, Berner and Cochran (1998) concede that there is "some evidence" for weathering beneath *S. vulcani* on old lava flows. However, with regard to the younger (100–3000 y old) flows they studied, there was little or no evidence for leaching of basalt on the micrometre scale. The most recent contributions have been from Brady *et al.* (1999) and Conrad *et al.* (1999). Brady *et al.* (1999) quantified the rates of weathering of olivine and plagioclase in Hawaiian basalts, some of which were colonized by *S. vulcani* and others that were lichen-free. Weathering rates were determined by digital analysis of backscattered SEM images of polished cross-sections of basalt surfaces. Brady *et al.* (1999) concluded that colonization by lichens enhances the weathering rate of plagioclase and olivine by a factor of 2–18 and that the enhancement was slightly greater for olivine. Conrad *et al.* (1999) examined <30 y old Hawaiian lava flows colonized by *S. vulcani*. Despite the lichens being young and small (most thalli were <2 mm in diameter) the basalt below them was visibly oxidized, with some concentrations of Cu and Zn, and olivine phenocrysts had been altered. Although much further work is clearly required in order to determine the mechanisms by which *S. vulcani* interacts with basalt on Hawaii, the weight of evidence does now indicate that colonization by this species enhances weathering rates.

### 5.9.2 *Rates determined from calcium oxalate accumulation*

A proxy for rates of biochemical weathering of Mg- and Ca-rich minerals, which are of most interest to palaeoclimate modellers, is the accumulation of Mg- and Ca-oxalate salts in the lichen's thallus. Equations below allow calculation of the volume of a Ca-rich precursor required to yield an observed volume of calcium oxalate salt (similar calculations can be performed for other oxalate minerals). These calculations assume that the Ca-rich precursor dissolves stoichiometrically and all Ca ions liberated into solution by biochemical weathering reprecipitate as Ca-oxalate within the thallus.

First, the number of Ca ions in every $mm^3$ of the oxalate and precursor ($n_{Ca}$) can be calculated using equation (5.3):

$$n_{Ca} = \left(\frac{1}{V_{cell}}\right) \times Z_{Ca} \tag{5.3}$$

where: $V_{cell}$ = Volume of a unit cell of oxalate or precursor ($mm^3$), $Z_{Ca}$ = The number of Ca ions in a unit cell of oxalate or precursor.

Next, it is necessary to calculate the volume of the Ca-rich precursor required to yield 1 $mm^3$ of oxalate (equation 5.4):

$$Vol_{precursor} = \frac{(n_{Ca})_{oxalate}}{(n_{Ca})_{precursor}} \tag{5.4}$$

where: $Vol_{precursor}$ = Volume of the Ca-rich precursor mineral in $mm^3$

Values of $V_{cell}$, $Z_{Ca}$, $n_{Ca}$ and $Vol_{precursor}$ for whewellite and weddellite and two possible Ca-rich precursors, calcite and dolomite, are listed in Table 5.2.

The following is an example calculation. Figure 5.9 is a CP-SEM image of a cross-sectional fracture surface of the crustose lichen *Trapelia placodioides* removed from a sandstone wall in the author's Edinburgh garden. A continuous layer of calcium oxalate, ~0.05 mm in cross-sectional thickness, occurs within the thallus. X-ray diffraction identified this oxalate as whewellite. The SEM imaging of the sandstone shows heavily-etched dolomite cements (Fig. 5.8), indicating that dolomite was one source of Ca; here we assume that it was the only source. Data in Table 5.2 show that 1 $mm^3$ of whewellite can be formed from 0.98 $mm^3$ of stoichiometric dolomite. In order to calculate the volume of dolomite precursor ($Vol_{dolomite}$) required to form the observed volume of whewellite, a modification of equation (5.4) is required:

$$Vol_{dolomite} = \frac{[(n_{Ca})_{whewellte} \times (t_{whewellite} \times f_{whewellite})]}{(n_{Ca})_{dolomite}} \tag{5.5}$$

TABLE 5.2 Data for use in calculating rates of biochemical weathering from measured volumes of calcium oxalate in the thallus

| | Whewellite $CaC_2O_4 \cdot H_2O$ | Weddellite $CaC_2O_4 \cdot 2H_2O$ | Calcite* $CaCO_3$ | Dolomite $CaMg(CO_3)_2$ |
|---|---|---|---|---|
| $V_{cell}$ ($mm^3$) | $8.749 \times 10^{-19}$ | $1.126 \times 10^{-18}$ | $3.68 \times 10^{-19}$ | $3.21 \times 10^{-19}$ |
| $Z_{Ca}$ | 8 | 8 | 6 | 3 |
| $n_{Ca}$ | $9.14 \times 10^{18}$ | $7.10 \times 10^{18}$ | $1.63 \times 10^{19}$ | $9.35 \times 10^{18}$ |
| $Vol_{calcite}$ | 0.56 | 0.44 | na | na |
| $Vol_{dolomite}$ | 0.98 | 0.76 | na | na |

na denotes not applicable
*Using the hexagonal unit cell
$Vol_{calcite}$ and $Vol_{dolomite}$ denote the volume of calcite and dolomite precursors, respectively, required to yield 1 $mm^3$ of oxalate

where: $t_{whewellite}$ = cross-sectional thickness of the whewellite layer in mm, and $f_{whewellite}$ = fraction of the whewellite layer that is porosity or filled by organics.

If we assume that $t_{whewellite}$ = 0.05 mm and $f_{whewellite}$ = 0.5 (giving 0.025 $mm^3{}_{whewellite}\ mm^{-2}{}_{lichen}$), $Vol_{dolomite}$ = 0.024 $mm^3\ mm^{-2}{}_{lichen}$. As only a

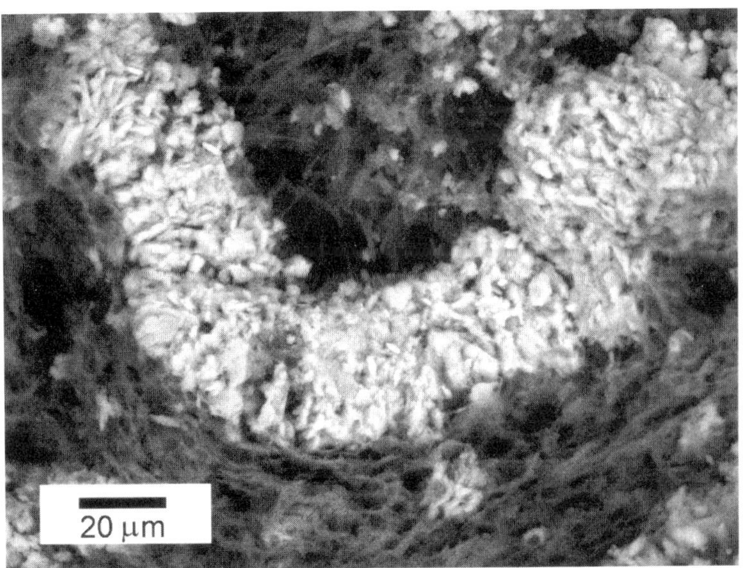


FIG. 5.9 Backscattered electron image (CP-SEM, 20 kV, 0.3 mbar) of a cross-sectional fracture surface of the thallus of *Trapelia placodioides* that has encrusted sandstone from the author's Edinburgh garden. A continuous layer of calcium oxalate (whewellite) up to 50 μm in thickness is distinguished by its considerably greater mean atomic number than the surrounding fungal hyphae.

portion of the wall rock is composed of dolomite, a final modification of equations 5.4 and 5.5 is required to calculate the depth to which it has been leached from the sandstone ($D_{\text{dolomite}}$, in mm):

$$D_{\text{dolomite}} = \frac{[(n_{\text{Ca}})_{\text{whewellte}} \times (t_{\text{whewellite}} \times f_{\text{whewellite}})]}{(n_{\text{Ca}})_{\text{dolomite}} \times f_{\text{dolomite}}} \tag{5.6}$$

where: $f_{\text{dolomite}}$ = fraction of the sandstone that contains dolomite.

If we assume that $f_{\text{dolomite}} = 0.1$ (i.e. 10% of the rock is dolomite cemented), equation 5.6 shows that the whewellite layer has formed by complete dissolution of dolomite from the sandstone to a depth of 0.244 mm. As the lichen has to be <101 y old (the author's house was constructed in 1898), the rate of biochemical leaching of dolomite from the sandstone must have been $\geqslant 0.0024$ mm $y^{-1}$ (note that the size of this lichen, 20 by 60 mm, which is smaller than others on the same wall, suggests that it may be considerably less than 101 y old). This specimen calculation has ignored many poorly-constrained variables such as the volume of dolomite in the sandstone wall rock, input of Ca from extraneous sources (e.g. runoff) and dissolution of other Ca-bearing minerals such as plagioclase feldspar, but does give an idea of the type of quantitative weathering rate data that may be acquired from oxalate accumulations.

Another example is provided by Seaward (1997). In one study he observed that a typical 10 mm diameter (~78.5 mm$^2$ in area) thallus of the lichen *Dirina massiliensis* forma *sorediata* on a $CaCO_3$-rich substrate had a layer of whewellite $1.288 \pm 0.064$ mm thick in its central part. This gives a total volume of whewellite of ~$101.2 \pm 4.9$ mm$^3$. Using data in Table 5.2, such a volume of whewellite will come from ~$56.7 \pm 2.7$ mm$^3$ (or $153.7 \pm 7.3$ mg) of calcite, indicating that this mineral has been biochemically leached to a uniform depth of ~$0.72 \pm 0.04$ mm below the whole thallus (note that Seaward (1997) calculated that 135 mg of calcite had been destroyed by each thallus, corresponding to a leaching depth of 0.63 mm).

### *5.9.3 Rates from thallus volume and erosion rate*

Wessels and Schoeman (1988) calculated the weathering rate of sandstone surfaces of known age by the endolithic lichen *Lecidea* aff. *sarcogynoides*. This lichen had been growing on the sandstone for $\leqslant 41$ y and the base of the cortex had reached a mean depth of 3.21 mm below the surrounding rock surface. Wessels and Schoeman suggested that quartz grains were incorporated into the base of the thallus as the sandstone's cement was removed by biochemical weathering and the grains were eventually lost with parts of the outer cortex (i.e. the lichen's action is a combination of biophysical and biochemical weathering and erosion). These authors calculated a weathering rate of 0.096 mm $y^{-1}$. In their most recent study, Wessels and Wessels (1995) concluded that *Lecidea* aff. *sarcogynoides* covering 0.69 m$^2$ of a wall it had colonized for 43 y had liberated $0.25 \times 10^{-2}$ $m^3$ of sand. This gives an average weathering rate of 0.084 mm $y^{-1}$. Wessels and Wessels (1995) did, however, emphasize that these rates were a combination of lichen-mediated processes and physical weathering, for example frost action.

McCarrol and Viles (1995) undertook a comparable study of the impact of the endolithic lichen *Lacedia auriculata* on the weathering rate of gabbro boulders lying on dated moraine ridges (oldest deposited in 1750). They inferred the dominant weathering mechanism to be biophysical by detaching and expelling flakes of rock. From measurements of the volume of the endolithic thalli, McCarrol and Viles (1995) suggested a minimum surface lowering rate of $120 \times 10^{-5}$ mm $y^{-1}$. This is considerably greater than the rate of weathering of lichen-free boulders of $2.9 \times 10^{-5}$ mm $y^{-1}$, which was determined by measurements of phenocryst relief (McCarrol, 1990).

### *5.9.4 Association of lichens with other organisms*

Although this review has concentrated on the action of lichens, in most environments they comprise one small part of a community of organisms that interact with each other in a complex manner. For example, Wessels and Wessels (1991) demonstrated that quartz grains were removed from a sandstone in South Africa by the combined action of an endolithic lichen and moth larvae. Quartz is initially loosened from the sandstone by lichen-mediated biochemical weathering. The grains are then eroded as the moth larvae graze the lichens and use quartz to construct their larval bags. The erosion rate was calculated to be 4.4 kg of quartz per hectare of sandstone per year, corresponding to an approximate rate of surface lowering of 0.017 mm $y^{-1}$.

### *5.9.5 Qualitative rate estimates*

From a TEM study of lichen-encrusted syenite, Barker and Banfield (1998) conclude "...geomicrobiological weathering is accelerated by several orders of magnitude compared to inorganic weathering". It is important to note that the work of Barker and Banfield (1996, 1998) is based on a relatively recently exposed (~90 y ago) rock surface. Once the freshly-exposed mineral surfaces have been coated with extracellular polymers and/or the most reactive parts of these surfaces (e.g. highly strained crystal structure) have been etched, weathering rates may well slow if new surfaces are not continually exposed by biophysical action or by physical weathering (e.g. freeze-thaw). This is echoed by Drever (1994), who concluded that biochemical weathering effects of lichens from excreted organic acids may be significant in the short term but will have little effect on long-term weathering rates. Drever (1994) attributes this in part to the relatively small volume of rock affected by any one lichen.

Schwartzman *et al.* (1997) compared the concentration of solutes in runoff from lichen-covered schists with runoff from identical rock surfaces that were apparently free of microbial activity at a locality in New Hampshire, USA. They found higher concentrations of Mg, Si, Ca, K and Na and lower concentrations of Al and Fe in runoff from the lichen-covered surfaces. These data were interpreted to demonstrate that lichen encrustation enhanced the weathering rate of schist. However, it is unclear whether biochemical processes were solely responsible for elevated solute fluxes or whether wind-borne dust trapped within the thallus or biophysically fragmented mineral grains beneath the thallus were being chemically weathered. Alternatively, organically-complexed chemical elements may have been leached.

### *5.9.6 Negative relative weathering rates*

Almost all studies of lichen-encrusted rock highlight the deleterious effects, but very few compare these effects with the impact of chemical and physical processes on identical but lichen-free surfaces. Processes and products of biophysical and biochemical weathering are especially easy to characterize because weathering products are retained in an organic matrix. By contrast, products of chemical and physical weathering are eroded from unencrusted surfaces and will not be available for examination.

This is exemplified in a study by Ariño *et al.* (1995) of weathering of a Roman sandstone pavement excavated in 1971. Most of the pavement, which is exposed to wind and rain, is lichen-free and has been significantly weathered by processes including salt crystallization and abraded by wind-borne particles. The portion of the pavement that is colonized by lichens shows features typical of biophysical and biochemical weathering including disaggregation and etching of quartz grains and calcium oxalate accumulations. However, Ariño *et al.* (1995) found that lichen-encrusted areas of the sandstone had undergone less intense weathering overall than lichen-free areas. This was attributed to the thallus protecting the sandstone from ingress of rainwater and runoff containing dissolved salts. Significantly, a ~0.1 mm thick patina of $CaCO_3$ and oxalate occur on some sandstone surfaces. These patinas are weathering products from previous generations of lichens and may further protect the sandstone from chemical and physical weathering. SaizJimenez (1999) has also stressed the protective role of oxalate patinas on building stones.

Benedict (1993) studied the retreat of granodiorite surfaces by measuring the relief of phenocrysts (mineralogy not stated). He found that rates of bedrock lowering had ranged from $1 \times 10^{-4}$ to $13 \times 10^{-4}$ mm $y^{-1}$ since deglaciation in the late Pleistocene, depending on snow cover. Maximum rates of retreat were found in areas that had shallow to moderate winter snow accumulation and 20–60% lichen coverage. Significantly, rates of postglacial weathering were lower on surfaces with even greater lichen coverage, indicating to Benedict (1993) that lichens have a bioprotective function by sheltering rock surfaces from large temperature variations, intercepting moisture and by binding mineral fragments.

## 5.10 Summary: the impact of lichen colonization on weathering rates

The paucity of detailed studies of lichen-mediated weathering of dated rock surfaces makes generalizations about their impact on weathering rates especially difficult. The conclusions below therefore represent a personal impression from the literature and unpublished work rather than firm conclusions from hard quantitative data.

The evidence that lichen-colonized rock surfaces are exposed to a variety of biophysical and biochemical weathering processes is overwhelming. Although much of the previous work on biochemical weathering has focused on the effects of oxalic acid and crystallization of oxalate salts, work by Barker and Banfield (1996, 1998) has demonstrated conclusively that extracellular polymers play a highly significant role in the dissolution of minerals, transport of ions and precipitation of weathering products. Biochemical weathering will be enhanced by the thermal insulating properties of the thallus, a greater residence time of water and active biophysical weathering, exposing fresh mineral surfaces.

In answer to the questions posed at the beginning of this chapter, I suggested that the weathering rate of Mg- and Ca-rich silicate minerals on late Precambrian and lower Palaeozoic land surfaces would have been significantly greater if these surfaces had been covered by algae, fungi and lichens than if they were bare. Thus, if present, microbes would have enhanced rates of removal of $CO_2$ from the atmosphere and correspondingly reduced global temperatures. This effect would have been largely a result of binding biophysically fragmented mineral grains and wind-borne dust within the thallus, exposing mineral surfaces to a range of acids and ligands. Although lichens are clearly effective agents of mineral weathering, as they bind and retain weathered mineral grains and weathering products, they will act to stabilize rock surfaces and protect them from erosion.

One of the main messages of this review is that quantitative studies are sorely needed before predictions of organically-mediated weathering rates can be made. This future work must use precisely dated rock surfaces and lichen-colonized and lichen-free parts of the surface must be characterized in tandem and using identical techniques. The different variables that can affect weathering rates (e.g. species and age of the lichen, rock type [specifically mineralogy, grain-size, porosity and permeability] and macro and micro-climate) must be progressively isolated by using careful sampling strategies. One obvious tactic will be to study the effects of highly substrate-specific (ornithocoprophilous) lichen species and compare their weathering effects with lichens that display little or no substrate specificity.

### Acknowledgements

I thank Jim Buckman (Department of Petroleum Engineering, Heriot-Watt University) for assistance with the ESEM, Ian Parsons (University of

Edinburgh) for reading an earlier version of this manuscript and Brian Coppins (Royal Botanic Gardens, Edinburgh) for invaluable discussions and lichen identification. I also thank Geoff Angel (Edinburgh) for assistance with the X-ray diffraction. This manuscript benefited greatly from reviews by Bill Barker, Janet Cotter-Howells and an anonymous reviewer.

## References

Adamo, P., Colombo, C. and Violante, P. (1997) Iron oxides and hydroxides in the weathering interface between Stereocaulon vesuvianum and volcanic rock. *Clay Miner.*, **32**, 453–61.

Ariño, X., Ortegacalvo, J.J., Gomezbolea, A. and SaizJimenez, C. (1995) Lichen colonization of the Roman pavement at Baelo-Claudia (Cadiz, Spain) – Biodeterioration vs bioprotection. *Sci. Total Environ.*, **167**, 353–63.

Ascaso, C. and Wierzchos, J. (1994) Structural aspects of the lichen–rock interface using backscattered electron imaging. *Botanica Acta*, **107**, 251–6.

Ascaso, C. and Wierzchos, J. (1995) Study of the biodeterioration zone between the lichen thallus and the substrate. *Cryptogamic Botany*, **5**, 270–81.

Ascaso, C., Galvan, J. and Ortega, C. (1976) The pedogenic action of *Parmelia conspersa*, *Rhizocarpon geographicum* and *Umbilicaria Pustulata*. *Lichenologist*, **8**, 51–171.

Ascaso, C., Wierzchos, J. and Castello, R. (1998) Study of the biogenic weathering of calcareous litharenite stones caused by lichen and endolithic microorganisms. *Int. Biodet. Biodeg.*, **42**, 29–38.

Banfield, J.F., Barker, W.W., Welch, S.A. and Taunton, A. (1999) Biological impact on mineral dissolution: Application of the lichen model to understanding mineral weathering in the rhizosphere. *Proc. Natl. Acad. Sci.*, **96**, 3404–11.

Barker, W.W. and Banfield, J.F. (1996) Biologically versus inorganically mediated weathering reactions: relationships between minerals and extracellular microbial polymers in lithobiontic communities. *Chem. Geol.*, **132**, 55–69.

Barker, W.W. and Banfield, J.F. (1998) Zones of chemical and physical interaction at interfaces between microbial communities and minerals: A model. *Geomicrobiol. J.*, **15**, 223–44.

Barker, W.W., Welch, S.A. and Banfield, J.F. (1997) Biochemical weathering of silicate minerals. Pp. 391–428 in: *Geomicrobiology: Interactions between Microbes and Minerals* (J.F. Banfield and K.H. Nealson, editors). Reviews in Mineralogy, **35**. Mineralogical Society of America, Washington D.C.

Benedict, J.B. (1993) Influence of snow on rates of granodiorite weathering, Colorado Front Range, USA. *Boreas*, **22**, 87–92.

Berner, R.A. (1992) Weathering, plants, and the long-term carbon cycle. *Geochim. Cosmochim. Acta*, **56**, 3225–31.

Berner, R.A. (1995) Chemical weathering and its effect on atmospheric $CO_2$ and climate. Pp. 565–583 in: *Chemical Weathering Rates of Silicate Minerals* (A. F. White and S.L. Brantley, editors). Reviews in Mineralogy, **31**. Mineralogical Society of America, Washington D.C.

Berner, R.A. and Cochran, M.F. (1998) Plant-induced weathering of Hawaiian basalts. *J. Sed. Res.*, **68**, 723–6.

Brady, P.V., Dorn, R.I., Brazel, A.J., Clark, J., Moore, R.B. and Glidewell, T. (1999) Direct measurement of the combined effects of lichen, rainfall and temperature on silicate weathering. *Geochim. Cosmochim. Acta*, **63**, 3293–300.

Brodo, I.M. (1973) Substrate ecology. Pp. 401–41 in: *The Lichens* (V. Ahmadjian and M.E. Hale, editors). Academic Press, London.

Chisholm, J.E., Jones, G.C. and Purvis, O.W. (1987) Hydrated copper oxalate, moolooite, in lichens. *Mineral. Mag.*, **51**, 715–8.

Cochran, M.F. and Berner, R.A. (1993*a*) Enhancement of silicate weathering rates by vascular land plants – Quantifying the effect. *Chem. Geol.*, **107**, 213–5.

Cochran, M.F. and Berner, R.A. (1993*b*) Reply to comments on 'Weathering, plants, and the long-term carbon cycle. *Geochim. Cosmochim. Acta*, **57**, 2147–8.

Cochran, M.F. and Berner, R.A. (1996) Promotion of chemical weathering by higher plants: Field observations on Hawaiian basalts. *Chem. Geol.*, **132**, 71–7.

Conrad, M.E., Stringfellow, W.T. and Lamble, G.M. (1999) Uptake and precipitation of metals from basalt by the lichen *Stereocaulon volcanii*. *Geol. Soc. Amer. Ann. Mtg.*, Abstract.

Cooks, J. and Otto, E. (1990) The weathering effects of the lichen *Lecidea* aff. *sarcogynoides* (Koerb.) on Magaliesberg quartzite. *Earth Surf. Proc. Landforms*, **15**, 491–500.

Dorn, R.I. (1995) Digital processing of back-scatter electron imagery: A microscopic approach to quantifying chemical weathering. *Geol. Soc. Amer. Bull.*, **107**, 725–41.

Drever, J.I. (1994) The effect of land plants on weathering rates of silicate minerals. *Geochim. Cosmochim. Acta*, **58**, 2325–32.

Easton, R.M. (1995) Lichens and rocks: a review. *Geosci. Canad.*, **21**, 59–76.

Edwards, H.G.M. (1997) FT Raman microscopy of lichen encrustations. *Microsc. Anal.*, **59**, 5–7.

Fry, E.J. (1924) A suggested explanation for the mechanical action of lithophytic lichens on rocks (shale). *Ann. Bot.*, **38**, 175–96.

Fry, E.J. (1927) The mechanical action of crustaceous lichens on substrate of shale, schist, gneiss, limestone and obsidian. *Ann. Bot.*, **41**, 437–60.

Gadd, G. (2000) Heterotrophic solubilization of metal-bearing minerals by fungi. Pp. 57–75 in: *Environmental Mineralogy: Microbial Interactions, Anthropogenic Influences, Contaminated Land and Waste Management* (J.D. Cotter-Howells, L.S. Campbell, E. Valsami-Jones and M. Batchelder, editors). Mineralogical Society Series, **9**. Mineralogical Society, London.

Galvan, J., Rodriguez, C. and Ascaso, C. (1981) The pedogenic action of lichens on metamorphic rocks. *Pedobiologica*, **21**, 60–73.

Hale, M.E. Jr. (1974) *The Biology of Lichens*. 2nd edition. Edward Arnold, London.

Jackson, T.A. (1993) Comment on 'Weathering, plants, and the long-term carbon cycle'. *Geochim. Cosmochim. Acta*, **57**, 2141–4.

Jackson, T.A. and Keller, W.D. (1970*a*) Comparative study of the role of lichens and inorganic processes in the chemical weathering of recent Hawaiian lava flows. *Amer. J. Sci.*, **269**, 446–66.

Jackson, T.A. and Keller, W.D. (1970*b*) Evidence for biogenic synthesis of an unusual ferric oxide mineral during alteration of basalt by a tropical lichen. *Nature*, **227**, 522–3.

Jones, D. and Wilson, M.J. (1985) Chemical activity of lichens on mineral surfaces – a review *Int. Biodet. Biodeg.*, **21**, 99–104.

Jones, D. and Wilson, M.J. (1986) Biomineralization in crustose lichens. Pp. 91–105 in: *Biomineralization in the Lower Plants and Animals* (B.S.C. Leadbetter and R. Riding, editors). Clarendon Press, Oxford.

Jones, D., Wilson, M.J. and Tait, J.M. (1980) Weathering of basalt by *Pertusaria corallina*. *Lichenologist*, **12**, 277–89.

Jones, D., Wilson, M.J. and McHardy, W.J. (1981) Lichen weathering of rock-forming minerals: application of scanning electron microscopy and microprobe analysis. *J. Microsc.*, **124**, 95–104.

Jongmans, A.G., van Breemen, N., Lundström, U., van Hees, P.A.W., Finlay, R.D., Srinivasan, M., Unestam, T., Giesler, R., Melkerud, P.-A. and Olsson, M. (1997) Rock-eating fungi. *Nature*, **389**, 682–3.

Lee, M.R. (1999) Organic-mineral interactions studied by controlled pressure SEM. *Microsc. Anal.*, **70**, 9–11.

Lee, M.R. and Parsons, I. (1999) Biophysical and biochemical weathering of lichen-encrusted granite: textural controls on organic-mineral interactions and deposition of silica-rich layers. *Chem. Geol.*, **161**, 385–97.

Lee, M.R., Hodson, M.E. and Parsons, I. (1998) The role of intragranular microtextures and microstructures in chemical and mechanical weathering: direct comparisons of experimentally and naturally weathered feldspars. *Geochim. Cosmochim. Acta,* **62**, 2771–88.

McCarroll, D. (1990) Differential weathering of feldspar and pyroxene in an arctic alpine environment. *Earth Surf. Proc. Landforms,* **15**, 641–51.

McCarroll, D. and Viles, H. (1995) Rock-weathering by the lichen *Lecidea auriculata* in an arctic alpine environment. *Earth Surf. Proc. Landforms,* **20**, 199–206.

Nash, T.H. III (1996*a*) Photosynthesis, respiration, productivity and growth. Pp. 88–120 in: *Lichen Biology* (T.H. Nash III, editor). Cambridge University Press.

Nash, T.H. III (1996*b*) Nutrients, elemental accumulation and mineral cycling. Pp. 136–53 in: *Lichen Biology* (T.H. Nash III, editor). Cambridge University Press.

Paradise, T.R. (1997) Disparate sandstone weathering beneath lichens, Red Mountain, Arizona. *Geografiska Annaler Series A – Physical Geography,* **79A**, 177–84.

Piervittotri, R., Salvadori, O. and Laccisaglia, A. (1994) Literature on lichens and biodeterioration of stonework. I. *Lichenologist,* **26**, 171–92.

Purvis, O.W. (1984) The occurrence of copper oxalate in lichens growing on copper sulphide-bearing rocks in Scandinavia. *Lichenologist,* **16**, 197–204.

Purvis, O.W., Gilbert, O.L. and James, P.W. (1985) The influence of copper mineralization on Acarospora smaragdula. *Lichenologist,* **17**, 111–6.

Purvis, O.W., Elix, J.A., Broomhead, J.A. and Jones, G.C. (1987) The occurrence of copper-nostictic acid in lichens from cupiferous substrata. *Lichenologist,* **19**, 193–203.

Purvis, O.W., Elix, J.A. and Gaul, K.L. (1990) The occurrence of copper psoromic acid in lichens from cupiferous substrata. *Lichenologist,* **22**, 345–54.

Richardson, D.H S. (1992) *Pollution Monitoring with Lichens.* The Richmond Publishing Co. Ltd, Slough, UK.

SaizJimenez, C. (1999) Biogeochemistry of weathering processes in monuments. *Geomicrobiol. J.,* **16**, 27–37.

Schwartzman, D.W. (1993) Comment on 'Weathering, plants, and the long-term carbon cycle'. *Geochim. Cosmochim. Acta,* **57**, 2145–6.

Schwartzman, D.W. and Volk, T. (1989) Biotic enhancement of weathering and the habitability of Earth. *Nature,* **340**, 457–60.

Schwartzman, D.W. and Volk, T. (1991) Biotic enhancement of weathering and surface temperatures on Earth since the origin of life. *Palaeogeog. Palaeoclimatol. Palaeoecol.,* **90**, 357–71.

Schwartzman, D.W., Aghamiri, R. and Bailey, S.W. (1997) Lichen weathering at Cone Pond, New Hampshire. *Seventh Annual V. M. Goldschmidt Conference*, p. 189.

Schwertmann, U. and Taylor, R.M. (1989) Iron oxides. Pp. 379–438 in: *Minerals in Soil Environments* (J.B. Dixon and S.B. Weed, editors). Soil Science Society of America, Madison, WI.

Seaward, M.D.R. (1988) Lichen damage to ancient monuments: a case study. *Lichenologist,* **20**, 291–4.

Seaward, M.R.D. (1997) Major impacts made by lichens in biodeterioration processes. *Int. Biodet. Biodeg,,* **40**, 269–73.

Stillings, L.L., Drever, J.I., Brantley, S.L., Sun, Y. and Oxburgh, R. (1996) Rates of feldspar dissolution at pH 3–7 with 0–8 mm oxalic acid. *Chem. Geol.,* **132**, 79–89.

Syers, J.K. (1964) *A study of soil formation on Carboniferous limestone with particular reference to lichens as pedogenic agents.* PhD thesis, Univ. Durham, UK.

Syers, J.K. and Iskandar, I.K. (1973) Pedogenic significance of lichens. Pp. 225–248 in: *The Lichens* (V. Ahmadjian and M.E. Hall, editors). Academic Press, London.

Syers, J.K., Birnie, A.C. and Mitchell, B.D. (1967) The calcium oxalate content of some lichens growing on limestone. *Lichenologist,* **3**, 409–14.

Viles, H. (1987) A quantitative scanning electron-microscope study of evidence for lichen weathering of limestone, Mendip Hills, Somerset. *Earth Surf. Proc. Landforms,* **12**, 467–73.

Welch, S.A. and Ullman, W.J. (1993) The effect of organic acids on plagioclase dissolution rates and stoichiometry. *Geochim. Cosmochim. Acta,* **57**, 2725–36.

Welch, S.A. and Ullman, W.J. (1996) Feldspar dissolution in acidic and organic solutions: Compositional and pH dependence of dissolution rate. *Geochim. Cosmochim. Acta,* **60**, 2939–48.

Welch, S.A., Barker, W.W. and Banfield, J.F. (1999) Microbial extracellular polysaccharides and plagioclase dissolution. *Geochim. Cosmochim. Acta,* **63**, 1405–19.

Wessels, D.C.J. and Schoeman, P. (1988) Mechanism and rate of weathering of Clarens sandstone by an endolithic lichen. *S. African J. Sci.,* **84**, 274–7.

Wessels, D.C.J. and Wessels, L.A. (1991) Erosion of biogenically weathered Clarens sandstone by Lichenophagous bagworm larvae (*Lepidoptera; Pyschidae*). *Lichenologist,* **23**, 283–91.

Wessels, D.C.J. and Wessels, L.A. (1995) Biogenic weathering and microclimate of Clarens sandstone in South Africa. *Cryptogamic Botany,* **5**, 288–98.

White, A.F. and Brantley, S.L. (1995) *Chemical Weathering Rates of Silicate Minerals.* Reviews in Mineralogy, **31**. Mineralogical Society of America, Washington D.C.

Wierzchos, J. and Ascaso, C. (1994) Application of back-scattered electron imaging to the study of the lichen–rock interface. *J. Microsc.,* **175**, 54–9.

Wierzchos, J. and Ascaso, C. (1996) Morphological and chemical features of bioweathered granitic biotite induced by lichen activity. *Clays Clay Miner.,* **44**, 652–7.

Wierzchos, J. and Ascaso, C. (1998) Mineralogical transformation of bioweathered granitic biotite, studied by HRTEM: Evidence for a new pathway in lichen activity. *Clays Clay Miner.,* **46**, 446–52.

Wilson, M.J. (1995) Interactions between lichens and rocks; a review. *Cryptogamic Botany,* **5**, 399–405.

Wilson, M.J. and Bayliss, P. (1987) Mineral nomenclature: glushinskite. *Mineral. Mag.,* **51**, 327–8.

Wilson, M.J. and Jones, D. (1983) Lichen weathering of minerals: implications for pedogenesis. Pp. 5–12 in: *Residual Deposits: Surface Related Weathering Processes and Materials* (R.C.L. Wilson, editor). Geological Society of London, Blackwell.

Wilson, M.J. and Jones, D. (1984) The occurrence and significance of manganese oxalate in *Pertusaria corallina. Pedobiologica,* **26**, 373–9.

Wilson, M.J., Jones, D. and Russell, J.D. (1980) Glushinskite, a naturally occuring magnesium oxalate. *Mineral. Mag.,* **43**, 837–40.

Wilson, M.J., Jones, D. and McHardy, W.J. (1981) The weathering of serpentine by *Lecanora atra. Lichenologist,* **13**, 167–76.

CHAPTER SIX

# Section 2: Anthropogenic influences on mineral interactions

L. S. CAMPBELL

*Telford Institute of Environmental Systems, (Environmental Resources), The University of Salford, Frederick Road, Salford M6 6PU, UK (E-mail: L.S.Campbell@salford.ac.uk)*

## 6.1 Introduction

The 'Anthropogenic Influences' section of this volume on Environmental Mineralogy is concerned with controls on mineral-environment interactions that are in some way influenced by human activity. Mineral-environment interactions include all types of mineral growth and decomposition, with associated chemical and isotopic signatures, and all other chemical reactions such as ion exchange and adsorption. The controls on these interactions are the physical, chemical and biological conditions that exist in the immediate vicinity of the mineral.

Many mineral-environment interactions occur naturally in response to natural processes of change, and the significance of an anthropogenic influence may simply be in the rate or the scale of change. Thus, for example, pyrite oxidation resulting in the liberation of protons (pH decrease), has occurred wherever pyritiferous rocks have been exposed to the atmosphere or to oxygenated waters (**Keith and Vaughan, chapter 7**). But it is often only where humans have opened new conduits in rocks, or have spatially concentrated gangue sulphides in heaps of high-porosity mine waste, that the scale of interactions has become significant to the wider environment. Alternatively, the significance of an anthropogenic influence may be in the combining of substances that would not normally be found together in nature, or it may be in the creation of reactive conditions.

Humans act as a rapid mode of transport, placing mineral, liquid and gaseous phases in close proximity to one another. Prior to the advent of the global environmental agenda, most anthropogenic influences have been unintentional, resulting in the vast array of contaminated or unbalanced

Campbell, L.S. (2000) Section 2: Anthropogenic influences on mineral interactions. Pp. 109–115 in: *Environmental Mineralogy: Microbial Interactions, Anthropogenic Influences, Contaminated Land and Waste Management* (J.D. Cotter-Howells, L.S. Campbell, E. Valsami-Jones and M. Batchelder, editors). Mineralogical Society Series, **9**. Mineralogical Society, London. ISBN 0 903056 20 8.

environments that exist today. However, it is in the relatively new field of geochemical engineering, where intentional interventions with mineral-environment interactions are growing, that vast opportunities exist not only for pollution remediation, but for long term sustainable management of resources and waste (Schuiling, 1998; Vriend and Zijlstra, 1998). Mineralogy has a crucial role to play in this, as exemplified by other papers in the current volume (see chapters 11–15 on 'Minerals in Contaminated Environments' and chapters 16–19 on 'Minerals and Waste Management').

## 6.2 Anthropogenic influences

The ways in which human activity can influence the immediate environment around a mineral can be grouped as follows: (1) exposure or changes to the atmosphere; (2) exposure or changes to aqueous solutions; (3) changes to the physical environment; (4) influences associated with the biosphere; and (5) juxtaposition of different mineral phases. Some of these categories are linked intimately with regard to certain geochemical processes such as weathering, but it is sometimes convenient to consider examples in the context of one dominant factor. The categories therefore, are suggested as a framework in which anthropogenic influences on mineral interactions can be rationalized. Examples are outlined for each of the categories, with indications as to the wider environmental issues associated with the mineral responses.

### *6.2.1 Exposure or changes to the atmosphere*

It is in the context of the cycling of two key elements, carbon and sulphur, that a wider understanding of mineral interactions due to atmospheric changes will be possible. Patterns of anthropogenic influence are becoming more and more familiar with headlines on $CO_2$ and global warming, and acid deposition as a result of $SO_2$ emissions from industrial activity. Increasingly, authors are evaluating the cycling of these mineralogically significant elements in global terms, but the extent to which atmospheric fluxes directly impact on minerals is less clear. In one such study, Varekamp and Thomas (1998) attempt to evaluate the volcanic and anthropogenic contributions to global weathering budgets. They estimate that the situation with regard to $CO_2$ and $SO_2$ is far from steady state, in that weathering (neutralization) is not keeping pace with emissions. Since S from atmospheric $SO_2$ tends to be destined to reside in the aqueous phase, it is easier to quantify the mineralogical impacts through sediment-water studies in specific locations, than it is for $CO_2$. Carbon dioxide reservoirs, exchanges and fluxes are very large, involving the atmosphere-ocean, ocean-sediment, ocean-biosphere, and atmosphere-biosphere, and it may well be that the climatic impacts will prove to be more dramatic than mineralogical impacts for anthropogenic influences on carbon cycling.

Of course, on a local scale (e.g. in mining operations), humans contribute to the fracturing and exposure of minerals to atmospheric oxygen, giving rise to weathering reactions on a scale that would not otherwise be significant. A contributary factor in mine waste mineral reactions must also be that we separate and remove potential buffering minerals in the processing of ores and waste rock.

### *6.2.2 Exposure or changes to aqueous solutions*

The abundance and distribution of aqueous solutions in the geosphere hardly requires introduction. One aspect of mineral-solution interaction of particular relevance to environmental mineralogy, but that also requires exposure to oxygen, is in acid rock drainage (ARD) due to sulphide oxidation. Keith and Vaughan (2000) review mechanisms of sulphide oxidation, from purely inorganic reactions to bacterially-induced mineral decomposition. They provide useful summaries of key conditions required for the formation of ARD. The emphasis is on the iron sulphides, notably pyrite, but some detailed attention is also given to galena, sphalerite, chalcopyrite and arsenopyrite.

The subsequent fate of ARD on interaction with surrounding rocks, particularly with respect to pH changes, is addressed in the contribution by **Jambor (chapter 8)**. He evaluates methods for assessing what is known as the acid- and neutralization-potential (AP, NP) of rocks subject to ARD. The role and significance of carbonate minerals in NP is stressed, thereby emphasizing the importance of mineralogical investigation alongside the commonly employed static (chemical) tests used for this purpose.

Aqueous solutions may be changed, anthropogenically, in other ways. For example, groundwater extraction can sometimes cause a saline influx, leading to solubilization and mobilization of mineral phases or adsorbed components in the host rock, and it can also lead to new precipitations that seal conduits. A further type of reaction that may occur as a consequence of a saline influx is ion-exchange in clay minerals. This can have serious consequences for the environment, because sodium will exchange with calcium in montmorillonite, resulting in an increase in basal layer spacing of the clay mineral (i.e. swelling).

Groundwater chemistry can be affected in other ways, even more directly. Agricultural and urban run-offs have practically no limits on their compositions, but at least some of them are relatively predictable. In high latitudes and high altitudes, halite is widely used as a de-icing agent on the roads. Some of these saline run-off fluids will find their way into surface sediments or groundwaters, and the mineralogical consequences may parallel those resulting from saline influx after excessive groundwater extraction.

For further, current examples of aqueous interactions with minerals, the reader is referred to the regularly produced volumes on *Water Rock Interaction* (e.g. Arehart and Hulston, 1998), which invariably contain environmental themes alongside purely geological studies.

### 6.2.3 *Changes to the physical environment*

Anthropogenic influences on the physical environment of a mineral are regarded as generally very limited. This is due to the fact that most human activities take place on the surface of the Earth in environments where both temperature and pressure remain within a relatively narrow range. Certainly, extremes of temperature and pressure are created by humans, but these tend to be on a small scale, and in controlled settings (e.g. industrial furnaces). However, one example affecting the wider environment is in the potential underground storage or disposal of radioactive waste, and the release of thermal energy due to radioactive decay. Tessier *et al.* (1998) experimentally examined thermal and pressure effects on activated clay minerals in relation to this issue, and found that changes in these physical conditions markedly affected the properties of the clays. Notable differences in microstructure and swelling pressure were observed between experiments conducted at 90°C and those conducted at 145°C.

In another example, Chipera and Bish (1997) undertook thermodynamic modelling for the clinoptilolite to analcime transformation in relation to the potential radioactive waste repository at Yucca Mountain (USA). The implications for this reaction occurring are two-fold. Firstly, the higher density of analcime relative to that of clinoptilolite means that the system would become hydrologically more open. Secondly, the advantage of the radionuclide-retardation properties of clinoptilolite would be compromised. Chipera and Bish (1997) found that clinoptilolite should remain stable at temperatures below 100°C, but that mineral and groundwater compositions were also important factors affecting its stability.

## 6.3 Influences associated with the biosphere

In chapters 1–5 of this volume, a great diversity of mineral-microbe interactions is described and evaluated. As noted previously, it is generally accepted that many of these interactions occur naturally in environments not affected by humans. However, humans have undoubtedly influenced the progression of geochemical, biological and mineralogical events at Mynydd Parys, Anglesey, as described in the contribution by **Jenkins *et al.* (chapter 9)**. The copper mining here has a protracted history. Although abundant iron sulphides have predictably led to acid mine drainage (AMD) at the site, there is an implication that the subsequent prolonged lack of disturbance throughout the twentieth century has enabled an extreme microbial ecosystem to develop. Devoid of further inputs (natural or anthropogenic), and essentially free of carbonates that might contribute to NP (Jambor, 2000), acidophillic microorganisms that utilize sulphur species for energy have been able to become well established at the Mynydd Parys site (Jenkins *et al.*, 2000). Mineralogically, it is compelling to speculate about the long term fate of sulphur in a comparable, but sterile system. Perhaps some of the answers lie in the advancing field of sulphide oxidation mechanisms (Keith and Vaughan, 2000), where research on the role of microorganisms has been heavily focused on the iron sulphides, as opposed to the Cu, As, Zn and Pb sulphides.

Methylation of heavy metals such as Hg and Pb is another microbially-associated reaction that is only partially understood. Experiments examining cinnabar solubility conclude that both the presence of S (Jay *et al.*, 2000; Paquette and Helz, 1997), and the presence of sulphate-reducing bacteria (Jay *et al.*, 2000) are significant factors in the mechanisms relating to mercury methylation. Branfireun *et al.* (1999) took the issue further in a field experiment, demonstrating that methyl mercury concentrations in peatlands increase with the addition of sulphate (and hence, sulphate-reducing bacteria). This study implies that the consequences of acid rain could be far more wide-ranging than previously thought. Further details on microbial interactions with minerals are given in chapters 1–5 of this volume and in Banfield and Welch (2000).

The isotopic composition of minerals can be affected, during mineral growth, by changes in the biosphere resulting directly from human activity. For example, Filippi *et al.* (1999) provided evidence for past isotopic disequilibrium in a Swiss lake, as recorded in the $\delta^{18}O$ and $\delta^{13}C$ signatures of ostracode calcite samples dating from Medieval times. The isotopic shifts were interpreted as reflecting a process of kinetic fractionation from increased biological productivity due to forest clearances and eutrophication. The question immediately arises, therefore, as to the undocumented scale of such influences in today's world of large scale agricultural practices.

### *6.3.1 Juxtaposition of different mineral phases*

It is assumed here, that either a gas or liquid (aqueous) phase is present in which reactions can take place, but that the dominant factor in these mineral-mineral reactions is their proximity to each other, and spatial concentration, as a result of human intervention. In the contribution by **Alves and Sequeira Braga (chapter 10)**, it is the concentrations and relative solubilities of halite and gypsum in various parts of the study sites, (historical granite buildings), that are strongly linked to building stone decay. Additional factors in the nature, intensity and distribution of secondary minerals and decay were found to be the amount of weathering in the granite (inherited from the *in situ* exposures), the presence of mortars (containing carbonates), and the presence of human excreta. Clearly then, the anthropogenic contributions to these interactions have been diverse and protracted.

Modern buildings and structures are also well known to be susceptible to deterioration due to mineral reactions, and in particular, the alkali-silica reaction (ASR) of concrete. Reaction mechanisms are discussed in a recent review by Fournier and Berube (2000), in which a broader term, alkali-aggregate reactivity (AAR), is used. Broekmans and Jansen (1998) highlighted mineralogical factors in concrete reactivity in their comparison of sandstone and chert aggregates. The sandstone was found to be more reactive than the chert in concrete, despite the disadvantage of a coarser grain-size (sandstone) relative to the microcrystallinity of the chert. The proposed reason was that the sandstone contained mineral constituents that are more susceptible to reaction (micas and clays), than the pure $SiO_2$ of the chert.

Agricultural uses of minerals (for example, calcite/limestone in soil amendments), is another vast area in which the proximity of different mineral phases plays a significant role in the interactions. However, it is more appropriate here to refer to the substantial literature on the complexities of soils, soil mineralogy, soil amendments and plants, than to attempt to select specific examples. A recent outline of mineralogy and soil processes can be found in Cotter-Howells and Paterson (2000).

## 6.4 Conclusion

These paragraphs serve to introduce the theme of anthropogenic influences on mineral interactions, to outline the scope, and to provide a context for the papers that follow. Documenting the mechanisms and occurrence of mineral-environment interactions is crucial in building an understanding of the ideal conditions necessary for the reactions to occur, and to identify where and how the human factor applies. Increasingly, we are raising our consciousness of how we affect the natural mineralogical environment; increasingly, this will impact on our ability to make responsible decisions for local and global environmental management.

## References

Alves, C.A.S. and Sequeira Braga, M.A. (2000) Influence of soluble salts on the decay of granitic buildings of Braga (NW Portugal). Pp. 181–99 in: *Environmental Mineralogy: Microbial Interactions, Anthropogenic Influences, Contaminated Land and Waste Management* (J.D. Cotter-Howells, L.S. Campbell, E. Valsami-Jones and M. Batchelder, editors). Mineralogical Society Series, **9**. Mineralogical Society, London.

Arehart, G.B. and Hulston, J.R. (1998) Water-rock interaction. *Proc. 9$^{th}$ Int. Symp. Water-Rock Interaction.* Balkema, Rotterdam.

Banfield, J.F. and Welch, S.A. (2000) Microbial controls on the mineralogy of the environment. Pp. 173–96 in: *Environmental Mineralogy* (D.J. Vaughan and R.A. Wogelius, editors). EMU Notes in Mineralogy, **2**. Eotvös University Press, Budapest.

Branfireun, B.A., Roulet, N.T., Kelly, C.A. and Rudd, J.W.M. (1999) In situ sulphate stimulation of mercury methylation in a boreal peatland: Toward a link between acid rain and methylmercury contamination in remote environments. *Global Biogeochemical Cycles,* **13,** 743–50.

Broekmans, M.A.T.M. and Jansen, J.B.H. (1998) Silica dissolution in impure sandstone; application to concrete. *J. Geochem. Expl. (Spec. Iss.),* **62,** 311–8.

Chipera, S.J. and Bish, D.L. (1997) Equilibrium modeling of clinoptilolite-analcime equilibria at Yucca Mountain, Nevada, USA. *Clays Clay Miner.,* **45,** 226–39.

Cotter-Howells, J.D. (2000) Section 1: Mineral-microbe interactions. Pp. 1–5 in: *Environmental*

*Mineralogy: Microbial Interactions, Anthropogenic Influences, Contaminated Land and Waste Management* (J.D. Cotter-Howells, L.S. Campbell, E. Valsami-Jones and M. Batchelder, editors). Mineralogical Society Series, **9**. Mineralogical Society, London.

Cotter-Howells, J.D. and Paterson, E. (2000) Minerals and soil development. Pp. 91–124 in: *Environmental Mineralogy* (D.J. Vaughan and R.A. Wogelius, editors). EMU Notes in Mineralogy, **2**. Eötvös University Press, Budapest.

Filippi, M.L., Lambert, P., Hunziker, J., Kubler, B. and Bernasconi, S. (1999) Climatic and anthropogenic influence on the stable isotope record from bulk carbonates and ostracodes in Lake Neuchatel, Switzerland, during the last two millennia. *J. Paleolimnol.*, **21**, 19–34.

Fournier, B. and Berube, M.A. (2000) Alkali-aggregate reaction in concrete: a review of basic concepts and engineering implications. *Canad. J. Civil Eng.*, 27, 167–91.

Jambor, J. (2000) The relationship of mineralogy to acid- and neutralization-potential values in ARD. Pp. 141–59 in: *Environmental Mineralogy: Microbial Interactions, Anthropogenic Influences, Contaminated Land and Waste Management* (J.D. Cotter-Howells, L.S. Campbell, E. Valsami-Jones and M. Batchelder, editors). Mineralogical Society Series, **9**. Mineralogical Society, London.

Jay, J.A., Morel, F.M.M. and Hemond, H.F. (2000) Mercury speciation in the presence of polysulfides. *Environ. Sci. Technol.*, **34,** 2196–200.

Jenkins, D.A., Johnson, D.B. and Freeman, C. (2000) Mynydd Parys: the mineralogy, microbiology and acid drainage of a metal sulphide mine. Pp. 161–79 in: *Environmental Mineralogy: Microbial Interactions, Anthropogenic Influences, Contaminated Land and Waste Management* (J.D. Cotter-Howells, L.S. Campbell, E. Valsami-Jones and M. Batchelder, editors). Mineralogical Society Series, **9**. Mineralogical Society, London.

Keith, C. and Vaughan, D.J. (2000) Mechanisms and rates of sulphide oxidation in relation to the problems of acid mine drainage. Pp. 117–39 in: *Environmental Mineralogy: Microbial Interactions, Anthropogenic Influences, Contaminated Land and Waste Management* (J.D. Cotter-Howells, L.S. Campbell, E. Valsami-Jones and M. Batchelder, editors). Mineralogical Society Series, **9**. Mineralogical Society, London.

Paquette, K.E. and Helz, G.R. (1997) Inorganic speciation of mercury in sulfidic waters: The importance of zero-valent sulfur. *Environ. Sci. Technol.*, **31,** 2148–53.

Schuiling, R.D. (1998) Geochemical engineering; taking stock. *J. Geochem. Expl. (Spec. Iss.),* **62,** 1–28.

Tessier, D., Dardaine, M., Beaumont, A. and Jaunet, A.M. (1998) Swelling pressure and microstructure of an activated swelling clay with temperature. *Clay Miner.,* **33,** 255–67.

Varekamp, J.C. and Thomas, E. (1998) Volcanic and anthropogenic contributions to global weathering budgets. *J. Geochem. Expl.,* **62,** 149–59.

Vriend, S.P. and Zijlstra, H.J.P. (editors) (1998) Geochemical engineering: Current applications and future trends. *J. Geochem. Expl. (Spec. Iss.),* **62**.

CHAPTER SEVEN

# Mechanisms and rates of sulphide oxidation in relation to the problems of acid rock (mine) drainage

C. N. KEITH[1] AND D. J. VAUGHAN[2]

[1] *Departments of Chemistry and Earth Sciences, Manchester University, Manchester M13 9PL, UK (E-mail: ckeith@fs1.ge.man.ac.uk)*

[2] *Department of Earth Sciences, Manchester University, Manchester M13 9PL, UK (E-mail: David.Vaughan@man.ac.uk)*

**ABSTRACT**

The aqueous oxidation of metal sulphide minerals in natural rocks, minewastes or mineworkings generates acidic waters, often containing elevated concentrations of toxic metals, and known as acid mine drainage (AMD) or, more generally, as acid rock drainage (ARD). Understanding the mechanisms and rates of oxidation of key sulphide minerals is the essential first stage in understanding the processes giving rise to ARD. In this chapter, our knowledge of the aqueous oxidation of the most important sulphide minerals (pyrite, pyrrhotite, galena, chalcopyrite, sphalerite, marcasite and arsenopyrite) is considered in the context of problems associated with ARD.

In certain cases, qualitative or semi-quantitative data concerning oxidation rates are available (for example, in tailings impoundments the sequence from most to least reactive is generally pyrrhotite > galena – sphalerite > pyrite – arsenopyrite > chalcopyrite) and a substantial body of data (some conflicting) exists concerning the products of oxidation. It is acknowledged that surface reaction control is the key to oxidation reaction mechanism. However, as reviewed here, the data and models currently available to describe the oxidation of particular sulphides do not, as yet, yield a consistent picture. Fundamental understanding of oxidation mechanisms remains sketchy, therefore, but the tools are now available to make progress in this field through *in situ* studies of oxidation processes at atomic resolution.

## 7.1 Introduction

The metal sulphide minerals, of which pyrite ($FeS_2$) is by far the most naturally abundant, are generally not stable in the Earth's atmosphere or when in contact with the (oxygenated) waters at or near the surface of the Earth. These minerals break down via oxygenation reactions and, when oxygenated waters are involved,

Keith, C.N. and Vaughan, D.J. (2000) Mechanisms and rates of sulphide oxidation in relation to the problems of acid rock (mine) drainage. Pp. 117–139 in: *Environmental Mineralogy: Microbial Interactions, Anthropogenic Influences, Contaminated Land and Waste Management* (J.D. Cotter-Howells, L.S. Campbell, E. Valsami-Jones and M. Batchelder, editors). Mineralogical Society Series, **9**. Mineralogical Society, London. ISBN 0 903056 20 8.

such reactions lead to the generation of acidic solutions. Major and minor (or trace) metals in the sulphide minerals are also released in these breakdown reactions; additionally, the acidic solutions may be active in dissolving other minerals, so that significant release of their components may also occur. The generation of acidic waters, often containing relatively high concentrations of metals or other toxic ions, by the breakdown of sulphides is commonly termed 'acid mine drainage' (AMD) because of the association with wastes derived from both metalliferous and coal mining. However, since the phenomenon can occur through the exposure to atmospheric weathering of rocks containing significant concentrations of metal (particularly iron) sulphides, even without mining activity, the term 'acid rock drainage' (ARD) is more generally appropriate.

In this chapter, the oxidation of key sulphide minerals is reviewed and the present state of knowledge critically evaluated. Understanding the mechanisms and rates of oxidation are the essential first stage in understanding the range of processes associated with ARD. Before considering the environmentally important sulphide minerals in this chapter, some background information on the problems associated with ARD is reviewed briefly.

## 7.2 Acid rock drainage: background to the problem

### *7.2.1 Principal source materials of ARD*

There are certain materials necessary for the generation of ARD, and their relative amounts in a system are important in determining the rate of ARD formation. Pyrite is the most common mineral involved in ARD generation (Gray, 1996), although there are other sulphide minerals which can be involved, as described below. Water and oxygen are also essential; these provide the oxidizing conditions needed for sulphide reactivity, as well as a transport mechanism for nutrients which supply the bacteria that may catalyse the oxidation reactions (as discussed below). The concentrations of oxygen in both the water and gas phases are critical, for without oxygen, none of the ARD reactions can proceed, and the bacteria die due to oxygen starvation (Salomons, 1995).

Acid rock drainage (ARD) is characterized by the presence of both fluids and fine particle solids; in particular the hydrated iron oxides often referred to as 'ochres'. The fluids are first to form, and the ochres form as a product of a secondary process of precipitation.

### *7.2.2 Fluids*

The fluids associated with ARD commonly exhibit:

(1) low pH, usually in the range 2.1–6.5 as shown in Fig. 7.1 (Salomons, 1995; Gray, 1996; Lin and Herbert, 1997; Monterroso and Macías, 1998);

(2) high Eh (Lin and Herbert, 1997) as also shown in Fig. 7.1 (from Gill, 1992);

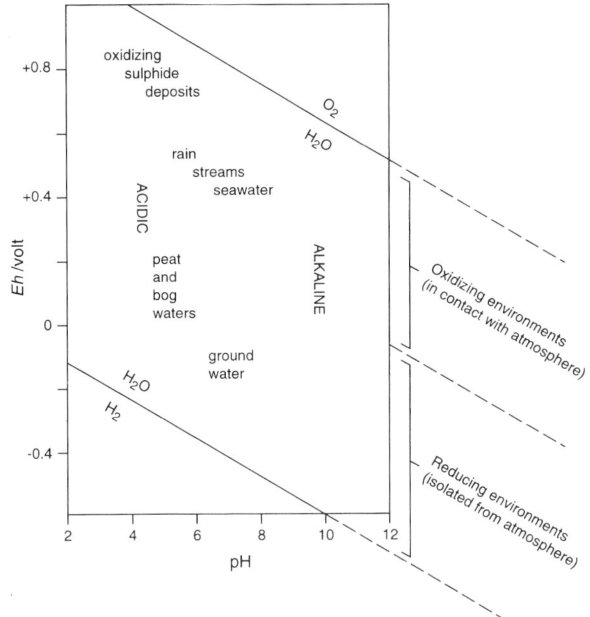


FIG. 7.1 Eh:pH diagram showing the stability ranges for various aqueous environments (after Gill, 1992).

(3) conductivities of 800–6500 μS $cm^{-1}$ and, hence, up to 100 times higher than the ambient groundwaters (Monterroso and Macías, 1998);

(4) elevated concentrations of particular elements and ions, especially Fe, sulphate and $H^+$ ions (Salomons, 1995; Monterroso and Macías, 1998), along with a wide range of metals leached out from the host rocks associated with the source of ARD notably Al, Mn and heavy metals such as Pb, Hg, Cd, etc. (Gray, 1996).

The acidity of these fluids promotes reactions with minerals in the host rocks; these fluids are often unstable, and react easily with the materials they encounter to form new compounds. Their constituent species are frequently altered, e.g. initially formed $Fe^{2+}$ reacts to form $Fe^{3+}$, a reaction which is responsible for the formation of ochres, and is itself an indication of system reactivity.

### *7.2.3 Fine particle solids ('ochres')*

Most 'ochres' are iron oxyhydroxides, and are formed as a result of the oxidation of ferrous iron in solution to ferric iron; this is accompanied by hydrolysis and partial precipitation as a solid. This process can be summarized by reactions such as:

$$4Fe^{2+} + O_2 + 4H^+ \xrightarrow{\text{oxidation}} 4Fe^{3+} + 2H_2O \tag{7.1}$$

$$Fe^{3+} + 3H_2O \xrightarrow{\text{hydrolysis}} FeOOH + H_2O + 3H^+ \tag{7.2}$$

$$FeOOH + H_2O + 3H^+ \rightarrow Fe(OH)_3 + 3H^+ \tag{7.3}$$

Other metals may also precipitate with the iron oxyhydroxides. For example, at pH <4.0, $Al^{3+}$ can dissolve from aluminosilicate minerals and, on mixing with higher pH waters, can precipitate as gibbsite (Banks *et al.*, 1997):

$$Al^{3+} + 3H_2O \rightarrow Al(OH)_3 + 3H^+ \tag{7.4}$$

### *7.2.4 Bacterial action in ARD generation*

The formation of ARD is promoted by the activity of certain bacteria; the bacterium which has been most studied is *Thiobacillus ferrooxidans* (Bierens de Haan, 1991; Sengupta, 1992; Salomons, 1995; Banks *et al.*, 1997; Belzile *et al.*, 1997). This is an obligate, acidophyllic chemotroph which is able to oxidize $Fe^{2+}$ and metal sulphides. Another bacterium of the same family, *T. thiooxidans*, cannot oxidize $Fe^{2+}$ but is able to oxidize both elemental sulphur and sulphide to sulphuric acid ($H_2SO_4$). This involves the reactions (Evangelou and Zhang, 1995):

$$S^o + 1.5O_2 + H_2O \rightarrow H_2SO_4 \tag{7.5}$$

$$S^{2-} + 2O_2 + 2H^+ \rightarrow H_2SO_4 \tag{7.6}$$

These bacteria thrive at pH values of between 1.5 and 3.0 and act to promote and/or catalyse the oxidation, hydrolysis and precipitation of dissolved iron (Banks *et al.*, 1997).

There are two principal mechanisms by which bacterial action can be classified: direct metabolic oxidation, and indirect metabolic oxidation (Evangelou and Zhang, 1995). As outlined below, using pyrite as an example, direct metabolic reactions require physical contact between bacteria and the sulphide mineral particles, whereas indirect metabolic reactions do not.

*7.2.4.1 Direct metabolic oxidation (DMO).* DMO is described by the reaction below (Torma, 1988, and references within):

$$FeS_2 + \tfrac{7}{2}O_2 + H_2O \xrightarrow{\text{bacteria}} Fe^{2+} + 2SO_4^{2-} + H^+ \tag{7.7}$$

This reaction is composed of two separate steps, the first of which is the dissolution of iron disulphide ($FeS_2$):

$$FeS_2 \rightarrow Fe^{2+} + S_2^{2-} \tag{7.8}$$

The second part of the reaction is the bonding of the released disulphide anion by bacterial enzymes, and subsequent oxidation to sulphate:

$$S_2^{2-} + 4O_2 \xrightarrow{\text{bacteria}} 2SO_4^{2-} \tag{7.9}$$

In theory, this process should continue until all the $FeS_2$ has been converted to sulphate. However, in practice the process is much more complicated than these reactions would suggest.

The behaviour of *T. ferrooxidans* is thought to be crucial in the perpetuation of the reactions (Konishi *et al.*, 1990); the bacteria use pyritic sulphur as a nutrient source, and in the process cause its oxidation to sulphate. Another possible mechanism proposed by Palencia *et al.* (1991) is the direct attack of the pyrite surface by oxidizing bacterial action. The reaction below shows the bacteria-catalysed oxidation of sulphur and iron by oxygen:

$$4FeS_2 + 15O_2 + 2H_2O \rightarrow 2Fe_2(SO_4)_3 + 2H_2SO_4 \qquad (7.10)$$

The association between *T. ferrooxidans* and pyrite surface oxidation has been studied extensively (e.g. Bennett and Tributsch, 1978; Konoshi *et al.*, 1990; Wakao *et al.*, 1982, 1983), and it has been concluded that the majority of activity occurs in the interaction zone between the bacteria and the surface of the pyrite crystal. This means that the surface structure of pyrite assumes an important role, and the influence of surface defects such as cracks and imperfections are of considerable importance.

*7.2.4.2 Indirect metabolic oxidation (IMO).* The indirect metabolic oxidation of pyrite involves chemical oxidation by $Fe^{3+}$ (Torma, 1988):

$$FeS_2 + Fe_2(SO_4)_3 \rightarrow 3FeSO_4 + 2S^o \qquad (7.11)$$

and:

$$2S^o + 6Fe_2(SO_4)_3 8H_2O \rightarrow 12FeSO_4 + 8H_2SO_4 \qquad (7.12)$$

These reactions produce ferrous iron and elemental sulphur ($S^o$), which are then oxidized by *T. ferrooxidans* to generate $Fe^{3+}$ and acidity. This oxidation takes place through the reactions outlined below, which can take place either at the pyrite surface or in interstitial fluids. Reaction 7.13 shows the first step: namely, oxidation of ferrous iron:

$$4FeSO_4 + O_2 + 2H_2SO_4 \rightarrow 2Fe_2(SO_4)_3 + 2H_2O \qquad (7.13)$$

Reaction 7.14 shows the oxidation of the elemental sulphur and formation of sulphuric acid:

$$2S^o + 3O_2 + 2H_2O \rightarrow 2H_2SO_4 \qquad (7.14)$$

The reactions above illustrate that bacteria do not act directly upon the surface of the sulphide (as in DMO), but regenerate $Fe^{3+}$ which then acts on the pyrite as an oxidizing agent. The bacteria themselves are not directly involved in the process of oxidation, instead they catalyse reactions to form other species which act as oxidizing agents.

However, there is still debate over the relative contributions of the DMO and IMO mechanisms to the process of pyrite oxidation. Silverman (1967) concluded that the two mechanisms are in operation at the same time. In contrast, Carranza (1983) reported that direct metabolism in the presence of *T. ferrooxidans* is the main mechanism by which pyrite oxidation occurs. This disagreement has not yet been resolved (Evangelou and Zhang, 1995).

Other bacteria which can be involved in the processes of sulphide oxidation include *Leptospirillum ferrooxidans,* and *Sulfobacillus thermosulfidooxidans* (Banks *et al.*, 1997).

Most mine waters (and other such waste waters) contain populations of such bacteria, but their activity and the exact species make-up are dependent upon the local environment (e.g. nutrient availability, oxygen concentration and pH) as well other ecological factors such as population density (Norris, 1990; Salomons, 1995; Banks *et al.*, 1997).

### *7.2.5 Key conditions necessary for ARD formation*

There are several key conditions necessary for the reactions which generate ARD to occur. These are principally associated with the source materials and the geochemical environment. These conditions are summarized below.

(1) The source must contain enough sulphide mineral materials to react and form acidic fluids at a rate which exceeds the neutralization capacity of any alkaline compounds (such as carbonates) contained in the system (Belzile *et al.*, 1997).

(2) The materials must be such that they allow the ingress of the water and air necessary to support chemical reactions, including those promoted by bacterial activity. Fine-grained materials, when compacted, limit the amount of oxygen input that can occur, whilst coarser-grained materials promote oxygen advection and diffusion, and so enable acid to be produced at a higher rate (Salomons, 1995). Smaller-grained materials also offer a larger reaction surface area.

(3) The climate must be such that there is sufficient rainfall and thus infiltration into the materials of the ARD system. This is to ensure a good water, oxygen and nutrient supply, as well as to enable acid water to move through the environment (Belzile *et al.*, 1997).

### *7.2.6 Potential sources of ARD*

Mining and quarrying can result in systems which have the potential to generate ARD and hence to release acidic leachates into the environment. Some of these systems include: waste rock or tailings dumps; ore stockpiles; underground or open pit mine workings; spent heap leach piles (Fornasiero *et al.*, 1992); and managed tailings schemes (Belzile *et al.*, 1997).

The reactivity of each system, and hence likelihood of the production of acid, is linked to the properties of each of the materials present, as well as the local geological conditions; for example, whether water has easy access to the material.

There are also examples of ARD generated without human activities such as mining or quarrying. A specific example of such a system occurs at Mam Tor, Derbyshire, UK. Here, a large landslip occurred some 3500 years ago and exposed pyrite-rich Namurian shales; part of a series of marine shales, calcareous mudstones, ironstones, iron-rich turbidites and micaceous sandstones (Cripps and Hird, 1991). In the period since the slope failure, fluids and ochres characteristic of ARD have formed. Vear and Curtis (1981) carried out research into the reactions occurring at depth within the landslip, and concluded that the waters emerging from the base were so strongly acidic and sulphate-rich that, for each litre of water, 1.4996 grams of pyrite would need to be oxidized to produce these conditions.

### *7.2.7 Environmental impact of ARD*

In their report into abandoned mines and water quality, the UK National Rivers Authority (1994) outlined some of the effects of acid rock drainage upon the biological systems of streams, lakes and rivers. These include: depletion of numbers of sensitive species, and a corresponding reduction in biodiversity; loss of spawning grounds for species which return to the same stretch of river or stream each year (e.g. salmon); fish and other species deaths, primarily due to metal poisoning (especially from aluminium contamination) and suffocation as dissolved oxygen is used in oxidation processes; damage to fish gills by low pH waters, decreasing their ability to absorb oxygen from the water; and an overall decline in both fish and other benthic populations in affected streams.

### *7.2.8 Other impacts of ARD*

Other impacts of ARD upon the wider environment include: potable water supplies – water abstraction points which draw water from supplies contaminated by ARD require a higher level of water treatment prior to distribution to customers, and the cost of this can be prohibitive; the visual impact of ARD can be spectacular, as the ochres and fluids are commonly highly coloured – often bright orange from the iron oxyhydroxides suspended in the fluids; the amenity value decreases when ARD pollutes a river or stream – the water becomes unsuitable for swimming, as well as fishing; and localized flooding can be caused by uncontrolled discharges from abandoned mines (National Rivers Authority, 1994).

## 7.3 Sulphide minerals: chemistry and reactivity, oxidation mechanisms, rates and controls

Although there are several hundred known metal sulphide minerals, only five are sufficiently abundant to be regarded as 'rock forming minerals' (Deer *et al.*, 1992). These are pyrite, pyrrhotite, galena, chalcopyrite and sphalerite. Of these five, two (pyrite and pyrrhotite) are responsible for most of the acidic drainage worldwide (Jambor and Blowes, 1998). In this chapter, the five major sulphide minerals will be considered, along with the dimorph of pyrite, marcasite, and the related As-bearing mineral, arsenopyrite. The emphasis will be on oxidation in aqueous fluids of the types encountered in ARD systems.

The key questions to be considered are: what are the reactivities of these sulphide minerals? By what mechanism(s) do they undergo oxidation, and what are the products of oxidation? What controls the rates and products of oxidation? That the overall controls of reactivity are complex and variable is well illustrated by comparison of relative reactivities amongst these commoner sulphides as reported from different laboratory and field studies and shown in Table 7.1 (after Jambor, 1994). These differences partly reflect the different conditions under which oxidation occurs (abiotic or biotic, gossan, waste rock, etc.) and there are evidently effects related to grain size and mineral associations. Despite those complexities, under the conditions prevalent in ARD systems, it appears that pyrite is generally more stable, and that pyrrhotite is highly reactive with galena and sphalerite falling somewhere

TABLE 7.1 Relative reactivity of sulphide minerals (after Jambor, 1994).

| Increasing order of resistance | Condition | Reference |
|---|---|---|
| sphalerite > galena > chalcopyrite > pyrite | Abiotic, calculated | 1 |
| marcasite > pyrrhotite > chalcopyrite > pyrite = arsenopyrite | Waste rock | 2 |
| pyrrhotite > chalcopyrite > fine pyrite > sphalerite > galena > coarse pyrite | Gossan | 3 |
| pyrrhotite > arsenopyrite > pyrite > chalcopyrite > sphalerite > galena | Laboratory, pH 2–6 | 4 |
| pyrrhotite > pyrrhotite-pyrite > pyrrhotite-arsenopyrite > arsenopyrite > pyrite > chalcopyrite > sphalerite > galena > chalcocite | – | 5 |
| pyrrhotite > chalcocite > tetrahedrite > galena > arsenopyrite > sphalerite > pyrite > marcasite > chalcopyrite | – | 6 |
| sphalerite > tetrahedrite group > chalcopyrite > Bi-Sb sulphosalts > galena > arsenopyrite > pyrite | Gossan | 7 |
| pyrite > chalcopyrite > galena > sphalerite | Air oxidation | 8 |

1. Sarveswara Rao *et al.* (1991)
2. Kwong and Ferguson (1990)
3. Andrew (1984)
4. Kakovsky and Kosikov (1975)
5. Flann and Lucaszewski (1970)
6. Brock *et al.* (1984), abridged
7. Boyle (1994)
8. Brion (1980), by X-ray photoelectron spectroscopy

between. Jambor (1994) noted that the sequence in tailings impoundments (most → least reactive) is usually:

pyrrhotite > galena – sphalerite > pyrite – arsenopyrite > chalcopyrite

Marcasite, as a metastable phase is also highly reactive. Details of the reactive behaviour of the individual sulphides will now be considered, with particular emphasis on pyrite and pyrrhotite as the dominant contributors to ARD.

### *7.3.1 Pyrite*

Simply because of its abundance, pyrite is by far the most important mineral contributing to ARD. Although the bulk crystal structure of pyrite is well known (see Vaughan and Craig, 1978), the surface atomic structure of pyrite has been subject to relatively limited studies thus far, due to problems in obtaining flat surfaces (Hochella, 1995; Eggleston *et al.*, 1996). However, a recent study of clean pyrite {100} surfaces generated by cleaving in vacuum and employing low energy electron diffraction (LEED) and scanning tunnelling microscopy (STM) methods marks a significant advance (Russo *et al.*, 1999*a*). These experiments, which were supported by computer modelling studies using quantum mechanical methods, suggest that the {100} surface undergoes very little 'relaxation' (i.e. adjustment of atom positions at the surface) and can be approximated by a simple termination of the bulk structure along a plane of cleaved Fe–S bonds. Although the atomic ('crystal') structures at the surface appears to differ little from the bulk, loss of co-ordination at the surface necessarily results in changes in the nature of bonding (or electronic structure). The calculations suggest that the highest energy occupied orbitals at the surface are iron 3d (specifically $3d^2_Z$) in character and lowest energy unoccupied orbitals of mixed Fe $3d^2_Z$ and sulphur 3p character. Since it is these orbitals that would be involved in reactions (such as oxidation) at the surface, the indications are that Fe sites are energetically more reactive than $S_2$ sites at the surface and will be involved in redox reactions leading to new surface species. However, other recent studies highlight the complexity of the problems associated with pyrite surface reactivity by identifying a variety of reactive sites and introducing the possibility that different mechanisms may be proceeding simultaneously across the surface. Thus, in addition to the work of Eggleston *et al.* (1996) and Russo *et al.* (1999*a,b*) for example, highlighting the role of surface iron atoms in oxidation, studies employing photoelectron spectroscopy have also identified important, reactive sulphur sites formed on breaking S–S bonds (Nesbitt *et al.*, 1998; Schaufuss *et al.*, 1998).

Given the importance of pyrite in environmental geochemistry and in mineral processing technology, it is not surprising that a large amount of work has been undertaken on the mechanisms and kinetics of pyrite oxidation. It is not possible to comprehensively discuss all aspects of this work here, and readers are referred to other review articles such as that of Evangelou and

Zhang (1995). However, the important point to be made is that key aspects of the mechanisms and kinetics of pyrite oxidation remain unresolved. This is firstly because of the large number of variables associated with the oxidation process of pyrite (and other sulphides); oxidation medium, pH, Eh, temperature and presence of certain bacteria can all significantly influence oxidation rates. For example, as illustrated in Fig. 7.2, the presence of bacteria can influence oxidation rates by many orders of magnitude (at low pH, *T. ferrooxidans* can accelerate the oxidation of $Fe^{2+}$ by more than $10^6$; Singer and Stumm, 1970). Secondly, the oxidation of pyrite to form sulphate requires the transfer of 8 electrons per S atom and must therefore involve several (as yet unresolved) reaction steps with S species of intermediate oxidation state. Overall, Biegler and Swift (1979) suggest that in pure water, pyrite oxidizes by a combination of the half reactions:

$$FeS_2 + 8H_2O \rightarrow Fe^{3+} + 2SO_4^{2-} + 16H^+ + 15e^- \quad (7.15)$$

$$FeS_2 \rightarrow Fe^{3+} + 2S + 3e^- \quad (7.16)$$

for which Eh and pH determine the dominant pathway; the sulphate route dominates under ambient conditions. At low pH, pyrite may be oxidized by $O_2$ and $Fe^{3+}$ (McKibben and Barnes, 1986) via reactions such as:

$$FeS_2 + \tfrac{7}{2}O_2 + H_2O \rightarrow Fe^{2+} + 2SO_4^{2-} + 2H^+ \quad (7.17)$$

$$FeS_2 + 14Fe^{3+} + 8H_2O \rightarrow 15Fe^{2+} + 2SO_4^{2-} + 16H^+ \quad (7.18)$$

The potential for electron transfer involved in sulphide oxidation reactions such as these is what causes them to be, in nature, largely catalysed by the activity of micro-organisms. As noted above *Thiobacillus ferrooxidans* accelerates the weathering of pyrite by catalysing the oxidation of $Fe^{2+}$ to $Fe^{3+}$ using free oxygen as an electron acceptor.

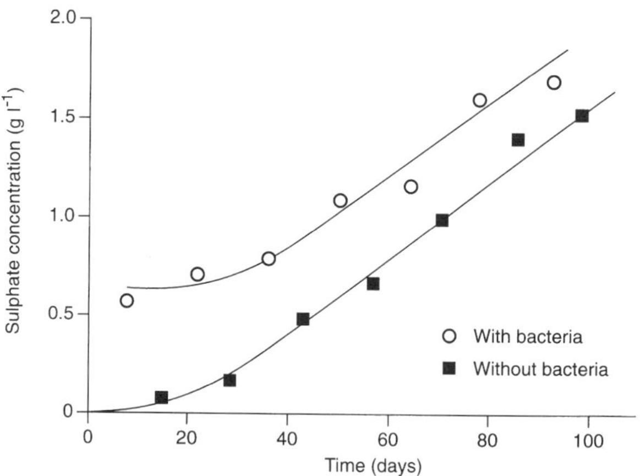


FIG. 7.2 Pyrite oxidation kinetics in minewaste at 21°C and pH 3 in the presence and absence of bacteria (after Scharer *et al.*, 1991, cited in Evangelou and Zhang, 1995).

The dynamic interactions between free oxygen, iron and dissolving pyrite can be shown schematically as in Fig. 7.3 (after Banks *et al.*, 1997). Here, pyrite is oxidized to sulphate by reaction A. Alternatively, reaction B shows the pyrite surface being oxidized directly by free oxygen, and sulphate being leached into solution. At step C, ferrous iron is oxidized, and may be precipitated as relatively insoluble iron hydroxide in step E. Alternatively, $Fe^{3+}$ may oxidatively dissolve pyrite in step D.

The earlier literature on pyrite oxidation has been reviewed in detail by Hiskey and Schlitt (1982), Lowson (1982) and Nordstrom (1982). These reviews and later papers universally acknowledge that any oxidation reaction mechanism must be surface reaction controlled. Thus McKibben and Barnes (1986) envisaged pyrite oxidation to occur at acid pH through one of two mechanisms. The first proposed mechanism, oxidation by dissolved oxygen, involves absorption of the latter onto the surface and subsequent oxidation by water to form $H_2O_2$. Pyrite oxidation is then driven by adsorbed hydrogen peroxide. The second proposed reaction pathway involves adsorption of Fe(III), electron transfer, dissociation of the surface complex, and desorption into solution. The last step was considered to be the rate-limiting one. Wiersma and Rimstidt (1984) measured pyrite dissolution rates at pH 2 and emphasized the importance of $Fe^{3+}$ in the control of oxidation. Moses *et al.* (1987) focused on the likely surface steps involved in the autocatalytic oxidation of pyrite by ferric iron (step D in Fig. 7.3). The proposed initial oxidation step is the movement of a hydroxyl radical from a hydrated ferric iron to a S site exposed at the mineral surface. By a subsequent series of hydroxyl transfers and dehydration steps, a thiosulphate ion is produced. This reaction hypothesis predicts that rates should depend on the availability of $Fe^{3+}$ and its affinity for sulphide species. Luther (1987) has applied molecular orbital theory to arrive at a very similar mechanism for pyrite dissolution.

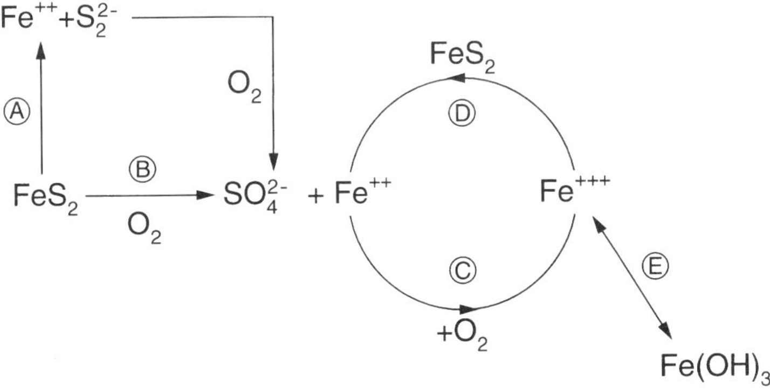


FIG. 7.3 Reaction pathways for pyrite oxidation (after Banks *et al.*, 1997).

Some of the more recent electrochemical studies of pyrite oxidation point to an initial surface oxidation step, followed by the surface association of hydroxyls onto $Fe^{2+}$ sites and subsequent oxidation of sulphide groups. On the other hand, cyclic voltammetry of pyrite in acidic aqueous electrolytes has shown very slow oxidation and reduction rates in the potential range −0.4 to 0.6 V (*vs.* Saturated Calomel Electrode; SCE; see Fig. 7.4) in conflict with thermodynamic predictions (Kelsall *et al.*, 1996). The rate limiting reaction was attributed to reversible electrochemical adsorption/desorption reactions of the type:

$$FeS_2 + H^+ + e^- \rightleftharpoons FeS_2H \text{ (ads)} \tag{7.19}$$

$$FeS_2 + H_2O + h^+ \rightleftharpoons FeS_2OH_2^+ \text{ (ads)} \tag{7.20}$$

A mechanism of pyrite oxidation was proposed that involves a complex sequence of series and parallel reaction steps, ultimately producing $Fe^{2+/3+}$ and $HSO_4^-/SO_4^-$ depending upon pH and applied potential. In the initial stages of reaction, at positive electrode potentials, the oxygen atom of water molecules is envisaged as adsorbing at the disulphide site; with increasing positive potential, one electron in the Π* orbital of $S_2^{2-}$ could be extracted, leading to interactions between those orbitals and the Π orbital of the water molecules.

At the start of this section, it was noted that only limited studies have thus far been undertaken of the surface structure and reactivity of pyrite at the level of atomic resolution. Furthermore, attempts to define individual reaction mechanisms at the surface have raised complex questions. However, the techniques now becoming available to study surface reactions at the molecular level directly, promise new insights. For example, following on from their studies of the structure of the {100} pyrite surface, Russo *et al.* (1999*b*) used STM, photoelectron spectroscopy and computer modelling methods to

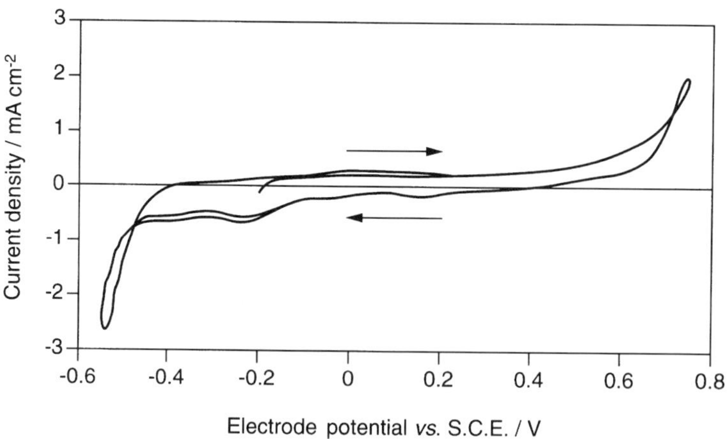


FIG. 7.4 Cyclic voltammogram of $FeS_2$ in 1 M HCl (0.3 $Vs^{-1}$) (after Kelsall *et al.*, 1996).

investigate interaction of gaseous $O_2$ and $H_2O$ and their mixtures with this surface under controlled (high vacuum) conditions. For such gaseous 'molecular' oxidants, it appears that $O_2$ sorbs dissociatively to surface Fe atoms and $H_2O$ sorbs molecularly to such sites and combined gases oxidize the surface more aggressively than does pure $O_2$.

### *7.3.2 Pyrrhotite*

Pyrrhotite is second in importance only to pyrite as a source of ARD. The name 'pyrrhotite' really refers to a family of iron sulphides of general formula $Fe_{1-x}S$ with structures based on that of NiAs, but in which up to ⅛ of the iron atoms are omitted leaving vacancies which may be ordered in various ways at lower temperatures (Vaughan and Craig, 1978). Although a large number of structure types is known, naturally occurring pyrrhotites are dominated by the monoclinic phase of composition $Fe_7S_8$ and a hexagonal form with a composition of ~$Fe_9S_{10}$.

Perhaps not surprisingly, given these vacancy-containing structures, pyrrhotite is amongst the most reactive of the sulphides (see Table 7.1). In general, rates of oxidation/breakdown are reported as being an order of magnitude or more faster than those of pyrite, and found to be independent of pH at 10°C ($2 < pH < 6$) but to show increasing pH dependence at higher temperatures.

Studies of pyrrhotite surfaces at atomic resolution are in their infancy but show the importance of S atom vacancies in the uppermost atomic layer of the {001} surface of monoclinic pyrrhotite, and of the role of surface step sites as the locus of oxidation reactions (Becker *et al.*, 1997). At a more general level of observation of surface reactions, X-ray photelectron spectroscopy (XPS) studies of monoclinic pyrrhotite (Pratt *et al.*, 1994*a,b*; Mycroft *et al.*, 1995) show that about one third of the iron in the surface region becomes $Fe^{3+}$, forming a layer containing '$Fe_2S_3$'. On further alteration, this gives way to a marcasite layer and, in turn, to an iron oxyhydroxide. This sequence, which is associated with Fe diffusion from the bulk, emulates what is commonly observed for natural samples (Jambor and Blowes, 1998).

There is disagreement over the relative susceptibility to oxidation of the monoclinic and hexagonal forms of pyrrhotite. In natural materials (e.g. Nickel *et al.*, 1977), monoclinic pyrrhotite is observed as an alteration product of the hexagonal form in the oxidation zone of ore deposits. However, laboratory studies using natural ARD fluids (Jones *et al.*, 1997) have been reported to show hexagonal pyrrhotite oxidizing more readily.

### *7.3.3 Galena*

The surface chemistry of galena has been much studied; for the dominant {001} cleavage or growth surface, experimental evidence from low energy electron diffraction (LEED) and scanning tunnelling microscopy (STM) suggests that the arrangement of atoms at the surface approximates to that represented by a truncation of the bulk solid (Hochella *et al.*, 1989; Eggleston and Hochella, 1990). Although experimental data show no significant movement of atoms at the surface compared with their positions in the bulk, quantum mechanical calculations do suggest a small amount of movement of the outermost atoms in the direction normal to the surface (Becker and Hochella, 1996). Inevitably, there will also be various defects at the surface associated with S and Pb vacancies and with steps between layers of atoms. Changes in electronic structure near steps, shown experimentally using STM (Laajalehto *et al.*, 1993; Eggleston and Hochella, 1994) typically lead to increased reactivity near step sites (Junta and Hochella, 1994). This has been confirmed by quantum mechanical calculations of the dissociation of water on the galena surface which show lack of reaction on a perfect {001} surface, but exothermic reaction at a step site (Wright *et al.*, 1999).

Reactivity of the galena surface has, in fact, been studied intensively in relation to understanding the mechanisms and kinetics of oxidation. Particularly for elucidation at stages beyond the very first oxidation reaction, electrochemical methods combined with spectroscopic techniques for surface analysis (especially X-ray and UV photoelectron spectroscopies) have provided a powerful means of investigation (e.g. Brion, 1980; Buckley and Woods, 1984; Buckley *et al.*, 1985; Laajalehto *et al.*, 1993; Fornasiero *et al.*, 1994). Understanding of the very first stages of oxidation and their detailed mechanisms at the molecular scale is much less advanced, but techniques centred on scanning tunnelling microscopy and spectroscopy (Hochella, 1990; Becker and Hochella 1996) and quantum mechanical calculation (see Becker *et al.*, 1997) are now yielding important insights. For example, STM images of clean PbS surfaces show S sites or, more precisely, the local density of valence band electron states above S atoms with predominantly S3p character (see Fig. 7.5). These lightly bound electrons are responsible for the tunnelling of electrons from the sample surface to the tip, and for electron transfer from the sample to an oxidizing agent such as $O_2$. Experimental images taken under oxidizing conditions show dark patches that grow with time of exposure, and that can be explained by oxygen withdrawing electrons from the S sites so that they are no longer available for the tunnelling process.

At more advanced oxidation stages, different oxidation products are observed at the galena surface depending upon overall oxidation conditions (pH, temperatures, etc.) and available reactants ($O_2$, $H_2O$, $CO_2$, etc.). These oxidation products include such species as elemental S, polysulphides, lead oxide, hydroxides, sulphates and carbonate, and have largely been identified from electrochemical measurements and peak fitting of X-ray photoelectron spectra (see Brion, 1980; Buckley and Woods, 1984, etc.).

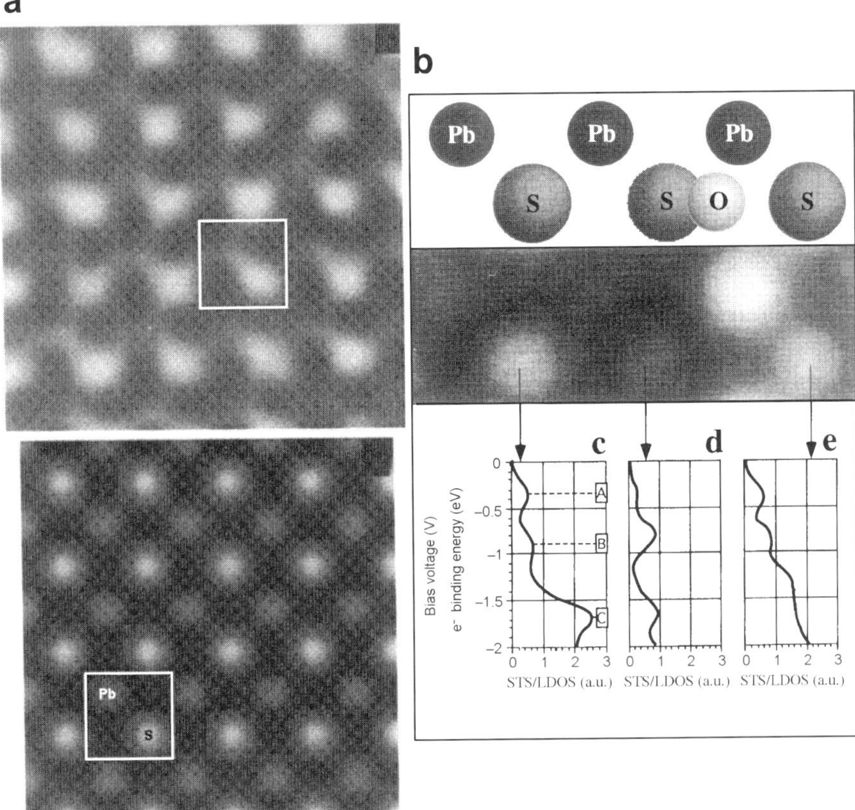


FIG. 7.5 (*a*) Experimental (upper) and calculated (lower) STM images for the {100} cleavage surface of galena. (*b*) Calculated STM images and STS data for the {100} cleavage surface of galena (after Becker and Hochella, 1996).

### 7.3.4 Sphalerite

The surface properties of pure synthetic ZnS phases have been particularly well studied because of their importance in the semiconductor industry; indeed, the (110) cleavage surface of sphalerite is the best understood of all semiconductor surfaces (Kahn, 1983). Detailed LEED studies have established that the (110) surface is reconstructed by movement of the Zn atoms inward (toward the bulk solid) and of the S atoms outward. Displacements of the cations and anions in the uppermost layer by $\ll 0.5$ Å and in the second layer by 0.1 Å compared to the bulk are involved (Duke, 1988). This reconstruction has been predicted using atomistic calculations by Wright *et al.* (1998) and discussed in terms of electronic structure by Harrison (1980) and Tossell and Vaughan (1993).

Although much is known about the pure ZnS surface, there have been relatively few studies of the oxidation of sphalerite. The XPS studies of Brion (1980) showed that sphalerite is more resistant to oxidation in air than pyrite, chalcopyrite or galena, and that surface oxidation of sphalerite leads directly to the formation of zinc sulphate. It is important to note, however, that field observations show galena to be more resistant than sphalerite. This is almost certainly due to the resistance to breakdown imparted by secondary coatings. In the case of galena, anglesite ($PbSO_4$) commonly forms a surface coating which is highly insoluble and shields the sulphide from contact with oxidizing agents. An equivalent secondary sulphate mineral does not exist for sphalerite, where sulphates are highly soluble and readily removed (Jambor and Blowes, 1998).

### *7.3.5 Chalcopyrite*

Chalcopyrite ($CuFeS_2$), although having a crystal structure that is essentially the same as that of sphalerite (Vaughan and Craig, 1978), has been less studied as regards the surface properties of the pure material. This is largely because good surfaces for study using techniques such as LEED or STM are not available; what appear to be single crystals of chalcopyrite are often polycrystalline material, and chalcopyrite has no good cleavage surface. Reconstructions of the kind observed in sphalerite are expected in chalcopyrite on the basis of crystal chemical arguments (Tossell and Vaughan, 1993), and this is now being confirmed by computer modelling studies using atomistic methods (Wright, in prep.). Further work in this area is currently being undertaken by the authors in collaboration with K.V. Wright.

Given how little is known about the surface structure of chalcopyrite at atomic resolution, it is not surprising that the initial oxidation process is not well understood. However, much more is known about the later stages of oxidation, including oxidation in aqueous solution. Studies employing electrochemical techniques in combination with XPS, Auger electron spectroscopy (AES) and X-ray absorption spectroscopy (XAS) (Yin *et al.*, 1995; Vaughan *et al.*, 1995) have enabled a sequence of oxidation reactions to be established. For example, considering a cyclic voltammogram for a chalcopyrite electrode in conjunction with XPS and XAS data has enabled the sequence of reactions shown in Fig. 7.6 to be proposed for oxidation under neutral or mildly alkaline conditions. Here, at potentials just above the rest potential, a monolayer of $Fe_2O_3/Fe(OH)_3$ is formed leaving Cu and S unoxidized in the original chalcopyrite structure as a metastable phase of $CuS_2$ stochiometry. Together, these phases inhibit further reaction (i.e. they cause passivation). With increasing potential these reactions continue, removing iron from deeper within the chalcopyrite and with solid state diffusion being the likely rate-controlling mechanism (Fig. 7.6). Above a critical (pH-dependent) potential, the passivating $CuS_2$ phase probably decomposes by the further reactions shown in Fig. 7.6. Under acid conditions, the metastable $CuS_2$ phase

a

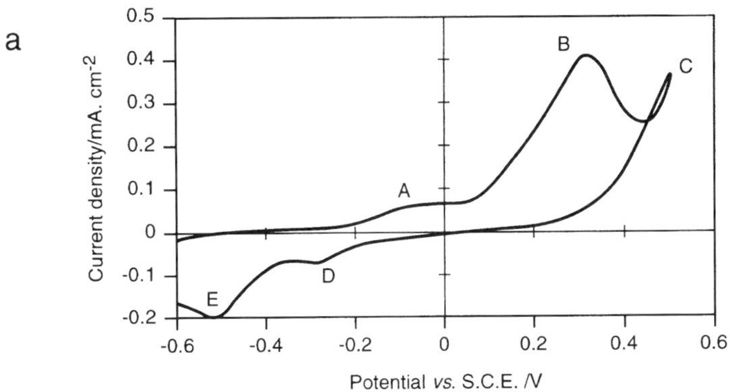


b

$$2CuFeS_2 + 3xOH^- \rightarrow 2CuFe_{1-x}S_2 + xFe_2O_3 + 3xH^+ + 6xe^-$$

$$2CuFeS_2 + 6OH^- \rightarrow 2CuS_2^* + Fe_2O_3 + 3H_2O + 6e^-$$

$$CuFeS_2 + 3OH^- \rightarrow CuS_2^* + Fe(OH)_3 + 3e^-$$

$$CuS_2^* + 2OH^- \rightarrow CuO + 2S + H_2O + 2e^-$$

$$CuS_2^* + 2OH^- \rightarrow Cu(OH)_2 + 2S + 2e^-$$

c

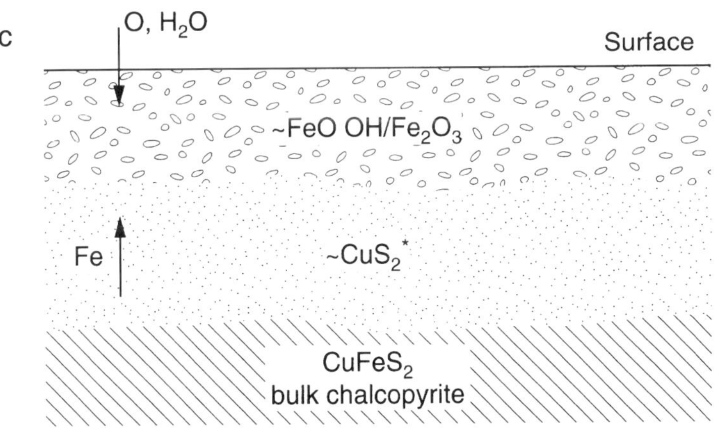


FIG. 7.6 (*a*) Cyclic voltammogram of chalcopyrite in 0.1 M $Na_2B_4O_7$ (pH 9.2) at 298 K. The potential was swept at 20 mV $s^{-1}$. Peaks A–E correspond to reactions at the electrode surface. (*b*) Oxidation reactions at the chalcopyrite surface corresponding to peaks A, B, C above. (*c*) Model showing development of oxidation products at the chalcopyrite surface (after Vaughan *et al.*, 1997).

again forms a passivating film at the surface, with diffusion of iron through that film being the rate-controlling mechanism (Yin *et al.*, 1995; Vaughan *et al.*, 1995). In this case, $Fe^{2+}$ ions or $Fe^{3+}$ ions at higher (more oxidizing) potentials are the products of chalcopyrite breakdown rather than iron oxides.

If an important control of oxidation rate is diffusion from the bulk to the surface (Fig. 7.6), it might be expected that variations in metal:S ratio influence oxidation rate. An interesting aspect of chalcopyrite crystal chemistry is the ability to accommodate additional metals in normally unoccupied interstitial sites. When ordered these give rise to the metal-rich 'stuffed derivatives' of chalcopyrite named haycockite, mooihoekite and talnakhite. Electrochemical studies of these phases have shown marked differences in oxidation rates with stochiometric chalcopyrite oxidizing the slowest, and haycockite and mooihoekite much more rapidly, suggesting that the interstitial metal atoms are more active and mobile, as might be expected (Vaughan *et al.*, 1995).

### *7.3.6 Arsenopyrite and marcasite*

Arsenopyrite (FeAsS) and marcasite ($FeS_2$) are both sulphides containing dianion groups and are both structurally related to pyrite. Although not as abundant as the other sulphides discussed above, arsenopyrite is included because of the particular problems associated with release of As into the environment, and marcasite is included because it can occur in significant quantities in coals and coal mining wastes which are major ARD sources in many countries. No studies of the detailed surface structures and properties of these phases have been undertaken at atomic resolution, but some information is available on the products and mechanisms of oxidation.

Richardson and Vaughan (1989) used XPS and AES to study arsenopyrite surfaces that had been treated with a variety of inorganic oxidants. It was concluded that arsenopyrite will oxidize rapidly in the low pH conditions that prevail in some mine wastes, and that the concentrations of sulphur on the surface layer can be influenced by the associated chemical conditions. Muir *et al.* (1993) found, using XPS and AES, that a one-day exposure of fresh arsenopyrite to air caused an increase in As and oxygen concentrations at the surface. In the upper 20–30 Å, ~35% of the As appeared as As(III) and As(V) oxides and only a minor amount of $Fe^{2+}$ was oxidized to $Fe^{3+}$. Sulphur remained unoxidized and buried beneath the surface oxides to a depth of ~100 Å. In water, however, oxidative dissolution was observed, and large amounts of $Fe^{3+}$ detected (presumably from iron oxyhydroxides) and S found as sulphide, polysulphide and minor sulphate.

Further studies by Nesbitt *et al.* (1995) using similar techniques essentially confirm these findings. In air-saturated water, Fe(III) oxyhydroxides are the dominant, Fe-containing surface product, $As^{5+}$, $As^{3+}$ and $As^{1+}$ are as abundant as $As^{1-}$, and sulphate can be detected on the surface. Nesbitt *et al.* (1995)

noted a striking similarity between the products of oxidation of arsenopyrite in air-saturated (distilled) water and in air, suggesting similar oxidation mechanisms. In air, As is most readily oxidized and S most slowly, whereas in air-saturated water, $As^{1-}$ and $Fe^{2+}$ oxidize at similar rates and much more rapidly than S. Studies of compositional variations in the near-surface region demonstrate that As diffuses from the interior of the mineral to the surface during oxidation; combined with the rapid oxidation of As to produce $As^{3+}$ and $As^{5+}$, this leads to release of significant amounts of arsenites and arsenates into solution. The arsenites ($AsO_3^{3-}$) are particularly toxic to biota. Although not yet confirmed experimentally, it is likely that similar effects will be observed for arsenian pyrite ($Fe(S,As)_2$) and löllingite ($FeAs_2$) oxidation.

The oxidation of marcasite can be described using the same overall reactions as for pyrite (e.g. reactions 7.17 and 7.18 above) but, as with pyrite, oxidation of the S component of marcasite to sulphate requires a transfer of 14 electrons. It must also involve intermediate steps, with products such as thiosulphate, sulphite and possibly polysulphides, not described in these overall reactions. In a recent study, Rinker *et al.* (1997) studied the mechanisms and rates of marcasite oxidation in acidic solutions (pH 3) at low temperatures. A marcasite surface exposed to an oxygenated HCl solution (pH 3.0) was studied using XPS and showed an increase in polysulphide species at the expense of disulphide; iron species present were not readily identified, although chloride and $OH^-$ were identified. Studies of compositional variation with depth revealed no formation of distinct zones after reaction. Experiments were also performed to determine the rates of aqueous oxidation reactions involving both cleaned and untreated (natural) marcasite samples exposed to oxygen-saturated HCl at pH 3.0 and at 25°C. Leach rates between these two samples varied by several orders of magnitude ($4.25 \times 10^{-5}$ mmol/m$^2$.s for cleaned and $2.2 \times 10^{-2}$ mmol/m$^2$.s for untreated marcasite). The breakdown of marcasite in air (commonly observed in museum specimens) has been relatively little studied and may offer further clues to understanding reaction pathways.

## 7.4 Concluding remarks

Earlier in this chapter, the key questions that need to be considered in relation to sulphide oxidation in the context of ARD were identified as being concerned with reaction rates, mechanisms and products, and the controls over rates and products. As is evident from the concise overview presented above, some qualitative or semi-qualitative knowledge of rates is available, and a substantial body of data (sometimes conflicting) exists concerning the products of oxidation. Fundamental understanding of mechanisms remains sketchy but the tools, both experimental and theoretical, are becoming available to enable real advances to be made in this area. In particular, as well as the capability to study clean surfaces under highly controlled (high vacuum) conditions and to subject them to oxidation reactions whilst making real time measurements at atomic resolution, it is now possible to conduct such high resolution studies with the surface in contact with a fluid phase. Such *in situ* measurements using a range of scanning probe (STM, etc.) techniques and methods employing synchrotron radiation to look at X-ray absorption and reflectivity at the oxidizing mineral/fluid interface will revolutionize our understanding of the processes at this interface. Gathering this information is the essential next step in understanding the controls of sulphide oxidation and generating predictive models of value in the control of acid rock drainage.

## Acknowledgements

CNK acknowledges EPSRC for support through a research studentship, and DJV acknowledges NERC for support through research grants GR3/10789 and GR3/11979.

## References

Andrew, R.L. (1984) Supergene alteration and gossan textures of base-metal ores in southern Africa. *Minerals Sci. Eng.*, **12**, 193–215.

Banks, D., Younger, P.L., Arnesen, R.-T., Iversen, E.R. and Banks, S.B. (1997) Mine-water chemistry: the good, the bad and the ugly. *Environ. Geol.*, **32,** 157–73.

Becker, U. and Hochella, M.F. Jr. (1996) The calculation of STM images, STS spectra and XPS peak shifts for galena: new tools for understanding mineral surface chemistry. *Geochim. Cosmochim. Acta*, **60**, 2413–26.

Becker, U., Hochella, M.F. Jr. and Vaughan, D.J. (1997) The adsorption of gold to galena surfaces: calculation of adsorption/reduction energies, reaction mechanisms, XPS spectra, and STM images. *Geochim. Cosmochim. Acta*, **61**, 3565–85.

Becker, U., Munz, A.W., Lennie, A.R., Thornton, G. and Vaughan, D.J. (1997) The atomic and electronic structure of the (001) surface of monoclinic pyrrhotite ($Fe_7S_8$) as studied using STM, LEED and quantum mechanical calculations. *Surf. Sci.*, **389**, 66–87.

Belzile, N., Maki, S., Chen, Y.-W. and Goldsack, D. (1997) Inhibition of pyrite oxidation by surface treatment. *Sci. Tot. Environ.*, **196**, 177–86.

Bennett, J.C. and Tributsch, H. (1978) Bacteria leaching patterns on pyrite crystal surface. *J. Bacteriol.*, **134**, 310–7.

Biegler, T. and Swift, D.A. (1979) Anoxic behaviour of pyrite in acid solutions. *Electrochim. Acta,* **24**, 415–20.

Bierens de Haan, S. (1991) A review of the rate of pyrite oxidation in aqueous systems at low temperature. *Earth Sci. Rev.*, **31**, 1–10.

Boyle, D.R. (1994) Oxidation of massive sulfide deposits in the Bathurst Mining Camp, New Brunswick. Pp. 535–50 in: *Environmental Geochemistry of Sulfide Oxidation* (C.N. Alpers and D.W. Blowes, editors) American Chemical Society, Symp. Series, **550**.

Brion, D. (1980) Etude par spectroscopy de photoelectrons de la degredation superficielle de $FeS_2$, $CuFeS_2$, ZnS and PbS a l'air dans l'eau. *Appl. Surf. Sci.*, **5**, 133–52.

Brock, T.P., Smith, D.W. and Madigan, M.T. (1984) *Biology of Micro-organisms*. Prentice-Hall, Englewood Cliffs, New Jersey.

Buckley, A.N. and Woods, R. (1984) An X-ray photoelectron spectroscopic study of the oxidation of galena. *Appl. Surf. Sci.,* **17**, 401–14.

Buckley, A.N., Hamilton, I.C. and Woods, R. (1985) Investigation of the surface oxidation of sulfide minerals by linear potential sweep voltammetry and X-ray photoelectron spectroscopy.

*Dev. Miner. Process*, **6**, 41–60.

Carranza, F. (1983) PhD thesis, Univ. Seville, Spain.

Cripps, J.C. and Hird, C.C. (1991) A guide to the landslide at Mam Tor. *Geoscientist*, **2**, 22–7.

Deer, W.A., Howie, R.A. and Zussman, J. (1992) *An Introduction to the Rock-forming Minerals*. Longman Scientific & Technical, Harlow, Essex, UK.

Duke, C.B. (1988) Atomic and electronic structure of tetrahedrally co-ordinated compound semiconductor interfaces. *J. Vac. Sci. Tech.*, **6**, 1957–62.

Eggleston, C.M. and Hochella, M.F. Jr. (1990) Scanning tunnelling microscopy of sulfide surfaces. *Geochim. Cosmochim. Acta*, **54**, 1511–7.

Eggleston, C.M. and Hochella, M.F. Jr. (1994) Atomic and electronic structure of PbS (100) surfaces and chemisorption-oxidation reactions. Pp. 201–20 in: *Environmental Geochemistry of Sulfide Oxidation* (C.N. Alpers and D.W. Blowes, editors). American Chemical Society, Symp. Series, **550**.

Eggleston, C.M., Ehrhardt, J.J. and Stumm, W. (1996) Surface structural controls on pyrite oxidation kinetics: and XPS-UPS, STM and modelling study. *Amer. Mineral.*, **81**, 1036–56.

Evangelou, V.P. and Zhang, Y.L. (1995) A review: pyrite oxidation mechanisms and acid mine drainage prevention. *Critical Rev. Environ. Sci. Tech.*, **25**, 141–99.

Flann, R.C. and Lukaszewski, G.M. (1970) The oxidation of pyrrhotite in ores and concentrates. Presentation at Austral. Inst. Mining Metall., Regional Mtg., Tennant Creek, Australia. Pp. 59–102 in: *Environmental Geochemistry of Sulfide Minewastes*. Short Course, **22**. Mineralogical Association of Canada.

Fornasiero, D., Ejit, V. and Ralston, J. (1992) An electrokinetic study of pyrite oxidation. *Coll. Surf.*, **62**, 63–73.

Fornasiero, D., Li, F., Ralston, J. and Smart, R. StC. (1994) Oxidation of galena surfaces. I X-ray photoelectron spectroscopic and dissolution kinetics studies. *J. Coll. Interf. Sci.*, **164**, 333–44.

Gill, R. (1992) *Chemical Fundamentals of Geology*. Chapman & Hall, London.

Gray, N.F. (1996) Field assessment of acid mine drainage contamination in surface and ground water. *Environ. Geol.*, **27**, 358–61.

Harrison, W. (1980) *Electronic Structure and the Properties of Solids*. W.H. Freeman & Co., San Francisco.

Hiskey, J.B. and Schlitt, W.J. (1982) Aqueous oxidation of pyrite. Pp. 55–74 in: *Interfacing Technologies in Solution Mining* (W.J. Schlitt and J.B. Hiskey, editors). Proc. 2nd SME-SPE International Soln. Mining Symposium, Denver.

Hochella, M.F. Jr. (1990) Atomic structure, microtopography, composition and reactivity of mineral surfaces.Pp. 87–132 in: *Interface Geochemistry* (M.F. Hochella, Jr. and A.F. White, editors). Reviews in Mineralogy, **23**. Mineralogical Society of America , Washington D.C.

Hochella, M.F. Jr. (1995) Mineral surfaces: their characterization and their chemical, physical and reactive nature. Pp. 17–60 in: *Mineral Surfaces* (D.J. Vaughan and R.A.D. Pattrick, editors). Mineralogical Society Series, **5**. Chapman & Hall, London.

Hochella, M.F., Jr., Eggleston, C.M., Elings, V.B., Parks, G.A., Brown, G.E. Jr., Wu, C.M. and Kjoller, K. (1989) Mineralogy in two dimensions: scanning tunnelling microscopy of semiconducting minerals with implications for geochemical reactivity. *Amer. Mineral.*, **74**, 1235–48.

Jambor, J.L. (1994) Mineralogy of sulfide-rich tailings and their oxidation products. Pp. 59–102 in: *Environmental Geochemistry of Sulfide Minewastes* (J.L. Jambor and D.W. Blowes, editors). Short Course, **22**. Mineralogical Association of Canada.

Jambor, J.L. and Blowes, D.W. (1998) Theory and applications of mineralogy in environmental studies of sulfide-bearing mine wastes. Pp. 367–402 in: *Modern Approaches to Ore and Environmental Mineralogy* (L.J. Cabri and D.J. Vaughan, editors). Short Course, **27**. Mineralogical Association of Canada.

Jones, R.A., Fox, D. and Zentilli, M. (1997) Relative rates of sulfide oxidation by chemical and microbial means: the role of mineralogy and texture in acid rock drainage from Meguma sulfide slates, Nova Scotia. *Geol. Assoc. Canad. Abst.*, **22**, p. 75.

Junta, J.L. and Hochella, M.F. Jr. (1994) Manganese (II) oxidation at mineral surfaces; a microscopic and spectroscopic study. *Geochim. Cosmochim. Acta*, **58**, 4985–99.

Kahn, A. (1983) Semiconductor surface structures. *Surface Sci. Reports*, **3**, 193–300.

Kakovsky, I. A. and Kosikov, Y.M. (1975) Study of the kinetics of oxidation of some sulfide minerals. *Obogashch. Rud.*, **20**, 18–21 (in Russian).

Kelsall, G.H., Yin, Q. Vaughan, D.J. and England, K.E.R. (1996) Electrochemical oxidation of pyrite ($FeS_2$) in acidic aqueous electrolytes. *Mineral and Metal Processing, Electrochem. Soc.*, **96**, 131–42.

Konishi, Y., Asai, S. and Sakai, H.K. (1990) Bacteria dissolution of pyrite by *Thiobacillus ferrooxidans*. *Bioprocess. Eng.*, **5**, 231–7

Kwong, Y.T.J. and Ferguson, K.D. (1990) Water chemistry and mineralogy at Mount Washington: implications to acid generation and metal leaching. Pp. 217–30 in: *Acid Mine Drainage: Designing for Closure* (J.W. Gadsby, J.A. Malick and S.J. Dat, editors). Bitech Publishers, Vancouver, British Columbia.

Laajalehto, K., Smart, R.St.C., Ralston, J. and Suoninen, E. (1993) STM and XPS investigation of reaction of galena in air. *Appl. Surf. Sci.*, **64**, 29–31.

Lin, Z. and Herbert, R.B. (1997) Heavy metal retention in secondary precipitates from a mine rock dump and underlying soil, Dalarna, Sweden. *Environ. Geol.*, **33**, 1–12.

Lowson, R.T. (1982) Aqueous oxidation of pyrite by molecular oxygen. *Chem. Rev.*, **82**, 461–97.

Luther, G.W. III (1987) Pyrite oxidation and reduction: molecular orbital theory considerations. *Geochim. Cosmochim. Acta,* **51**, 3193–9.

McKibben, M.A. and Barnes, H.L. (1986) Oxidation of pyrite in low temperature acidic solutions: rate laws and surface textures. *Geochim. Cosmochim. Acta*, **50**, 1509–20.

Monterroso, C. and Macías, F. (1998) Drainage waters affected by pyrite oxidation in a coal mine in Galicia (NW Spain): composition and mineral stability. *Sci. Tot. Environ.*, **216**, 121–32.

Moses, C.O., Nordstrom, D.K., Herman, J.S. and Mills, A.L. (1987) Aqueous pyrite oxidation by dissolved oxygen and by ferric iron. *Geochim. Cosmochim. Acta,* **51**, 1561–71.

Muir, I.J., Nesbitt, H.W. and Pratt, A.R. (1993) Surface oxidation of arsenopyrite upon exposure to air and water. *Geol. Assoc. Canada – Min. Assoc. Canada Meeting Abstr.,* p. 73.

Mycroft, J.R., Nesbitt, H.W., and Pratt, A.R. (1995) X-ray photoelectron and Auger electron spectroscopy of air oxidised pyrrhotite: distribution of oxidised species with depth. *Geochim. Cosmochim. Acta*, **59**, 721–33.

National Rivers Authority (1994) *Abandoned Mines and the Water Environment.* HMSO, London.

Nesbitt, H.W., Muir, I.J. and Pratt, A.R. (1995) Oxidation of arsenopyrite by air and air-saturated, distilled water, and implications for mechanism of oxidation. *Geochim. Cosmochim. Acta*, **59**, 1773–86.

Nesbitt, H.W., Bancroft, G.M., Pratt, A.R. and Scaini, M.J. (1998) Sulfur and iron surface states on fractured pyrite surfaces. *Amer. Mineral.*, **83**, 1067–76.

Nickel, E.H., Allchurch, P.D., Mason, M.G. and Wilmshurst, J.R. (1977) Supergene alteration at the Perseverance nickel deposit, Agnew, Western Australia. *Econ. Geol.*, **72**, 184–203.

Norris, P.R. (1990) Acidophyllic bacteria and their activity in mineral sulfide oxidation. Pp. 3–27 in: *Microbial Mineral Recovery* (H.L. Ehrlich and C.L. Brierly, editors). McGraw-Hill, New York.

Nordstrom, D.K. (1982) Aqueous pyrite oxidation and consequent formation of secondary iron minerals. Pp. 3–56 in *Acid Sulfate Weathering* (L.R. Hossaer, J.A. Kittrick and D.F. Faming, editors). Soil Society of America, Madison, WI.

Palencia, I.R., Wan, W. and Miller, J.D. (1991) The electrochemical behaviour of a semiconducting natural pyrite in the presence of bacteria. *Metall. Trans.*, **B22**, 765–74.

Pratt, A.R., Muir, I.J. and Nesbitt, H.W. (1994*a*) X-ray photoelectron and Auger electron spectroscopic studies of pyrrhotite and mechanism of air oxidation. *Geochim. Cosmochim. Acta*, **58**, 827–41.

Pratt, A.R., Nesbitt, H.W. and Muir, I.J. (1994*b*) Generation of acids in mine waste: oxidative leaching of pyrrhotite in dilute $H_2SO_4$ solutions (pH 3.0). *Geochim. Cosmochim. Acta*, **58**, 5147–59.

Richardson, S. and Vaughan, D.J. (1989) Arsenopyrite: a spectroscopic investigation of altered surfaces. *Mineral. Mag.*, **53**, 223–9.

Rinker, M.J., Besbitt, H.W. and Pratt, A.R. (1997) Marcasite oxidation in low-temperature acidic (pH 3.0) solutions: mechanism and rate laws. *Amer. Mineral.*, **82**, 900–12.

Russo, K.M., Becker, U. and Hochella, M.F. Jr. (1999*a*) Atomically resolved electronic structure of pyrite {100} surfaces: an experimental and theoretical investigation with implications for reactivity. *Amer. Mineral.*, **84**, 1535–48.

Russo, K.M., Becker, U. and Hochella, M.F. Jr. (1999*b*) The interaction of pyrite {100} surface with $O_2$ and $H_2O$: fundamental oxidation mechanisms. *Amer. Mineral.*, **84**, 1549–61.

Salomons, W. (1995) Environmental impact of metals derived from mining activities: processes, predictions, prevention. *J. Geochem. Expl.*, **52**, 5–23.

Sarveswara Rao, R.K., Das, R.P. and Ray, H.S. (1991) Study of leaching of multimetal sulphides through an interdisciplinary approach. *Mineral. Process. Extr. Metall. Rev.*, **7**, 209–33.

Schaufuss, A.G., Nesbitt, H.W., Kartio, I., Laajalehto, K., Bancroft, G.M. and Szargan, R. (1998) Reactivity of surface chemical states on fractured pyrite. *Surf. Sci.*, **411**, 321–28.

Sengupta, M. (1992) *Environmental Impacts of Mining: Monitoring, Restoration and Control.* Lewis Publishers, Boca Raton, FL.

Silverman, M.P. (1967) Mechanism of bacterial pyrite oxidation. *J. Bacteriol.*, **94 (4)**, 1046–51.

Singer, P.C. and Stumm, W. (1970) Acid mine drainage: rate determining step. *Science*, **167**, 1121–3.

Torma, A.E. (1988) Leaching of metals. *Biotechnology*, **6B**, 367–90. VCH Verlagsgesellschaft, Weinheim, Germany.

Tossell, J.A. and Vaughan, D.J. (1993) Bisulfide complexes of zinc and cadmium in aqueous solution: calculation of structure, stability, vibrational and NMR spectra and of speciation on sulfide mineral surfaces. *Geochim. Cosmochim. Acta*, **57**, 1935–45.

Vaughan, D.J. and Craig, J.R. (1978) *Mineral Chemistry of Metal Sulfides*. Cambridge University Press.

Vaughan, D.J., England, K.E.R., Kelsall, G.H. and Yin, Q. (1995) Electrochemical oxidation of chalcopyrite ($CuFeS_2$) and the related metal-enriched derivatives $Cu_4Fe_5S_8$, $Cu_9Fe_9S_{16}$ and $Cu_9Fe_8S_{16}$. *Amer. Mineral.*, **80**, 725–73.

Vaughan, D.J., Becker, U., Wright, K. (1997) Sulfide mineral surfaces: theory and experiment. *Int. J. Mineral Proc.*, **51**, 1–14.

Vear, A. and Curtis, C. (1981) A quantitative evaluation of pyrite weathering. *Earth Surface Process. Landforms*, **6**, 191–8.

Wakao, N., Mishina, M., Sakurai, Y. and Shiota, H. (1982) Bacterial pyrite oxidation I. The effect of the pure and mixed cultures of *Thiobacillus ferrooxidans* and *Thiobacillus thiooxidans* on release of iron. *J. Gen. Appl. Microbiol.*, **28**, 331–43.

Wakao, N., Mishina, M., Sakurai, Y. and Shiota, H. (1983) Bacterial pyrite oxidation II. The effect of various organic substances on release of iron from pyrite by *Thiobacillus ferrooxidans*. *J. Gen. Appl. Microbiol.*, **29**, 177–85.

Wiersma, C.L. and Rimstidt, J.D. (1984) Rates of reaction of pyrite and marcasite with ferric iron at pH 2. *Geochim. Cosmochim. Acta,* **48**, 85–92.

Wright, K., Watson, G.W., Parker, S.C. and Vaughan, D.J. (1998) Simulation of the structure and stability of sphalerite (ZnS) surfaces. *Amer. Mineral.*, **83**, 141–6.

Wright, K., Hillier, I.H., Vaughan, D.J. and Vincent, M. (1999) Cluster models of the dissociation of water on the surface of galena (PbS). *Chem. Phys. Lett.*, **299** 527–31.

Yin, Q., Kelsall, G.H., Vaughan D.J. and England, K.E.R. (1995) Atmospheric and electrochemical oxidation of the surface of chalcopyrite ($CuFeS_2$). *Geochim. Cosmochim. Acta,* **59**, 1091–100.

CHAPTER EIGHT

# The relationship of mineralogy to acid- and neutralization-potential values in ARD

J. L. JAMBOR

*Leslie Research and Consulting, 316 Rosehill Wynd, Tsawwassen, British Columbia, Canada V4M 3L9 (E-mail: jlj@wimsey.com)*

**ABSTRACT**

Static tests are the most widely used method to assess the net acid-generating potential of rocks at prospective mine sites. The results influence mine plans in which the safe disposal of wastes, such as tailings and waste rocks, is an environmental requirement in most jurisdictions. Static tests are chemical tests that are intended to provide predictions of the extent to which the minerals in representative samples will react to produce acidity, or neutralize acidity, during weathering in potential ARD (acid rock drainage) scenarios. Weathering, however, is not an instantaneous process that affects all minerals equally. Consequently, if static tests are to be interpreted meaningfully, the rates of weathering of the common gangue minerals need to be taken into account. Although the amount of neutralization contributed by an individual mineral during the vigorous reactions in most laboratory static tests is an important measure, it is commonly assumed that this amount of neutralization is directly correlative with that accessible during natural weathering. Static tests by definition exclude mineral-dissolution kinetics, but these vectors are intrinsic to the interpretation and application of the results of static tests. Experimental dissolution rates of silicate and aluminosilicate minerals are rapid relative to normal weathering rates, but most of these minerals react slowly relative to the rapidity at which acid is generated by the oxidation of iron sulphides. Most silicates or aluminosilicates therefore contribute to the attenuation of acidity only after ARD has already been established. For environmental assessments, it is suggested that carbonate contents (calcite and dolomite, but with siderite excluded) provide a more realistic assessment of whether rocks have neutralization potential adequate for the prevention of ARD.

## 8.1 Introduction

In environmental assessments of potential mines and in the disposal of mine wastes, there is currently a significant dependency on the results from determinations of acid-producing potential and neutralization-potential values. These values are intended to provide an indication of the behaviour of a rock during weathering. Hence, the tests are commonly referred to as an acid–base

Jambor, J.L. (2000) The relationship of mineralogy to acid- and neutralization-potential values in ARD. Pp. 141–159 in: *Environmental Mineralogy: Microbial Interactions, Anthropogenic Influences, Contaminated Land and Waste Management* (J.D. Cotter-Howells, L.S. Campbell, E. Valsami-Jones and M. Batchelder, editors). Mineralogical Society Series, **9**. Mineralogical Society, London. ISBN 0 903056 20 8.

account. The fundamental answers sought in these determinations are whether the effects of natural weathering will result in the minerals of a specific rock sample oxidizing or otherwise altering to produce acid or base. The answers are commonly sought by means of relatively simple chemical tests that involve no direct mineralogical input. The results of these tests can have a profound impact on decisions that affect the environment, including predictions of post-mining water quality (Skousen *et al.*, 1997). Although the tests are commonly stated to be no more than an initial screening process prior to more in-depth studies, the screening process is itself a decision point. Thus, the results do influence mine planning with respect to the permitting process and to waste-disposal options, including restoration by topsoils or acceptable substitutes at sites disturbed by surface mining. Considerable effort has therefore been expended to evaluate existing procedures and to develop alternative tests. The focus in this paper is not on these alternatives, all of which by bulk methods attempt to probe by chemical means the mineralogical behaviour of a sample during weathering; rather, the focus is on the common gangue minerals in mineral deposits, and how these minerals might behave in the acidic conditions that typify acid rock drainage (ARD) or, alternatively, acid mine drainage (AMD). Acid rock drainage is generally considered to exist if aqueous effluents have a pH of 5–5½ or less.

## 8.2 Static and kinetic tests

Whether a rock has the potential to be acid generating or, conversely, has the capacity to consume acidity, is determined by static and kinetic tests. Static tests represent an attempt to measure the acid-producing potential or neutralization potential of a rock, or sample, but the measure does not include consideration of the rate at which 'weathering' reactions occur. Kinetic tests are used to simulate the rate factor, typically by monitoring the amount of acidity and ion release to leachates that ensue from numerous cycles of wetting and drying a crushed sample. Most kinetic tests are conducted in controlled conditions in a laboratory, using an apparatus such as a column or a humidity cell that holds a small weight of sample, but tests involving large samples exposed to field conditions are also carried out. Both in static and in kinetic tests, the approach is chemical, and in few jurisdictions is it mandatory to have detailed mineralogical work done on the samples. The consequence is that little is known about the mineral alterations or consumptions that take place in these tests; reactions are inferred, but are rarely confirmed. This lack of confirmation is a concern especially where silicate-mineral reactions are thought to make an important contribution to the results obtained in static tests. The procedures used in static tests are therefore reviewed briefly, with the objective of relating the results to mineralogy. Several alternative methods for determining acid potential and neutralization potential have been proposed. Detailed analytical procedures for the most commonly used methods are given in Sobek *et al.* (1978) and MEND (1991*a,b*), and alternative methods are discussed by O'Shay *et al.* (1990), Lapakko (1994), Miller *et al.* (1994), Skousen *et al.* (1997), and many others. The performance of the most commonly used methods has been tested and compared most recently by Adam *et al.* (1997), Lawrence and Wang (1996, 1997), Skousen *et al.* (1997), White *et al.* (1999), and Jambor *et al.* (2000).

## 8.3 Acid-producing potential

The acid-producing potential (AP) of a sample is determined by measuring its S content. The theoretical basis for an AP number is the assumption that all of the acid-producing potential in a sample is present as $FeS_2$ (pyrite), all of which is oxidized by the reaction

$$FeS_2 + 15/4O_2 + 7/2H_2O \rightarrow Fe(OH)_3 + 2SO_4^{2-} + 4H^+ \quad (8.1)$$

Thus the equivalency is $2H^+$ for each S, and the $2H^+$ can be neutralized by the reaction

$$CaCO_3 + 2H^+ \rightarrow Ca^{2+} + CO_2 + H_2O \quad (8.2)$$

Ideally, therefore, one mole of S is equivalent to one mole of $CaCO_3$. As AP numbers are traditionally expressed as kg of $CaCO_3$ per tonne of material:

$$AP = \left(\frac{x}{100}\right) \times 1000\ \text{kg} \times \left(\frac{\text{molecular wt. } CaCO_3}{\text{atomic wt. S}}\right) \quad (8.3)$$

$$AP = S\% \times 31.25\ \text{kg}\ CaCO_3/\text{t} \quad (8.4)$$

where $\frac{x}{100}$ is the percentage of S, and the atomic weight of sulphur and the mole weight of $CaCO_3$ have been taken as 32 and 100, respectively (i.e. 100/32 = 3.125). The numerous assumptions in equations 8.1 to 8.4 are outlined in Morin (1990) and MEND (1991*a*). Although the use of $S_{total}$ apparently works well in some jurisdictions, in others the practice is to do S speciation so that sulphates are distinguished and eliminated, with $S_{sulphide}$ alone used to calculate the AP number.

More than 99% of sulphide-related acid generation from mineral deposit wastes worldwide is from the oxidation of pyrite and pyrrhotite. Although other sulphide minerals can generate acidity (e.g. Jennings *et al.*, 2000), other than in extremely rare cases the principal importance of other sulphide, arsenide and antimonide minerals is that their dissolution contributes metals and semi-metals to ARD, and these dissolved species are commonly of greater environmental concern than the associated acidity. Whether pyrite or pyrrhotite, or both, are present is not discriminated in static tests, but the distinction is important in terms of the rate of acid generation. Pyrrhotite in waste settings has been observed to alter much more rapidly than pyrite (Blowes and Jambor, 1990; Jambor, 1994; Jambor and Blowes, 1998), and in laboratory studies the reaction rates of hexagonal pyrrhotite have been found to be up to 100 times faster than those of pyrite at 25°C and atmospheric oxygen concentrations (Nicholson, 1994; Nicholson and Scharer, 1994). Unless carbonate minerals are associated with a pyrrhotite-dominant assemblage, ARD will be generated because, regardless of the static-test neutralization potential values, it is unlikely that any of the common silicate gangue and aluminosilicate minerals react rapidly enough to prevent acidity.

## 8.4 Sulphide resistance and persistence

The rate of alteration of sulphide minerals in mine-waste settings is affected by what may be termed the 'persistence' factor. For example, Emmons (1917) noted, from his field observations of oxidized mineral deposits, that galena and enargite persisted as relicts after all other sulphides had been destroyed. He also observed that these relict sulphides persisted because they occurred as cores within anglesite and arsenic oxide, respectively. The lower aqueous solubility of the oxidized rims had clearly impeded alteration of the core sulphides. Thus Emmons' (1917) order of sulphide reactivity contrasted markedly with the order obtained in laboratory studies (Table 8.1).

In mine-waste settings, the two principal wastes are tailings and waste rock (spoils). The two differ substantially in that waste-rock piles are more heterogeneous in terms of local mineral associations, are more varied in particle size and in the non-comminuted grain sizes of the individual minerals, and are more diverse insofar as none of the ore minerals has been removed. These properties increase the chance of forming a diverse suite of oxidation products during weathering, and increase the possibility that galvanic interactions (Sato, 1992; Kwong, 1995) may influence reactivity rates among the sulphides.

Tailings represent a much more homogeneous medium, one from which the majority of the ore minerals has been removed. In most sulfide-bearing tailings, the observed general order of increasing resistance to alteration was

TABLE 8.1 Comparison of field and laboratory reaction rates of sulphide minerals.

| Reactivity | Field (Emmons, 1917) | Laboratory reaction with Fe(III) (Emmons, 1917)[1] | | (Rimstidt *et al.*, 1994)[2] | |
|---|---|---|---|---|---|
| High | Sphalerite | Pyrrhotite | 4.3 | | |
| | Pyrrhotite | Arsenopyrite | 1.7 | Arsenopyrite | $1.7 \times 10^{-6}$ |
| | Chalcopyrite | Galena | 1.5 | Galena | $1.6 \times 10^{-6}$ |
| | Pyrite | Enargite | 1.1 | Pyrite | $2.7 \times 10^{-7}$ |
| | Galena | Marcasite | 0.9 | Marcasite | $1.5 \times 10^{-7}$ |
| | Enargite | Pyrite | 0.8 | Sphalerite | $7.0 \times 10^{-8}$ |
| | | Sphalerite | 0.5 | Chalcopyrite | $9.6 \times 10^{-9}$ |
| Low | | Chalcopyrite | 0.3 | Covellite | $9.1 \times 10^{-9}$ |

[1] $m_{Fe^{3+}} = 10^{-1}$; numbers refer to relative rates (Rimstidt *et al.*, 1994)
[2] rate in mol $m^{-2}$ $s^{-1}$; $m_{Fe^{3+}} = 10^{-3}$, pH = 2.5 at 25°C

suggested (Jambor, 1994) to be pyrrhotite → galena-sphalerite → pyrite-arsenopyrite → chalcopyrite → magnetite. On the basis of recent additional observations, the order is suggested to be pyrrhotite → galena → sphalerite → bornite → arsenopyrite → pyrite → chalcopyrite → magnetite. There is no 'right' or 'wrong' to this order (see also the listing in Plumlee, 1999), which does not follow the laboratory-determined data in Table 8.1. What is needed are more observations, accompanied by compositional data, of oxidizing sulphides in tailings impoundments to gain insights as to why the differences occur. Thus, for example, galena is now recognized to have a low resistance to oxidation, but has a high persistence in some non-laboratory settings. In a further example, the possibility that the substantial difference between field *vs.* laboratory reactivity of sphalerite may be related to the mineral's Fe or trace-element contents requires investigation.

## 8.5 Neutralization potential

The most widely used method for determining the neutralizing potential (NP) of mine-waste samples is that of Sobek *et al.* (1978). A sample is tested with HCl to determine the intensity of effervescence, or 'fizz rating' (nil, slight, moderate or strong), and the rating governs the amount and concentration of HCl added to react with the sample, which is then heated to almost boiling, diluted, briefly boiled, and subsequently titrated with NaOH to determine the amount of acid that has been consumed. The vigorous reaction in the acidification stage typically consumes some silicate minerals (Sherlock *et al.*, 1995; Jambor *et al.*, 2000), thus resulting in NP values that are high relative to those obtained at lower temperatures. To temper the high-temperature digestion, one of several proposed modified methods (commonly referred to as the B.C. Modified Technique and distinguished here as the Lawrence method, after the principal protagonist, R.W. Lawrence) has been extensively tested in British Columbia (Lawrence and Wang, 1996, 1997) even though the Sobek method is recommended in that jurisdiction's regulatory guidelines (Price *et al.*, 1997). In the Lawrence method, the HCl digestion is for 24 h at ambient temperatures (25–35°C) and the back titration is to pH 8.3 rather than 7.

Lawrence and Wang (1997) observed that some Sobek NP values are substantially higher than Lawrence NP values for the same sample (Fig. 8.1*a*). The latter values are higher, but closer to, carbonate NP values obtained by determining the inorganic C content of the sample, and calculating the C as $CaCO_3$ (Fig. 8.1*b*). The consensus seems to be that, for sulphide-bearing wastes from metalliferous ore deposits, the Sobek test overestimates what is termed 'available NP' or 'readily accessible NP'; thus, there is a tacit inclination to compensate for this presumed 'overestimation' by incorporating a kinetic factor into the results of the static tests.

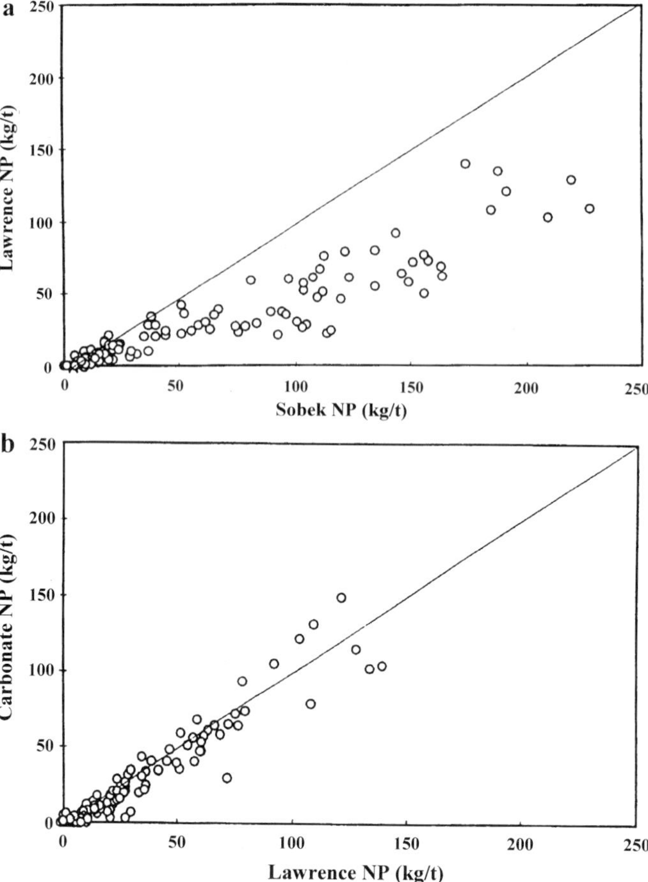


FIG. 8.1 Comparison of NP values for waste-rock and tailings samples associated with metalliferous deposits. (*a*) Sobek NP *vs*. Lawrence NP; (*b*) Lawrence NP *vs*. carbonate NP calculated from inorganic C. Data are from Lawrence and Wang (1996). Scales are kg of $CaCO_3$ equivalent per tonne of material.

## 8.6 Weathering rates of minerals

The recognition that kinetic factors must be taken into consideration if NP values, whether Lawrence or Sobek, are to be meaningful in real-life scenarios for metalliferous ore-deposit wastes has led to an examination of the weathering rates of silicate minerals so that these rates can be incorporated to adjust NP values. Thus Kwong (1993) assigned numerical values to the relative rates of reactivity in Sverdrup's (1990) six mineral groups as shown in Table 8.2. Kwong's (1993) rate values have been used in recent studies (Li, 1997; Lawrence and Scheske, 1997) to adjust NP values to ones thought to reflect more realistically the NP available during the weathering of sulphide-bearing mine wastes. Kwong's (1993) rate values have also been used in normative calculations in an attempt to adjust measured NP to 'effective NP' (Lawrence and Scheske, 1997). The use of normative calculations, however, is not recommended here because: (1) the net effect is to add to the reliance on chemical analyses independent of mineralogical observations, and (2) the introduction of normative minerals represents a step away from, rather than closer to, the hydrous assemblages that are characteristic of ARD alteration mineralogy. Paktunc (1999) has also criticized such normative usage as having "only tenuous relevance to mineralogy" in terms of neutralization potential.

The observation by Goldich (1938) that the principal rock-forming silicates are increasingly resistant to weathering as one progresses down Bowen's (1928) reaction series for magmatic minerals has more or less stood the test of time. Modern studies to quantify the rates of silicate-mineral weathering have mainly involved controlled laboratory experiments on the dissolution rates of individual minerals, as well as quantification of weathering rates in soils. The results provide useful insights for the interpretation of NP values.

Table 8.3 shows the relative rates of mineral weathering in rocks (Ollier, 1984), soil (Allen and Hajec, 1989), as well as the relative order used in modelling soil weathering rates by the computer program PROFILE (Sverdrup and Warfinge, 1993, 1995; Hodson *et al.*, 1997). Results from laboratory-determined dissolution rates under far-from-equilibrium conditions at pH 5 are shown in Tables 8.4 and 8.5. Table 8.4 is a slightly abbreviated version of that in Lasaga and Berner (1998); the rate for near-end-member anorthite, based on data by Fleer (1982) as noted in Lasaga (1984) and Lasaga *et al.* (1994), has been omitted because it is considered to be too high relative to rates determined for other Ca-rich plagioclase. Moreover, Stillings and Brantley (1995), for example, observed at pH 3 a consistent, straight-line increase in dissolution rates of the plagioclase feldspars as An content increased to, and including, $An_{76}$. Even if rates increase at Ca contents higher than $An_{76}$ (see the discussion in Blum and Stillings, 1995), this is of little consequence to most potential ARD settings because few occurrences of plagioclase with compositions exceeding $An_{76}$ are known in sulphide-rich deposits.

Solid solutions are prevalent in many of the rock-forming minerals, and the results from NP tests of a series of mineral standards may not be precisely transferable to NP results for a specific mine site. The positions of garnet and apatite in the weathering series of Tables 8.2 and 8.3 exemplify some of the difficulties in attempting to develop a universally applicable weathering order. Both minerals are of considerable importance in the study of weathering rates, especially in soils and in soil development, but neither mineral plays a prominent role in ARD related to metalliferous deposits. Although garnet is abundant in many skarn deposits, these typically also contain abundant carbonates.

Dryden and Dryden (1946) observed that weathering of garnet from schists in eastern U.S.A. was more rapid than that of hornblende, and Sverdrup (1990),

TABLE 8.2 Grouping of minerals according to their acid-neutralizing capacity[1].

| Group name | Typical minerals | Relative reactivity |
|---|---|---|
| 1. Dissolving | calcite, aragonite, dolomite, magnesite, brucite | 1.0 |
| 2. Fast weathering | anorthite, nepheline, olivine, garnet, jadeite, leucite, spodumene, diopside, wollastonite | 0.6 |
| 3. Intermediate weathering | epidote, zoisite, enstatite, 'hypersthene', augite, hedenbergite, hornblende, glaucophane, tremolite, actinolite, anthophyllite, serpentine, chrysotile, talc, chlorite, biotite | 0.4 |
| 4. Slow weathering | albite, 'oligoclase', 'labradorite', vermiculite, montmorillonite, gibbsite, kaolinite | 0.02 |
| 5. Very slow weathering | K-feldspar, muscovite | 0.01 |
| 6. Inert | quartz, rutile, zircon | 0.004 |

[1] After Sverdrup (1990) and Kwong (1993). Relative reactivity on the basis of 100% mineral abundance, i.e. a monomineralic soil. For 'oligoclase', read albite $Ab_{90-70}$, and for 'labradorite' read anorthite $Ab_{50-30}$. Note the several recent adjustments for various minerals as given in Table 8.3

Tejan-kella *et al.* (1991) and others placed garnet within the fast-weathering group of minerals (Table 8.2). Pettijohn (1941), however, observed garnet to be a persistent mineral in geologically older sediments; persistence, which reflects both physical and chemical stability, was ranked in the order olivine → augite → hornblende → apatite → biotite → garnet, i.e. Goldich's (1938) trend, with apatite and garnet added. Comparison with the ranking in Table 8.3 shows the considerable variation in the placement of garnet in the weathering series. In recent studies of soil weathering rates (Sverdrup and Warfinge, 1995; Hodson *et al.*, 1997; Läng, 2000), garnet has been shifted from the 'fast-weathering' group to the 'intermediate' group. Part of the difficulty in attempting to classify garnet in this manner is that garnet is a group name currently applicable to 15 mineral species. Although only a few are common rock-forming minerals, little is known about the effects that their solid solutions have with regard to weathering resistance. Velbel (1984, 1993) reported that individual species may behave quite differently during weathering. In ARD conditions, it would be expected that grossular would be more susceptible than pyrope to dissolution, but if so, the magnitude of the difference is not known (see also the discussion in Velbel, 1999). Apatite is another example of how compositions apparently can have a profound effect on weathering rates; hydroxylapatite is reported to be unstable under leaching conditions, whereas fluorapatite is relatively more stable and can persist as a detrital mineral (Loughnan, 1969).

From a practical viewpoint, however, the primary concern in ARD-related studies is the behaviour of the minerals in the lower parts of Bowen's reaction

TABLE 8.3 Rates of mineral weathering.

| Rock weathering[1] | Soil weathering[2] | PROFILE[3] |
|---|---|---|
| Olivine | | Calcite |
| Actinolite | Apatite | Apatite |
| Diopside | Biotite | Chlorite |
| Enstatite | Pyroxene | Biotite |
| Augite | Chlorite | Anorthite |
| Zoisite | Hornblende | Garnet |
| Hornblende | Ca plagioclase | Hornblende |
| Epidote | Na plagioclase | Pyroxene |
| Magnetite | K-feldspar | Plagioclase $Ab_{90-70}$ |
| Ilmenite | Muscovite | Plagioclase $Ab_{100-90}$ |
| Apatite | Epidote | Epidote |
| Biotite | Garnet | K-feldspar |
| Garnet | | Muscovite |
| Muscovite | | Vermiculite |

[1] Abridged from Ollier (1984)
[2] Allen and Hajec (1989) in Hodson *et al.* (1997)
[3] PROFILE version 3.2 (Hodson *et al.*, 1997)

series, and the behaviour of the sulphide-associated gangue minerals such as the carbonates, the feldspar, chlorite and mica groups, talc, stilpnomelane, and for coal deposits, kaolinite and smectite. For Mg-Fe solid solutions in Bowen's series, higher Mg contents have been observed to decrease the stability of olivine and orthopyroxene (Gílason and Anórsson, 1993), and experimental data summarized in Brantley and Chen (1994) also indicate a higher dissolution rate for enstatite than for ferroan enstatite. In laboratory experiments by Kalinowski and Schweda (1996), phlogopite ($Mg \gg Fe$) showed K/Si release and a greater susceptibility to vermiculitization than did 'biotite' ($Mg > Fe$; actually ferroan phlogopite). This Mg-Fe relationship is not

TABLE 8.4 Mean lifetime of a 1 mm crystal at 25°C and pH 5.

| Mineral | Lifetime (year) | Mineral | Lifetime (year) |
|---|---|---|---|
| Calcite | 0.43 | Albite | 575,000 |
| Wollastonite | 79 | Microcline | 921,000 |
| Forsterite | 2300 | Epidote | 923,000 |
| Diopside | 6800 | Muscovite | 2,600,000 |
| Enstatite | 10,100 | Kaolinite | 6,000,000 |
| Sanidine | 291,000 | Quartz | 34,000,000 |

From Lasaga and Berner (1998)

TABLE 8.5 Relative dissolution rates of non-sulphide minerals[1].

| | $\frac{\text{Rate}}{\text{Rate for calcite}} \times 10^5$ | | | $\frac{\text{Rate}}{\text{Rate for calcite}} \times 10^5$ |
|---|---|---|---|---|
| Calcite | 100,000 | Plagioclase | $An_{76}$ | 0.25 |
| Dolomite | 6,000 | | $An_{46}$ | 0.12 |
| Forsterite | 4 | | $An_{13}$ | 0.02 |
| Diopside | 1.4 | | $An_0$ | 0.02 |
| Enstatite | 1 | Sanidine | | 0.03 |
| Talc | 0 | Microcline | | 0.01 |
| Chrysotile | 0.06 | | | |
| Biotite | 0.01–0.03 | | | |
| Phlogopite | 0.02 | | | |
| Chlorite | 0.02 | | | |
| Kaolinite | 0.006–0.02 | | | |
| Muscovite | 0.006 | | | |
| Montmorillonite | 0.002 | | | |
| Quartz | 0.0005 | | | |

[1] In laboratory experiments at pH 5, far from equilibrium. Rates relative to calcite were converted from data in Drever and Clow (1995) and from Nagy (1995), using in the latter, the rates relative to muscovite.

constant, however; e.g. Wogelius and Walther (1992) determined that the dissolution rate of Fe-rich olivine is greater than that of Mg-rich olivine, and in oxygenated conditions both increases and decreases in weathering rates of olivines and pyroxenes have been reported (Hoch *et al.*, 1996). Where decreased rates of Fe release or lower rates of dissolution have been observed in oxygenated conditions, the result has been attributed to the formation of an oxidized Fe-bearing protective surface (Schott and Berner, 1985; Siever and Woodford, 1979; Chen and Brantley, 2000). In the ARD regime, the rapid oxidation of sulphides releases Fe which is subsequently hydrolysed, and Fe oxyhydroxides typically coat all minerals, including quartz. These oxyhydroxides may not have an intimate crystallographic relationship with the host mineral, but their abundance and pervasiveness suggests that the rinds are likely to impede weathering.

## 8.7 Relationships to ARD

As discussed most recently by Malmström *et al.* (2000), experimental dissolution rates determined in laboratory reactors may be up to hundreds of times faster than field-determined weathering rates. Nevertheless, the data in Tables 8.4 and 8.5 indicate the recalcitrance of the silicates and aluminosilicates relative to the dissolution rates of calcite and dolomite. Particularly informative are the rates for the feldspar minerals, as these almost invariably are assumed to contribute appreciable neutralization potential in ARD situations, and by analogy the feldspars have also been assumed to contribute to NP in both the Sobek and Lawrence methods. From the perspective of sulphide-bearing, metalliferous ore deposits, dissolution rates of albite and K-feldspar are one of the primary concerns, but it is evident from Table 8.5 that these minerals, although commonly invoked, would be unlikely to contribute significant NP either during or long after the onset of acid generation at a mine site. Where zoned plagioclase is present, moreover, the average dissolution rate would probably be skewed because of the inhibiting effect of the outermost zones, which are typically the most sodic and therefore also the least reactive. Therefore, if the dissolution rates are indicative of 'effective NP' (i.e. that which is immediately available and is readily accessible), the feldspars should be largely discounted as a neutralization source other than over the long term, many years after mine-site closure. As is indicated below, plagioclase contributes only a small amount of NP in the Sobek and Lawrence tests, and the mineral does not seem to be sufficiently reactive to prevent or significantly modify ARD in field settings, except on a geological time scale.

Examination of Tables 8.3, 8.4 and 8.5 also suggests that none of the common silicate minerals associated with most sulphide-bearing ore deposits is likely to retard, in a meaningful, practical way, the onset of acid generation. Once ARD conditions are established, typically in the pH range 3 to 5, reactions of the common silicate and aluminosilicate minerals become less clear. In laboratory experiments to determine dissolution rates, great care is taken to prevent the formation of secondary minerals because their presence can notably slow dissolution. Complete, or even substantial, armouring is not necessary to have an impact on the experimental weathering rates. In natural ARD settings, extensive coatings of Fe oxyhydroxides on silicates are the norm, thereby extending the uncertainties arising from the extrapolation of laboratory dissolution data. The trend, however, is clearly toward reduced rather than accelerated weathering rates. Thus, if the primary purpose of AP and NP tests is the determination of whether the wastes will have a net acid-generating potential during mine operation and after closure, it can be argued that the most realistic assessment is that of carbonate NP, where $NP_{carb}$ is that derived from calcite and dolomite-ankerite. Siderite and the solid-solution $Fe^{2+}$ component in other carbonates are generally deducted from $NP_{carb}$ because it has been shown that oxidation of Fe-bearing carbonates in natural, non-laboratory settings may produce neutral to slightly acid rather than alkaline solutions (Postma, 1983; Morrison *et al.*, 1990; MEND, 1991*a*). At the high temperatures utilized in the Sobek NP test, however, the reaction of dissolved siderite is initially to neutralize acidity, thereby giving overstated NP values (O'Shay *et al.*, 1990; Frisbee and Hossner, 1995; Skousen *et al.*, 1997). The details of the theoretical reactions involved in arriving at a net allocation of zero NP for siderite are given in Skousen *et al.* (1997); whether the allocation is appropriate is important in environmental studies, and the issue merits further investigation in real or realistically-simulated field conditions.

Iron sulphides oxidize rapidly, but numerous field observations have confirmed that the resulting acidity can be neutralized by the associated carbonates. In the absence of carbonates, however, ARD conditions almost inevitably ensue. In British Columbia, environmental guidelines pertaining to mine wastes suggest that, in the initial screening, $S_{sulphide}$ = 0.3 wt.% may mark the boundary between potentially acid-generating wastes and those expected not to be acid-generating (Price *et al.*, 1997; Soregaroli and Lawrence, 1997). Unless carbonate is present, however, our experience is that 0.3% S is slightly too high a limit for most rocks associated with metalliferous deposits (highly mafic to ultramafic rocks may be an exception, but specific data are not available for these).

It is remarkable that NP determinations have been done on tens of thousands, and perhaps hundreds of thousands of samples, but until recently (White *et al.*, 1999; Jambor *et al.*, 2000) individual mineral species had not been tested to determine their reactivity, and hence NP contribution, to standard static tests. For example, the answer to whether albite ($An_0$) or calcic albite contributed NP under the conditions of the Sobek and Lawrence tests, and to what extent in one test *vs.* the other, was not known. That silicate–aluminosilicate alteration does occur in ARD conditions is not questioned; kinetic tests of crushed samples typically show an initial rapid release of ions prior to attainment of steady-state conditions, presumably because of the initial presence of fine-grained (<1 μm) particles produced during the crushing, and the presence of fresh surfaces. Further evidence of aluminosilicate dissolution is the high pore-water contents of characteristic elements, such as Al, in the vadose zone of sulphide-bearing tailings impoundments. Even where the sulphide contents are relatively low, the commonly detected presence of jarosite attests to aluminosilicate alteration to provide the K that is typically the predominant *A*-site cation in the accompanying precipitates of secondary jarosite. Aluminosilicate reactions may therefore help in attenuating acidic conditions, but sulphide reactions are too rapid for aluminosilicates to prevent the development of acidic conditions.

To date, mineralogical studies of field sites have been focused mainly to establish what happens to the primary sulphide minerals, and to determine the nature and distribution of the secondary minerals that have formed as a consequence of sulphide oxidation. Silicate–aluminosilicate alteration is a much less conspicuous feature at these sites, at which the maximum age of weathering for most has been ~50 years in a temperate climate. Nevertheless, at the sulphide-rich tailings impoundment of the Heath Steele mine, New Brunswick, in which the gangue assemblage includes muscovite, chlorite, talc, stilpnomelane and albite, the chlorite showed X-ray diffraction evidence of having been depleted in the low-pH zone (Jambor *et al.*, 1992). At Sudbury, Canada, the Nickel Rim and Copper Cliff tailings impoundments, and tailings in 180 tonne outdoor lysimeters, all show that biotite-phlogopite in the low-pH vadose zone has undergone *in situ* alteration to hydrobiotite, a mixed-layer phyllosilicate that contains vermiculite interlayers (Jambor and Owens, 1993; Jambor *et al.*, 1999; McGregor *et al.*, 1998; Shaw *et al.*, 1998; Johnson *et al.*, 2000). Alteration in the lysimeters occurred within a three-year exposure to oxidation. In these tailings, and in others, trioctahedral mica typically provides the source of K for secondary jarosite precipitates. These results may seem to be at variance with the low dissolution rates determined for some of the phyllosilicates in laboratory experiments (Table 8.5), but the results are in accord with the rapidity of K release. Compositional variations of the minerals may also play a role, and for chlorite it has been reported that higher Fe contents increase the dissolution rate by up to three orders of magnitude relative to the rate in Table 8.5 (May *et al.*, 1995, in Nagy, 1995). In acidic conditions near room temperature, biotite characteristically undergoes a large and rapid release in K (Acker and Bricker, 1992; Malström and Banwart, 1997), which signals the onset of alteration even though surface area and colour do not change. Rapid alteration of biotite-phlogopite was detected in dissolution experiments by Kalinowski and Schweda (1996) and Murakami and Utsunomiya (1998). Although Murphy *et al.* (1998) calculated rates of biotite weathering that are significantly slower than laboratory rates, indications are that the values for biotite and phlogopite in Table 8.5 do not reflect the reactivity of the mineral in ARD settings because of the importance of incongruent cation release. Thus, the rate of biotite weathering in ARD is more in accord with that reported for weathering in soils (Table 8.3).

Table 8.6, which is condensed from Jambor *et al.* (2000), illustrates the magnitude of the NP values that are obtained from some common rock-forming minerals. The NP values were obtained using standard Sobek and Lawrence test protocols. Bearing in mind that the NP value for calcite is 1000 tonnes of $CaCO_3$ equivalent per tonne of mineral, and that each of the tested minerals was, or approximated, 100% of each test sample, it is evident that the silicates and aluminosilicates do not contribute large amounts of NP. The specifications for maximum grain size differ in the Sobek and Lawrence tests, and much of the variation between the test results for the two methods is related to grain size (Jambor *et al.*, 2000).

It is evident from Table 8.6 that, relative to calcite, most granitoid rocks have little inherent neutralization potential in terms of attenuating ARD. White *et al.* (1999) observed that fresh granitoid rocks commonly contain accessory calcite in amounts up to 3 g/kg. This amount, however, would be insufficient to prevent ARD if the rocks also contained appreciable amounts of sulphides. One of the previously mentioned 180 tonne lysimeters at Sudbury contains tailings with 0.35 wt.% $S_{sulphide}$ (Shaw *et al.* 1998; Jambor *et al.*, 1999); although the tailings are acid-generating, the experiment has shown that it may take several years before wastes with these low sulphide contents eventually yield acidic effluents. At slightly lower sulphide contents, the acid-generating sites are increasingly widely disseminated rather than pervasive; the attainment of overall acidity for a large mass may be slow enough that reactions on a

TABLE 8.6 Results of NP tests on individual minerals[1].

| Mineral | Sobek NP | Lawrence NP | Mineral | Sobek NP | Lawrence NP |
|---|---|---|---|---|---|
| Feldspar group | | | Mica group | | |
| Anorthite $An_{93}$ | 10.7 | 10.5 | Muscovite | 0.3 | 1.7 |
| Anorthite $An_{50}$ | 2.6 | 3.9 | Phlogopite | 2.7 | 3.9 |
| Albite $Ab_{80}$ | 0.6 | 0.5 | Phlogopite | 8.5 | 10.3 |
| Albite $Ab_{98}$ | 0.5 | – | | | |
| K-feldspar | 1.4 | 1.3 | Chlorite group | | |
| K-feldspar | 1.0 | 0.8 | Clinochlore | 10.3 | 14.3 |
| K-feldspar | 1.3 | 3.3 | Clinochlore | 0.8 | – |
| Microcline | 0.5 | 2.8 | | | |
| | | | Talc, serpentine | | |
| Pyroxene group | | | Antigorite | 15.1 | 25.8 |
| Enstatite | 3.2 | 25.8 | Lizardite | 16.1 | 37.3 |
| Diopside | 4.5 | 5.6 | Talc | 1.7 | 3.1 |
| Hedenbergite | 6.6 | 11.7 | | | |
| Augite | 4.6 | 23.0 | Clay minerals | | |
| | | | Kaolinite | 0.0 | 0.0 |
| Amphibole group | | | Montmorillonite | 13.8 | 13.6 |
| Tremolite | 5.2 | 3.8 | | | |
| Pargasite | 4.4 | 17.5 | Olivine (forsterite) | 23.9 | 42.0 |

[1] Values for Sobek and Lawrence tests are in tonnes of $CaCO_3$ equivalent per tonne of mineral
Repetition of a mineral name, such as K-feldspar, indicates that samples from different localities were tested

micro-environmental scale become significant, thereby allowing the silicate–aluminosilicate minerals to interact at the disseminated acid-generating sites. Such a scenario would be marked by eventual elevation of the concentrations of Si and Al in the pore waters. An alternative possibility, for which such a pore-water signature would be absent, would be the frequent periodic flushing (related to local climatic conditions) of the soluble sulphates that typically develop in the initial stages of sulphide oxidation, thereby lessening the cumulative buildup of acidity and also diluting the overall sulphate content of the frequently flushed effluents. The question of what happens in low-sulphide mine wastes is environmentally important, but it is a question for which we do not yet have the answers.

## 8.8 Conclusions

Although the role of static tests has been downplayed in environmental assessments in some jurisdictions, these tests nevertheless are the most widely employed initial method of assessing the potential for ARD. Thousands of static tests are performed annually, and with each result there is an inference as to how the sample will react when weathered. Carbonate minerals are an accessible and effective source for neutralizing the acid in ARD. In the absence of carbonates, however, recourse is made to inferred silicate and aluminosilicate consumption and attendant neutralization capacity. Rarely will these siliceous minerals prevent the onset of ARD in rocks containing >0.3 wt.% S, especially if the S is present as pyrrhotite, because the rate of sulphide oxidation is more rapid than silicate-mineral alteration/dissolution. In other words, silicate minerals in most wastes are ineffective attenuators of acidity until ARD has already been established. Silicate minerals are not inert, and they do contribute to attenuation of the acidity of solutions, thereby tempering the overall ARD problem over the long term. Although feldspars are the most abundant labile minerals in natural weathering, they do not seem to play a significant role in ameliorating ARD in the short terms applicable to mined wastes. For many of the common silicate and aluminosilicate gangue minerals, and especially their solid-solution variants, the scarcity of data on how much neutralization each contributes to standard NP tests has been a huge gap in our knowledge about acid–base accounting and its extrapolation to environmental scenarios.

Adoption of carbonate NP as the standard NP measure for metalliferous deposits would have significant repercussions for the mining industry. Many of the rocks currently categorized as ‘possibly’ or ‘likely’ acid generators would shift to the acid-generating category, thus adding to the costs to ensure an environmentally benign disposal of these wastes. There is clearly a need for increased mineralogical research pertaining to the interpretation of the results from static tests, as well as kinetic tests, and both the mining industry and regulators should at least heed the questions that have been raised.

**Acknowledgements**

This paper was presented in January 1999 at Aberdeen, Scotland, as the Hallimond Lecture of the Winter Meeting of the Mineralogical Society of Great Britain & Ireland. The original manuscript has been updated by incorporating references to papers published in 1999 and 2000, and by the addition of Table 8.6 *et sequitur*, but little else has been changed. In the original Acknowledgements the Society was thanked “for the opportunity to present this review and its somewhat controversial opinions and conclusions”. The opinions and conclusions now seem less controversial than when originally expressed, but this may be due to familiarity or delusion. I thank W.A. Price of the British Columbia Ministry of Energy and Mines, D.W. Blowes of the University of Waterloo, two anonymous referees, and Editor Linda Campbell for numerous helpful suggestions.

## References


Acker, J.G. and Bricker, O.P. (1992) The influence of pH on biotite dissolution and alteration kinetics at low temperature. *Geochim. Cosmochim. Acta*, **56**, 3073–92.

Adam, K., Kourtis, A., Gazea, B. and Kontopoulos, A. (1997) Evaluation of static tests used to predict the potential for acid drainage generation at sulphide mine sites. *Trans. Inst. Mining Metall.*, **106**, A1–8.

Allen, B.I. and Hajeck, B.F. (1989) Mineral occurrences in soil environments. Pp. 199–278 in: *Minerals in Soil Environments* (J.B. Dixon and S.B. Weed, editors). Soil Science Society of America, Madison, WI.

Blowes, D.W. and Jambor, J.L. (1990) The pore-water geochemistry and the mineralogy of the vadose zone of sulfide tailings, Waite Amulet, Québec, Canada. *Appl. Geochem.*, **5**, 327–46.

Blum, A.E. and Stillings, L.L. (1995) Feldspar dissolution kinetics. Pp. 291–351 in: *Chemical Weathering Rates of Silicate Minerals* (A.F. White and S.L. Brantley, editors). Reviews in Mineralogy, **31**. Mineralogical Society of America, Washington D.C.

Bowen, N.L. (1928) *The Evolution of the Igneous Rocks*. Dover Publications Inc., New York.

Brantley, S.L. and Chen, Y. (1994) Chemical weathering rates of pyroxenes and amphiboles. Pp. 119–72 in: *Chemical Weathering Rates of Silicate Minerals* (A.F. White and S.L. Brantley, editors). Reviews in Mineralogy, **31**. Mineralogical Society of America, Washington D.C.

Chen, Y. and Brantley, S.L. (2000) Dissolution of forsteritic olivine at 65°C and $2 < pH < 5$. *Chem. Geol.*, **165**, 267–81.

Drever, J.I. and Clow, D.W. (1995) Weathering rates in catchments. Pp. 463–83 in: *Chemical Weathering Rates of Silicate Minerals* (A.F. White and S.L. Brantley, editors). Reviews in Mineralogy, **31**. Mineralogical Society of America, Washington D.C.

Dryden, L. and Dryden, C. (1946) Comparative rates of weathering of some common heavy minerals. *J. Sed. Petrol.*, **16**, 91–6.

Emmons, W.H. (1917) The enrichment of ore deposits. *U.S. Geol. Surv. Bull.*, **625.**

Fleer, V.N. (1982) *The dissolution kinetics of anorthite ($Ca_2Al_2Si_2O_8$) and synthetic strontium feldspar ($SrAl_2Si_2O_8$) in aqueous solutions at temperatures below 100°C, with applications to the geological disposal of radioactive nuclear wastes.* PhD Thesis, Pennsylvania State Univ., University Park, PA (in Lasaga, 1984).

Frisbee, N.M. and Hossner, L.R. (1995) Siderite weathering in acidic solutions under carbon dioxide, air, and oxygen. *J. Environ. Qual.*, **24**, 856–60.

Gílason, S.R. and Anórsson, S. (1993) Dissolution of primary basaltic minerals in natural waters: saturation state and kinetics. *Chem. Geol.*, **105**, 117–35.

Goldich, S.S. (1938) A study in rock weathering. *J. Geol.*, **46**, 17–58.

Hoch, A.R., Reddy, M.M and Drever, J.I. (1996) The effect of iron content and dissolved $O_2$ on dissolution rates of clinopyroxene at pH 5.8 and 25°C: preliminary results. *Chem. Geol.*, **132**, 151–6.

Hodson, M.E., Langan, S.J. and Wilson, M.J. (1997) A critical evaluation of the use of the profile model in calculating mineral weathering rates. *Water, Air, Soil Pollut.*, **98**, 79–104.

Jambor, J.L. (1994) Mineralogy of sulfide-rich tailings and their oxidation products. Pp. 59–102 in: *Environmental Geochemistry of Sulfide Mine-wastes* (J.L. Jambor and D.W. Blowes, editors). Short Course **22**. Mineralogical Association of Canada.

Jambor, J.L. and Blowes, D.W. (1998) Theory and applications of mineralogy in environmental studies of sulfide-bearing mine wastes. Pp. 367–401 in: *Modern Approaches to Ore and Environmental Mineralogy* (L.J. Cabri and D.J. Vaughan, editors). Short Course **27**. Mineralogical Association of Canada.

Jambor, J.L. and Owens, D.R. (1993) *Mineralogy of the tailings impoundment at the former Cu-Ni deposit of Nickel Rim Mines Ltd., eastern edge of the Sudbury Structure, Ontario.* CANMET Report MSL 93-4 (CF), Natural Resources Canada, Ottawa.

Jambor, J.L., Owens, D.R. and Blowes, D.W. (1992) *Examination of possible dissolution in the low-pH zone of the Heath Steele old tailings impoundment, Bathurst–Newcastle area, New Brunswick.* CANMET Report MSL 92-83 (IR), Natural Resources Canada, Ottawa.

Jambor, J.L., Groat, L.A., Shaw, S.C., Blowes, D.W. and Hanton-Fong, C.J. (1999) Mineralogical reactions in lysimeters containing tailings with different sulfide contents. Pp. 1381–90 in: *REWAS '99 – Global Symposium on Recycling, Waste Treatment and Clean Technology* (I. Gaballah, J. Hager and R. Solozabal, editors). TMS, Warrendale, PA.

Jambor, J.L., Dutrizac, J.E. and Chen, T.T. (2000) Contribution of specific minerals to the neutralization potential in static tests. *Proceedings of the Fifth International Conference on Acid Rock Drainage,* **1,** 551–65.

Jennings, S.R., Dollhopf, D.J. and Inskeep, W.P. (2000) Acid production from sulfide minerals using hydrogen peroxide weathering. *Appl. Geochem.*, **15**, 235–43.

Johnson, R.H., Blowes, D.W., Robertson, W.D. and Jambor, J.L. (2000) The hydrogeochemistry of the Nickel rim tailings impoundment, Sudbury, Ontario. *J. Contam. Hydrol.*, **41**, 49–80.

Kalinowski, B.E. and Schweda, P. (1996) Kinetics of muscovite, phlogopite, and biotite dissolution and alteration at pH 1–4, room temperature. *Geochim. Cosmochim. Acta*, **60**, 367–85.

Kwong, Y.T.J. (1993) *Prediction and prevention of acid mine drainage from a geological and mineralogical perspective.* Project **1.32.1**, MEND, Natural Resources Canada, Ottawa.

Kwong, Y.T.J. (1995) Influence of galvanic sulfide oxidation on mine water chemistry. Pp. 477–83 in: *Proceedings of Sudbury '95, Mining and the Environment* (T.P. Hynes and M.C. Blanchette, editors). CANMET, Natural Resources Canada, Ottawa.

Lång, L. (2000) Heavy mineral weathering under acidic soil conditions. *Appl. Geochem.*, **15**, 415–23.

Lapakko, K.A. (1994) Evaluation of neutralization potential determinations for metal mine waste and a proposed alternative. Pp. 129–37 in: *Proceedings of the International Land Reclamation and Mine Drainage Conference*, **USBM SP 06A-94,** Vol. **1**.

Lasaga, A.C. (1984) Chemical kinetics of water–rock interactions. *J. Geophys. Res.*, **89**, 4009–25.

Lasaga, A.C. and Berner, R.A. (1998) Fundamental aspects of quantitative models for geochemical cycles. *Chem. Geol.*, **145**, 161–75.

Lasaga, A.C., Soler, J.M., Ganor, J., Burch, T.E. and Nagy, K.L. (1994) Chemical weathering rate laws and global geochemical cycles. *Geochim. Cosmochim. Acta*, **58**, 2361–86.

Lawrence, R.W. and Scheske, M. (1997) A method to calculate the neutralization potential of mining wastes. *Environ. Geol.*, **32**, 100–6.

Lawrence, R.W. and Wang, Y. (1996) *Determination of neutralization potential for acid rock drainage prediction.* Draft Report, MEND Project **1.16.1**, Natural Resources Canada, Ottawa.

Lawrence, R.W. and Wang, Y. (1997) Determination of neutralization potential in the prediction of acid rock drainage. Pp. 451–64 in: *Proceedings of the Fourth International Conference on Acid Rock Drainage.* MEND, Natural Resources Canada, Ottawa.

Li, M. (1997) Neutralization potential versus observed mineral dissolution in humidity cell tests for Louvicourt tailings. Pp. 151–64 in: *Proceedings Fourth International Conference on Acid Rock Drainage.* MEND, Natural Resources Canada, Ottawa.

Loughnan, F.C. (1969) *Chemical Weathering of the Silicate Minerals.* Elsevier, New York.

Malström, M. and Banwart, S. (1997) Biotite dissolution at 25°C: The pH dependence of dissolution rate and stoichiometry. *Geochim. Cosmochim. Acta*, **61**, 2779–99.

Malström, M.E., Destouni, G., Banwart, S.A. and Str'mberg, B.H.E. (2000) Resolving the scale-dependence of mineral weathering rates. *Environ. Sci. Tech.*, **34**, 1375–8.

McGregor, R.G., Blowes, D.W., Jambor, J.L. and Robertson, W.D. (1998) Mobilization and attenuation of heavy metals in a tailings impoundment near Subdbury, Ontario. *Environ. Geol.*, **36**, 305–19.

MEND (1991*a*) New methods for determination of key mineral species in acid generation predicting by acid–base accounting. Project **1.16.1c**, report by Norecol Environmental

Consultants Ltd., MEND, Natural Resources Canada, Ottawa.

MEND (1991*b*) Acid rock drainage prediction manual. Project **1.16.1b**, report by Coastech Research Inc., MEND, Natural Resources Canada, Ottawa.

Miller, S.D., Jeffery, J.J. and Donohue, A. (1994) Developments in predicting and management of acid forming mine wastes in Australia and Southeast Asia. Pp 177–84 in: *Proceedings of the International Land Reclamation and Mine Drainage Conference,* **USBM SP 06A**, Vol. **1**.

Morin, K.A. (1990) Problems and proposed solutions in predicting acid drainage with acid–base accounting. Pp. 93–107 in *Acid Mine Drainage – Designing for Closure*. Geological Association of Canada/Mineralogical Association of Canada Annual Meeting, Vancouver. BiTech Publishers, Vancouver, B.C.

Morrison, J.L., Scheetz, B.E., Stickler, D.W., Williams, E.G., Rose, A.W., Davis, A. and Parizek, R.R. (1990) Predicting the occurrence of acid mine drainage in the Alleghenian coal-bearing strata of western Pennsylvania; an assessment by simulated weathering (leaching) experiments and overburden characterization. *Geol. Soc. Amer. Spec. Paper*, **248**, 87–99.

Murakami, T. and Utsunomiya, S. (1998) Comparison of biotite dissolution in the laboratory and in the field by high-resolution transmission electron microscopy. *Mineral. Mag.*, **62A**, 1044–5.

Murphy, S.M., Brantley, S.L., Blum, A.E., White, A.F. and Dong, H. (1998) Chemical weathering in a tropical watershed, Luquillo Mountains, Puerto Rico: II Rate and mechanism of biotite weathering. *Geochim. Cosmochim. Acta*, **62**, 227–43.

Nagy, K.L. (1995) Dissolution and precipitation kinetics of sheet silicates. Pp. 173–233 in: *Chemical Weathering Rates of Silicate Minerals* (A.F. White and S.L. Brantley, editors). Reviews in Mineralogy, **31**. Mineralogical Society of America, Washington D.C.

Nicholson, R.V. (1994) Iron-sulfide oxidation mechanisms: laboratory studies. Pp. 163–83 in: *Environmental Geochemistry of Sulfide Mine-wastes* (J.L. Jambor and D.W. Blowes, editors). Mineralogical Association of Canada Short Course, Vol. **22**.

Nicholson, R.V. and Scharer, J.M. (1994) Laboratory studies of pyrrhotite oxidation kinetics. Pp. 14–30 in: Environmental Geochemistry of Sulfide Oxidation (C.N. Alpers and D.W. Blowes, editors). *ACS Symp. Ser.* **550**. American Chemical Society, Washington, D.C.

Ollier, C. (1984) *Weathering*, 2nd edition. Longman, New York.

O'Shay, T., Hossner, L.R. and Dixon, J.B. (1990) A modified hydrogen peroxide oxidation method for determination of potential acidity in pyritic overburden. *J. Environ. Qual.*, **19**, 778–82.

Paktunc, A.D. (1999) Discussion of "A method to calculate the neutralization potential of mining wastes" by Lawrence and Scheske. *Environ. Geol.*, **38**, 82–4.

Pettijohn, F.J. (1941) Persistence of heavy minerals and geologic age. *J. Geol.*, **49**, 610–25.

Plumlee, G.S. (1999) The environmental geology of mineral deposits. Pp. 71–116 in: *The Environmental Geochemistry of Mineral Deposits* (G.S. Plumlee and M.S. Logsdon, editors). Reviews in Economic Geology, **6A**. Society of Economic Geologists Inc.

Postma, D. (1983) Pyrite and siderite oxidation in swamp sediments. *J. Soil. Sci.,* **34**, 163–82.

Price, W.A., Morin, K. and Hutt, N. (1997) Guidelines for the prediction of acid rock drainage and metal leaching for mines in British Columbia: part II. Recommended procedures for static and kinetic testing. Pp. 15–30 in: *Proceedings of the Fourth International Conference on Acid Rock Drainage*. MEND, Natural Resources Canada, Ottawa.

Rimstidt, J.D., Chermak, J.A. and Gagen, P.M. (1994) Rates of reaction of galena, sphalerite, chalcopyrite, and arsenopyrite with Fe(III) in acidic solutions. Pp. 1–13 in: *Environmental Geochemistry of Sulfide Oxidation* (C.N. Alpers and D.W. Blowes, editors). ACS Symp. Ser. **550**. American Chemical Society, Washington, D.C.

Sato, M. (1992) Persistency-field Eh–pH diagrams for sulfides and their application to supergene oxidation and enrichment of sulfide ore bodies. *Geochim. Cosmochim. Acta*, **56**, 3133–56.

Schott, J. and Berner, R.A. (1985) Dissolution mechanisms of pyroxenes and olivines during weathering. Pp. 35–53 in: *The Chemistry of Weathering* (J.I. Drever, editor). Reidel,

Dordrecht, The Netherlands.

Shaw, S.C., Groat, L.A., Jambor, J.L., Blowes, D.W., Hanton-Fong, C.J. and Stuparyk, R.A. (1998) Mineralogical study of base metal tailings with various sulfide contents, oxidized in laboratory columns and field lysimeters. *Environ. Geol.*, **33**, 209–17.

Sherlock, E.J., Lawrence, R.W. and Poulin, P. (1995) On the neutralization of acid rock drainage by carbonate and silicate minerals. *Environ. Geol.*, **25**, 43–54.

Siever, R. and Woodford, N. (1979) Dissolution kinetics and the weathering of mafic minerals. *Geochim. Cosmochim. Acta*, **43**, 717–24.

Skousen, J., Renton, J., Brown, H., Evans, P., Leavitt, B., Brady, K., Cohen, L. and Ziemkiewicz, P. (1997) Neutralization potential of overburden samples containing siderite. *J. Environ. Qual.*, **26**, 673–81.

Sobek, A.A., Schuller, W.A., Freeman, J.R. and Smith, R.M. (1978) *Field and laboratory methods applicable to overburdens and mine soils*. Report **EPA-600/2-78-054**, U.S. Environmental Protection Agency.

Soregaroli, B.A. and Lawrence, R.W. (1997) Waste rock characterization at Dublin Gulch: a case study. Pp. 633–45 in: *Proceedings of the Fourth International Conference on Acid Rock Drainage*. MEND, Natural Resources Canada, Ottawa.

Stillings, L.L. and Brantley, S.L. (1995) Feldspar dissolution at 25°C and pH 3: reaction stoichiometry and the effect of cations. *Geochim. Cosmochim. Acta*, **59**, 1483–96.

Sverdrup, H.U. (1990) *The Kinetics of Base Cation Release Due to Chemical Weathering.* Lund University Press, Lund, Sweden.

Sverdrup, H. and Warfvinge, P. (1993) Calculating field weathering rates using a mechanistic geochemical model PROFILE. *Appl. Geochem.*, **8**, 273–83.

Sverdrup, H. and Warfvinge, P. (1995) Estimating field weatheirng rates using laboratory kinetics. Pp. 485–541 in: *Chemical Weathering Rates of Silicate Minerals* (A.F. White and S.L. Brantley, editors). Reviews in Mineralogy, **31**. Mineralogical Society of America, Washington D.C.

Tejan-Kella, M.S., Chittleborough, D.J. and Fitzpatrick, R.W. (1991) Weathering assessment of heavy minerals in age sequences of Australian sandy soils. *Soil Sci. Soc. Amer. J.*, **55**, 427–38.

Velbel, M.A. (1984) Natural weathering mechanisms of almandine garnet. *Geology*, **12**, 631–4.

Velbel, M.A. (1993) Formation of protective surface layers during silicate-mineral weathering under well-leached, oxidizing conditions. *Amer. Mineral.*, **78**, 405–14.

Velbel, M.A. (1999) Bond strength and the relative weathering rates of simple orthosilicates. *Amer. J. Sci.*, **299**, 679–96.

White, A.F., Bullen, T.D., Vivit, T.D., Schulz, M.S. and Clow, D.W. (1999) The role of disseminated calcite in the chemical weathering of granitoid rocks. *Geochim. Cosmochim. Acta*, **63**, 1939–53.

White, W.W. III, Lapakko, K.A. and Cox, R.L. (1999) Static-test methods most commonly used to predict acid-mine drainage: practical guidelines for use and interpretation. Pp. 325–38 in: *The Environmental Geochemistry of Mineral Deposits* (G.S. Plumlee and M.S. Logsdon, editors). Reviews in Economic Geology, **6A**. Society of Economic Geologists Inc.

Wogelius, R.A. and Walther, J.V. (1992) Olivine dissolution kinetics at near-surface conditions. *Chem. Geol.*, **97**, 101–12.

CHAPTER NINE

# Mynydd Parys Cu-Pb-Zn Mines: mineralogy, microbiology and acid mine drainage

D. A. Jenkins[1], D. B. Johnson[2] and C. Freeman[2]

*Schools of [1]Agricultural & Forest Science, and [2]Biological Sciences, University of Wales, Bangor, Gwynedd LL57 2UW, UK (E-mail: d.a.jenkins@bangor.ac.uk)*

**ABSTRACT**

Mynydd Parys in Anglesey is the site of two famous mines which dominated copper production in the early industrial revolution, and is still under investigation for possible future deep mining. It is now a valuable research site for its ore deposits, being an excellent example of the volcanic-associated massive sulphide ('VMS') type, its primary and post-mining mineralogy, surface water geochemistry, and related specialized microbiology and problems of acid mine drainage, as well as for its industrial and archaeological record. This chapter discusses the general background to these different geochemical aspects and their specific occurrence and intricate interdependence on Mynydd Parys as well as the problems posed by the management of such a site.

## 9.1 Introduction

Mynydd Parys in the Isle of Anglesey, North Wales (Fig. 9.1), hosts a major polymetallic sulphide deposit that is unique in the UK. Exploitation of this orebody can be traced back 3500 years to the Bronze Age, but reached its zenith at the end of the eighteenth century when the mines dominated the world's copper markets. Deep mining had ceased by the end of the nineteenth century, although smaller scale production of copper by precipitation continued into the early twentieth century. The deposits are currently being re-evaluated by Anglesey Mining plc, with the intention of re-opening the mines for Zn, Cu and Pb should financial conditions permit. Although not particularly large, compared to many VMS deposits, or diverse in its ore mineralogy, as in the Cornish mines, Mynydd Parys is a 'neat', well defined and preserved entity which provides scope for research in a variety of disciplines.

Being remote, this important site has remained relatively undisturbed and provides not only a wealth of archaeological and industrial history, but also a

Jenkins, D.A., Johnson, D.B. and Freeman, C. (2000) Mynydd Parys Cu-Pb-Zn Mines: mineralogy, microbiology and acid mine drainage. Pp. 161–179 in: *Environmental Mineralogy: Microbial Interactions, Anthropogenic Influences, Contaminated Land and Waste Management* (J.D. Cotter-Howells, L.S. Campbell, E. Valsami-Jones and M. Batchelder, editors). Mineralogical Society Series, **9**. Mineralogical Society, London. ISBN 0 903056 20 8.

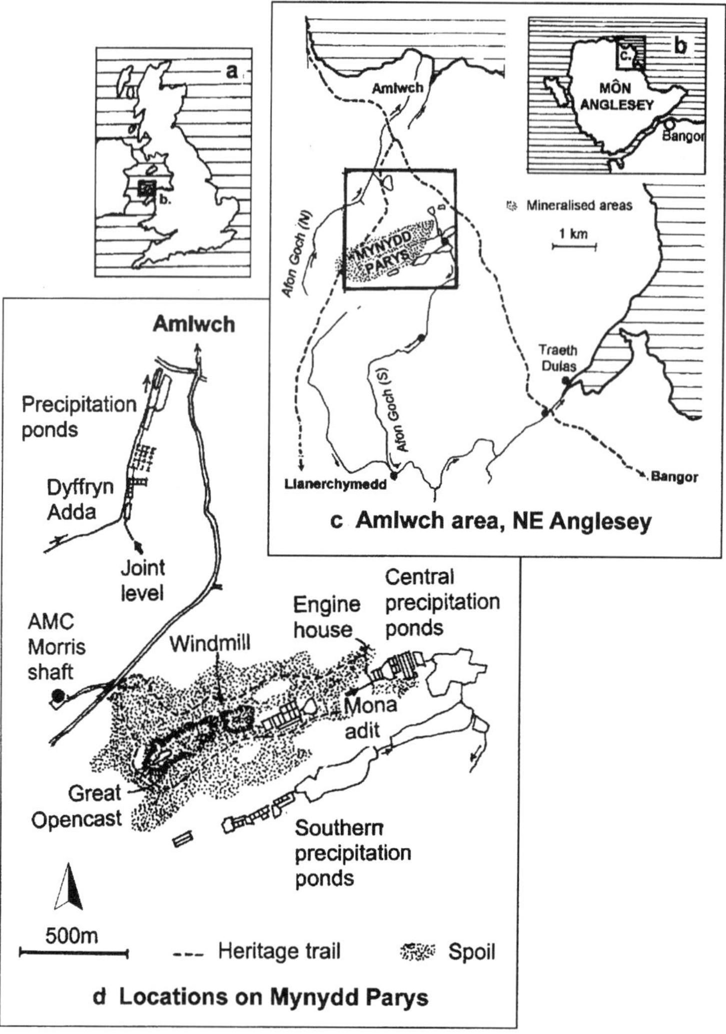


FIG. 9.1 Location map of Mynydd Parys.

remarkable geochemical environment. The sulphide anomaly that gave rise to the mines is now expressed in extreme values of acidity and heavy metal concentrations of surface waters, in a consequent suite of unusual secondary and post-mining minerals, and in a specialized microbiology and macro-flora. As such it is a prime example of the features that characterize abandoned metal mines in the UK generally, the mountain already hosting a number of geological and biological SSSIs as well as scheduled industrial monuments. It therefore merits conservation, management and presentation to provide a teaching, research and visitor facility, and the Amlwch Industrial Heritage Trust has been set up to achieve this end in consultation with the Countryside Council for Wales, the Environment Agency, Cadw and Anglesey Mining plc. This paper first summarizes the general geological setting and history of the mines, and then deals specifically with the surface geochemistry and mineralogy, the microbiology and the potential treatment of AMD, and with the general inter-relationships between these aspects.

## 9.2 Geology and mining history

The sulphide deposits occur within an Ordovician–Silurian volcanic-sedimentary sequence, overlying a metamorphic Precambrian basement. It forms the low (147 m OD) but prominent feature of Mynydd Parys on the Anglesey landscape. Geological relations and lithogeochemical data indicate that this sequence, whose outcrops have been proved over an area extending for 6 km ENE–WSW, formed in an extensional tectonic environment on a continental margin (Tennant, 1999). The bimodal volcanic sequence comprises rhyolitic lavas and pyroclastics and minor basaltic lavas together with intercalated shales, and includes a massive quartz breccia now interpreted to be a product of hydrothermal silicification. Interpretation of the area's complex structure is difficult due to the paucity of fossils, to the devitrification of the volcanics, and to their extensive and repeated hydrothermal alteration. This alteration was followed by low-grade metamorphism, associated with Acadian tectonism that resulted in steep dips to the N, reverse faulting/thrusting and N–S faulting. Earlier interpretations involved a tight overturned syncline (e.g. Greenly, 1919; Pointon and Ixer, 1980), but this structural model has now been re-evaluated in the light of detailed rhyolite chemostratigraphy (Tennant, 1999; Barrett *et al.*, 1999).

The mineralization at Mynydd Parys is of volcanic-associated massive sulphide ('VMS'; Franklin, 1993) type, predominantly stratiform and precipitated from hydrothermal solutions in a marine environment. Fine-grained VMS deposits, dominated by Zn and known locally as 'bluestone', are concentrated at the lower (southern) boundary of the volcanic sequence and show close spatial relationships to dome-like bodies of rhyolite, whose emplacement redeposited some of the sea-floor sulphides as debris flows. Those at the upper (northern) boundary are characterized by disseminated Cu-dominated sulphides whose subsequent remobilization also led to Cu-rich vein deposits. The ore mineralogy comprises chalcopyrite, galena and sphalerite, although a range of other minerals has been recorded in trace quantities (Table 9.1*a*: Pointon and Ixer, 1980; Southwood and Bevins, 1995). These ore minerals are accompanied by abundant pyrite.

Historically, this deposit provided an estimated $2.6 \times 10^6$ tonnes of ore averaging ~5% Cu, i.e. an overall production of $0.13 \times 10^6$ tonnes of copper (Manning, 1959). Several parallel ENE–WSW lodes were recorded and much

TABLE 9.1 Provisional list of recorded minerals from Mynydd Parys (after Lentin, 1800; Greenly, 1919; Pointon and Ixer, 1980; Southwood and Bevins, 1995; etc.).

(*a*) Ore and gangue minerals

| Mineral | Formula | Mineral | Formula |
|---|---|---|---|
| Bismuth | $Bi$ | Bismuthinite | $Bi_2S_3$ |
| Copper | $Cu$ | Bornite | $Cu_5FeS_4$ |
| Gold | $Au$ | Bournonite | $CuPbSbS_3$ |
| Silver | $Ag$ | Chalcocite | $Cu_2S$ |
| Sulphur | $S$ | Chalcopyrite | $CuFeS_2$ |
| | | Covellite | $CuS$ |
| Cuprite | $Cu_2O$ | Galena | $PbS$ |
| Minium | $Pb_3O_4$ | Jordanite | $Pb_{14}(As,Sb)_6S_{23}$ |
| Tenorite | $CuO$ | Kobellite | $Pb_{22}Cu_4(Bi,Sb)_{30}S_{29}$ |
| | | Pyrite | $FeS_2$ |
| Anhydrite | $CaSO_4$ | Pyrrhotite | $Fe_{1-x}S$ |
| | | Sphalerite | $ZnS$ |
| Ankerite | $Ca(Mg,Fe)(CO_3)_2$ | Spionkopite | $Cu_{39}S_{28}$ |
| Dolomite | $CaMg(CO_3)_2$ | Tennantite | $(Cu,Fe,Ag,Zn)_{12}As_4S_{13}$ |
| Malachite | $Cu_2(CO_3)(OH)_2$ | Tetrahedrite | $(Cu,Ag,Fe,Zn)_{12}As_4S_{13}$ |
| Siderite | $FeCO_3$ | | |

(*b*) Secondary, post-mining, minerals
(N.B. Fe″ = $Fe^{2+}$, Fe = $Fe^{3+}$; (?) yet to be confirmed by XRDA)

| Mineral | Formula | Mineral | Formula |
|---|---|---|---|
| Allophane | $Al_2O_3/SiO_2/H_2O$ | Goethite | $\alpha$-FeO(OH) |
| Alunogen (?) | $Al_2(SO_4)_3 \cdot 17H_2O$ | Gunningite | $ZnSO_4 \cdot H_2O$ |
| Anglesite | $PbSO_4$ | Gypsum (selenite) | $CaSO_4 \cdot 2H_2O$ |
| Antlerite | $Cu_3(SO_4)(OH)_4$ | Halotrichite | $Fe''Al_2(SO_4)_4 \cdot 22H_2O$ |
| Aragonite | $CaCO_3$ | Hydronium jarosite (?) | $(H_3O)_2Fe_6(SO_4)_4(OH)_{12}$ |
| Baryte | $BaSO_4$ | Jarosite | $K_2Fe_6(SO_4)_4(OH)_{12}$ |
| Basaluminite | $Al_4(SO_4)(OH)_{10} \cdot 5H_2O$ | Melanterite | $Fe''SO_4 \cdot 7H_2O$ |
| Brochantite | $Cu_4(SO_4)(OH)_6$ | Pisanite | $(Cu,Fe'')(SO_4) \cdot 7H_2O$ |
| Calcite | $CaCO_3$ | Pyromorphite (?) | $Pb_5(PO_4)_3Cl$ |
| Chalcanthite | $CuSO_4 \cdot 5H_2O$ | Römerite (?) | $Fe''Fe_2(SO_4)_4 \cdot 14H_2O$ |
| Copiapite | $Fe''Fe_4(SO_4)_6(OH)_2 \cdot 20H_2O$ | Siderotil | $Fe''SO_4 \cdot 5H_2O$ |
| Coquimbite | $Fe_2(SO_4)_3 \cdot 9H_2O$ | Rozenite | $Fe''SO_4 \cdot 4H_2O$ |
| Fibroferrite | $Fe(SO_4)(OH) \cdot 5H_2O$ | | |

ore lay at shallow depths and was exploited in the late eighteenth century by the spectacular opencasts for which Mynydd Parys is renowned – the Great Opencast is up to 380 m long, 170 m wide and 35 m deep. Later underground extraction occurred progressively northwards at increasing depths up to 300 m, and recorded workings extend over 20 km in length. Associated with this activity are extensive surface deposits of spoil totalling several million tonnes, much of which is strongly pyritic.

Mining on Mynydd Parys has a long history. It has now been established that copper ore was mined in the Early Bronze Age ~3500 years BP (Jenkins, 1995). Prehistoric spoil has been located on the surface and workings discovered over 20 m deep underground, as well as possible evidence in the geochemical stratigraphy of sediment cores downstream from Mynydd Parys (Timberlake and Jenkins, 2000). There is also circumstantial evidence of Roman activity in the form of inscribed copper ingots found near the mountain. However, local tradition has the first major discovery of copper dated to 2nd March 1768, with the subsequent development of two mines, Mona mine to the East and Parys mine to the West (Rowlands, 1966). Under the astute management of the major industrialist Thomas Williams (Harris, 1964), these two mines were to dominate world copper production in the 1780s, their copper sheathing Nelson's fleet at the time of Trafalgar. A work force of over 1600 was employed on the mountain at this time, and Amlwch briefly became the second largest town in Wales. In the nineteenth century, development of deep mining took place under Cornish management, and Mynydd Parys enjoyed a second though lesser phase of productivity.

In common with many British mines, however, underground production slumped in the third quarter of the nineteenth century with the discovery of cheaper overseas sources of copper. Mining effectively ceased in the 1880s, although recovery of dissolved copper, as described below, continued on into the twentieth century. The orebody has since been subject to a number of reappraisals involving extensive deep coring. During 1988–90 the present lessees, Anglesey Mining plc, sank a new 300 m shaft with 1.5 km of exploratory driving and estimated remaining reserves at $6.5 \times 10^6$ tonnes, averaging 10% Zn+Cu+Pb (Anglesey Mining plc, 1998). When favourable economic conditions permit, they plan to recommence deep mining, mainly for Zn, which would add a further chapter to the 3500 year history of the mining on Mynydd Parys.

## 9.3 Surface mineralogy and geochemistry

### *9.3.1 Pyrite oxidation*

The inherent instability of sulphide ore minerals leads to their breakdown in the moist, oxidizing conditions that characterize most weathering environments. However, the precise mechanisms involved in the oxidation of the sulphide, ultimately to sulphate, and the release of the metal ions (and where relevant, e.g. $Fe^{2+} \rightarrow Fe^{3+}$, their oxidation also) are complex and have been the subject of much recent study (e.g. Evangelou, 1995; Keith and Vaughan, 2000). In the case of pyrite, the dominant sulphide in the spoil at Mynydd Parys and many other mines, the overall process (equation 9.1) and its components (equation 9.1a–c) can be summarized by the following 'balanced' reactions, although these give no indication of reaction rates:

$$4FeS_2 + 15O_2 + 14H_2O \rightarrow 4Fe(OH)_3 + 8SO_4^{2-} + 16H^+ \quad (9.1)$$

$$2FeS_2 + 7O_2 + 2H_2O \rightarrow 2Fe^{2+}_{aq} + 4SO_4^{2-} + 4H^+ \quad (9.1a)$$

$$4Fe^{2+}_{aq} + O_2 + 4H^+ \rightarrow 4Fe^{3+}_{aq} + 2H_2O \quad (9.1b)$$

$$Fe^{3+}_{aq} + 3H_2O \rightarrow Fe(OH)_3 + 3H^+ \quad (9.1c)$$

An important additional step (equation 9.2) is now known to be the oxidation of pyrite at low pH values (<3) by the oxidized $Fe^{3+}{}_{aq}$ itself (and at pH 3.5 by $Fe(OH)^{2+}{}_{aq}$) which may be summarized overall as:

$$FeS_2 + 14Fe^{3+}{}_{aq} + 8H_2O \rightarrow 15Fe^{2+}{}_{aq} + 2SO_4^{2-} + 16H^+ \qquad (9.2)$$

Pyrite breakdown can therefore continue as long as $Fe^{3+}{}_{aq}$ is being regenerated. As discussed in more detail below in terms of the intermediate stages, this is where microbial mediation is now known to be a crucial rate-determining factor in that such organisms as *Thiobacillus ferrooxidans* carry out this reaction with great efficiency and rapidity at low pH conditions (Evangelou, 1995). There is also, however, a 'galvanic' effect of relevance on Mynydd Parys whereby pyrite in contact with chalcopyrite will be effectively protected whilst promoting the preferential dissolution of the chalcopyrite due to its lower 'rest potential' (Torma, 1988).

From an environmental point of view, the significant feature of the oxidation of sulphides such as pyrite is the copious production of protons – e.g. $4H^+$ per mole $FeS_2$ for example in equation 9.1 and, in equation 9.2, $16H^+$ per mole $FeS_2$. This results in the development of some of the most acidic environments known with values of pH 2 or less being recorded. This is shown in the general context of surface environments in Fig. 9.2. These values are well below the 4.3 buffered by the '$Al^{3+}/Al(OH)_3$' system operating in 'normal' acid mineral soils; they may even exceed the buffering capacity of the '$Fe^{3+}/Fe(OH)_3$' system operating under oxidizing conditions (Eh >790 mV) with values ~pH 2–3. Accompanying the acidity is the production of abundant sulphate anions (2 per mole $FeS_2$) and, typically, the precipitation of orange-brown ferric hydroxide as 'ochre'.

### 9.3.2 *Mynydd Parys hydrochemistry*

This extreme geochemical situation is well illustrated at Mynydd Parys. The results of some typical water analyses from both above and below ground (Table 9.2) fall into different categories. The major colourless effluent from flooded workings (Mona adit) has a low Eh and pH (2.3), and contains ferrous iron at high levels (500 μg/ml); in other exposed ponds the pH is still low but the high 'dissolved' (i.e. <0.2 μm filter) iron content is now in the ferric form, giving the water a deep orange colour. In underground pools extreme values of pH (2.0) and iron content (2000 μg/ml = 50 mM) have been recorded. In other seepages from the workings abundant ochre has already been deposited raising the pH to 3.6 and giving correspondingly lower iron concentrations (5 μg/ml). As the surface water course is followed down the Afon Goch ('Red River') some 12 km to the sea, a progressive and predictable decrease in concentrations is evident due to dilution and precipitation/adsorption. The values for other metals (e.g. Al, Cd, Co, Cu, Mn, Ni, Zn) generally show a parallel pattern. However, the behaviour of Pb is different in that it is

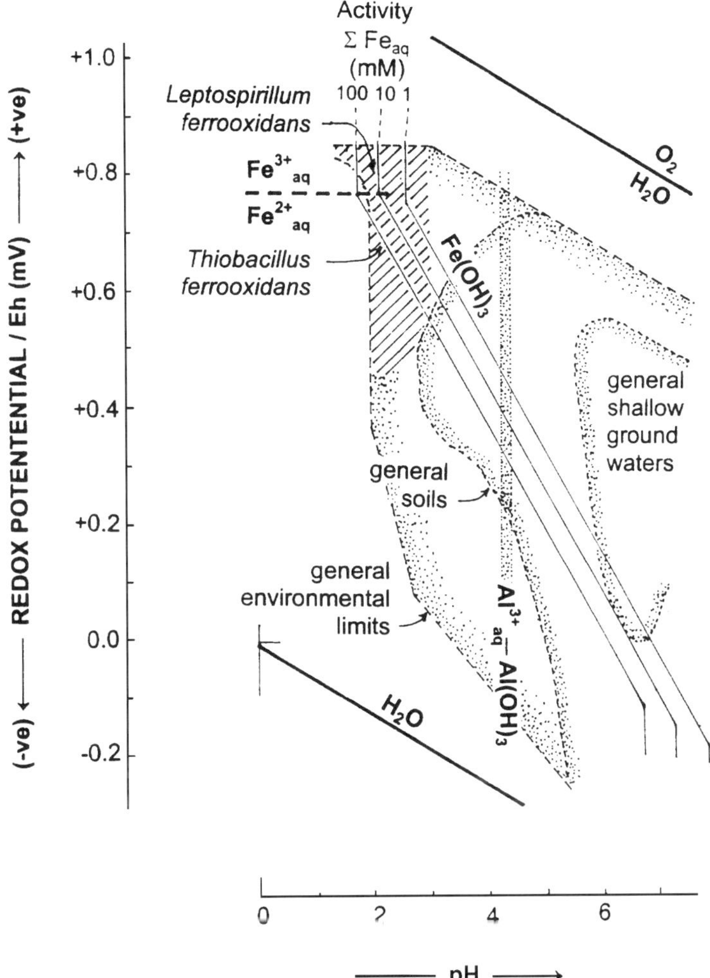


FIG. 9.2 Environments in terms of Eh–pH at Mynydd Parys (hatched) at 25°C and 1 atm *P* (excluding S species), showing Fe activities and the Al buffer system, in relation to 'normal' environments (after Baas Becking *et al.*, 1960).

precipitated (presumably as $PbSO_4$) with the ochres. So also is the behaviour of As which is depleted through adsorption of the dissolved arsenate onto the ochre; otherwise As can achieve values of several μg/ml in pools lacking ochre. A small seasonal fluctuation in concentrations has also been detectable corresponding to dilution by rainfall.

Dissolved copper is of especial interest since a remarkably efficient system was developed by the eighteenth century miners at Mynydd Parys in particular

TABLE 9.2 Representative analyses of selected 'soluble' (<0.2 μm) elements in water samples from Mynydd Parys 1998–9.

| Sampling sites | pH | Cond (mS) | Mg | Al | Ca | Co | Mn | Fe | Cu | Zn | As | Cd | Pb |
|---|---|---|---|---|---|---|---|---|---|---|---|---|---|
| | | | | | | | values in μg/ml | | | | | | |
| (a) Underground | | | | | | | | | | | | | |
| 16fNW brown pool | 2.1 | 9.45 | 146.00 | 326.00 | 80.90 | 0.611 | 9.050 | 2070.00 | 145.000 | 68.100 | 3.450 | 0.133 | 0.067 |
| 16fW blue pool | 2.9 | 1.11 | 20.30 | 18.60 | 20.50 | 0.038 | 1.120 | 12.30 | 10.200 | 23.800 | 0.005 | 0.063 | 0.072 |
| 20fN water table | 2.7 | 2.10 | 48.80 | 61.20 | 29.20 | 0.096 | 1.820 | 77.00 | 28.300 | 15.200 | 0.130 | 0.038 | 0.045 |
| (b) Surface Sites (see Fig. 9.1*d*) | | | | | | | | | | | | | |
| Great Opencast | 2.4 | 9.87 | 21.50 | 36.50 | 19.60 | 0.121 | 1.480 | 358.00 | 22.600 | 27.200 | 0.527 | 0.074 | 0.304 |
| Maria shaft pool | 2.0 | 1.91 | 11.90 | 47.80 | 49.90 | n.d. | 0.466 | 1610.00 | 70.900 | 63.400 | 4.240 | 0.117 | 0.990 |
| Pearl opencast | 3.8 | 0.53 | 22.30 | 9.38 | 3.63 | 0.067 | 0.744 | 1.72 | 12.300 | 19.900 | 0.005 | 0.074 | 0.598 |
| Mona adit | 2.3 | 3.70 | 60.20 | 81.60 | 27.20 | 0.244 | 7.000 | 431.00 | 39.100 | 44.600 | 0.236 | 0.129 | 0.193 |
| Dyffryn Adda adit | 3.5 | 0.72 | 28.00 | 5.66 | n.d. | 0.041 | 3.650 | 5.61 | 1.030 | 3.640 | n.d. | <0.001 | 0.002 |
| (c) Afon Goch (South) (see Fig. 9.1*c*) | | | | | | | | | | | | | |
| Penysarn dam - 0 km | 2.4 | 1.80 | 27.50 | 16.00 | 29.90 | 0.056 | 6.870 | 54.40 | 6.150 | 21.300 | 0.006 | 0.040 | 0.245 |
| Capel Parc - 6 km | 2.7 | 0.17 | 15.00 | 8.27 | 24.70 | 0.017 | 1.730 | 14.10 | 3.060 | 5.960 | 0.002 | 0.010 | 0.022 |
| City Dulas - 10 km | 3.9 | 0.04 | 7.97 | <0.01 | 27.10 | <0.001 | <0.001 | <0.01 | <0.001 | 1.910 | 0.002 | <0.001 | <0.001 |
| Traeth Dulas - 11km | 4.5 | 1.86 | 11.90 | <0.01 | 30.40 | <0.001 | <0.001 | <0.01 | <0.001 | 0.064 | 0.002 | <0.001 | <0.001 |

cond = conductivity; n.d. = no data available
ICP-AES results from Thomas (2000) and Butler (2000)
For details of sampling sites, analytical equipment and technique, reproducibilities, etc. see Thomas (2000)

to reclaim it. Water was pumped to the top of the mountain and allowed to drain down through the spoil ('sparging') and join drainage from the underground workings in numerous purpose-built, brick-lined ponds around the mountain. Imported scrap iron was then added to the ponds, the iron dissolving and metallic copper precipitating as a sludge that was then readily smelted. By this procedure it was possible to reduce copper levels from 70 μg/ml (comparable to present effluent levels) to 15 μg/ml to give an annual production of the order of 30 tonnes of copper (Manning, 1959). Nevertheless it has been calculated that there is currently an annual discharge to the sea of some 18 and 40 tonnes of Cu and Zn respectively from Mynydd Parys (Boult and Curtis, 1994).

### *9.3.3 Post-mining mineralogy*

The mineralogy relating to this oxidizing sulphide environment is predictably dominated by a variety of rare, but interesting, sulphate minerals in addition to the ferric hydrous oxides ('limonite'). The expanding list of these 'post-mining' minerals (or 'supragossan' – Ixer, 1999) that have so far been identified from Mynydd Parys is given in Table 9.1*b*, and adds to our expanding knowledge of such minerals in Wales (e.g. Mason and Rust, 1997). The characteristic mineral both above and below ground is an earthy yellow jarosite, possibly including the hydronium and other varieties although this has yet to be confirmed by quantitative chemical analysis. The abundant ochre, which forms spectacular stalactites/stalagmites underground and blocks some of the drainage adits, is generally amorphous, although a weak goethite peak is sometimes evident in XRD traces; no schwertmannite has been detected but ferrihydrite is invoked in stream deposits (Boult and Curtis, 1994). There is a range of ferric sulphates, some of which are relatively uncommon, such as fibroferrite, coquimbite and copiapite, which may be found in overhangs and recesses and other protected sites in the two major opencasts. The ferrous sulphate melanterite can contain significant quantities of Cu to give the distinctive blue-green variety pisanite (Bor, 1950), again characteristic of parts of the underground mines where pyrite is present with subordinate chalcopyrite.

However, the commonest Cu-product coating chalcopyrite exposures underground is the bright green hydroxy-sulphate, antlerite (Fig. 9.3*a*), a mineral previously unrecorded in Wales (Bevins, 1994). It is often overlying the blue-green mineral brochantite indicating a drop in pH, the conditions determining their relative stabilities (chalcanthite $<$ pH 2.7 $<$ antlerite $<$ pH 4.5 $<$ brochantite) having been presented by Pollard *et al.* (1992). The generally rare aluminium hydroxy-sulphate, basaluminite (Fig. 9.3*b*), is also a common efflorescence on passage walls underground, together with allophane, again reflecting the extreme acidity. Surprisingly, high-pH niches have also been found underground containing secondary calcite, aragonite and malachite,

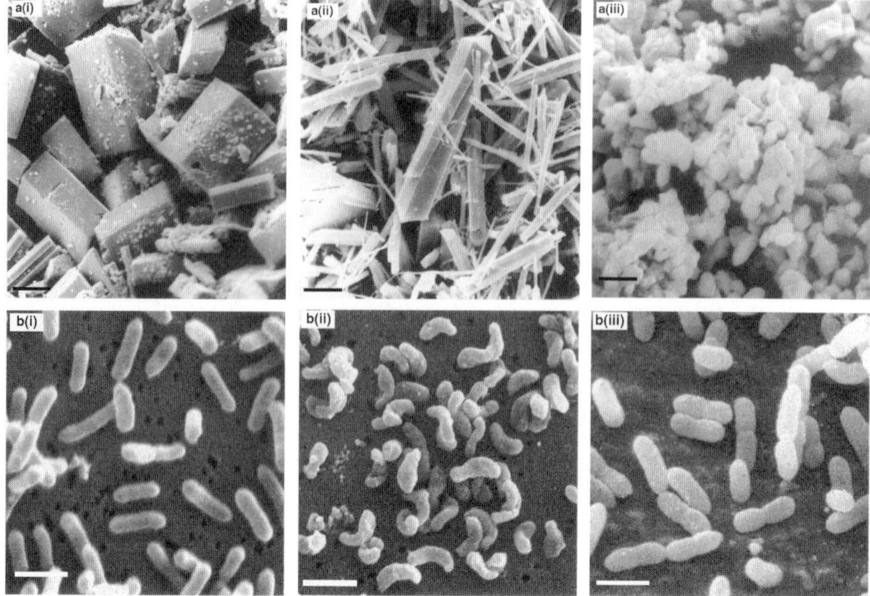


FIG. 9.3 (*a*) Post-mining minerals from Mynydd Parys (scale bar is 2 μm for each): (i) antlerite; (ii) bassaluminite; and (iii) gunnigite. (*b*) Iron-oxidizing bacteria indigenous to the Afon Goch, Mynydd Parys (scale bar is 1 μm for each): (i) *Thiobacillus ferrooxidans;* (ii) *Leptospirillum ferrooxidans*; and (iii) *Ferromicrobium acidophilus*.

presumably deriving from rare, localized carbonate gangue minerals. Mynydd Parys is of course the type locality of the lead sulphate anglesite (Southwood and Bevins, 1995) which occurs as small lustrous rhombohedra; apparently it was originally abundant in a 2 m thick weathered cap on exposures of the 'bluestone' but is now only rarely to be found in the tips. A unique occurrence of a previously unrecorded hydrated zinc sulphate mineral, gunningite, as a rare, thin, white coating on sphalerite has recently been destroyed, presumably in the 'cleaning' of an exposure of the ore vein. This highlights the need for geological conservation in this much-visited site.

## 9.4 Biology

### *9.4.1 Macrobiology*

The extreme conditions (pH, metal content, etc.) of the surface and sub-surface waters, and of the weathered regolith ('soil') at Mynydd Parys have a major bearing on the biodiversity of indigenous life-forms. Acid-tolerant plants, such as *Calluna* spp. have established sporadically over parts of the landscape, although there are islands of increased macrophyte biodiversity around derelict buildings where lime-containing mortar has sustained some pH amelioration of the surrounding 'soil'. However, much of the mine surface currently remains devoid of plant life. By contrast, in the Bronze Age evidence in the form of leaves, acorns and fronds preserved in underground deposits indicates that oak and bracken were common and that the original ore exposure and relatively small-scale mining had yet to influence the ecology significantly (Jenkins, 1995). The most obvious effects today are expressed in the lichen flora where an *Acarosporion sinopicae* community colonizes spoil, including *Rhizocarpon furfurosum, Lecanora epanora* and possibly a previously unrecognized *Fuscidea* sp. A distinctive community including the green copper-rich ecotype of *Acarospora smaragdula* and the rare *Psilolechia leprosa* also colonizes copper-rich mortar on old mine buildings (Purves, 1991). For these reasons eight sites on the mountain have been designated as Sites of Special Scientific Interest (SSSIs). Within the acidic (pH 2–2.5) streams draining the mine, green filamentous streamer-like growths of *Euglena mutabilis* (unicellular algae) are commonplace.

### *9.4.2 Microbiology*

Microbial life at Mynydd Parys is dominantly acidophillic (i.e. displaying pH optima for growth at 3 or below). The biodiversity of indigenous acidophiles is considerable, and many of the mesophilic acidophiles that have been described in the literature may be found at the site (Johnson, 1998). Of particular note are the mineral oxidizing chemolithotrophic bacteria that, by accelerating the oxidative dissolution of sulphide minerals, are directly involved in acid genesis and metal solubilization in the waters and 'soils' at Mynydd Parys. The detailed mechanism (relative to equations 9.1 and 9.2) whereby these bacteria catalyse the oxidation of sulphide minerals and the reactions involved in the process have both been the subject of considerable debate (e.g. Schippers *et al.*, 1996). The primary role of mineral-oxidizing bacteria is the regeneration of ferric iron, which is the major oxidant of sulphide minerals in low pH environments, and involves the production of transient thiosulphate species:

$$FeS_2 + 6Fe(H_2O)_6^{3+} + 3H_2O \rightarrow Fe^{2+} + S_2O_3^{2-} + 6Fe(H_2O)_6^{2+} + 6H^+ \qquad (9.3)$$

Iron oxidizing acidophilic bacteria are therefore acknowledged to have the most central role in mineral oxidation though they are not necessarily the dominant microorganisms (numerically) in mineral leaching environments. Reaction 9.3 can occur under anoxic or aerobic conditions, though the subsequent microbial oxidation of ferrous iron does require molecular oxygen. The thiosulphate formed in reaction 9.3 is unstable in acidic environments (particularly in the presence of ferric iron) and is oxidized to tetrathionate ($S_4O_6^{2-}$) which, in turn, decomposes (in the presence of pyrite) to form elemental S, traces of tri- and pentathionate, and sulphate (Schippers *et al.*, 1996). The various polythionates, and elemental S, formed during the oxidation of pyrite and other metal sulphides, are used as energy sources by some chemolithotrophic acidophiles, thereby generating sulphuric acid, e.g.:

$$S_4O_6^{2-} + 3.5O_2 + 3H_2O \rightarrow 4SO_4^{2-} + 6H^+ \qquad (9.4)$$

Iron- and sulphur-oxidizing autotrophic bacteria are important primary producers in acidic, sulphide-rich ecosystems, such as at Mynydd Parys, and are the sole carbon-fixing biota in subterranean zones. A significant proportion of this fixed C is lost to the environment *via* exudation and cell lysis. Organic carbon may also originate from other sources, such as leachates from rotting wooden roof supports. Although levels of dissolved organic C in Mynydd Parys ecosystems are relatively low (typically <20 mg/l), acidophilic heterotrophic bacteria, as well as fungi and yeasts, are well represented in surface waters. Direct interaction has been observed between jarosite and bacteria underground at Mynydd Parys in the form of pale brown gelatinous deposits, but antlerite and basaluminite have so far proved to be sterile, presumably because of the toxic levels of Cu and Al respectively. This mineral-bacteria interaction is a topic of continuing research.

Both of the two most well-characterized acidophilic iron-oxidizing chemolithotrophs (*Thiobacillus ferrooxidans* and *Leptospirillum ferrooxidans,* Fig. 9.3*b*) have been isolated from surface waters at Mynydd Parys, at numbers varying between $10^3$ and $10^4$/ml in the Afon Goch (Walton and Johnson, 1992). The major factor determining distribution appears to be redox potential, with *T. ferrooxidans* dominating the zones of the stream near to the discharge adit, where ferrous:ferric iron ratios are high, and redox potentials relatively low (< +770 mV), and *L. ferrooxidans* more numerous in downstream zones where the opposite conditions exist. Heterotrophic iron-oxidizing bacteria ('*Ferromicrobium acidophilus*') and the S-oxidizing chemolithotroph *Thiobacillus thiooxidans* have also been isolated from the Afon Goch using enrichment cultures (Johnson and Roberto, 1997). The diversity of *Acidiphilium*-like heterotrophic bacteria was noted by Walton and Johnson (1992) to increase in the Afon Goch with distance from the discharge adit.

Whilst these acidophilic bacteria do not catalyse iron oxidation, most isolates are able to bring about the dissimilatory reduction of ferric iron (using $Fe^{3+}$ in place of oxygen as a terminal electron acceptor) and have been implicated in the biogeochemical cycling of iron in extremely acidic environments, and in the re-mobilization of Fe from ferric-rich sediments (Johnson *et al.*, 1993; Johnson and Bridge, 1997). Novel acidophilic heterotrophic bacteria that use sulphate as terminal electron acceptor have recently been isolated from sediment samples taken from the Afon Goch and elsewhere (Sen and Johnson, 1999). Unlike previously characterized sulphate-reducing bacteria (SRB), which are highly sensitive to even mild acidity (pH <5.5), these isolates are active at extremely low pH. Due to their abilities to generate alkalinity and to remove, as highly insoluble sulphides, metals which may be present in acidic discharge waters, these bacteria have considerable potential for remediating AMD and related waste waters, as discussed below.

A summary of the various redox transformations of Fe, S and C catalysed by the acidophilic microflora present in the surface waters and sediments of the Afon Goch is shown in Fig. 9.4. In the O-rich surface waters, oxidation of ferrous iron and reduced inorganic sulphur compounds is favoured. This is catalysed predominantly by chemo-autotrophic bacteria (*L. ferrooxidans* and *Thiobacillus* ssp.) which fix dissolved carbon dioxide, with a minor contribution from the chemo-organotroph '*F. acidophilus*' (which evolves $CO_2$). Further carbon mineralization results from the activities of other heterotrophic bacteria such as *Acidiphilium* ssp. There is a very sharp gradation of dissolved oxygen concentrations in the sediments in the Parys mine streams.

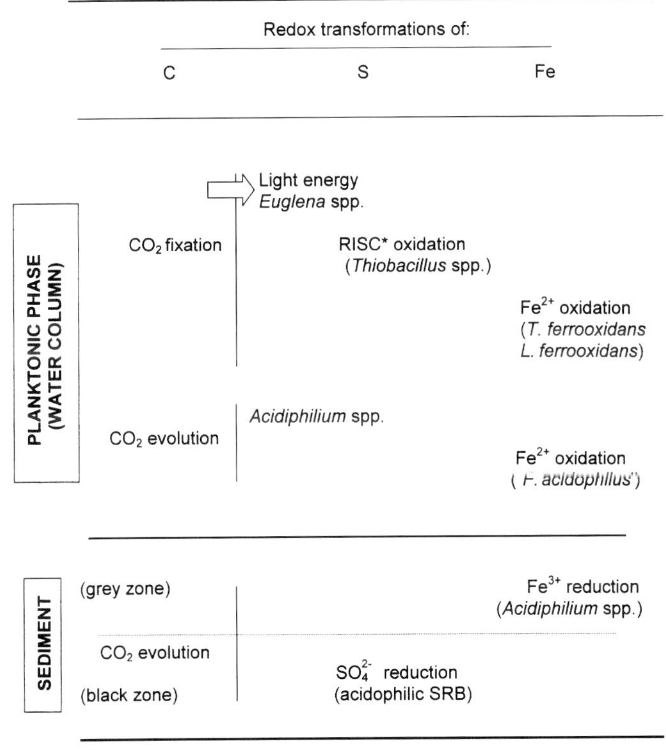


FIG. 9.4 Microbial transformations of C, S and Fe in surface and sediment waters in the Afon Goch, Mynydd Parys. In the oxygen-containing surface water (variable depth, ~1–10 cm) oxidation of both ferrous iron and reduced inorganic sulphur compounds occur, while within the sediment, oxygen concentrations fall sharply, resulting in visible stratification associated with iron reduction (upper grey zone) and sulphate reduction (lower black zone).

The surface oxidized (orange-brown) layer overlies a grey layer in which the main redox transformation is reduction of ferric iron by *Acidiphilium* ssp. now using ferric iron in place of oxygen as terminal electron acceptor. Below this layer is a distinctive black sulphidic zone within which sulphate is reduced to sulphide by a variety of sulphate-reducing bacteria, including the novel acid-tolerant strains referred to earlier.

## 9.5 Acid mine drainage

### *9.5.1 Background*

Although acidic streams can arise naturally, as in the thermal springs of Yellowstone National Park, USA, today these acid environments are far more likely to be found in association with mining activities. Mining, specifically for metals as at Mynydd Parys but also for coal, disturbs large volumes of geological material. Where this is brought to the surface as spoil it is exposed to the sub-aerial environment causing the oxidation of minerals which were previously in a reduced state (Wiggering, 1993). Mining similarly allows the introduction of oxygen to the deep geological environment. The result is ‘acid-mine drainage’ (AMD) which is highly acidic, typically with a pH of $<3$, and contains elevated levels of such metals as Al, Fe and Zn, as well as sulphate at levels typically $>3000$ mg/l (Lyew *et al.*, 1994). The production of AMD, as described in sections 9.3 and 9.4 above, is controlled by a number of factors including: sulphide (particularly pyrite) mineral content; morphology of the strata; effective exposed surface area; availability of oxygen; concentration of ferric ions; and bacterial activities and concentrations.

Where the water levels in a mine are kept stable by pumping, little pyrite oxidation occurs below the water level and few metals are leached from above, resulting in mine water which is relatively non-aggressive environmentally (Banks *et al.,* 1997). Active pyrite oxidation will, however, continue to occur in the unsaturated zone and if pumps are turned off at the cessation of mining, the rising water level will leach out a large amount of heavy metals resulting in a highly acidic and contaminated solution (NRA, 1994). The same applies to spoil tips where precipitation and rising groundwater cause the oxidized pyrite to form a polluted solution. This is the situation at Mynydd Parys where pumping and deep mining ceased over a century ago. Overall, the sulphuric acid and iron sulphates produced under aerobic conditions are mobilized by rising water levels and these create the potential problem of AMD (NRA, 1994).

In the UK more than 100 streams and rivers are affected in some way by AMD, killing not only plant life but also affecting vertebrate and invertebrate organisms, as well as killing the majority of microbiological species present in non-polluted waters (Johnson, 1995). Thus AMD has both direct and indirect biological, chemical, physical and ecological effects on organisms comprising the community structure of the ecosystem (Gray, 1996). Acid mine drainage is a multi-factor pollutant with its main impacts varying within and between ecosystems. These impacts/factors include acidity, salinization, metal toxicity and sedimentation processes. Prime examples include the Red River and River Carnon in Cornwall (NRA, 1995), the Avoca mines, Co. Wicklow, in the Irish Republic (Gallagher *et al.*, 1998), and the Afon Goch at Mynydd Parys.

### *9.5.2 Treatment of AMD*

Before the 1980s all AMD was treated by chemical procedures. This involved the addition of basic chemicals which neutralize acidity and promote the formation of metal solids, primarily metal hydroxides. Treated slurries were then directed into sedimentation basins where the metal-rich solids settled, leaving a low metal content discharge. Additional chemical and mechanical supplements were also used at this time to increase the kinetics of the chemical reactions and sludge settling (Hedin, 1997). The most widely used method of treating AMD is the direct addition of lime ($Ca(OH)_2$,/$CaCO_3$) to the water, the overall result being a rise in pH when metals and sulphate are removed. However this system has the disadvantage of forming a sludge from the precipitation of metal hydroxides and gypsum. The sludge is a hazardous waste in itself, as the metals within could easily remobilize should the water become re-acidicified. As such, it has to be collected and disposed of in a suitable manner (Lyew *et al.*, 1994). Thus, although cheap and easy, liming is a short-term solution which in fact generates its own problems. It is also limited in scale, and would be impractical and environmentally undesirable on a site such as Myndd Parys.

Over the past 15 years, passive treatment using a biological process based on the activity of sulphate-reducing bacteria in constructed wetlands, natural wetlands, with limestone and organic substances, has gained increasing acceptance. Constructed wetlands represent a complex artificial ecosystem, the biological and physical constituents of which interact to provide a mechanical and biogeochemical 'filter' capable of removing many types of contaminants from waters (Mitsch and Gosselink, 1993). The observations during the early 1980s in the USA of the improvement in mine water quality once it had flowed through a wetland system provided the move towards passive treatment technologies. Interest has increased following studies of natural wetlands subject to AMD which show a substantial decline in sulphate and iron concentrations and a concomitant rise in pH in the effluent. These include Tub Run Bog in northern West Virginia which receives AMD from a number of deep mines in the area. The wetland shows a rise in pH from 3 to 5.5, a reduction in sulphate concentrations from 250 to 10 mg/l and a decrease in iron concentrations from 50 to 2 mg/l (Wieder and Lang, 1982). Surveys show that treatment efficiencies for H, Fe, Al, Mn and sulphate of at least 68, 67, 81, 48, 34 and 8% respectively, can be obtained (Wieder, 1989).

Passive treatment of mine waters using constructed wetland technology has only been applied in earnest in the UK since 1992 following the outburst of mine water from the Wheal Jane tin mine in Cornwall. That episode caused the Environment Agency (then the NRA) to initiate a review of mine water treatment in the UK (Younger, 1997). The review resulted in the adoption of constructed wetlands as a possible element of long run remediation of pollution at Wheal Jane. These systems tend to incorporate a number of different cells to treat different aspects of AMD pollution. In general, anaerobic compost cells are put in place to enable sulphate reduction while aerobic cells encourage precipitation of metals and adjust the pH. The wetland aerobic cells in the Wheal Jane treatment removed 63–74% of Fe, the anaerobic cells removed 45–86% of Zn and the pH was reduced by dosing the systems with lime (Hamilton *et al.*, 1997).

The hydrological conditions are probably the single most important determinant of the establishment and maintenance of wetland processes (Mitsch and Gosselink, 1993). The hydrology influences the soil and nutrients and these in turn influence the biota established within the systems. The biological interactions provide the negative feedback loop and effect the flow rate and retention of the waters, therefore having an influence on the hydrological cycle. The rate of flow, storage volume and retention times all determine the length of time that the water spends in the wetland. They also determine the opportunity for interaction between water-borne substances and the wetland ecosystem, and it is this interaction that creates an opportunity for the water to be cleaned and filtered by the numerous chemical, physical and biological processes.

It could be argued that one of the most important of these interactions is due to the sulphate-reducing bacteria (SRB). Dissimilatory sulphate-reduction is the dominant terminal process in the anaerobic degradation of biomass present in sulphate-rich habitats. Sulphate-reducing bacteria reduce sulphate to sulphide, and sulphide complexes readily with metals to precipitate sulphides, thereby removing metals from the water and reducing sulphate concentrations. The process of sulphate reduction also consumes protons and the carbon metabolism of the SRB produces carbonate alkalinity, consequently the pH of the water is increased (Lyew *et al.*, 1994). Therefore SRB have the potential to be highly valuable in the treatment of AMD. However, surveys in the USA have shown that treatment efficiencies can be highly variable, with some systems even becoming sources of metal pollution (Wieder, 1989).

On Mynydd Parys many of the copper and ochre precipitation ponds have, upon disuse, effectively developed into natural wetland systems that can be investigated *in situ*. These areas can provide ideal sites for field studies on AMD remediation. They would also allow the construction of model systems near to laboratory facilities so that critical variables affecting the hydrology of wetlands or the physiology of the SRB could be manipulated for investigations into optimizing treatment performance.

## 9.6 Conclusions

Since their zenith in the eighteenth century and revival in the nineteenth century, the mines on Mynydd Parys have remained relatively undisturbed. Much of the post-mining mineralogy and microbiology can be seen to have developed remarkably rapidly within decades. Nevertheless, there are also areas which have not been disturbed for centuries and, underground, even millennia, providing potential for research on the time factor in these systems. This survival is providential and mainly due to the isolation of the area and its inherently low agricultural productivity such that 'reclamation' has been deemed uneconomic. As a result we now have a remarkably rich and well-preserved historic and scientific resource, and one that consequently offers considerable potential in a range of research disciplines. Primarily, these arise from the geochemical anomaly that produced the ore body and the consequent extreme acidities and heavy metal levels. They are manifested in an intriguing geology and biology and, in particular, in the biogeochemical interactions between minerals and bacteria, as well as in the valuable industrial archaeology.

However, this multi-disciplinary potential also poses distinct problems in management. The different historical, biological and geological aspects are all of a quality that has resulted in official recognition by the Countryside Council for Wales as SSSIs (four geological and six biological) and as scheduled ancient monuments (five) by Cadw. There is, therefore, a need for an integrated management scheme to balance all these diverse, but interlinked and sometimes incompatible, interests, and to this end the Amlwch Industrial Heritage Trust was established in 1997. Many such metal mining sites in the UK have already been lost through 'reclamation' and Mynydd Parys is therefore an important part of a diminishing resource illustrating the biological, geological and environmental features of abandoned metal mines (Jenkins and Johnson, 1994). The site now merits recognition and protection for its unique scientific value.

## Acknowledgements

The authors wish to acknowledge the encouragement and valuable comments of Mr I. Cuthbertson and Dr S. Tennant of Anglesey Mining plc., and of Dr R. Ixer and Prof. D. Taylor Smith.

## References

Anglesey Mining plc. (1998) *Annual Report*. Anglesey Mining plc, Amlwch, Anglesey.

Baas Becking, L.G.M., Kaplan, I.R. and Moore, D. (1960) Limits of the natural environment in terms of pH and oxidation-reduction potentials. *J. Geol.*, **68,** 243–84.

Banks, D., Younger, P.L., Arnesen, R.T., Iversen, R.E. and Banks, S.B. (1997) Mine water chemistry: the good, the bad and the ugly. *Environ. Geol.*, **32**, 157–74.

Barrett, T.J., Tennant, S.C. and Maclean, W.H. (1999) *Geology and mineralisation of the Parys Mountain Polymetallic Sulphide deposit, Wales, U.K.* Unpublished Report for Anglesey Mining, Wales, by Ore Systems Consulting, Vancouver.

Bevins, R. (1994) *A Mineralogy of Wales.* National Museum of Wales, Cardiff.

Bor, L. (1950) Pisanite from Parys Mountain, Anglesey. *Mineral. Mag.*, **29**, 63–7.

Boult, S. and Curtis, C.D. (1994) The predictability of metal flux in a stream heavily polluted by acid mine-drainage. Pp. 227–36 in: *Trace Substances, Environment and Health* (C.R. Cothern, editor). Science Reviews, Northwood, UK.

Butler, S. (2000) *Acid mine drainage in the Afon Goch (South), Anglesey.* Unpublished MSc thesis, Univ. Wales, Bangor.

Evangelou, V.P. (1995) *Pyrite Oxidation and its Control.* CRC Press, Boca Raton, FL.

Franklin, J.M. (1993) Volcanic-associated massive sulphide deposits. Pp. 315–34 in: *Mineral Deposit Modelling* (R.V. Kirkham, W.D. Sinclair and R.I. Thorpe, editors). Geological Association of Canada, Spec. Pap. **40**.

Gallagher, V., O'Connor, P., Good, J., Kilkenny, B., O'Suilleabhain, D. and Precott, T. (1998) Environmental problems and rehabilitation trials at Avoca mine, Ireland *Chron. Rech. Minière,* **533**, 51–69.

Gray, N.F. (1996) Environmental impact and remediation of acid mine drainage; a management problem. *Environ. Geol.*, **30**, 62–71.

Greenly, E. (1919) *The Geology of Anglesey.* HMSO, London.

Hamilton, Q.U.I., Lamb, H.M., Hallet, C. and Proctor, J.A. (1997) Passive treatment systems for the remediation of acid mine drainage. In: *Mine Water Treatment using Wetlands; Proc. CIWEM Natl. Conf., 1997* (P.L Younger, editor).

Harris, J.R. (1964) *The Copper King.* Liverpool University Press.

Hedin, R.S. (1997) Passive mine water treatment in the eastern United States. In: *Mine Water Treatment Using Wetlands; Proc. CIWEM Natl. Conf. 1997* (P.L. Younger, editor).

Ixer, R. (1999) The role of ore geology and ores in the archaeological provenancing of metals Pp. 43–51 in: *Metals in Antiquity* (S.M.M. Young, A.M. Pollard, P. Budd and R. Ixer, editors). BAR International Series, **792**.

Jenkins, D.A. (1995) Mynydd Parys copper mines. *Archaeol. Wales,* **35**, 35–6.

Jenkins, D.A. and Johnson, D.B. (1994) Abandoned metal mines: a unique mineralogical and microbiological resource. *J. Russell Soc.*, **5,** 40–4.

Johnson, D.B. (1995) Acidophilic microbial communities; Candidates for bioremediation of acidic mine effluents. *Int. Biodet. Biodeg.*, **35,** 41–58.

Johnson, D.B. (1998) Biodiversity and ecology of acidophilic microorganisms. *FEMS Microbiol. Ecol.*, **27**, 307-17.

Johnson, D.B. and Bridge, T.A.M. (1997) The role of microbial dissimilatory reduction processes in the bioremediation of metal-rich, acidic drainage waters. In: *Environmental Biotechnology: Part 1* (H. Verachtert and W. Verstraete, editors). Proc. Int. Symp., Ostend, Belgium. Technologisch Instituut, Belgium.

Johnson, D.B. and Roberto, F.F. (1997) Biodiversity of acidophilic bacteria in mineral leaching and related environments. Pp. 3–10 in: *IBS Biomine '97 Conf. Proc.* Australian Mineral Foundation, Glenside, Australia.

Johnson, D.B., McGinness, S. and Ghauri, M.A. (1993) Biogeochemical cycling of iron and sulfur in leaching environments. *FEMS Microbiol. Rev.*, **11**, 63-70.

Keith, C.N. and Vaughan, D.J. (2000) Mechanisms and rates of sulphide oxidation in relation to the problems of acid rock (mine) drainage. Pp. 117–39 in: *Environmental Mineralogy: Microbial Interactions, Anthropogenic Influences, Contaminated Land and Waste Management* (J.D. Cotter-Howells, L.S. Campbell, E. Valsami-Jones and M. Batchelder, editors). Mineralogical Society Series, **9**. Mineralogical Society, London.

Lentin, A.G.L. (1800) *Briefe über die Insel Anglesea, vorzuglich über das dasige Kupfer-Berwerk und die dazu gehorigen Schmelzwerke und Fabriken.* Leipzig.

Lyew, D., Knowles, R. and Sheppard, J. (1994) The biological treatment of acid mine drainage under continuous flow conditions in a reactor. *Trans. Inst. Chem. Eng.*, **72**, 42–7.

Manning, W. (1959) *The future of non-ferrous mining in GB and Ireland.* Institution of Mining and Metallurgy, London.

Mason, J.S. and Rust, S.A. (1997) The mineralogy of Ystrad Einion mine, Dyfed, Wales. *U.K. J. Mines Minerals*, **18,** 33–6.

Mitsch, W.J. and Gosselink, J.G. (1993) *Wetlands.* Van Nostrand Reinhold, New York.

NRA (1994) Abandoned mines and the water environment. *Water Quality Series*, **14**.

NRA (1995) *Wheal Jane, Cornwall – an example of acid mine drainage amelioration in the UK.* APEM.

Pointon, C.R. and Ixer, R.A. (1980) Parys mountain mineral deposit, Anglesey, Wales: geology and ore mineralogy. *Trans. Inst. Mining Metall., Sect. B*, 143–55.

Pollard, M., Thomas, R.G. and Williams, P.A. (1992) The stabilities of antlerite $Cu_3SO_4(OH)_4.2H_2O$: their formation and relation to other copper (II) sulphate minerals. *Mineral. Mag.*, **56**, 359–65.

Purvis, O.W. (1991) Parys Mountain – lichen survey. *Unpublished report to the NCC.*

Rowlands, J.R. (1966) *The Copper Mountain* (Llangefni).

Sen, A.M. and Johnson, D.B. (1999) Acidophilic sulphate-reducing bacteria: candidates for bioremediation of acid mine drainage. Pp. 709-18 in: *Biohydrometallurgy and the Environment Toward the Mining of the 21st Century* (R. Amils and A. Ballester, editors). Process Microbiology, **9A**. Elsevier, Amsterdam.

Schippers, A., Jozsa, P.G. and Sand, W. (1996) Sulfur chemistry in bacterial leaching of pyrite. *Appl. Environ. Microbiol.*, **62**, 3424-31.

Southwood, M. and Bevins, R. (1995) Parys Mountain: the type locality for anglesite *U.K. J. Mines Minerals*, **15**, 11–7.

Tennant, S.C. (1999) *Volcanic stratigraphy and lithogeochemistry at the Parys Mountain polymetallic sulphide deposit, North Wales.* PhD thesis, Univ. Wales, Cardiff.

Thomas, R. (2000) *The implications of dewatering Parys Mountain: a study in trace metal pollution.* MSc thesis, Univ. Wales, Bangor.

Timberlake, S. and Jenkins, D.A. (2000) Prehistoric mining: geochemical evidence from sediment cores at Mynydd Parys, Anglesey. Pp. 193–9 in: *Archaeological Sciences 1997* (A. Millward, editor). BAR vol., Archaeopress.

Torma, A.E. (1988) Leaching of metals Pp. 367–99 in: *Biotechnology*, **6B** (H.J. Rehm and G. Reed, editors). Weinheim, Germany.

Walton, K.C. and Johnson, D.B. (1992) Microbiological and chemical characteristics of an acidic stream draining a disused copper mine. *Environ. Pollution*, **76**, 169–75.

Wieder, R.K. (1989) A survey of constructed wetlands for acid coal mine drainage treatment in the Eastern United States. *Wetlands,* **9**, 299–315.

Wieder, R.K. and Lang, G.E. (1982) Modification of Acid Mine drainage in a freshwater wetland. In: *Proc. Symp. Wetlands of the Unglaciated Appalachian Region, USA* (B.R. McDonald, editor).

Wiggering, H. (1993) Sulfide oxidation – an environmental problem within colliery dumps. *Environ. Geol.,* **22**, 99–105.

Younger, P.L. (1997) The future of passive minewater treatment in the UK; a review from the Wear catchment. In: *Mine Water Treatment Using Wetlands. Proc. CIWEM Natl. Conf. 1997* (P.L Younger, editor).

CHAPTER TEN

# Decay effects associated with soluble salts on granite buildings of Braga (NW Portugal)

C. A. S. ALVES AND M. A. SEQUEIRA BRAGA*

*CCA/CT, Univ. Minho, 4700-320 Braga, Portugal*
*(E-mail: casaix@dct.uminho.pt)*

**ABSTRACT**

Granite buildings of Braga show extensive saline contamination, mainly related to anthropogenic activities. Gypsum is the salt associated with the most intense and widespread decay, affecting both internal and external surfaces of buildings. More soluble salts such as thenardite, niter and halite are associated with decay of wall paint. Soluble salts also affect mortars, mostly by the disruption of mortar joints, but, in addition, there is evidence of chemical attack. The zoned distribution of decay indicates that salt concentration is strongly linked to decay.

## 10.1 Introduction

Soluble salts are one of the main weathering agents on the Earth's surface. Besides its contribution to landscape evolution, salt weathering affects both old and new man-made constructions all over the world, endangering cultural heritage and recent structures (Goudie and Viles, 1997). The presence of soluble salts in buildings can derive from several geogenic and anthropogenic sources (Arnold and Zehnder, 1989).

Salt attack on building materials can be modelled by theoretical equations and simulated by laboratory experiments. However, the application of these theoretical and laboratory studies to the variability and complexity of field conditions is a controversial matter. In field studies there are several important uncertainties, namely in relation to salt sources, salt concentration, migration paths of solutions, etc. However, with field studies it is possible to compare decay features and patterns associated with different salt settings, and thus to evaluate the decay impact of different salts.

* Corresponding author

Alves, C.A.S. and Sequeira Braga, M.A. (2000) Decay effects associated with soluble salts on granite buildings of Braga (NW Portugal). Pp. 181–199 in: *Environmental Mineralogy: Microbial Interactions, Anthropogenic Influences, Contaminated Land and Waste Management* (J.D. Cotter-Howells, L.S. Campbell, E. Valsami-Jones and M. Batchelder, editors). Mineralogical Society Series, **9**. Mineralogical Society, London. ISBN 0 903056 20 8.

After a review of the main theoretical principles, laboratory results and field observations concerning salt weathering of building materials, this chapter goes on to characterize the salts affecting selected monuments in Braga (NW Portugal), aiming to: (1) identify sources of saline pollution and to show their link with human activities; and (2) describe and compare decay features and patterns associated with different salts and, therefore, to evaluate, in field conditions, the weathering effects of salts.

## 10.2 Salt weathering on buildings: a review

### *10.2.1 Theoretical principles*

Several aspects have been considered in order to understand the damaging effects of soluble salts.

*10.2.1.1 Development of pressures.* Pressures developed by salts are the result of crystallization, hydration and differential thermal expansion. Table 10.1 displays some of the equations that have been proposed for the calculation of these pressures and the ranking of the different pressure developed by the formation of some common salts.

*10.2.1.2 Salt crystallization position.* Soluble salts inflict more damage when they are formed inside the affected materials than when they are on the surface. The crystallization position is determined by the balance between water gained (by capillary rise or from the rain) and water lost through drying (Wendler *et al.*, 1990; Jeannette and Hammecker, 1992); by surface characteristics (Hammecker, 1993); by salt solubility (Arnold, 1982; Hammecker, 1993, 1995) and by the influence of salt mixtures (Zehnder, 1996). Hammecker (1993, 1995) proposed a model of saline solutions inside a capillary, calculating that the presence of very soluble salts would lower the water activity of solutions and promote a higher crystallization position in the capillary, while the presence of less soluble salts would not affect water activity. Less soluble salts, namely gypsum, will tend to crystallize inside the stone.

*10.2.1.3 Cyclical transformations.* Salts that experience several transformation cycles (dissolution-crystallization or hydration-dehydration) with variations in environmental conditions produce repeated crystallization and hydration pressures. Aires-Barros and Maurício (1996), based on environmental data, attempted to forecast salt transformation cycles on different parts of selected monuments and, thus, to evaluate the different weathering potential of these monuments.

*10.2.1.4 Chemical attack.* Soluble salts can inflict chemical weathering effects on building materials (Lea, 1970; Bernabé *et al.*, 1995; Goudie and Viles, 1997). Generally, salt solutions can, by the increase of ionic force, increase the solubility of minerals. Alkaline salts react with aggregates in

TABLE 10.1 Equations for the calculation of pressures developed by soluble salts and ranking of some of the most common salts.

| Mechanism | Equation | Ranking of pressures developed |
|---|---|---|
| Crystallization | Corren's equation (Winkler and Singer, 1972)<br>$P = RT\ln(C/Cs)/Vm$<br><br>R: gas constant of ideal gas;<br>$T$: temperature;<br>$C/Cs$: supersaturation of solution;<br>$V$m: molecular volume of solid salt | Halite (NaCl) > nitratine ($NaNO_3$) > niter ($KNO_3$) > thenardite ($Na_2SO_4$) > gypsum ($CaSO_4 \cdot 2H_2O$) > trona ($Na_3H(CO_3)_2*2H_2O$) > aphthitalite ($K_3Na(SO_4)_2$) > hexahydrite ($MgSO_4 \cdot 6H_2O$) > epsomite ($MgSO_4 \cdot 7H_2O$) > mirabilite ($Na_2SO_4 \cdot 10H_2O$) |
| | Fitzner-Snethlage's equation, as referred to in LaIglesia *et al.* (1997)<br>$P = 2\sigma(1/r - 1/R)$<br>$\sigma$: salt-solution interfacial tension;<br>$r$: radii of small pore;<br>$R$: radii of coarse pore | Aphthitalite > epsomite > hexahydrite > mirabilite > thenardite (from LaIglesia *et al.*, 1997) |
| Hydration | Winkler and Wilhelm (1970)<br>$P = \frac{2.3nRT}{V_h - V_a}\log\left(\frac{P_w}{P_{w'}}\right)$<br><br>$n$: water moles gained by hydration;<br>R: gas constant of ideal gas;<br>$T$: temperature;<br>$V_h$: volume of hydrate;<br>$V_a$: volume of original salt before hydration;<br>$P_w$: vapour pressure of water at temperature $T$;<br>$P_{w'}$: vapour pressure of hydrate salt at temperature $T$ | Bassanite ($CaSO_4 \cdot \frac{1}{2}H_2O$) to gypsum > thenardite to mirabilite > hexahydrite to epsomite |

mortars and concrete and can contribute to silicate mineral weathering (by increasing the pH).

### *10.2.2 Laboratory crystallization tests*

Goudie and Viles (1997) gathered data on crystallization tests performed with several soluble salts in diverse building stones and mortars, finding some variability in the damage ranking of soluble salts. However, $Na_2SO_4$ was consistently amongst the most damaging, while salts like NaCl and gypsum, which, from the previous discussion, should be particularly dangerous, consistently produced the least damage. These studies also questioned the importance of differential thermal expansion of soluble salts in causing damage. Environmental conditions and solution concentration play important roles in the crystallization tests, with greater decay for greater drying temperatures and more concentrated solutions (Alonso *et al.*, 1987; Goudie and Viles, 1997).

### *10.2.3 Previous studies on buildings*

In the study of soluble salts affecting buildings, the decay of building stones is systematized (Jeannette, 1992; Winkler, 1994) into a series of decay forms (also known as 'pathologies'). These include: efflorescences – superficial crystallization of salts on the surface of the stones; black crusts – aggregates of gypsum crystals mixed with atmospheric particles from urban and industrial pollution that form deposits on the stone surface; scales and flakes – quasi-planar stone fragments that detach themselves from the stone surface, scales being larger and thicker than flakes; and granular disintegration – loss of grains from the stone surface.

From the studies of soluble salts in buildings, the most remarkable aspect is the widespread presence of gypsum in all stone types, latitudes and environmental conditions. Gypsum occurs mostly as a superficial crystallization (black crusts) and in the detachment surface of scales and flakes. The association of gypsum with scales and flakes has been linked with the mineral's low solubility and stone exposure conditions, namely, extreme drying or cyclical wetting-drying (Wendler *et al.*, 1990; Jeannette, 1992). In contrast, Rivas *et al.* (1994) found a different situation in a granite monument in NW Spain, since, in the north façade, the frequency of scales was greater than in the sunnier south façade (where gypsum formed mainly efflorescences).

Field studies of monument decay usually deal with scaling of stone façades, without any reference to the occurrence of scales or flakes in interior stones, and Wendler *et al.* (1990) stated that scales did not occur in sheltered places.

In general, the association of scales and flakes with other soluble salts is seldom registered, e.g. the presence of scales associated with halite and calcium chloride in places where these salts were used for de-icing (Winkler and Singer, 1972; Winkler, 1994).

The development of granular disintegration has been linked to the presence of more soluble or more hygroscopic salts than gypsum (Porto *et al.*, 1991; Hammecker, 1993; Silva *et al.*, 1996; Begonha *et al.*, 1996). However, Silva *et al.* (1996) considered that the presence of more soluble salts would increase gypsum solubility, favouring the crystallization of this mineral nearer to the surface and the development of granular disintegration.

Soluble salts also affect other building materials, including wall paint (Arnold and Zehnder, 1989; Zehnder, 1996), mortar and concrete (Lea, 1970; Novak and Colville, 1989).

Another aspect to be considered in order to understand salt weathering of buildings is the spatial distribution of salts and associated decay features.

Arnold (1982) found that, in walls affected by rising damp, soluble salts were distributed in zones according to solubility, with more pronounced decay in the lower zones where the less soluble salts (like gypsum) crystallize. However, gypsum has been found at the upper zone of rising damp, when associated with more soluble salts. This has been explained by solubility increase due to ionic force increase (Zehnder, 1996).

## 10.3 Granite buildings of Braga: case studies

The city of Braga has a rich building heritage and granite has been and is a widely-used building material. From the wide range of monuments in Braga, three structures were investigated:

(1) the architectural complex of the Largo do Paço, consisting of several sections built during the Medieval, Renaissance and Baroque periods (Atanásio and Nunes, 1976), was once the official residence of the archbishops of Braga and is currently occupied by the University of Minho;

(2) the Biscainhos Museum, formerly part of the Biscainhos Palace, built by an important noble family in the 17–18th centuries (Almeida d'Eça, 1990); and

(3) the Assembleia Distrital building, formerly part of the Biscainhos Palace, but now considered a different building for administrative reasons and for its past use as a salt storehouse.

In these buildings, granite is the principal material used for walls, pavements, arches, columns, window frames and doors, sculptures and fountains. Previous studies (Alves *et al.*, 1996; Alves, 1997) showed that the granites used in the buildings exhibit different degrees of weathering inherited from the quarries and outcrops. It was verified that, in the buildings, decay is more intense and extensive where stones had inherited the greatest degree of alteration from the quarry (indicated by colour and by the presence of secondary weathering minerals like kaolinite and gibbsite). The different susceptibility to decay is explained by the consequences of natural weathering (affecting granite massifs) in the physical properties of the rocks, namely by the development of a well interconnected porous media that allows the penetration of saline solutions into the granite stones. In the buildings, stones with different degrees of weathering (inherited from the quarry) are distributed widely throughout several sections of the buildings, which allows comparison and discussion of decay effects of a diverse range of soluble salts.

In addition to granite, other building materials used in these buildings include lime mortars, with, in some cases, a more recent covering of cement mortar. Some walls are covered by a mortar rendering, finished with paint, leaving the granite exposed only in windows, doors and at the base of walls. Tiles are also found, either in panels, where a typical Portuguese tile, the 'azulejo', is used, or on the floor.

## 10.4 Characterization of soluble salts

Samples of the main decay forms identified on the buildings (efflorescences, granular disintegration, scales and flakes) were studied by X-ray diffraction (XRD), optical microscopy (following the method described in Arnold, 1984), scanning electron microscopy (SEM) with an EDS analysis system and chemical analysis of the aqueous extract by atomic absorption spectrometry (AAS) and ion chromatography.

### *10.4.1 Efflorescences*

Efflorescences are found on stone, mortar joints, mortar renderings and tiles. They are frequent and widespread in sheltered parts of the buildings. Several salts were identified either on granite or on other building materials (Figs 10.1 and 10.2). A common feature for most of the efflorescences is the presence of gypsum (generally in small amounts) associated with several other salts.

Niter is the most frequent and widespread salt in the Biscainhos Museum and is almost the only mineral found in the efflorescences that occur in a ground floor room of the Medieval section (Largo do Paço complex).

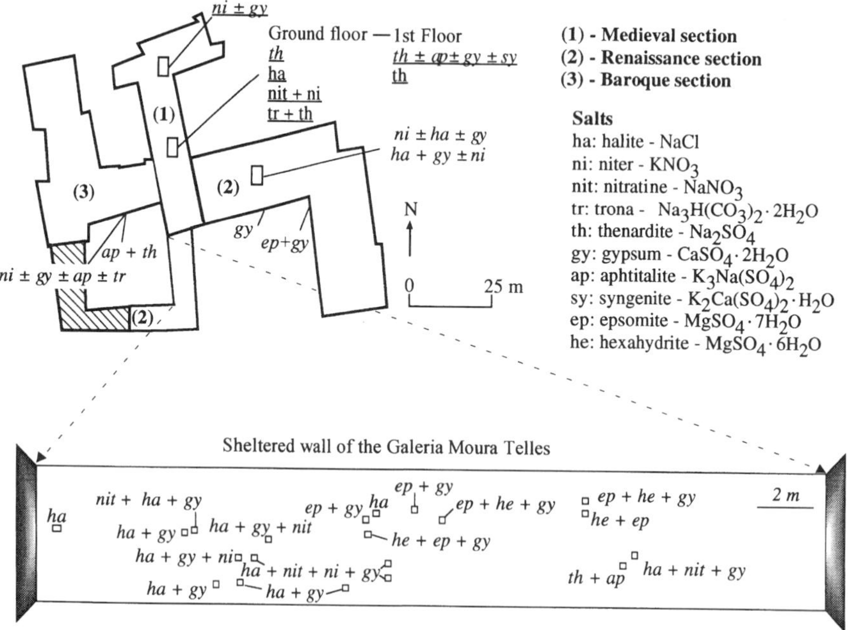


FIG. 10.1 Distribution of salts in efflorescences in the Largo do Paço complex and detailed study of the sheltered wall area of the Galeria Moura Telles. Minerals in efflorescences on granite stones are in *italics*; minerals in efflorescences on other building materials (mortars and tiles) are underlined; and minerals in efflorescences on both granite stones and other building materials are underlined and in *italics*.

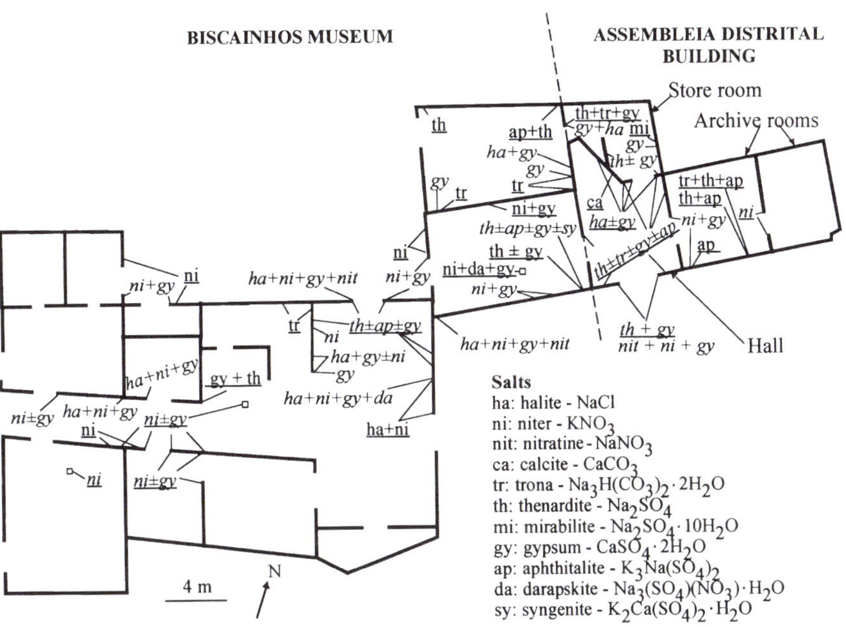


FIG. 10.2 Distribution of salts in efflorescences in the Biscainhos Museum and in the Assembleia Distrital building. Symbols as in Fig. 10.1.

Halite is a major component of efflorescences in the Assembleia Distrital building (Fig. 10.2), as expected from an old salt storehouse, but its occurrence is mostly confined to the passage between the hall and the storeroom, where supposedly the salt was kept, and to the stair and stone handrail linking the ground floor to the first floor. In the ground floor area there are efflorescences of trona and thenardite (sometimes with minor aphthitalite) and the presence of these minerals extends to the adjacent zones of the Biscainhos Museum. In the entrance hall of the Assembleia Distrital building, there are also occurrences of nitratine and niter, which becomes the main salt in the efflorescences of the east portion of the Assembleia Distrital building.

Alkaline sulphates and carbonates are dominant in the efflorescences of the Medieval section of the Largo do Paço complex, with thenardite as the most frequent mineral. There are also occurrences of halite and nitratine (besides niter) on the ground floor, and minor amounts of syngenite on the first floor.

Halite, niter and nitratine dominate the efflorescences in the Renaissance section of the Largo do Paço complex, where Mg sulphates are also found, especially in the upper part of the sheltered wall of the Galeria Moura Telles (Fig. 10.1).

Usually, the same minerals occur on the granite and mortar joints. The most striking aspect of the efflorescences on the mortar renderings is the scarcity of salt species: niter in the Biscainhos Museum; halite and niter in the

Assembleia Distrital building (plus one occurrence of calcite efflorescence); and thenardite in the Largo do Paço complex (only on the first floor of the Medieval section). The occurrence of salt efflorescences in tiles is very rare, and the most remarkable example is the large occurrence of trona in an 'azulejos' panel of the Biscainhos Museum.

### *10.4.2 Scales and flakes*

In the buildings studied, scales and flakes occur on several interior and exterior surfaces, with grossly similar frequency, challenging the statements and observations of some authors (Wendler *et al.*, 1990; Hammecker, 1993), which indicated or suggested that scales and flakes do not form in sheltered areas or building interiors. Scales and flakes, in these buildings, are generally found in areas near the floor (up to 1 m) but occasionally they have been observed at greater distances from the floor. The distribution of scales and flakes in the Assembleia Distrital building is different, being found mainly at heights >2 m.

Samples of scales and flakes from sheltered areas of the Largo do Paço complex and Biscainhos Museum are characterized by the presence of gypsum aggregates on the detachment surfaces (Fig. 10.3), regardless of the diversity

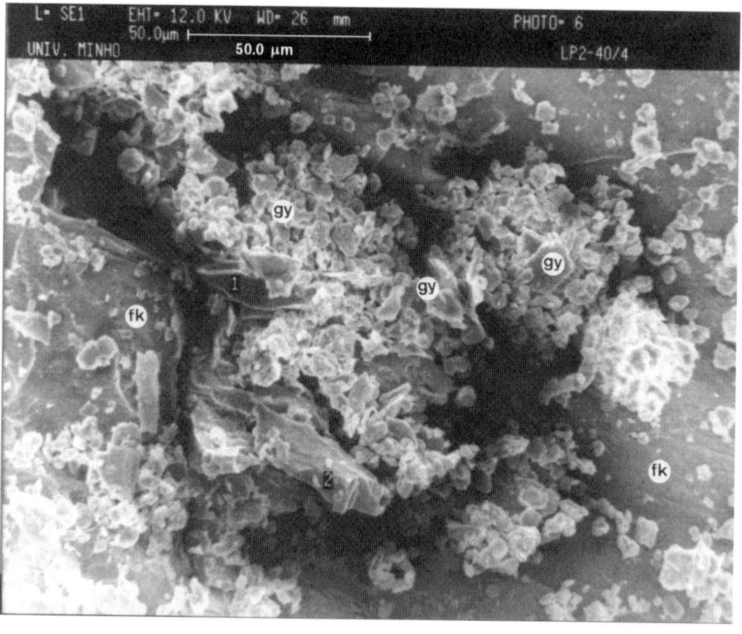


Fɪɢ. 10.3 SEM images of the detachment surface of scales from the Largo do Paço complex. Gypsum (gy) on alkaline feldspar (fk) surfaces and filling its fractures. (1,2) feldspar fragments associated with gypsum crystallization.

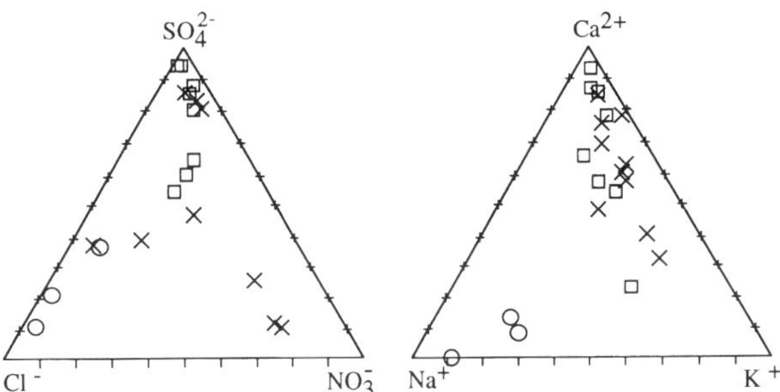


FIG. 10.4 Proportions of the main anions and cations (in mEq) in the aqueous extract of scales and flakes samples from the Largo do Paço complex and the Biscainhos Museum: sheltered (□) and unsheltered (×) locations; and from the Assembleia Distrital building (○).

of salts present in the efflorescences. Chemical analyses of the aqueous extracts of these samples show a trend towards the $SO_4^{2-}$ and $Ca^{2+}$ apexes (Fig. 10.4 – □). It is much rarer to find gypsum aggregates in the samples from unsheltered areas of these buildings and there is a major scattering of the aqueous extract composition (Fig. 10.4 – ×).

The main salt in scales and flakes from inside the Assembleia Distrital building is halite, on the detachment surface (Fig. 10.5). The aqueous extract

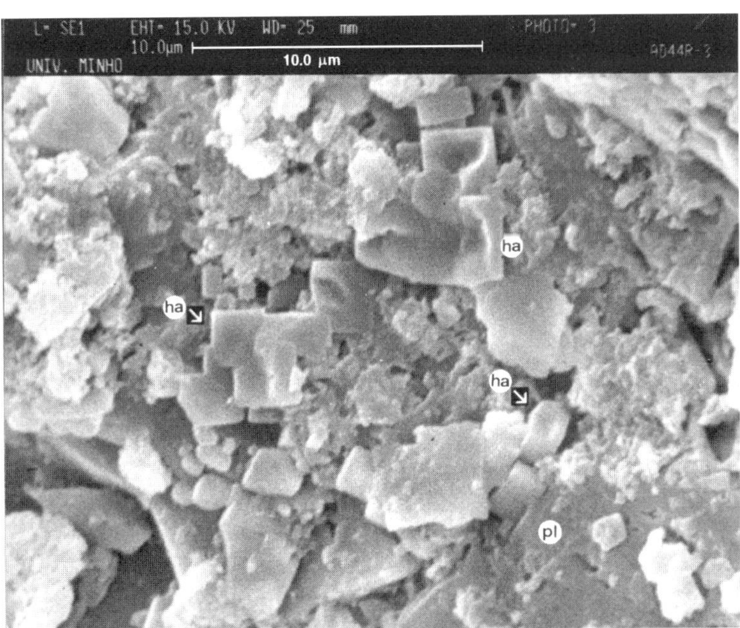


FIG. 10.5 SEM images of detachment surfaces of flakes from the Assembleia Distrital building. Halite (ha) on plagioclase (pl) crystal.

compositions of these samples show a trend towards the $Na^+$ and $Cl^-$ apexes (Fig. 10.4 – ○).

### 10.4.3 *Granular disintegration*

In the buildings studied, granular disintegration on interior and exterior surfaces is most frequent above the zone of scaling and flaking. The salt mineralogy of granular disintegration products is difficult to determine, because of interference of granite minerals. However, in some samples, gypsum was identified. The great dispersion of the aqueous extract composition (Fig. 10.6) indicates that several soluble salts are found with this decay form. Most of the samples plot in the central part of the diagram. However, some of them plot very near the $Ca^{2+}$ and $SO_4^{2-}$ apexes, suggesting that gypsum is associated with the development of granular disintegration. A sample from the Assembleia Distrital building, plots near the $Na^+$ and $Cl^-$ apexes, reflecting the influence of halite. The other sample from the Assembleia Distrital building was collected near the door of the hall (away from the area of halite concentration) and has a composition nearer the average values.

### 10.4.4 *Soluble salts zoning*

Samples of different decay forms that occur on the same stones were collected in order to understand the migration of soluble salts and to assess possible related decay effects. Alves and Sequeira Braga (1994) had previously found a zoning of soluble salts associated with the zoning of decay forms in the Biscainhos Museum.

Stones that presented efflorescences and granular disintegration in three columns of the Biscainhos Museum (profiles T, X and Y) and a stone that

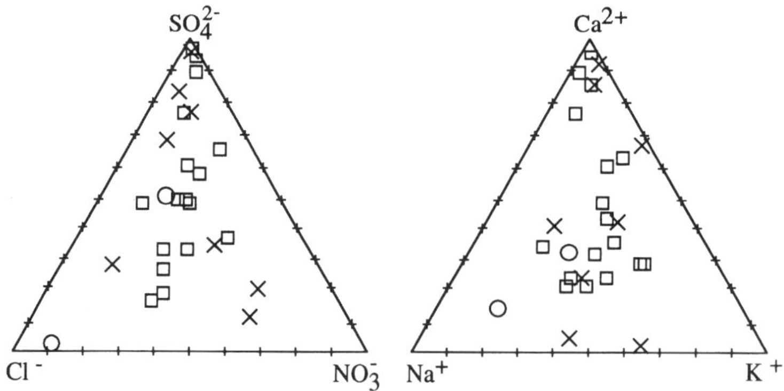


FIG. 10.6 Proportions of the main anions and cations (in mEq) in the aqueous extract of granular disintegration samples from the Largo do Paço complex and the Biscainhos Museum: sheltered (□) and unsheltered (×) locations; and from the Assembleia Distrital building (○).

presented efflorescences and scales in a door jamb of the same building (profile Z) were sampled, allowing a study of horizontal profiles and the distribution of soluble salts in decay forms. Each profile corresponds to one stone where several pathologies occur. Niter is the main mineral in the efflorescences. Besides niter, gypsum was found (in minor amounts) in the efflorescences of profile Z and halite and gypsum in the efflorescences of profile T. Gypsum was found in the detachment surface of scales of profile Z. The results of the aqueous extracts analyses (Fig. 10.7) show that, in each profile, the composition of the soluble salts that crystallize inside the stone (associated with scales and granular disintegration) is not the same as the soluble salts that crystallize on the surface (forming efflorescences). It is evident that there is an enrichment of $Ca^{2+}$ and $SO_4^{2-}$, corresponding to greater amounts of gypsum.

Samples were collected, at different heights, from stones in columns of the Largo do Paço complex (profiles A, B and C) and in a door jamb of the Biscainhos Museum (profile D) in order to study the vertical distribution of soluble salts. Samples of the same profile were collected from the same stone. For each profile, there is a decrease in the milliequivalent proportions of $Ca^{2+}$ and $SO_4^{2-}$ with the increasing height levels of the samples (Fig. 10.8). The wt.% results plotted against height (Fig. 10.9) show that there is a marked decrease of $SO_4^{2-}$ and $Ca^{2+}$ (Fig. 10.9*a,f*) with height and a dispersion of results for the other ions, with the exception of $Cl^-$ and $Na^+$, which generally tend to increase with height. The vertical zoning of the $SO_4^{2-}$ and $Ca^{2+}$ seems more related to the height of the samples than to the decay forms. This vertical zoning is related to the distribution of gypsum, which is a less soluble salt than the others most frequently found in these buildings (halite, niter or thenardite) and, hence, is concentrated in areas nearer the ground, in accordance with the solubility model of Arnold (1982).

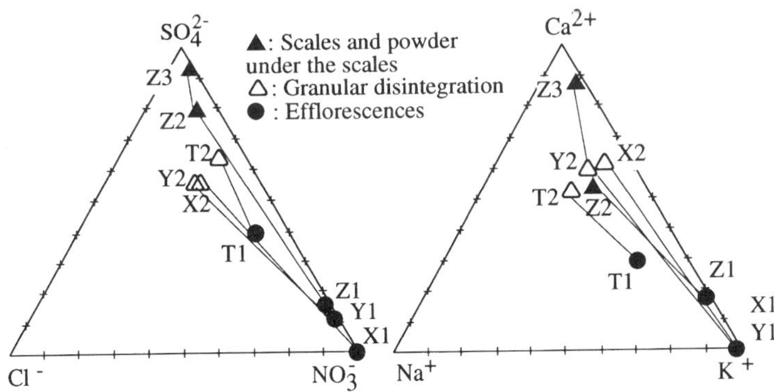


FIG. 10.7 Representation of the aqueous extract analyses of the stone decay forms in $SO_4^{2-}$-$Cl^-$-$NO_3^-$ and $Ca^{2+}$-$Na^+$-$K^+$ diagrams (mEq proportions). The letters refer to the horizontal profiles and the numbers to the sample positions in the same stone block of each profile.

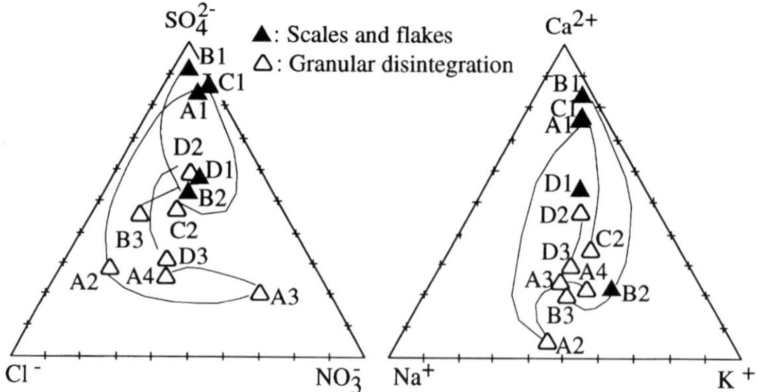


FIG. 10.8 Representation of the aqueous extract analyses of the stone decay forms in $SO_4^{2-}$-$Cl^-$-$NO_3^-$ and $Ca^{2+}$-$Na^+$-$K^+$ diagrams (mEq proportions). The letters refer to the vertical profiles and the numbers to the height of the sample positions. The numbers increase with increasing height from the floor.

### *10.4.5 Discussion of salt systems and pollution sources*

As has already been shown by Arnold and Zehnder (1989) and Arnold (1996), the type of soluble salts and their distribution allow the discussion of salt systems and pollution sources affecting buildings.

The ubiquity of gypsum for the several decay forms and for diverse buildings seems to point to a background salt system common to all the buildings and linked to ground solutions, since the presence of gypsum is more frequent in the zones near the ground. This salt system may result from the combination of several geogenic and anthropogenic sources, since $Ca^{2+}$ and $SO_4^{2-}$ are common ions in rainwater, groundwater, mortars and human and animal waste, etc.

The formation of abundant quantities of niter has been linked to the decomposition of organic matter, e.g. from old stables (Arnold and Zehnder, 1989; Arnold, 1996). The distribution and abundance of niter in the Biscainhos Museum and in a room of the Largo do Paço complex, reflect organic contributions from past use of these edifices and rooms, i.e. from domestic animals, whose presence is registered in the Biscainhos Museum.

In the Assembleia Distrital building, the presence of the halite efflorescences is clearly due to the old salt storehouse. The evolution to trona and thenardite around the area of concentration of halite efflorescences seems to relate to the interaction between solutions from the salt deposit and the mortars. This is supported by the type of soluble salts, their zoned distribution, the intense decay of the mortars and the presence of calcite efflorescences, suggesting the mobilization of compounds from the mortars. Gurrera *et al.* (1994) referred to the formation of thenardite, trona and

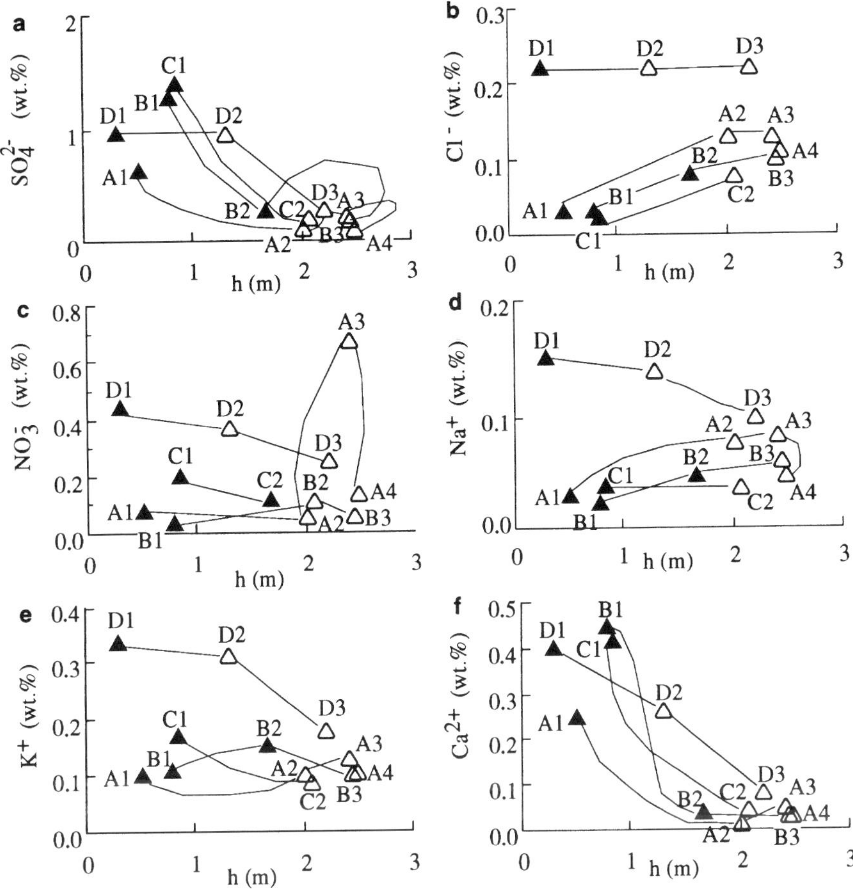


FIG. 10.9 Plot of wt.% of main ions *vs.* height, h (in m) in vertical profiles. The letters and numbers are as indicated in Fig. 10.8.

aphthitalite as a consequence of the reaction between sea-spray and cement compounds. This salt system contaminates the adjacent rooms of the Biscainhos Museum, as is clearly indicated by the distribution of the salts (Fig. 10.2). The presence of niter in the east portion of the Assembleia Distrital building reflects what must have been a common salt system in the old Biscainhos Palace. Darapskite and nitratine occur in zones of contact between the salt system characterized by niter and the salt system originating in the old salt storehouse.

The formation of trona efflorescences in the 'azulejos' panel of the Biscainhos Museum is also linked to the interaction between the saline solutions and the mortar. This is evident from the active formation of this carbonate and from other factors observed during the conservation work performed on the 'azulejos' panel. Part of the 'azulejos' panel suffered binding problems. During the conservation work it was observed that the mortar used to lay the *'azulejos'* lost most of the lime (which led to the problems with the 'azulejos' panel).

The dominance of alkaline sulphates and carbonates in most of the medieval section of the Largo do Paço complex seems to be linked to the use of cement mortar in restorations performed in the 20th century. The use of cement mortars has been postulated by several authors as an important source of alkaline sulphates and carbonates (Figg *et al.*, 1976; Halle *et al.*, 1978; Arnold and Zehnder, 1989).

In the Largo do Paço complex, the presence of halite, with nitratine and niter, appears to be associated with past and present pollution by human excreta, which is evident in the sheltered wall of the Galeria Moura Telles. Several authors (Lea, 1970; Arnold and Zehnder,1989; Martin *et al.*, 1992) connected the formation of Mg sulphates in monuments with the presence of Mg compounds in either the stones or the mortars. However, the very limited spatial distribution of Mg sulphates, their appearance near the sites of pollution by excreta and, particularly, their zoned distribution in the sheltered wall of the Galeria Moura Telles (Fig. 10.1), seem to indicate that this same salt system could be responsible for the formation of the Mg sulphates. It should be noted that hexahydrite-like crystals have been found in studies of calculi in the urinary system (Gibson, 1974).

## 10.5 Decay effects of soluble salts in granite buildings of Braga: a discussion

The analysis of the decay features present in the buildings together with their associated salt systems, allows the following discussion of soluble salts effects on different building materials.

### *10.5.1 Granite*

This discussion considers the salts that have a wider distribution in the buildings: gypsum, halite, niter and thenardite.

*10.5.1.1 Gypsum.* Gypsum is the salt most frequently associated with the deterioration of stone. It has the greatest weathering impact, as observed under field conditions. The decay effects of gypsum can be explained by its tendency to crystallize inside the stone and by its high hydration pressure (see Table 10.1).

The tendency for gypsum to crystallize inside the stone is explained by its low solubility (compared with other salts considered here) but is also linked to the wide distribution of a salt system with geochemical characteristics that favour gypsum crystallization. This salt system pervades the buildings, namely in the near ground zones, and justifies the widespread decay effect of gypsum in interior and exterior locations. The formation of scales and flakes inside the buildings cannot be linked to exceptional climatic conditions since it was verified that conditions are rather mild (Alves, 1997).

The tendency for gypsum to concentrate near the ground and inside the stones is confirmed by the geochemical studies of horizontal and vertical profiles. Besides scales and flakes, gypsum can also be associated with the development of granular disintegration, as shown by the mineralogical and geochemical results.

Efflorescences with gypsum as the main mineral are very rare, and, when they occur, the stone surface is well preserved. Even if hydration pressure can contribute to the decay action of this salt, it is clear that the decisive factor is the formation of gypsum inside the stone.

*10.5.1.2 Halite.* The previous considerations are reinforced by the contrast between the decay effects of halite in the Assembleia Distrital building and in other buildings.

Inside the Assembleia Distrital building, halite appears to be responsible for the development of scales and flakes, a situation that is not observed in the other buildings. Humidity balance and the concentration of solutions explain the distribution of decay forms in this building (Fig. 10.10). In the zone of high halite concentration (the corner near the stairs to the first floor), efflorescences dominate near the floor, where capillary rise favours the crystallization of halite on the stone surface, which remains well preserved. With increasing height, capillary supply decreases, and when it is exceeded by drying rates, salt crystallization inside the stone's porous media is favoured, therefore favouring occurrences of granular disintegration, scales and flakes at higher portions. Similar considerations of humidity balance explain the intense decay of edges, which have a greater ratio of evaporation surface to capillary supply surface (Hammecker, 1993). There is a decrease in damage intensity around these sites, as is seen in the frame of the door between the hall and the storeroom (Fig. 10.10). In the stone used as the lintel, the decay clearly diminishes from the extremity near the column to the opposite extremity. The decay of stones in the door jambs is also illustrative: the door jamb near the column is very damaged while the opposite one presents some efflorescences but has a well preserved surface. In the rest of the Assembleia Distrital building, there are frequent and abundant occurrences of efflorescences of diverse minerals (especially thenardite, trona and niter) but the decay is occasional and much less marked.

In accordance with the calculations of Hammecker (1993, 1995), the effects of the more soluble halite are concentrated at a greater height than is usually observed for gypsum-related decay forms.

In the Biscainhos Museum and in the Largo do Paço complex, the weathering aspects associated with halite are less impressive and very dispersed. In some interior locations, dissolution-crystallization cycles of

FIG. 10.10 Distribution of decay features in the hall of the Assembleia Distrital building. White arrow shows where the salt was kept. (1) efflorescences on stone; (2) scales; (3) accumulation of falling powdery material; (4) decay of mortar rendering; (5) flakes and granular disintegration on the lintel stone causing its thinning towards the column; and (6) wall paint peeling off.

halite were observed, but their effect appears to be a very slight granular disintegration, which does not increase with the several cycles observed, since these cycles occurred on the surface of the stone. In stones exposed to the sun, granular disintegration associated with halite efflorescences can be more intense but never as intense as inside the Assembleia Distrital building.

*10.5.1.3 Thenardite and niter.* In areas with frequent efflorescences of thenardite or niter, the granite stones show no more physical decay than in the rest of the buildings. The occurrence of thenardite and mirabilite in the Assembleia Distrital building indicates that Na sulphate could experience cycles of hydration-dehydration, but, as was already mentioned, the decay observed in the Assembleia Distrital building clearly diminishes in the zone where these salts are found.

### *10.5.2 Other building materials*

Wall paint is particularly susceptible to disruption by the formation of salt efflorescences in the underlying wall rendering. This decay aspect is associated with more soluble salts like niter, thenardite and halite, while gypsum is absent.

In the Biscainhos Museum, niter has a remarkable damaging effect on wall paint, linked to the widespread distribution of this mineral and to its solubility, that allows the remobilization of the solutions through the porous media of the rendering. The decay action of niter is still active since the formation of niter efflorescences is cyclically observed (Alves and Sequeira Braga, 1994).

Thenardite attack on wall paint in the Largo do Paço is confined to windows and walls affected by infiltration of rain waters and is linked to the use of modern cement mortars. The effect of thenardite efflorescences, associated with the use of cement mortars and the migration of moisture through them, is a process that continues to affect wall paint of very recent buildings.

In the 'azulejos' panel of the Biscainhos Museum, the frequent and recurrent crystallization of trona was observed in the fissures of the 'azulejos', without increasing the glaze decay, since the crystallization occurs on the surface.

Mortar wall rendering and joints did not show evidence of physical decay associated with the efflorescences of several salts. However, it is observed that the erosion of cement mortar joints and the patterns of this erosion accompany the patterns of stone decay, indicating that soluble salts contribute to the erosion of the cement mortar joints. This is well illustrated in the Assembleia Distrital building, where the high saline concentrations lead to the erosion of the mortar joints (and, in this case, also of the mortar rendering), in areas with high halite concentrations (Fig. 10.10). Besides the physical effect of salt crystallization, the concentrated saline solutions seem to cause chemical weathering in the mortar as evidenced by the calcite efflorescences in the Assembleia Distrital building as well as by the recurrent trona formation in the Biscainhos Museum 'azulejos' and the aspects observed in the underlying mortars.

## 10.6 Conclusions

It is clear that the main factor that determines the weathering of stone and other building materials in the studied granite buildings of Braga is the intensity of saline pollution, i.e. the concentration of saline solutions in relation to the solubility of the salts. Indeed, for the most active weathering agent (gypsum), the decisive factors are its low solubility (in relation to the other salts) and the geochemical abundance of Ca and sulphate ions in the pollution sources, which makes a salt system that pervades the buildings and that favours the crystallization of gypsum inside the stone in the near ground zones of the building, either inside or outside. More soluble salts need higher concentrations of solutions to favour intense crystallization, and, when that is achieved, the decay effects can be severe, as seen in the Assembleia Distrital building. But, these more soluble salts affect other building materials not affected by gypsum, namely wall paint where decay is enhanced by the higher solubility that allows the remobilization of the solutions. The results of laboratory crystallization tests have only limited applicability to these buildings, since the conditions are very different. Anthropogenic sources are the most important components of saline pollution, not only from car pollution but also from organic waste, the use of cement mortars and the mismanagement of salt deposits. The pernicious effects of saline pollution proceed since, once the materials are impregnated with soluble salts, moisture movement can lead to repeated remobilization. Saline contamination events may, therefore, act for many centuries. The mobility of these weathering agents makes them even more dangerous, and, as exemplified in the case of the Biscainhos Museum and the Assembleia Distrital building, the pollution from one place can disseminate throughout the neighbouring area.

## Acknowledgements

This research was supported financially by the EC Environment Programme (STEP-CT90-0101 project), and by FCT (Portugal), ((a) in the form of a PhD scholarship for the first author; (b) through the PRAXIS XXI – project n° 2/2.1/CSH/254/95); and (c) by the CCA-CT/FCT R&D Contract Program). Thanks are due to Dr Linda S. Campbell, Dr Bernard Smith and an anonymous reviewer for comments, suggestions and corrections which improved the manuscript, and to Dr Orlanda Marina de Nóbrega Correia and Dr Lillian Santos Reis for their help in correcting the English.

## References

Aires-Barros, L. and Maurício, M. (1996) Chronology, probability estimations and salt efflorescence occurrences forecasts on monumental building stones surfaces. *8th Int. Congr. Deterior. Conserv. Stone*, Berlin, 497–511.

Almeida d'Eça, T. (1990) *Guia - Roteiro do Museu dos Biscainhos*. IPPC.

Alonso, F.J., Ordaz, J., Valdeon, L. and Esbert, R.M. (1987) Revisión critica del ensayo de cristalización de sales. *Materiales de Construcción*, **37**, 53–60.

Alves, C.A.S. (1997) *Estudo da deterioração de materiais graníticos aplicados em monumentos da cidade de Braga. Implicações na Conservação do Património Construído*. PhD thesis, Univ. Minho, Portugal.

Alves, C.A.S. and Sequeira Braga, M.A. (1994) Niter and gypsum and their decay effects in a granitic monument of Braga (Portugal). *16th Gen. Meeting Int. Mineral. Assoc.*, Pisa, **9**.

Alves, C.A.S., Sequeira Braga, M.A. and Hammecker, C. (1996) Water transfer and decay of granitic stones in monuments. *C.R. Acad. Sci. Paris*, **323**, série IIa, 397–402.

Arnold, A. (1982) Rising damp and saline minerals. *4th Int. Cong. Deterior. Preserv. of Stone Objects*, Louisville, 11–28.

Arnold, A. (1984) Determination of mineral salts from monuments. *Studies in Conservation*, **29**, 129–38.

Arnold, A. (1996) Origin and behaviour of some salts in context of weathering on monuments. *E.C. Res. Workshop: Origin, Mechanisms and Effects of Salts on Degradation of Monuments in Marine and Continental Environments*, Bari, 133–9.

Arnold, A. and Zehnder, K. (1989) Salt weathering on monuments. *1st Int. Symp.: The Conserv. of Monuments in the Mediterranean Basin*, Bari, 31–58.

Atanásio, M.M. and Nunes, H.B. (1976) *Conjunto arquitectónico do Largo do Paço. Obras de adaptação 1974/1975*. Univ. Minho, Portugal.

Begonha, A., Sequeira Braga, M.A. and Gomes da Silva, F. (1996) Rainwater as a source of the soluble salts responsible for stone decay in the granitic monuments of Oporto and Braga –

Portugal. *8th Int. Congr. Deterior. Conserv. Stone*, Berlin, 481–7.

Bernabé, E., Bromblet, P. and Robert, M. (1995) Rôle de la cristallisation du natron dans la désagrégation sableuse d'un monument granitique en Bretagne. *C.R. Acad. Sci. Paris*, **320**, série IIa, 571–8.

Figg, J., Moore, A.E. and Gutteridge, W.A. (1976) On the occurrence of the mineral trona ($Na_2CO_3.NaHCO_3{\cdot}2H_2O$) in concrete deterioration products. *Cem. Con. Res.*, **6**, 691–6.

Gibson, R.I. (1974) Descriptive human pathological mineralogy. *Amer. Mineral.*, **59**, 1177–82.

Goudie, A. and Viles, H. (1997) *Salt Weathering Hazards*. John Wiley & Sons.

Gurrera, M.A., Raventos, X.D., Bou, V.E., Perez, J.L.P., Viñas, R.R., and Horta, A.V. (1994) Degradation forms and weathering mechanisms in the Berà Arch (Terragona, Spain). *3rd. Int. Symp.: The Conservation of Monuments in the Mediterranean Basin,* Venice, 673–9.

Halle, R., Carin, V., Hvala, M. and Crnkovic, B. (1978) A Discussion of the paper 'On the occurrence of the mineral trona ($Na_2CO_3.NaHCO_3{\cdot}2H_2O$) in concrete deterioration products'. *Cem. Con. Res.*, **8**, 251–4.

Hammecker, C. (1993) *Importance des transferts d'eau dans la dégradation des pierres en oeuvre*. PhD thesis, ULP.

Hammecker, C. (1995) The Importance of the Petrophysical Properties and External Factors in the Stone Decay on Monuments. *PAGEOPH*, **145**, 337–61.

Jeannette, D. (1992) Morphologie et nomenclature des altérations. *La conservation de la pierre monumentale en France*. CNRS, 51–72.

Jeannette, D. and Hammecker, C. (1992) Facteurs et mécanismes des altérations. *La conservation de la pierre monumentale en France*. CNRS, 73–81.

LaIglesia, A., Gonzalez, V., López-Acevedo, V. and Viedma, C. (1997) Salt crystallization in porous construction materials. I Estimation of crystallization pressure. *J. Cryst. Growth*, **177**, 111–8.

Lea, F.M. (1970) *The Chemistry of Cement and Concrete*. E. Arnold.

Martin, L., Bello, M.A. and Martin, A. (1992) The efflorescences of the cathedral of Almeria (Spain). *7th Int. Cong. Deterior. Conserv. Stone*, Lisbon, 869–73.

Novak, G.A. and Colville, A.A. (1989) Efflorescent mineralogical assemblages associated with cracked and degraded residential concrete foundations in southern California. *Cem. Conc. Res.*, **19**, 1–6.

Porto, M.C., Silva, B. and Rodrigues, J.D. (1991) Agents and forms of weathering in granitic rocks used in monuments. Pp. 439–42 in: *Europ. Symp.: Sci. Technol. Europ. Cultural Heritage, Bologna*. Butterworth-Heinemann Ltd, London.

Rivas, T., Lamas, B.P. and Silva, B. (1994) Plaque-shedding by granite in the Monastery of San Martin Pinario (Santiago de Compostela, NW Spain). *3rd. Int. Symp.: The Conservation of Monuments in the Mediterranean Basin,* Venice, 737–41.

Silva, B., Rivas, T. and Prieto, B. (1996) Relation between type of soluble salts and decay forms in granitic coastal churches in Galicia (NW Spain). *E.C. Res. Workshop: Origin, Mechanisms and Effects of Salts on Degradation of Monuments in Marine and Continental Environments, Bari*, 183–90.

Wendler, E., Klemm, D.D. and Snethlage, R. (1990) Contour scaling on building facades – Dependence on stone type and environmental conditions. *Adv. Workshop Anal. Method. Invest. Damaged Stone, Pavia*.

Winkler, E.M. (1994) *Stone in Architecture. Properties, Durability*. Springer-Verlag, Berlin.

Winkler, E.M. and Singer, P.C. (1972) Crystallization pressure of salts in stone and concrete. *Geol. Soc. Amer. Bull.*, **83**, 3509–14.

Winkler, E.M. and Wilhelm, E.J. (1970) Salt burst by hydration pressures in architectural stone in urban atmosphere. *Geol. Soc. Amer. Bull.*, **81**, 567–72.

Zehnder, K. (1996) Gypsum efflorescences in the zone of rising damp. Monitoring of slow decay process caused by crystallizing salts on wall paintings. *8th Int. Cong. Deterior. Conserv. Stone*, Berlin, 1669–78.

CHAPTER ELEVEN

# Section 3: Minerals in contaminated environments

E. VALSAMI-JONES

*Department of Mineralogy, The Natural History Museum, Cromwell Road, London SW7 5BD, UK (E-mail: evj@nhm.ac.uk)*

## 11.1 Introduction

High concentrations of toxic substances in the environment may be the result of natural processes (e.g. hydrothermal processes). However, it is mainly human activities, particularly industrial operations and mining, that have scarred the planet with a plethora of contaminated environments. As we are becoming increasingly aware of the potentially permanent damage inflicted upon the Earth's surface, and the need to ameliorate this damage, we are also developing a precise scientific understanding of environmental processes, including those that either release or lock up pollution. An essential component of the pollution jigsaw puzzle is the substrate that sustains it; a large part of this substrate is made of minerals. It is thus somewhat surprising that only in the last decade has Environmental Mineralogy been recognized as a research field in its own right.

## 11.2 Mineral-pollutant interactions

Minerals can play different roles in controlling pollutant behaviour: (1) they may be the source of the pollutant, potentially releasing it into the environment via dissolution; (2) they may control the pollutant through surface interactions: either a reversible surface attraction (adsorption) or a more permanent process of heterogeneous (i.e. induced by the mineral surface) precipitation; (3) they may nucleate homogeneously (i.e. directly from solution), to form a stable phase, capturing the pollutant within the newly-formed mineral structure; or (4) they may sequester or release the pollutant from their structure by ion exchange with the solution.

These reactions are mostly controlled by the mineral surface, which is the interface with air, water, organic/inorganic molecules and (micro-)organisms. Importantly, mineral surface properties are substantially different from those of

Valsami-Jones, E. (2000) Section 3: Minerals in contaminated environments. Pp. 201–205 in: *Environmental Mineralogy: Microbial Interactions, Anthropogenic Influences, Contaminated Land and Waste Management* (J.D. Cotter-Howells, L.S. Campbell, E. Valsami-Jones and M. Batchelder, editors). Mineralogical Society Series, **9**. Mineralogical Society, London. ISBN 0 903056 20 8.

the bulk mineral; so much so that one may consider the surface as a separate entity. Mineral surfaces represent the disruption of the 3-dimensional configuration of the crystal lattice, resulting in changes to the co-ordination chemistry around the ions exposed on the surface. The consequence is a surface electrostatic charge, when the surface ionizes in contact with water. Internal crystal chemistry imperfections and surface adsorption also contribute to the surface charge. This charge, combined with topographic heterogeneity, makes surfaces reactive. Reactivity is further influenced by the pH of the aqueous phase, temperature, solution composition, particle size and crystallinity. Specific details of how each of these variables affects the environmental role of minerals are beyond the scope of this introduction, but the principles of mineral reactivity are discussed in Stumm and Morgan (1996). Recent advances in the field are presented in a collection of papers on research in progress (Stipp *et al.*, 1999).

Environmental Mineralogy does not always consider minerals in the strict sense of the term; poorly crystalline or amorphous phases are often important. Poor crystallinity is common in phases forming on Earth's surface, and it is perhaps due to the difficulties in identifying and characterizing such phases that minerals in surface environments are not as well understood as some of their more crystalline counterparts.

Recent advances in analytical techniques, including techniques specific for studying mineral surfaces, have contributed significantly to creating the field of Environmental Mineralogy. The interested reader may wish to refer to Wogelius and Vaughan (2000) for a recent review. However, a couple of techniques are worthy of mention here, as they revolutionized our ability to observe minerals in ways not previously expected. Scanning tunnelling microscopy (STM) and atomic force microscopy (AFM) provided a capability for the *in situ* imaging of mineral surfaces on the atomic scale, in air, vacuum or even in contact with fluids. X-ray absorption spectroscopy (XAS), which relies upon high energy X-rays from synchrotron light sources, orders of magnitude brighter than conventional laboratory sources, provide the means for studying the short-range structural and chemical environment of atoms at very low concentrations, in crystalline or amorphous solids, and also in liquids and gases. This has enabled us to observe for the first time pollutants adsorbed onto mineral surfaces, and study their interactions.

Finally, important recent developments in theoretical simulations of mineral surface interactions with pollutants have emerged as a result of ever-improving computing facilities (for a review, see Tossell, 1995).

## 11.3 Environmental minerals

Certain groups of minerals are worthy of specific mention, due to their greater environmental significance and their role in influencing the flux of pollutants in contaminated environments. This role may be due to one of the following properties: (1) High solubility of a pollutant host (e.g. carbonate minerals), which will result in the release of the pollutant. (2) Low solubility of a pollutant host, which is desirable in order to reduce the availability of a pollutant. A good example here is the sparingly soluble phosphates, which are discussed further below. (3) Large surface area, which is important since release or uptake is often surface controlled. A key group of minerals here are Fe oxyhydroxides, often poorly crystalline phases, forming very small crystallites and possessing, as a consequence, exceptionally high surface areas. Cornell and Schwertmann (1996) have reviewed this important group of minerals. (4) Layered crystal structure, a property that, together with high surface areas, characterizes clay minerals. This structure of clay minerals gives them a unique ability to intercalate pollutants between layers. This is also discussed further below. (5) Open structures, a property characteristic of zeolites, which may act as molecular sieves and trap a range of pollutants. A discussion on zeolites follows in chapter 17 (Dyer, 2000).

## 11.4 Environmental Mineralogy and contaminated environments

The previous group of papers (chapters 7–10) on 'Anthropogenic Influences' in the main describes minerals as sources of pollution. Chapters 12–15 focus on the role of minerals in containing pollutants in contaminated environments, using four paradigms. **Hudson-Edwards (chapter 12)** discusses the role of Mn oxides in controlling heavy metal distribution in river sediments; their significance becomes apparent when considering their disproportionate accumulation of metals such as Pb and Zn in relation to their low relative abundance. This effect is due to a combination of large surface area, availability of surface sites and favourable surface charge of the Mn-oxides.

The role of minerals in stabilizing contaminants is well recognized. Clays, arguably more than any other group of minerals, can undergo natural or artificial transformations, which allow the intercalation of a wide range of inorganic and organic contaminants within their structure. This is discussed by **Dubbin (chapter 13)** who focuses on the mechanisms of intercalation of contaminants in expanding layer silicates. Ironically, the intercalation of organic contaminants in clays may have the dual role of both immobilizing them, but also preventing them from degrading into less toxic phases. One could envisage that by manipulating clay properties further, it might be possible in the future to create designer clays, tailor-made for specific pollutants or specific environmental applications.

**Ragnarsdottir and Charlet (chapter 14)** describe the uranium cycle on the Earth's surface. The mining of uranium for use in nuclear reactors and in weapons, has generated an unwanted stockpile of radioactive waste, the disposal of which may plague humanity for many thousands of years. In terms of its Environmental Mineralogy, uranium poses an interesting problem, being very stable under reducing conditions, but becoming soluble and mobile in the presence of oxygen, due to kinetically favourable surface oxidation reactions. Understanding uranium mineral stability may ultimately lead to better nuclear waste repository designs.

Environmental Mineralogy does not only contribute to the understanding of the distribution of contaminants. It also empowers environmental scientists with solutions derived from the understanding of mineral stability and mineral surface reactivity. An example of this is given by **Hodson *et al.* (chapter 15)** who discuss the potential of phosphate precipitation to immobilize metal contaminants in soils. Dissolution of a phosphate mineral source and subsequent precipitation of a new highly insoluble metal phosphate phase could be the key to a financially viable *in situ* remediation process.

## References

Cornell, R.M. and Schwertmann, U. (editors) (1996) *The Iron Oxides. Structure, Properties, Reactions, Occurrence and Uses*. VCH Verlagsgesellschaft mbH, Weinheim.

Dyer, A. (2000) Applications of natural zeolites in the treatment of nuclear wastes and fall-out. Pp. 315–64 in: *Environmental Mineralogy: Microbial Interactions, Anthropogenic Influences, Contaminated Land and Waste Management* (J.D. Cotter-Howells, L.S. Campbell, E. Valsami-Jones and M. Batchelder, editors). Mineralogical Society Series, **9**. Mineralogical Society, London.

Dubbin, W.E. (2000) Intercalation of organic and inorganic contaminants by expanding layer silicates. Pp. 227–44 in: *Environmental Mineralogy: Microbial Interactions, Anthropogenic Influences, Contaminated Land and Waste Management* (J.D. Cotter-Howells, L.S. Campbell, E. Valsami-Jones and M. Batchelder, editors). Mineralogical Society Series, **9**. Mineralogical Society, London.

Hodson, M.E., Valsami-Jones, E. and Cotter-Howells, J.D. (2000) Metal phosphates and remediation of contaminated land. Pp. 291–311 in: *Environmental Mineralogy: Microbial Interactions, Anthropogenic Influences, Contaminated Land and Waste Management* (J.D. Cotter-Howells, L.S. Campbell, E. Valsami-Jones and M. Batchelder, editors). Mineralogical Society Series, **9**. Mineralogical Society, London.

Hudson-Edwards, K.A. (2000) Heavy-metal bearing Mn oxides in river channel and floodplain sediments. Pp. 207–26 in: *Environmental Mineralogy: Microbial Interactions, Anthropogenic Influences, Contaminated Land and Waste Management* (J.D. Cotter-Howells, L.S. Campbell, E. Valsami-Jones and M. Batchelder, editors). Mineralogical Society Series, **9**. Mineralogical Society, London.

Ragnarsdottir, K.V. and Charlet, L. (2000) Uranium behaviour in natural environments Pp. 245–89 in: *Environmental Mineralogy: Microbial Interactions, Anthropogenic Influences, Contaminated Land and Waste Management* (J.D. Cotter-Howells, L.S. Campbell, E. Valsami-Jones and M. Batchelder, editors). Mineralogical Society Series, **9**. Mineralogical Society, London.

Stipp, S.L.S., Brady, P.V., Ragnarsdottir, K.V. and Charlet, L. (editors) (1999) Geochemistry in aqueous systems. A special issue in honour of Werner Stumm. *Geochim. Cosmochim. Acta*, **63**, 2891–3497.

Stumm, W. and Morgan, J.J. (1996) *Aquatic Chemistry. Chemical Equilibria and Rates in Natural Waters*. John Wiley & Sons, New York.

Tossell, J.A. (1995) Mineral surfaces: theoretical approaches. Pp. 61–86 in: *Mineral Surfaces* (D.J. Vaughan and R.A.D. Pattrick, editors). Mineralogical Society Series, **5**. Chapman & Hall, London.

Wogelius, R.A. and Vaughan, D.J. (2000) Analytical, experimental and computational methods in environmental mineralogy. Pp. 7–87 in: *Environmental Mineralogy* (D.J. Vaughan and R.A. Wogelius, editors). EMU Notes in Mineralogy, **2**. Eötvös University Press, Budapest.

CHAPTER TWELVE

# Heavy metal-bearing Mn oxides in river channel and floodplain sediments

K. A. HUDSON-EDWARDS

*School of Earth Sciences, Birkbeck College, University of London, London WC1E 7HX, UK (k.hudson-edwards@geology.bbk.ac.uk)*

**ABSTRACT**

Mn oxides are one of the most significant groups of substances which control the distribution of heavy metals in river sediments. This occurs because of their favourable structures, sizes and surface areas. In the Tyne, Tees and Yorkshire Ouse river basins in northeast England, heavy metal-bearing, X-ray amorphous Mn oxides are common (forming up to 15 modal % of heavy metal-bearing grains) though not always abundant phases in river channel and floodplain sediment. Manganese oxides, however, contain significantly more Pb (up to 23 wt.%) than do Fe oxides, and also contain Zn, Cd and Cu (up to 19 wt.%, 0.5 wt.% and 0.8 wt.%, respectively). Within floodplain alluvium, Mn oxides play a major role in controlling the post-depositional redistribution of heavy metals. This is shown by authigenic, Pb-, Zn-, Cd- and Cu-bearing Mn oxides in alluvial profiles at Blagill and Prudhoe in the Tyne basin, and overbank alluvial sediments at Myton-on-Swale and York in the Yorkshire Ouse basin containing multiple horizons of X-ray amorphous Mn-Ba oxides, some of which contain up to 1.0 wt.% Zn. The overbank alluvial sediments at Myton-on-Swale contain the Ba (-Mn) silicate verplanckite [$Ba_2(Mn,Fe,Ti)Si_2O_6(OH)_2 \cdot 3H_2O$], although this contains negligible (<0.1 wt.%) amounts of Zn, Pb, Cr, Ni and Co. Manganese oxides play an environmentally significant role in sequestering heavy metals, as shown by contents of up to 39–74% of total sediment-borne Pb, and 10–35% of total Zn, held by Mn oxides in north-east England fluvial sediments. Lower proportions of readily removable Pb and Zn (0.8–18% of total Pb, and 3–26% of total Zn) under ambient conditions suggest that the Mn oxides may be long-term sinks for heavy metals in the fluvial environment.

## 12.1 Introduction

Rivers are the primary arteries for the transportation and redistribution of metals at the Earth's surface (Graf, 1990). Within river systems, fine-grained (<2 mm) sediment is the principal carrier of metals, and once deposited along the rivers, these metal-bearing particles may reside within floodplain and other

Hudson-Edwards, K.A. (2000) Heavy metal-bearing Mn oxides in river channel and floodplain sediments. Pp. 207–226 in: *Environmental Mineralogy: Microbial Interactions, Anthropogenic Influences, Contaminated Land and Waste Management* (J.D. Cotter-Howells, L.S. Campbell, E. Valsami-Jones and M. Batchelder, editors). Mineralogical Society Series, **9**. Mineralogical Society, London. ISBN 0 903056 20 8.

deposits for tens to hundreds of years, where they may impact on soil, surface and ground water quality (Graf, 1994; Lewin and Macklin, 1987; Miller, 1997). Despite the fact that Mn oxides are not abundant in river systems, they play a vital role in the cycling of metals in these environments. This chapter reviews the literature on Mn oxides and their role in heavy metal distribution in river sediments, and illustrates this with studies on heavy metal-bearing Mn oxides in river channel and floodplain sediment in northeast England which have been affected by several centuries of base metal mining.

## 12.2 Mineralogy and heavy metal affinity of Mn oxides

Manganese oxides are common in both freshwater and marine deposits (Brooks, 1968; Jenne, 1968; Cronan and Thomas, 1970; Varentsov and Grasselly, 1980*a,b*; Nicholson *et al.*, 1997), and in soils (McKenzie, 1989), where they occur as suspended materials, sediments, coatings, nodules and concretions. There is a vast number of Mn oxides, hydroxides and oxyhydroxides (hereafter referred to simply as 'Mn oxides'), and in many of these, substitution of $Mn^{2+}$ and $Mn^{3+}$ for $Mn^{4+}$ occurs extensively (McKenzie, 1989). Within these, all of the Mn ions may be reduced or oxidized without changing structural position, $OH^-$ and $O^{2-}$ may occupy the same structural positions, and ions such as $Fe^{2+}$, $Li^+$, $Ba^{2+}$, $Ca^{2+}$, $Mg^{2+}$ and $Al^{3+}$ may be included in the structures (Taylor *et al.*, 1964). Mn oxides have until recently been relatively inadequately characterized, mainly because of their poor crystallinity and occurrence as mixtures. Some of the major naturally-occurring Mn oxides include pyrolusite [$\beta$-$MnO_2$], ramsdellite [$MnO_2$], nsutite [$\gamma$-$MnO_2$], romanèchite [$BaMn_9O_{16}(OH)_4$], birnessite [$(Na_{0.7}Ca_{0.3})Mn_7O_{14}\cdot 8H_2O$], todorokite [$(Na,Ca,K,Ba,Mn^{2+})_2Mn_4O_{12}\cdot 3H_2O$], hollandite [$Ba_2Mn_8O_{16}$], and lithiophorite [$(Al,Li)MnO_2(OH)_2$] (Burns and Burns, 1979; McKenzie, 1989). Most Mn oxides either possess a layer or a tunnel structure, formed from single, double, or wider chains of $MnO_6$ octahedra, linked into a framework that encloses tunnels through the structures (McKenzie, 1989).

Heavy metal adsorption on Mn oxides has been studied by many authors (McKenzie, 1967, 1970, 1972*a,b*, 1978, 1979, 1980*a,b*, 1981, 1983; Anderson *et al.*, 1973; Loganathan and Burau, 1973; Gadde and Laitinen, 1974; Murray, 1975*a,b*; Burns, 1976; Loganathan *et al.*, 1977; Hem, 1978; Means *et al.*, 1978; Murray and Dillard, 1979; Eary and Rai, 1987; Hem *et al.*, 1989; Moore *et al.*, 1990; Nicholson and Eley, 1997). This work has shown that the Mn oxides have strong but differing adsorption affinities for the heavy metals. For example, Loganathan and Burau (1973), Loganathan *et al.* (1977) and Murray (1975*a*) showed that $\delta$-$MnO_2$ adsorbs Co and Zn, while McKenzie (1979) reported that birnessite adsorbs Pb, Cu, Mn and Zn. In general, the affinity of Mn oxides for metals decreases in the order Pb > Cu > Co > Zn > Ni (Murray, 1975*b*; McKenzie, 1980*a*). Lead in particular is strongly adsorbed by natural and synthetic Mn oxides (Taylor and McKenzie, 1966; Norrish, 1975;

Davis *et al.* 1993; Thomson and Nicholson, 1998), and adsorbs preferentially to synthetic Mn oxides over synthetic Fe oxides (McKenzie, 1980*a*).

The Mn oxides are favourable hosts for heavy metals for several reasons: (1) they have structural sites which are ideal for heavy metal adsorption (Jenne, 1968; Burns and Burns, 1979; Förstner and Wittmann, 1979; Giovanoli and Balmer, 1981; Turner and Buseck, 1981; Post *et al.*, 1982; Manceau and Combes, 1988; Post and Bish, 1988); (2) they have small particle sizes and large surface areas; and (3) their surface charge is generally strongly negative due to a very low point of zero charge (Murray, 1974; Glasby, 1984). Adsorption, cation exchange, and coprecipitation are the major mechanisms for uptake of metals by Mn oxides.

The Mn oxides are investigated using a number of mineralogical and geochemical methods. The former include microscopic and spectroscopic approaches such as scanning force microscopy (SFM), X-ray photoelectron spectroscopy (XPS), auger electron spectroscopy (AES), scanning electron microscopy (SEM), X-ray diffraction (XRD), and electron microprobe. Selective chemical extractions are also used to estimate amounts of Mn oxide-associated heavy metal, particularly in sediments (e.g. Robinson, 1981; Evans and Davies, 1994), and to show how their mobility may change with respect to variations in ambient physicochemical environments. For example, hydroxylamine hydrochloride is often used to dissolve 'easily-reducible' elemental fractions which are generally represented by Mn oxides, but poorly crystallized Mn oxides are also thought to be broken down during weaker extractions using reagents such as $MgCl_2$ which removes 'exchangeable' fractions (cf. Tessier *et al.*, 1979).

## 12.3 Mn oxides in river sediments

Lind *et al.* (1987) reviewed the formation of Mn oxides in natural aqueous systems. The Mn oxides tend to form at redox fronts where water containing dissolved, reduced Mn, encounters dissolved $O_2$ or other oxidizing agents. This is particularly common at the sediment-water interface, as shown by Nowlan *et al.* (1983), who demonstrated that concretionary Mn oxides formed at an interface between oxygenated stream water and reduced Mn-bearing sediment porewater, and in groundwater systems. In groundwater-sediment systems, oxidizing fronts may occur near the water table if water moving downward as recharge differs in redox properties from the main groundwater body, or along flow paths due to permeability changes. When Mn and O concentrations reach saturation, the oxidation and precipitation reactions occur at favourable sites on solid surfaces, especially pre-existing Mn oxide surfaces (Wilson, 1980). The Mn oxide mineralogy depends on temperature, solute composition and concentrations with one or more of several metastable Mn solid phases initially precipitating, later altering to more stable minerals by disproportionation, protonation or structural rearrangement (Lind *et al.*, 1987).

Manganese originally present as highly insoluble forms may become relatively soluble when pH and/or Eh changes occur, enabling relocation in response to diffusion gradients and water movement (Krauskopf, 1957; Morgan and Stumm, 1964; Hem, 1972). Manganese in the reduced state is mobile, and Mn (III) and Mn (IV) oxides are more soluble than Fe (III) oxides (Garrels and Christ, 1965; Stumm and Morgan, 1970). As the reduced Mn is oxidized, it tends to precipitate around extant nuclei, resulting in the often-noted concentric arrangement within concretions. The Mn oxides may also be dissolved by photoreduction and natural organic compounds (Stone and Morgan, 1984).

In both polluted and unpolluted river systems, Mn oxides have been shown to play an important role in the cycling of heavy metals such as Pb, Zn, Cu and Co, and of other potentially toxic elements such as As (Jenne, 1968; Whitney, 1975; Carpenter *et al.*, 1975, 1978; Chao and Theobald, 1976; Nowlan, 1976; Carpenter and Hayes, 1978, 1980; Filipek *et al.*, 1981; Brook and Moore, 1988; Gann and Lopez, 1992; Lind and Hem, 1993; Evans and Davies, 1994). Whitney (1975), for example, suggested that the main control on the distribution of Pb, Zn and Cd in stream sediments was adsorption by Mn oxide coatings. In another stream sediment study, Nowlan (1976) showed that Ba, Cd, Co, Ni, Tl and Zn were strongly associated with Mn oxides, and Cu, Mo, Pb and Sr were weakly scavenged by Mn-Fe oxides. Hydrous Mn oxide coatings on pebbles in a stream in Virginia, USA, were demonstrated to adsorb Cu, Pb and Zn from solution by Robinson (1981). Birnessite has been identified in stream deposits, but most of the Mn oxides are X-ray amorphous (Potter and Rossman, 1979; Robinson, 1981).

The recognition that Mn oxides play an important role in sequestering heavy metals has resulted in their use as a sample medium for stream sediment geochemical surveys, particularly for ore deposit exploration (e.g. Carpenter *et al.*, 1975; Chao and Theobald, 1976; Nowlan, 1976), but more recently, fluvial Mn oxides have been investigated in a more environmental context. Andrews (1987) studied bed sediment in the Clark Fork River, Montana, USA, into which 100 million tons of As-, Cd-, Cu-, Pb- and Zn-bearing mine tailings material was supplied, and found that these elements were associated with ferromanganese sediment coatings. Scanning electron microscopy and X-ray elemental mapping showed that Cu and Zn were primarily associated with Mn-rich material on quartz and feldspar grain surfaces. In a study of stream sediments downstream of two landfill sites in Missouri, Mantei and Foster (1991) demonstrated that Co, Mn and Ba were concentrated in the chemically, operationally defined Mn oxide phase.

The formation, stability and distribution of heavy metal-bearing Mn oxides in river sediments can depend on the geomorphology, biology and amounts of Mn and Fe in the river system. Evans and Davies (1994), for example, investigated the chemical partitioning of Pb and Zn in contaminated stream sediments from the River Ystwyth in mid-Wales, UK, and showed that confinement of the channel by bedrock created very turbulent flow conditions which favoured the precipitation of Mn(IV) oxides on the surface of sand-sized particles (0.063–2 mm). Chemical phase-specific extraction work showed that Pb from this particle size was preferentially adsorbed to these Mn oxides. Von Gunten *et al.* (1991) showed that heavy metal-bearing Mn (hydr)oxide stability in the Perialpine Belt of central Europe was associated with temperature-related annual variations in microbiological activity in river water and bed sediments. Each summer, under anoxic conditions, Mn (hydr)oxides dissolved, and Cu, Zn and Cd were mobilized, whereas in the winter, relatively higher concentrations of $O_2$ and $NO_3^-$ resulted in the precipitation of Mn and other redox-sensitive elements, together with Zn and Cd. Pettine *et al.* (1992) investigated the dissolved and particulate transport of As and Cr in the Po River in Italy and found that the relative distribution of the oxidized and reduced forms of these elements was strongly dependent on the particulate concentrations of Mn and Fe.

After deposition, Mn oxides in alluvial sediment play a significant role in controlling the long-term stability and potential remobilization of metals in floodplains. Bradley and Cox (1987) demonstrated that in just 100 years significant amounts of Cd were translocated down-profile within floodplain sediments in Staffordshire, England, in response to a change in its chemical partitioning from an 'exchangeable' to an 'Fe/Mn oxide' form. This probably occurs because of the sensitivity of Mn (and Fe) oxides to changes in redox and pH (Krauskopf, 1957; Morgan and Stumm, 1964; Hem, 1972).

## 12.4 Mn oxides in river channel and floodplain sediment in northeast England

Much of the literature on heavy-metal-bearing Mn oxides in river sediments has described heavy metal-Mn oxide associations, but few papers focus on their petrography and potential environmental significance in these environments. This section of the chapter thus supplements and enhances the existing literature by (1) describing the chemistry and petrography of Mn oxides in channel and floodplain sediments from several northeast England river basins which drain Pb-Zn orefields of the Northern Pennines; (2) describing their affinity for heavy metals, especially Pb, Zn, and to a lesser extent, Cd, Cu, Cr, Ni; and (3) inferring and discussing their modes of formation and environmental significance. This section draws on results published previously in Hudson-Edwards *et al.* (1996, 1997*a,b*, 1998, 1999*b*), and also presents new data on Mn-Ba oxides and a Ba (-Mn) silicate mineral in floodplain alluvium in Yorkshire.

### *12.4.1 Area of study*

The Tyne, Tees and Yorkshire Ouse basins drain the most productive areas of the Northern Pennine and Yorkshire Dales lead-zinc-fluorite-baryte orefield in northeast England. Galena [PbS], sphalerite [ZnS], and locally, chalcopyrite

[$CuFeS_2$] are the major sulphide ore minerals, and occur with a wide range of gangue minerals such as pyrite [$FeS_2$], fluorite [$CaF_2$], baryte [$BaSO_4$], ankerite [$Ca(Mg,Fe,Mn)(CO_3)_2$], siderite [$FeCO_3$] and calcite [$CaCO_3$]. *In situ* oxidation of the orebodies has also resulted in the formation of secondary minerals such as cerussite [$PbCO_3$], anglesite [$PbSO_4$], smithsonite [$ZnCO_3$], hydrozincite [$ZnCO_3{\cdot}3Zn(OH)_2$] and malachite [$Cu_2CO_3(OH)_2$] (Dunham and Wilson, 1985; Dunham, 1990). The basins are underlain, from west to east, by Carboniferous to Cretaceous sedimentary rocks which include limestones, shales, sandstones, thin coals, marls, mudstones and chalks.

Lead mining was first documented in the Yorkshire Dales during Roman times with evidence of smelting in Nidderdale, Yorkshire, at the end of the first century AD (Jennings, 1967). Mining-related activities such as smelting, and dumping of mining material on land and into rivers contributed to river pollution from at least Medieval times onwards (e.g. Hudson-Edwards *et al.*, 1999*a*), peaking from the mid-eighteenth century until the present day (Hudson-Edwards *et al.*, 1999*b*). At least 30 $km^2$ of the Tyne, Tees and Yorkshire Ouse basins have been directly affected by historic metal mining (Macklin *et al.*, 1997), and $\sim 4.0 \times 10^5$ tonnes of Pb and $2.0 \times 10^5$ tonnes of Zn are stored in alluvial units dated to <250 years in the Yorkshire Ouse basin alone (Hudson-Edwards *et al.*, 1999*b*).

### *12.4.2 Methodology*

All of the studies have integrated a variety of field and laboratory techniques. At representative floodplain reaches, valley floor morphology was mapped onto aerial photographs, and representative floodplain and terrace units exposed and sampled both by mechanical excavator and percussion drilling. Samples of contemporary river overbank material were also collected using a stainless steel scoop and pooling material from an area of $\sim 10\ m^2$.

The sediment samples were air dried, disaggregated and sieved to pass through a 2 mm aperture. Samples were analysed for major (e.g. Fe, Mn) and trace elements (e.g. Pb, Zn, Cu, Cd, Cr) either by XRF spectrometry or by atomic absorption spectrophotometry following nitric acid digestion. Selected samples were analysed by both methods to check inter-analytical precision and accuracy and were found to be within 10% for the elements of interest. To isolate heavy metal-bearing minerals for better identification, high density minerals were separated from bulk samples using methods described by Henley (1977) and Cotter-Howells (1993). Black nodules and black Mn oxides were hand-picked using a binocular microscope and examined separately by XRD, SEM and electron microprobe (see below). Polished thin-sections of the hand-picked materials were prepared at the Department of Earth Sciences, University of Manchester, in order to avoid dissolution of the Mn oxides.

Analyses for Fe, Mn, Pb, Zn, Cu and Cd on a phase-specific, operationally-defined basis were carried out using a method adapted from Chao (1972) and

Tessier *et al.* (1979). This method partitions sediment-associated contaminant metal into the following fractions: (1) exchangeable: 1 M $MgCl_2$ shaken for 1 h, solid:solution ratio 1:10; (2) specifically adsorbed: 1 M $CH_3COONa$ (pH 5; $CH_3COOH$), shaken for 5 h, solid:solution ratio 1:10; (3) easily reducible: 0.1 M $NH_2OH{\cdot}HCl$ (0.01 M $HNO_3$), solid:solution ratio 1:20; (4) moderately reducible: 0.4 M $NH_2OH{\cdot}HCl$ (25% v/v $CH_3COOH$) 99°C, solid:solution ratio 1:20; (5) oxidizable: 30% v/v $H_2O_2$, 0.02 M $HNO_3$, 85°C, solid:solution ratio 1:20; 3.2 M $NH_4OAc$ (20% v/v $HNO_3$) added, shaken for 30 min; (6) residual: 25% v/v $HNO_3$, heated for several hours to dryness, leached in 10% v/v $HNO_3$. Manganese oxides are dissolved mainly in the easily reducible fraction, but the most reactive (and possibly most amorphous) Mn oxides are extracted during the exchangeable and specifically adsorbed procedures.

The Mn oxide and other mineralogical analysis was carried out by optical and electron microscopy (Jeol JSM 6400 scanning electron microscope (SEM) combined with energy dispersive X-ray spectrometry (EDX) and equipped with a Link Analytical backscattered electron detector; Cameca CAMEBAX electron microprobe) and confirmed in many cases by XRD (Philips PW1730 instrument with Cu-$K\alpha$ radiation at 40 kV/20 mA operating conditions; Philips PW1710 instrument fitted with a graphite monochromator, with Cu-$K\alpha$ radiation at 40 kV/30 mA operating conditions). Analytical data was obtained using polished thin-sections both on the SEM and the CAMECA microprobe. Operating conditions for the SEM were 15 kV accelerating voltage and 1.5 nA incident specimen current, and a standardless Link Analytical eXL energy dispersive analysis system with a ZAF4-FLS deconvolution/recalculation package was used to obtain the data. The counting time was 45 s for each analysis. Operating conditions for the Cameca microprobe were 20 kV accelerating voltage and 10 nA incident specimen current, and the Oxford Instruments (Link Analytical) SPECTA software was used to facilitate analysis. Modal abundance of heavy metal-bearing minerals was estimated using a grain-counting method based upon the work of Davis *et al.* (1993). The grain area of individual Pb-, Zn-, Cd- and Cu-bearing grains was measured using a Link image-processing-analysis program on the SEM. Frequencies of occurrence of the heavy metal-bearing grains in each sample were determined by summing the total area of all grains and dividing the area for each grain by the total area counted.

### *12.4.3 Mn oxides in Tyne and Tees river channel and overbank sediment*

Pb-, Zn-, Cd- and Cu-bearing Mn oxides are ubiquitous in river channel and overbank sediment from the Rivers Tyne and Tees, forming up to 15 modal % of the heavy metal-bearing mineralogy (Hudson-Edwards *et al.*, 1997*a*). The Mn oxides are often intimately associated with Fe oxides, but discrete Mn oxides also occur. All of the Mn oxides were X-ray amorphous, and contain up to 23 wt.% Pb, 19 wt.% Zn, 0.5 wt.% Cd and 0.8 wt.% Cu. Those Mn oxides which form on and pseudomorph sphalerite and galena (see below) generally contain the largest amounts of heavy metal.

Chemical extraction data for the Tyne and Tees river sediments suggest that the Mn oxides are dissolved mainly during the easily reducible extraction (Fig. 12.1; Table 12.1). The data presented in Table 12.1 suggest that 39–74% of the total Pb, and 10–35% of the total Zn, is bound to this easily reducible, Mn oxide, phase. This reflects the affinity of Pb for Mn oxides (cf. Taylor and McKenzie, 1966; Norrish, 1975; Davis *et al.* 1993; Thomson and Nicholson, 1998). By contrast, Zn, Cd and Cu are found in proportionate amounts mainly in the moderately-reducible (Fe oxide) fraction. Although Pb has been shown to

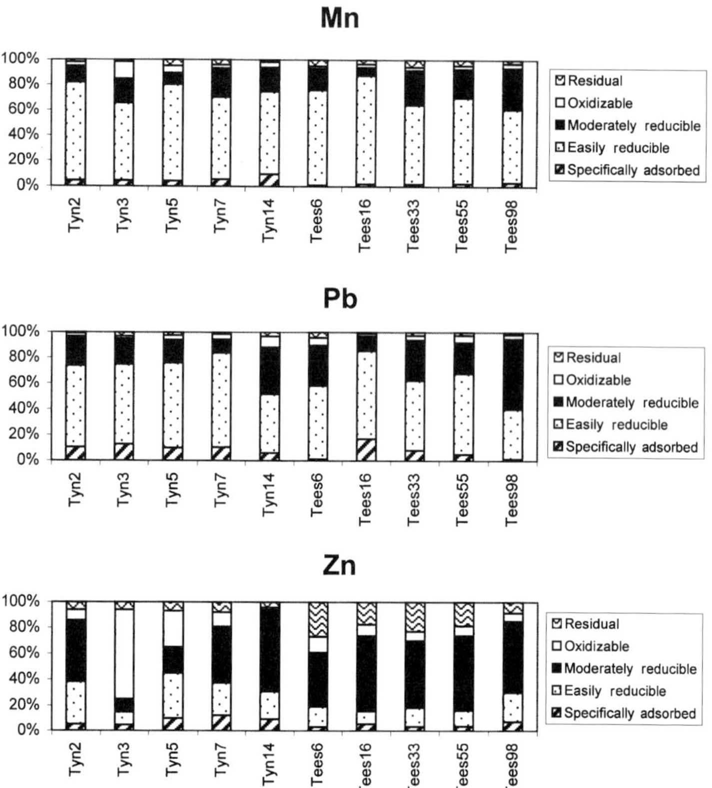


FIG. 12.1 Sequential extraction chemical data for Tyne and Tees overbank river sediments, normalized to 100%. Mn oxides are extracted mainly in the 'easily reducible' phase. A large proportion of the sediment-borne Pb is also extracted in this phase, reflecting the affinity of Mn oxides for Pb. By contrast, a smaller proportion of the Zn (and Cu and Cd, not shown) is extracted during the 'easily reducible' stage, with a larger proportion of this heavy metal associated with the 'moderately reducible' phase represented by Fe oxides. The data for this figure are listed in Table 12.1.

TABLE 12.1 Chemical extraction data for Tyne and Tees river channel sediments.

| Element/ Extraction | Tyn2 | Tyn3 | Tyn5 | Tyn7 | Tyn14 | Tees6 | Tees16 | Tees33 | Tees55 | Tees98 |
|---|---|---|---|---|---|---|---|---|---|---|
| Mn | | | | | | | | | | |
| Specifically adsorbed | 51 | 130 | 100 | 65 | 56 | 42 | 59 | 71 | 65 | 34 |
| Easily reducible | 890 | 1900 | 2000 | 790 | 390 | 3100 | 2600 | 2400 | 2000 | 650 |
| Moderately reducible | 150 | 590 | 240 | 280 | 120 | 710 | 200 | 1100 | 680 | 370 |
| Oxidizable | 36 | 420 | 150 | 33 | 25 | 84 | 81 | 96 | 78 | 40 |
| Residual | 26 | 58 | 130 | 50 | 11 | 210 | 100 | 220 | 130 | 32 |
| Pb | | | | | | | | | | |
| Specifically adsorbed | 180 | 490 | 230 | 63 | 18 | 21 | 1000 | 300 | 97 | 10 |
| Easily reducible | 1100 | 2400 | 1500 | 450 | 140 | 860 | 4200 | 2000 | 1200 | 280 |
| Moderately reducible | 400 | 780 | 410 | 63 | 110 | 470 | 700 | 1200 | 470 | 400 |
| Oxidizable | 45 | 72 | 81 | 25 | 25 | 91 | 130 | 130 | 110 | 25 |
| Residual | 14 | 130 | 62 | 10 | 10 | 65 | 57 | 94 | 44 | 13 |
| Zn | | | | | | | | | | |
| Specifically adsorbed | 60 | 1500 | 560 | 150 | 44 | 45 | 52 | 41 | 28 | 25 |
| Easily reducible | 400 | 3400 | 2000 | 310 | 100 | 240 | 100 | 190 | 100 | 80 |
| Moderately reducible | 570 | 3500 | 1100 | 540 | 310 | 640 | 580 | 640 | 460 | 190 |
| Oxidizable | 100 | 23200 | 1600 | 140 | 10 | 190 | 87 | 90 | 62 | 22 |
| Residual | 73 | 2000 | 400 | 100 | 20 | 410 | 170 | 290 | 150 | 30 |

All values in mg/kg

have a greater affinity than Zn, Cd or Cu for amorphous Fe oxides (Gadde and Laitinen, 1974; Kinniburgh *et al.*, 1976), the apparent affinity of Tyne and Tees Zn, Cd and Cu for amorphous Fe oxides may be due to the incorporation of Pb into Mn oxides, which would allow Zn, Cd and Cu to fill adsorption sites on Fe oxides. Manganese oxides (particularly $MnO_2$) tend to be negatively charged under the normal pH conditions of surface waters (>5), and the iso-electric point for Fe oxides is typically pH 6–7 (Stumm and Morgan, 1981). The apparent association of Zn, Cd and Cu with Fe oxides, and of Pb with Mn oxides, suggests that the pH of the local environment was probably within the range of 5–7, which is supported by measurements of sediment pH for the Tyne and Tees sediments (Hudson-Edwards *et al.*, 1996, 1997*a*).

As in other river systems (e.g. Andrews, 1987), many of the Tyne and Tees heavy metal-bearing Mn oxides occur as coatings on quartz (Fig. 12.2*a*), illite, Fe and other Mn oxide grains (Fig. 12.2*b*), and as discrete particles (Fig. 12.2*c*). These Mn oxides are delicate structures, which are interpreted to form by *in situ* precipitation, along with co-precipitation or adsorption of the heavy metals, at the sedimentary deposition site. The Tyne and Tees Mn oxides also replace and occur as pseudomorphs of sphalerite, cerussite, hemimorphite, chlorite, siderite (Fig. 12.2*d*) and augite. The pseudomorphs are interpreted to form by dissolution of the host phase, and precipitation of the Mn oxide, utilizing available elements from the host. These probably formed by *in situ* chemical weathering either at the deposition site or upstream during storage in mine-waste tips or other alluvium, followed by downstream transport of the weathering products.

In both the Rivers Tyne and Tees a downstream decrease in the abundance of Mn oxides relative to Fe oxides has been recorded (Hudson-Edwards *et al.*, 1996, 1997*a*). The reason for this is not fully understood, but may be due to the higher solubility of Mn oxides under low pH and Eh conditions relative to Fe oxides. In the River Tees, peaks in the abundances of Mn oxides occur in downstream alluvium, and may be due to factors such as channel morphology (cf. Evans and Davies, 1994) or the availability of particulate Mn.

### *12.4.4 Mn oxides in Tyne floodplain alluvium*

In overbank sedimentary profiles in the River Tyne, Mn oxides play a major role in the redistribution and storage of Pb, Zn, Cd and Cu (Hudson-Edwards *et al.*, 1998). Sediments from the Blagill profile, located upstream and near the former mining areas, exhibit very high metal contents (280–13000 mg/kg Pb, 790–38000 mg/kg Zn, 13–160 mg/kg Cd, 17–230 mg/kg Cu). Down-profile declines in moderately reducible fractions of Pb, Zn, Cd and Mn were thought to be related to the breakdown of Mn-bearing Fe oxides or discrete, crystalline Mn oxides. Further down-profile, relatively high proportions of exchangeable Pb, Zn, Cd and Mn, and specifically adsorbed Mn, together with petrographic evidence for authigenic, heavy metal-bearing Mn oxides, suggested that recent

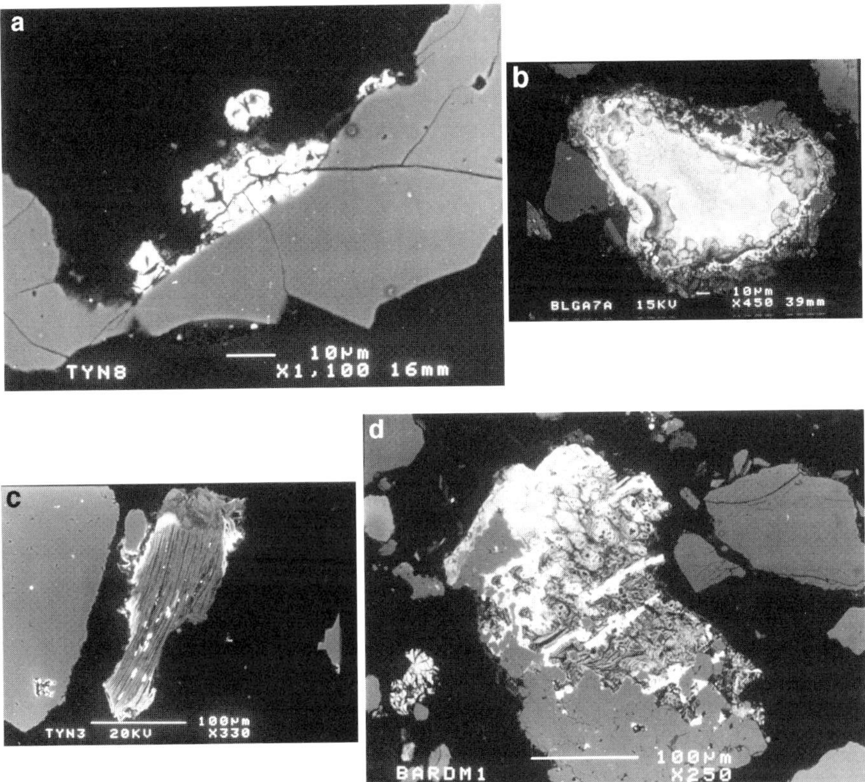


Fig. 12.2 Back-scattered image scanning electron photomicrographs: (*a*) Pb-bearing Mn oxide (white) growth on a quartz (grey) grain within Tyne overbank sediment; (*b*) Pb- and Zn-bearing Mn oxide (bright grey core of grain), partially resorbed and overgrown by bulbous Pb- and Zn-bearing Fe oxides, and overgrown by Pb-bearing Mn oxides (white outer rim); (*c*) very fine-grained Pb- and Zn-bearing Mn oxide grains (white) within and on the edges of organic matter in Tyne river channel sediment; (*d*) pseudomorphing of siderite (grey) by finely banded Pb-bearing Mn oxide (white) and Fe oxide (grey) minerals.

metal-bearing Mn oxide precipitation had occurred, possibly incorporating some of the released Pb, Zn, Cd and Mn (Hudson-Edwards *et al.*, 1998). Proportions of the total Pb and Zn released are ~7–18% and 15–26%, respectively. The evidence for the breakdown and subsequent reprecipitation of Pb-, Zn-, Cd- and Cu-bearing Mn oxides in the Blagill profile was explained by fluctuating water table conditions described in geomorphological studies of the area (Macklin, 1986). During a rise in water table levels, Mn oxides would dissolve by the reduction of Mn (III, IV) to Mn (II), releasing Pb, Zn, Cd, and Cu. When the water table falls, some of the mobilized, reduced Mn is oxidized and re-precipitated as amorphous Mn oxides which could adsorb or co-precipitate the contaminant metals.

The Prudhoe profile, located 81 km downstream of the mining areas, also exhibits high metal sediment-borne contents (610–2300 mg/kg Pb, 720–2700 mg/kg Zn, 1.6–8.0 mg/kg Cd, 11–42 mg/kg Cu), and the heavy metal-bearing mineralogy is dominated by Mn and Fe oxides. At Prudhoe, relatively high levels of exchangeable Pb, Zn, Cd and Mn in the upper part of the profile were thought to be related to the breakdown of heavy metal-bearing Mn oxides in the presence of abundant organic material. Secondary redistribution of the heavy metals was corroborated by the presence of authigenic Mn-Pb-Zn oxides within charcoal fragments in this part of the profile. Further down-profile, peaks in easily reducible Pb, Mn and Zn suggested that the heavy metals leached from the upper part of the profile may have accumulated in secondary Mn oxides (Hudson-Edwards *et al.*, 1998). Calculations using the sequential extraction data in Hudson-Edwards *et al.* (1998) suggest that 0.8–2% of the total Pb, and 3–7% of total Zn, are mobilized from the upper part of the Prudhoe profile.

### *12.4.5 Mn-Ba oxides in floodplain sediments of the Yorkshire Ouse basin*

Mn- and Ba-rich horizons occur in floodplain alluvium at Myton-on-Swale (SE 4304 6600, Fig. 12.3) and York (SE 6055 5100; Fig. 12.4) in the Yorkshire Ouse basin. The enrichments in Mn and Ba are also associated with enrichments in sediment-borne Zn, Ni, Cu, Co and depletions in Cr (Fig. 12.3). At Myton-on-Swale, the main Mn-Ba-rich horizon occurs in sediments lying between 2.65 and 2.71 m below the present-day surface, and dated to between 5450–4990 cal. years BP and 670–530 cal. years BP. Lesser horizons occur at 1.00–1.52 m and 5.00–5.49 m below the present-day surface. Within the Mn-Ba-rich horizons, discrete Mn-Ba phases with plate-like morphologies overgrow and cement clusters of detrital sediment grains comprising mixtures of quartz, albite, muscovite, kaolinite and illite. X-ray diffraction investigations of these minerals show that verplanckite [$Ba_2(Mn,Fe,Ti)Si_2O_6(OH)_2 \cdot 3H_2O$] is present at Myton-on-Swale, and SEM and microprobe analysis suggest that Mn-Ba oxides are present.

There are at least two distinct Mn-Ba-rich horizons in the sediment cores at York (Fig. 12.4), occurring between 8.5 and 10.5 m below the present-day surface, in sediments dated to at least 940–1160 cal. years BP. In contrast to the Mn-Ba horizons at Myton-on-Swale, those at York are dominated by cm-diameter black to rusty-black nodules composed of successive generations of finely laminated to rounded, interlaminated Fe- and Mn-Ba oxides, as shown by SEM-EDX analysis. The XRD analysis did not reveal the presence of any particular Mn oxides, but abundant quartz and amorphous Fe oxide were present.

Electron microprobe analysis has shown that some of the Myton-on-Swale and York Mn-Ba oxides contain up to 1.0 wt.% Zn, but analyses for Pb, Cr, Ni and Co were below detection limits of 0.1 wt.%. The Ba (-Mn) silicate (verplanckite) in the Myton-on-Swale core was analysed, but contained <0.1 wt.% Pb, Cr, Ni and Co.

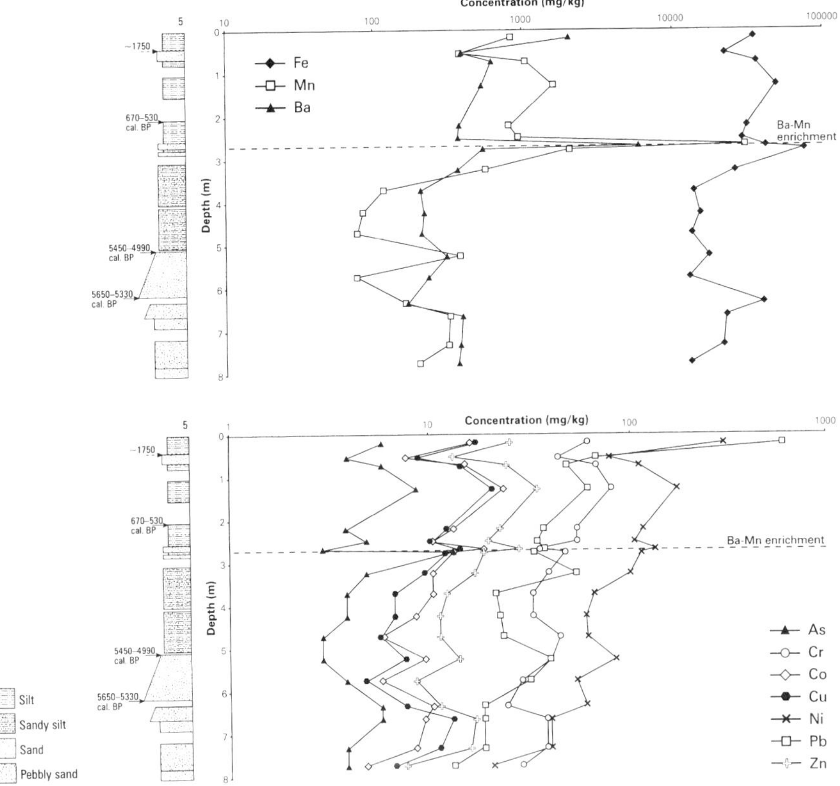


Fig. 12.3 Sediment-borne concentrations of Fe, Mn, Ba, As, Cr, Co, Cu, Ni, Pb and Zn for Myton-on-Swale core 5. The midpoints of the most prominent Mn-Ba enrichment zone, which contains discrete Mn-Ba oxides and the Ba (±Mn) silicate verplanckite [$Ba_2(Mn,Fe,Ti)$-$Si_2O_6(OH)_2 \cdot 3H_2O$] is shown as a dashed horizontal line. Other Ba-Mn enrichment zones occur at 1.00–1.52 m and 5.00–5.49 m.

Pearson correlation coefficients for Mn and the other analysed elements (Table 12.2) show a very strong correlation between Mn and Ba throughout the sediment columns (0.97, Myton-on-Swale; 0.96, York), suggesting that much of the Mn within the sediments is bound to Ba. Nicholson (1992) has reported the Mn-Ba association as a characteristic of freshwater Mn deposits. The coefficient for Mn and Zn in the Myton-on-Swale core is also high (0.82; Table 12.2), reflecting the association of Zn with the Mn-Ba oxides described above.

Despite the different modes of occurrence and chemistries of the Myton-on-Swale and York Mn-Ba oxides, in both locations they occur at considerable depths below the present-day surface and control the distribution of Zn, and

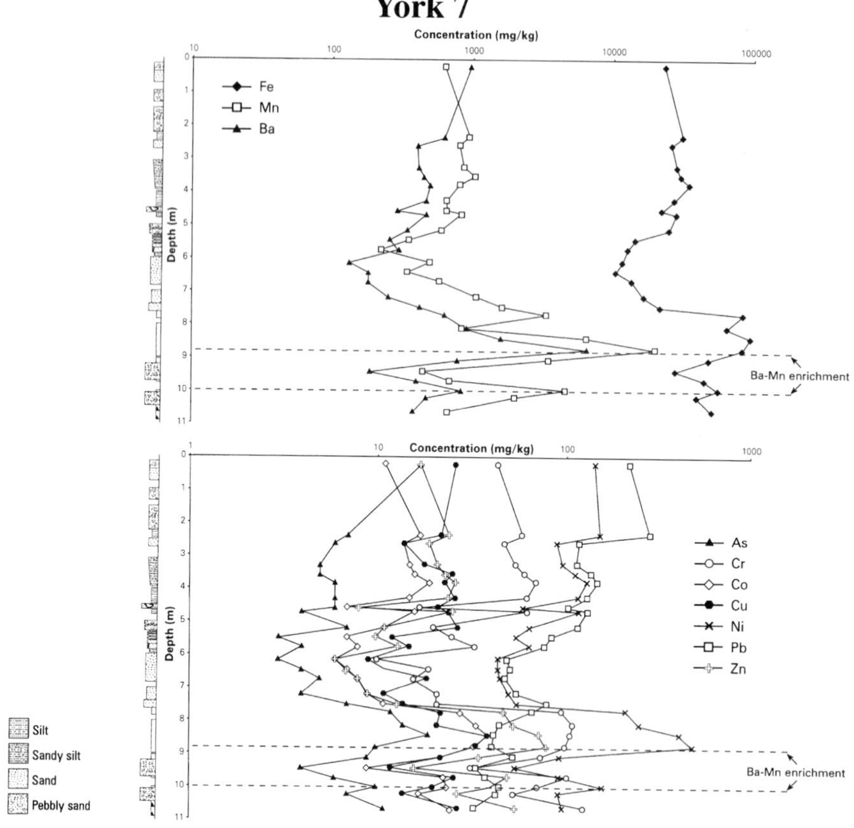


FIG. 12.4 Sediment-borne concentrations of Fe, Mn, Ba, As, Cr, Co, Cu, Ni, Pb and Zn for York core 7. The midpoints of the most prominent Mn-Ba enrichment zones which contain black nodules are shown as dashed horizontal lines, but other minor enrichment zones occur at 4.61–4.79 m and 7.63–7.85 m. The prominent Mn-Ba enrichment zones are at least as old as 940–1100 cal. years B.P., as indicated by a dated piece of included wood occurring at 7.57–7.60 m.

probably Ni, Cu and Co. The cement-like and concretionary morphologies suggest that the Mn-Ba oxides are probably authigenic, and the occurrence of several horizons of the Mn-Ba oxides suggests that they may have formed in response to pH and Eh changes related to changes in water-table levels (cf. Hem *et al.*, 1989). Indeed, the existence of several Ba-Mn enrichment horizons (Figs 12.3, 12.4), the correlations between Mn and Ba (Table 12.2) and the cyclicity in the Mn and Ba geochemistries (Figs 12.3, 12.4), suggests that these water-table levels may have fluctuated over time, and thus that the enrichment horizons represent palaeo-oxidation fronts. This is supported by rapid aggradation and frequent flooding at Myton and York during the Holocene (Macklin *et al.*,

TABLE 12.2 Selected Pearson correlation coefficients for Myton-on-Swale and York core sediments.

| Element | Correlation coefficient with Mn Myton-on-Swale core 5 | Correlation coefficient with Mn York core 7 |
|---|---|---|
| Fe | 0.65 | 0.27 |
| Ba | 0.97 | 0.96 |
| As | 0.35 | 0.14 |
| Cr | 0.41 | 0.09 |
| Co | 0.57 | 0.45 |
| Cu | 0.50 | 0.27 |
| Ni | 0.69 | 0.42 |
| Pb | −0.25 | −0.04 |
| Zn | 0.82 | 0.20 |

2000), probably causing relative changes in water-table level within the alluvial sediment piles. The Mn, Ba and heavy metal cations may have been derived from the downward percolation of interstitial soil/sediment water, or from groundwater. This is corroborated by studies near York (Jennings, 1974) which have described spring waters which contain up to 68 mg $l^{-1}$ Ba.

The Mn-Ba horizons are important in that they may represent both barriers to heavy metal migration from groundwater or from downward percolating soil/sediment water, and an indication of fossil water-table levels. In an analogous study, Carlos *et al.* (1993) identified several tunnel-structure, fracture-lining Mn oxides in silicic tuff at Yucca Mountain, Nevada, U.S.A, and suggested that they were probably the most significant minerals for the retardation of some radionuclides in a waste repository.

The Mn-Ba oxides at Myton-on-Swale, however, may have caused mobilization of Cr. The main Mn-Ba horizon shows a depletion in Cr (Fig. 12.3), suggesting that either the primary sediment-borne Cr concentration is relatively low at this horizon, or that the Mn oxides oxidized Cr (III) to Cr (VI), greatly increasing the mobility of Cr (cf. Eary and Rai, 1987).

## 12.5 Conclusions and environmental significance

Manganese oxides are one of the most significant groups of substances which control the distribution of heavy metals in river channel and floodplain sediment. Although the Mn oxides do not form a large proportion of the total heavy metal-bearing lode in these sediments, they do contain a disproportionately large amount of heavy metal, particularly Pb and Zn (in Mn-Ba oxides). The Mn oxides also appear to play possibly a more significant role than Fe oxides in the redistribution and storage of heavy metals within floodplain sediment profiles, which in turn appears to be largely related to fluctuations in water table levels.

The Mn oxides also play an environmentally significant role in the sequestering of heavy metals and in reducing their mobility. Chemical extraction data suggest that a large proportion of the total Pb and Zn (39–74% and 10–35%, respectively) in Tyne basin river sediments is bound to Mn oxides. The extraction data and accompanying evidence for metal mobility within floodplain overbank profiles suggest that only 0.8–18% of total Pb, and 3–26% of total Zn are readily removed under ambient conditions. These lower proportions suggest that the Mn oxides may be a significant sink for heavy metals, reducing their potential effects on organisms which live in and utilize the fluvial environment.

## Acknowledgements

This paper summarizes the results of work funded by both NERC and NSERC (Canada), carried out at the Department of Earth Sciences, University of Manchester, The School of Geography, University of Leeds, the Department of Mineralogy, Natural History Museum (NHM), London, and the School of Earth Sciences, Birkbeck College, University of London. Assistance from D. Plant and T. Hopkins (Manchester Electron Microprobe Facility), P. Lythgoe (Manchester), J. Spratt (NHM) and S. Hirons (Birkbeck) is gratefully acknowledged. L. Wright (Leeds) prepared many of the diagrams. S. Edwards is thanked for reading an early version of the manuscript, and the continued inspiration and support from M.G. Macklin in the field of fluvial geomorphology is especially appreciated. K. Nicholson, B. Dubbin and an anonymous reviewer are thanked for their incisive and helpful comments which greatly improved the manuscript.

## References

Anderson, B.J., Jenne, E.A. and Chao, T.T. (1973) The sorption of silver by poorly crystallized manganese oxides. *Geochim. Cosmochim. Acta*, **37**, 611–22.

Andrews, E.D. (1987) Longitudinal dispersion of trace metals in the Clark Fork River, Montana. Pp. 179–91 in: *Chemical Quality of Water and the Hydrologic Cycle* (R.C. Averett and D.M. McKnight, editors). Lewis Publishers Inc., Chelsea, Michigan.

Bradley, S.B. and Cox, J.J. (1987) Heavy metals in the Hamps and Manifold valleys, north Staffordshire, U.K.: partitioning of metals in floodplain soils. *Sci. Tot. Environ.*, **65**, 135–53.

Brook, E.J. and Moore, J.N. (1988) Particle-size and chemical controls of As, Cd, Cu, Fe, Mn, Ni, Pb, and Zn in bed sediment from the Clark Fork River, Montana (USA). *Sci. Total Environ.*, **76**, 247–66.

Brooks, D.B. (1968) Deep sea manganese nodules: from scientific phenomenon to world resource. *Nat. Res. J.*, **8**, 401–23.

Burns, R.G. (1976) The uptake of cobalt into ferromanganese nodules, soils, and synthetic manganese (IV) oxides. *Geochim. Cosmochim. Acta*, **40**, 95–102.

Burns, R.G. and Burns, V.M. (1979) Manganese oxides. Pp. 1–46 in: *Marine Minerals* (P.H. Ribbe, editor). Reviews in Mineralogy, **6**. Mineralogical Society of America, Washington D.C.

Carlos, B.A., Chipera, S.J., Bish, D.L. and Craven, S.J. (1993) Fracture-lining manganese oxides minerals in silicic tuff, Yucca Mountain, Nevada, U.S.A. *Chem. Geol.*, **107**, 47–69.

Carpenter, R.H. and Hayes, W.B. (1978) Precipitation of iron, manganese, zinc, and copper on clean ceramic surfaces in a stream draining a polymetallic sulfide deposit. *J. Geochem. Explor.*, **9**, 31–37.

Carpenter, R.H. and Hayes, W.B. (1980) Annual accretion of Fe-Mn oxides and certain associated metals in a stream environment. *Chem. Geol.*, **29**, 249–59.

Carpenter, R.H., Pope, T.A. and Smith, R.L. (1975) Fe-Mn oxide coatings in stream sediment geochemical surveys. *J. Geochem. Explor.*, **4**, 349–63.

Carpenter, R.H., Robinson, G.D. and Hayes, W.B. (1978) Partitioning of manganese, iron, copper, zinc, lead, cobalt and nickel in black coatings on stream boulders in the vicinity of the Magruder mine, Lincoln County, Georgia. *J. Geochem. Explor.*, **10**, 75–89.

Chao, T.T. (1972) Selective dissolution of manganese oxides from soils and sediments with acidified hydroxylamine hydrochloride. *Soil Sci. Soc. Am. Proc.*, **36**, 764–8.

Chao, T.T. and Theobald, P.K., Jr. (1976) The significance of secondary iron and manganese oxides in geochemical exploration. *Econ. Geol.*, **71**, 1560–9.

Cotter-Howells, J. (1993) Separation of high-density minerals from soil. *Sci. Tot. Environ.*, **132**, 93–8.

Cronan, D.B. and Thomas, R.L. (1970) Ferromanganese concretions in Lake Ontario. *Canad. J. Earth Sci.*, **7**, 1346–9.

Davis, A., Drexler, J.W., Ruby, M.V. and Nicholson, A. (1993) Micromineralogy of mine waste in relation to lead bioavailability. *Environ. Sci. Tech.*, **27**, 1415–25.

Dunham, K.C. (1990) Geology of the Northern Pennine Orefield. Vol. 1. Tyne to Stainmore. 2nd Ed. *Econ. Mem. Br. Geol. Surv.*, HMSO. London, Sheets 19 and 25 and Parts of 13, 24, 26, 31, 32 (England and Wales).

Dunham, K.C. and Wilson, A.A. (1985) Geology of the Northern Pennine Orefield, Volume 2 Stainmore to Craven. *Econ. Mem. Br. Geol. Surv.*, Sheets, 40, 41, 50, and parts of 31, 32, 51, 60, and 61.

Eary, L.E. and Rai, D. (1987) Kinetics of chromium (III) oxidation to chromium (VI) by reaction with manganese dioxide. *Environ. Sci. Tech.*, **21**, 1187–93.

Evans, D. and Davies, B.E. (1994) The influence of channel morphology on the chemical partitioning of Pb and Zn in contaminated river sediments. *Appl. Geochem.*, **9**, 45–52.

Filipek, L.H., Chao, Y.T. and Carpenter, R.H. (1981) Factors affecting the partitioning of Cu, Zn and Pb in boulder coatings and stream sediments in the vicinity of a polymetallic sulphide deposit. *Chem. Geol.*, **33**, 45–64.

Förstner, U. and Wittmann, G.T.W. (1979) *Metal Pollution in the Aquatic Environment.* Springer, Berlin.

Gadde, R.R. and Laitinen, H.A. (1974) Studies of heavy metal adsorption by hydrous iron and manganese oxides. *Anal. Chem.*, **46**, 2022–6.

Gann, A. and Lopez, E. (1992) Complex Fe-Mn oxide coatings on boulders in a tropical river. Pp. 557–60 in: *Proc. 7th Int. Symp. Water-Rock Interaction.* Balkema, Rotterdam.

Garrels, R.M. and Christ, C.L. (1965) *Solutions, Minerals and Equilibria.* Harper and Row, New York.

Giovanoli, R. and Balmer, B. (1981) A new synthesis of hollandite. A possibility for immobilization of nuclear waste. *Chimia*, **35**, 53–5.

Glasby, G.P. (1984) Manganese in the marine environment. *Oceanogr. Mar. Biol. Ann. Rev.*, **22**, 169–94.

Graf, W.L. (1990) Fluvial dynamics of thorium-230 in the church rock even, Puerco River, New Mexico. *Ann. Assoc. Amer. Geogr.*, **80**, 327–42.

Graf, W.L. (1994) *Plutonium and the Rio Grande: Environmental Change and Contamination in the Nuclear Age.* Oxford University Press, New York.

Hem, J.D. (1972) Chemical factors that influence the availability of iron and manganese in aqueous systems. *Geol. Soc. Amer. Bull.*, **83**, 443–50.

Hem, J.D. (1978) Redox processes at surfaces of manganese oxide and their effects on aqueous metal ions. *Chem. Geol.*, **21**, 199–218.

Hem, J.D., Lind, C.J. and Roberson, C.E. (1989) Coprecipitation and redox reactions of manganese oxides with copper and nickel. *Geochim. Cosmochim Acta*, **53**, 2811–22.

Henley, K.J. (1977) Improved heavy liquid separation at fine particle sizes. *Amer. Mineral.*, **62**, 377–81.

Hudson-Edwards, K.A., Macklin, M.G., Curtis, C.D. and Vaughan, D.J. (1996) Processes of formation and distribution of Pb-, Zn-, Cd-, and Cu-bearing minerals in the Tyne basin, northeast England: implications for metal-contaminated river systems. *Environ. Sci. Tech.*, **30**, 72–80.

Hudson-Edwards, K., Macklin, M. and Taylor, M. (1997*a*) Historic metal mining inputs to Tees river sediment. *Sci. Tot. Environ.*, **194/195**, 437–45.

Hudson-Edwards, K.A., Macklin, M.G., Curtis, C.D. and Vaughan, D.J. (1997*b*) Characterisation of Pb-, Zn-, Cd-, and Cu-bearing phases in river sediments by combined geochemical and mineralogical techniques. CD-ROM File 016.PDF in *Contaminated Soils. Proc. 3$^{rd}$ Int. Conf. on Biogeochemistry of Trace Elements, Paris*. French Ministry of the Environment.

Hudson-Edwards, K.A., Macklin, M.G., Curtis, C.D. and Vaughan, D.J. (1998) Chemical remobilisation of contaminant metals within floodplain sediments in an incising river system: implications for dating and chemostratigraphy. *Earth Surf. Proc. Landforms*, **23**, 671–84.

Hudson-Edwards, K.A., Macklin, M.G., Finlayson, R. and Passmore, D.G. (1999*a*) Medieval lead pollution in the river Ouse at York, England. *J. Archaeol. Sci.*, **26**, 809–19.

Hudson-Edwards, K.A., Macklin, M.G. and Taylor, M.P. (1999*b*) 2000 years of sediment-borne heavy metal storage in the Yorkshire Ouse basin, NE England. *Hyd. Proc.*, **13**, 1087–102.

Jenne, E.A. (1968) Controls on Mn, Fe, Co, Ni, Cu, and Zn concentrations in soils and water: the significant role of hydrous Mn and Fe oxides. *Adv. Chem. Ser.*, **73**, 337–88.

Jennings, B. (editor) (1967) *A History of Nidderdale*. Advertiser, Huddersfield.

Jennings, B. (1974) *A History of the Wells and Springs of Harrogate*. Harrogate Corporation Department of Conference and Resort Services, Harrogate.

Kinniburgh, D.G., Jackson, M.L. and Syers, J.K. (1976) Adsorption of alkaline earth, transition and heavy metal cations by hydrous oxide gels of iron and aluminum. *Soil Sci. Soc. Amer. J.*, **40**, 796–9.

Krauskopf, K.B. (1957) Separation of manganese from iron in sedimentary processes. *Geochim. Cosmochim. Acta*, **12**, 61–84.

Lewin, J. and Macklin, M.G. (1987) Metal mining and floodplain sedimentation in Britain. Pp. 1009–27 in: *International Geomorphology 1986*, Part 1 (V. Gardiner, editor). Wiley, Chichester.

Lind, C.J. and Hem, J.D. (1993) Manganese minerals and associated fine particulates in the streambed of Pinal Creek, Arizona, USA – A mining related acid drainage problem. *Appl. Geochem.*, **8**, 67–80.

Lind, C.J., Hem, J.D. and Roberson, C.E. (1987) Reaction products of manganese-bearing waters. Pp. 271–301 in: *Chemical Quality and the Hydrologic Cycle* (R.C. Averett and D.M. McKnight, editors). Lewis Publishers Inc., Chelsea, Michigan.

Loganathan, P. and Burau, R.G. (1973) Sorption of heavy metals by a hydrous manganese oxide. *Geochim. Cosmochim. Acta*, **37**, 1277–93.

Loganathan, P., Burau, R.G. and Fuerstenau, D.W. (1977) Influence of pH on the sorption of $Co^{2+}$, $Zn^{2+}$ and $Ca^{2+}$ by a hydrous manganese oxide. *Soil Sci. Soc. Amer. J.*, **41**, 57–62.

Macklin, M.G. (1986) Channel and floodplain metamorphosis in the River Nent, Cumberland. Pp. 13–19 in: *Quaternary River Landforms and Sediments in the Northern Pennines, England* (M.G. Macklin and J. Rose, editors). B.G.R.G./Q.R.A Field Guide, London.

Macklin, M.G., Hudson-Edwards, K.A. and Dawson, E.J. (1997) The significance of pollution from historic metal mining in the Pennine orefields on river sediment contaminants fluxes to the North Sea. *Sci. Tot. Environ.*, **194/195**, 391–7.

Macklin, M.G., Taylor, M.P., Hudson-Edwards, K.A. and Howard, A.J. (2000) Holocene environmental change in the Yorkshire Ouse basin and its influence on river dynamics and sediment fluxes to the coastal zone. Pp. 87–96 in: *Holocene Land-Ocean Interaction and Environmental Change around the North Sea* (I. Shennan and J. Andrews, editors). Special Publication, **166**. Geological Society, London.

Manceau, A. and Combes, J.M. (1988) Structure of Mn and Fe oxides and oxides: A topological approach by EXAFS. *Phys. Chem. Miner.*, **15**, 283–95.

Mantei, E.J. and Foster, M.V. (1991) Heavy metals in stream sediment: effects of human activities. *Environ. Geol. Water Sci.*, **18**, 95–104.

McKenzie, R.M. (1967) The sorption of cobalt by manganese minerals in soils. *Aust. J. Soil Res.*, **5**, 235–46.

McKenzie, R.M. (1970) The reaction of cobalt with manganese dioxide minerals. *Aust. J. Soil Res.*, **8**, 97–106.

McKenzie, R.M. (1972*a*) The manganese oxides in soils – a review. *Zeits. Pflanzenernaehr. Bodenkd.*, **131**, 221–42.

McKenzie, R.M. (1972*b*) The sorption of some heavy metals by the lower oxides of manganese. *Geoderma*, **8**, 29–35.

McKenzie, R.M. (1978) The effect of two manganese dioxides on the uptake of Pb, Co, Ni, Cu and Zn by subterranean clover. *Aust. J. Soil Res.*, **16**, 209–14.

McKenzie, R.M. (1979) Proton release during adsorption of heavy metal ions by a hydrous manganese dioxide. *Geochim. Cosmochim. Acta*, **43**, 1855–7.

McKenzie, R.M. (1980*a*) The adsorption of lead and other heavy metals on oxides of manganese and iron. *Aust. J. Soil Res.*, **18**, 61–73.

McKenzie, R.M. (1980*b*) The manganese oxides in soils. Pp. 259–69 in: *Geology and Geochemistry of Manganese*, vol. 1 (I.M. Varentsov and Gy. Grasselly, editors). Publishing House of Hungary Academy of Science, Budapest.

McKenzie, R.M. (1981) The surface charge on manganese dioxides. *Aust. J. Soil Res.*, **19**, 41–50.

McKenzie, R.M. (1983) The adsorption of molybdenum on oxide surfaces. *Aust. J. Soil Res.*, **21**, 505–13.

McKenzie, R.M. (1989) Manganese oxides and hydroxides. Pp. 439–65 in: *Minerals in Soil Environments* (2nd edition) (J.B. Dixon and S.B. Weed, editors). SSSA Book Series, **1**, Soil Science Society of America, Madison, WI.

Means, J.L., Crerar, D.A., Borcsik, M.P. and Duguid, J.O. (1978) Adsorption of Co and selected actinides by Mn and Fe oxides in soils and sediments. *Geochim. Cosmochim. Acta*, **42**, 1763–73.

Miller, J.R. (1997) The role of fluvial geomorphic processes in the transport and storage of heavy metals from mine sites. *J. Geochem. Explor.*, **58**, 101–18.

Moore, J.N., Walker, J.R. and Hayes, T.H. (1990) Reaction scheme for the oxidation of As(III) to As(V) by birnessite. *Clays Clay Miner.*, **38**, 549–55.

Morgan, J.J. and Stumm, W. (1964) Colloid-chemical properties of manganese dioxide. *J. Coll. Sci.*, **19**, 347–59.

Murray, J.W. (1974) The surface chemistry of hydrous manganese oxide. *J. Coll. Interf. Sci.*, **46**, 357–71.

Murray, J.W. (1975*a*) The interaction of metal ions at the manganese dioxide-solution interface. *Geochim. Cosmochim. Acta*, **39**, 505–19.

Murray, J.W. (1975*b*) The interaction of cobalt with hydrous manganese dioxide. *Geochim. Cosmochim. Acta*, **39**, 635–7.

Murray, J.W. and Dillard, J.G. (1979) The oxidation of cobalt (II) adsorbed on manganese dioxide. *Geochim. Cosmochim. Acta*, **43**, 781–7.

Nicholson, K. (1992) Contrasting mineralogical-geochemical signatures of manganese oxides: guides to metallogenesis. *Econ. Geol.*, **87**, 1253–64.

Nicholson, K. and Eley, M. (1997) Geochemistry of manganese oxides: metal adsorption in freshwater and marine environments. Pp. 309–26 in: *Manganese Mineralization: Geochemistry and Mineralogy of Terrestrial and Marine Deposits* (K. Nicholson, J.R. Hein, B. Bühn and S. Dasgupta, editors). Spec. Publ., **119**. Geological Society, London.

Nicholson, K., Hein, J.R., Bühn, B. and Dasgupta, S. (editors) (1997) *Manganese Mineralization: Geochemistry and Mineralogy of Terrestrial and Marine Deposits*. Spec. Publ., **119**. Geological Society, London.

Norrish, K. (1975) Geochemistry and mineralogy of trace elements. Pp. 55–81 in: *Trace Elements in Soil-Plant-Animal Systems* (D.J. Nicholas and A.R. Egan, editors). Waite Agricultural Research Institute Jubilee Symposium. Academic Press Ltd., New York.

Nowlan, G.A. (1976) Concretionary manganese-iron oxides in streams and their usefulness as a sample medium for geochemical prospecting. *J. Geochem. Explor.*, **6**, 193–210.

Nowlan, G.A., McHugh, J.B. and Hessin, T.D. (1983) Origin of concretionary Mn-Fe oxides in stream sediments of Maine, U.S.A. *Chem. Geol.*, **38**, 141–56.

Pettine, M., Camusso, M. and Martinotti, W. (1992) Dissolved and particulate transport of arsenic and chromium in the Po River (Italy). *Sci. Tot. Environ.*, **119**, 253–80.

Post, J.E. and Bish, D.L. (1988) Rietveld refinement of the todorokite structure. *Amer. Mineral.*, **73**, 861–9.

Post, J.E., Von Dreele, R.B. and Buseck, P.R. (1982) Symmetry and cation displacements in hollandites: structure refinements of hollandite, cryptomelane and priderite. *Acta Crystallogr.*, **B38**, 1056–65.

Potter, R.M. and Rossman, G.R. (1979) Mineralogy of manganese dendrites and coatings. *Amer. Mineral.*, **64**, 1219–26.

Robinson, G.D. (1981) Trace metal adsorption potential of phases comprising black coatings on stream pebbles. *J. Geochem. Explor.*, **17**, 205–19.

Stone, A.T. and Morgan, J.J. (1984) Reduction and dissolution of manganese (III) and manganese (IV) oxides by organics: 2. Survey of the reactivity of organics. *Environ. Sci. Technol.*, **18**, 617–24.

Stumm, W. and Morgan, J.J. (1970) *Aquatic Chemistry*. Wiley, New York.

Stumm, W. and Morgann, J.J. (1981) *Aquatic Chemistry*, 2nd edition. Wiley, New York.

Taylor, R.M. and McKenzie, R.M. (1966) The association of trace elements with manganese minerals in Australian soils. *Aust. J. Soil Res.*, **4**, 29–39.

Taylor, R.M., McKenzie, R.M. and Norrish, K. (1964) The mineralogy and chemistry of manganese in some Australian soils. *Aust. J. Soil Res.*, **2**, 235–48.

Tessier, A., Campbell, P.G.C. and Bisson, M. (1979) Sequential extraction procedure for the speciation of particulate trace metals. *Anal. Chem.*, **51**, 844–51.

Thomson, B. and Nicholson, K. (1998) Multi-metal adsorption on manganese oxides: the effect of lead on the adsorption process and its implications in the formation of manganese minerals. In: *River and Estuarine Pollution, Fossil Fuel and Environmental Quality, Society for Environmental Geochemistry and Health, 16$^{th}$ European Conference*. University of Derby.

Turner, S. and Buseck, P.R. (1981) Todorokites: a new family of naturally occurring manganese oxides. *Science*, **212**, 1024–7.

Varentsov, I.M. and Graselly, G. (1980*a*) *Geology and Geochemistry of Manganese; Manganese Deposits on Continents*. E. Schweizerbart'sche Verlag, Stuttgart, Germany.

Varentsov, I.M. and Graselly, G. (1980*b*) *Geology and Geochemistry of Manganese; Manganese on the Bottom of Recent Basins*. E. Schweizerbart'sche Verlag, Stuttgart, Germany.

Von Gunten, H.R., Karametaxas, G., Krähenbühl, U., Kuslys, M., Giovanoli, R., Hoehn, E. and Keil, R. (1991) Seasonal biogeochemical cycles in riverborne groundwater. *Geochim. Cosmochim. Acta*, **55**, 3597–609.

Whitney, P.R. (1975) Relationship of manganese-iron oxides and associated heavy metals to grain size in stream sediments. *J. Geochem. Explor.*, **4**, 251–63.

Wilson, D.E. (1980) Surface and complexation effects on the rate of Mn(II) oxidation in natural waters. *Geochim. Cosmochim. Acta*, **44**, 1311–7.

CHAPTER THIRTEEN

# Intercalation of organic and inorganic contaminants by expanding layer silicates

W. E. DUBBIN

*Department of Mineralogy, The Natural History Museum, London SW7 5BD, UK (b.dubbin@nhm.ac.uk)*

## ABSTRACT

Expanding layer silicates have been proposed as sorbents for both organic and inorganic contaminants. With specific surfaces that may exceed 800 $m^2/g$ and interlayer distances as large as 30 Å, montmorillonite in particular possesses a high capacity to sorb large amounts of pollutants with diverse morphologies. The intercalation of hydrolytic polymers of Al, Cr, Cu, Fe, Mn and Ni by smectite has been widely reported and reduces the bioavailability of these metals in both terrestrial and aquatic ecosystems. Hydroxy-Al interlayers possess the greatest order and, when present with other hydrolysed metals in the interlayer, confer a greater order and stability to the coprecipitated intercalate. Organic contaminants may also be effectively sorbed by expandable layer silicates. Cationic organic molecules may be intercalated through simple exchange reactions. Non-polar organic contaminants may be effectively sorbed only by those clays that have been modified to possess an organophilic character.

## 13.1 Introduction

Expanding layer silicates are a major component of many earth-surface environments, particularly those in temperate regions, and those that are poorly drained. The high permanent negative charge and large surface area of these minerals make them effective sorbents for both organic and inorganic cations. Moreover, the interlayer environment of these clays provides the sorbed ions with a degree of protection from remobilization and microbial degradation not offered by the various other inorganic sorbents common in soils and sediments. Such protection may reduce the bioavailability of sorbed contaminants but, in the case of organic pollutants, delay their degradation to more benign forms.

Intercalation as used in this review is defined as the sorption of organic molecules or polymeric metals onto the internal surfaces of expanding layer silicates. This sorption is accompanied by an expansion of the interlayer space and a concomitant increase in basal spacing. It is acknowledged, however, that

Dubbin, W.E. (2000) Intercalation of organic and inorganic contaminants by expanding layer silicates. Pp. 227–244 in: *Environmental Mineralogy: Microbial Interactions, Anthropogenic Influences, Contaminated Land and Waste Management* (J.D. Cotter-Howells, L.S. Campbell, E. Valsami-Jones and M. Batchelder, editors). Mineralogical Society Series, **9**. Mineralogical Society, London. ISBN 0 903056 20 8.

non-swelling clays such as the 1:1 layer silicates halloysite and kaolinite may also form interlamellar complexes. These latter complexes have been prepared in the laboratory as a way to differentiate kaolinite group minerals from chlorite or serpentine. This chapter begins with a brief review of expanding 2:1 layer silicates, then discusses the intercalation of these clays with both organic and inorganic contaminants.

## 13.2 Structure and properties of expandable layer silicates

The basic structure of layer silicates has been presented elsewhere (Brindley and Brown, 1980; Bailey, 1988; Dixon and Weed, 1989). Expandable layer silicates, i.e. those of the smectite and vermiculite groups, as well as many mixed-layered clays, are characterized by a positive charge deficit arising from isomorphous substitution in the octahedral and/or tetrahedral sheets. Smectites are defined as those expandable layer silicates that have a charge deficit ranging from 0.5 to 1.2 per unit cell (Bailey, 1980). In montmorillonite, which is a dominant component of the smectite group, the majority of this charge originates from the octahedral sheet. The need for electroneutrality results in the complexation of cations on the montmorillonite surface. The nature and concentration of these sorbed cations, as well as their mode of complexation, determine the distance between the individual 2:1 layers. The complexation of a simple hydroxy-metal polymer on montmorillonite is shown schematically in Fig. 13.1. With specific surfaces that may exceed 800 $m^2/g$ (Van Olphen and Fripiat, 1979) and interlayer distances as large as 30 Å (Rausell-Colom and Serratosa, 1987), montmorillonite possesses a high capacity to sorb large

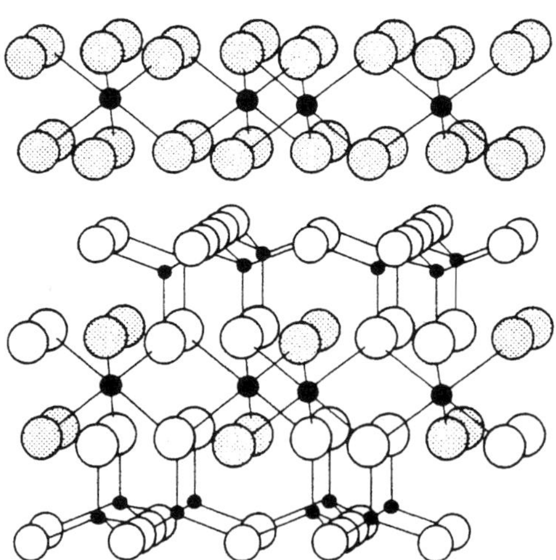

FIG. 13.1 Ball and stick representation of a 2:1 layer with a hydroxide interlayer (From Sposito, 1984).

amounts of both organic and inorganic counter-cations of diverse morphologies.

Functional groups in clay minerals may be defined as chemically reactive molecular units that are located at the particle margin (Sposito, 1984). Phyllosilicates possess two principal kinds of functional groups. The first type consists of the various hydroxyl groups at the edges of the octahedral and tetrahedral sheets. Of greater importance in studies concerning the intercalation of organic and inorganic contaminants are the functional groups on the basal surface of the tetrahedral sheets. The basal oxygen atoms of the silicon tetrahedra collectively comprise the siloxane surface. As a result of rotation of the silicon tetrahedra to fit the octahedral sheet, the siloxane oxygen atoms lie in two distinct planes that are separated by a distance of 0.2 Å. When viewed along the *c*-axis the siloxane oxygen appear to form hexagonal cavities but, because the oxygen atoms lie in two separate planes, the cavities are more accurately described as ditrigonal (Fig. 13.2). In the absence of a positive charge deficit in the 2:1 layer, the ditrigonal cavities serve only as very soft Lewis bases. With isomorphous substitution, however, the ditrigonal cavities become increasingly stronger Lewis bases. The Lewis base character of the ditrigonal cavities is determined not only by the magnitude of the layer charge, but also by its location. If substitution occurs at a single site in the octahedral sheet, the resulting negative charge will be distributed mainly over 10 siloxane oxygen atoms (Sposito, 1984). However, if the substitution occurs in the tetrahedral sheet, the excess negative charge will be distributed over only three oxygen atoms. Therefore, for an equivalent amount of substitution, charge originating from the tetrahedral sheet imparts a much greater Lewis base character to the siloxane oxygen.

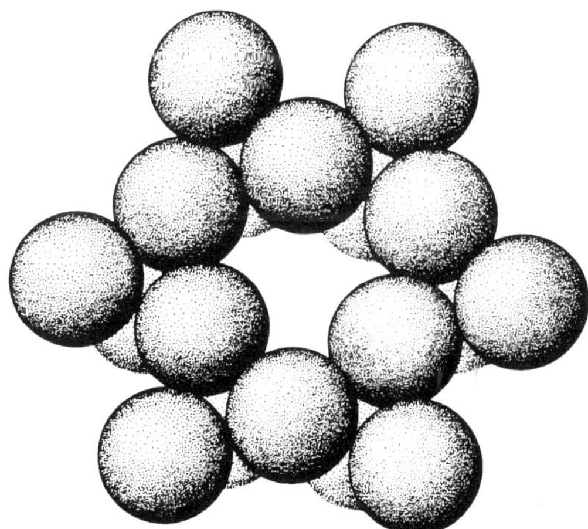

FIG. 13.2 The siloxane ditrigonal cavity in a tetrahedral sheet of a layer silicate as viewed along the *c*-axis (From Sposito, 1984).

## 13.3 Intercalation of inorganic contaminants

Hydroxy interlayers of many cations have been prepared in the laboratory or observed in natural environments. Specifically, hydroxy interlayers of Al (Keren, 1980; Barnhisel and Bertsch, 1989; Sakurai and Huang, 1998), Bi (Yamanaka *et al.*, 1980), Cr (Brindley and Yamanaka, 1979; Carr, 1985; Dubbin and Goh, 1995), Cu (Bassett, 1958), Fe (Herrera and Peech, 1970; Indraratne *et al.*, 1999), Mg (Brindley and Kao, 1980), Mn (Dubbin and Goh, 1999), Ni (Yamanaka and Brindley, 1978; Ghesquiere *et al.*, 1982), Si (Sterte and Shabtai, 1987; Lou and Huang, 1988), and Zr (Yamanaka and Brindley, 1979) have been reported. Of all the intercalated hydroxy polymers that have been studied, those of Al have received the most attention. The diverse and exhaustive studies of hydroxy-Al interlayers have been justified by the abundance and potential toxicity of Al, and have been made feasible by the ease with which these interlayers form. The present section of this review will focus on the intercalation of hydrolytic polymers of Al, Cr and other environmentally significant inorganic contaminants.

### *13.3.1 Cation hydrolysis and polymerization*

Cation hydrolysis is an important process in both terrestrial and aquatic environments. The hydrolysis of metal cations, leading to their precipitation as hydroxides, decreases their solubility and, consequently, their bioavailability. Depending on the element, this may be deleterious or beneficial. Hydroxy polymers of metals, particularly Al and Fe, also play an important role in enhancing soil clay aggregation (El Swaify and Emerson, 1975).

The ionic potential (IP) of a cation, defined as the cation's valence to radius ratio, may be used to identify those cations which hydrolyse extensively at pH 7. If the IP is 30 to 95 $nm^{-1}$, a cation is able to repel protons from a solvating water molecule strongly enough to form a hydrolytic species, but not so strongly that an oxyanion forms (Sposito, 1994). For example, the IP values of $Al^{3+}$ and $Cr^{3+}$ are 56 and 48 $nm^{-1}$, respectively, well within the range of hydrolysing cations. Clearly, hydrolytic species of these two metals will be abundant at pH values normally encountered in soil.

The hydrolysis of $Al^{3+}$ has been studied extensively (Hsu, 1989). Hydrolysis begins with the loss of a proton from the hexaaqua Al complex and proceeds stepwise with the loss of additional protons:

$$Al(H_2O)_6^{3+} + H_2O \rightarrow Al(OH)(H_2O)_5^{2+} + H_3O^+ \quad (13.1)$$
$$Al(OH)(H_2O)_5^{2+} + H_2O \rightarrow Al(OH)_2(H_2O)_4^{+} + H_3O^+ \quad (13.2)$$
$$Al(OH)_2(H_2O)_4^{+} + H_2O \rightarrow Al(OH)_3(H_2O)_3^{0} + H_3O^+ \quad (13.3)$$

The pK values for the first, second and third hydrolysis reactions of $Al^{3+}$ have been determined to be 5, 10 and 17, respectively (Nordstrom and May, 1989). The hydrolysis of $Cr^{3+}$ can likewise be described with a series of equilibrium reactions. The corresponding hydrolysis constants for $Cr^{3+}$, while not known as precisely as those for $Al^{3+}$, have been found to be comparable (Baes and Mesmer, 1976; Rai *et al.*, 1987).

The kinetics of mononuclear hydrolysis are rapid for both $Al^{3+}$ and $Cr^{3+}$. The specific rate constant for the loss of a proton from $Al(H_2O)_6^{3+}$ was determined to be $1.1 \times 10^5\ s^{-1}$ (Fong and Grunwald, 1969) while that for $Cr(H_2O)_6^{3+}$ was found to be $1.4 \times 10^5\ s^{-1}$ (Rich *et al.*, 1969). With respect to the thermodynamics and kinetics of hydrolysis, therefore, $Al^{3+}$ and $Cr^{3+}$ behave similarly.

The rates of formation of polynuclear complexes of $Al^{3+}$ and $Cr^{3+}$ are significantly different. Polymers of Al form much more rapidly than those of Cr. The fundamental reason for this difference lies in the different rates of ligand substitution for these two metals. As shown in Fig. 13.3, the exchange of coordinated water molecules with bulk water in a hexaaqua $Al^{3+}$ complex is several orders of magnitude faster than the exchange in a $Cr^{3+}$ complex. Similar exchanges are required for polymerization to occur. For example, during the dimerization process, each hydrolysed cation must shed a single water molecule. The bound hydroxyls can then serve as bridging ligands. Because a water molecule is displaced, however, the hydrolysed cation must temporarily adopt five-fold coordination. For metals with $d^3$ configuration, such as $Cr^{3+}$, the energy required in going from octahedral to trigonal bipyramidal geometry is prohibitively high. This high energy barrier can be explained in terms of ligand field theory. In an octahedral field, the three d electrons of $Cr^{3+}$ are in their ground state configuration. That is, all three electrons are unpaired and in the lower energy $t_{2g}$ orbitals. Deviation from octahedral geometry forces electrons to be either spin-paired or promoted to higher energy orbitals. Because both the ligand field splitting energy, $\Delta_O$, and the pairing energy are high, the formation of a trigonal bipyramidal intermediate is

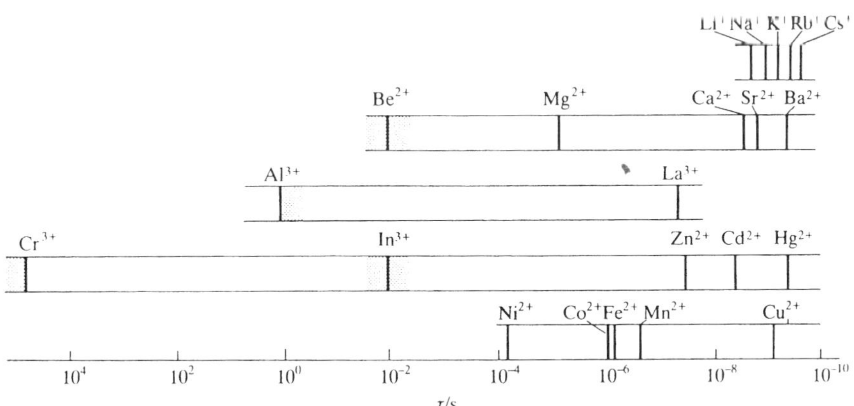


FIG. 13.3 Typical lifetimes for the exchange of water molecules in hexaaqua complexes (from Shriver *et al.*, 1990).

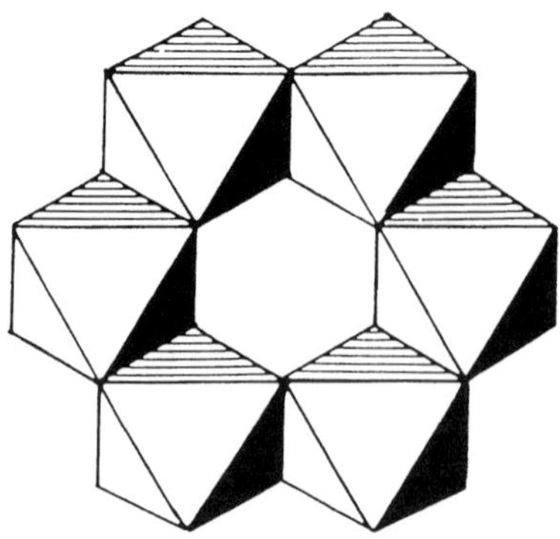

FIG. 13.4 Illustration of the ring structure formed by the aluminium hexamer, $[Al_6(OH)_{12}(H_2O)_{12}]^{6+}$ (From Schutz *et al.*, 1987).

highly unfavoured. The end result of the slow ligand displacement around $Cr^{3+}$ is that its polymerization is extremely slow. Unless the hydroxy-Cr polymers are allowed to age for months or years, they will remain small and poorly ordered.

### *13.3.2 Aluminium intercalation*

The fate of Al in the environment is of wide interest due to its toxicity to both terrestrial and aquatic organisms. The basic building block of the most widely accepted structure for hydroxy-Al interlayers is the hexamer, $[Al_6(OH)_{12}(H_2O)_{12}]^{6+}$ (Fig. 13.4) (Hsu and Bates, 1964). Intercalation of layer silicates with these hexameric units yields a basal spacing of approximately 14 Å, similar to that of chlorite. Growth of the polymer, much of which occurs within the interlayer spaces, is believed to be initiated by deprotonation of the edge-group water molecules. The polynuclear units can then coalesce as hydroxide bridges are formed. The growth of the hydroxy-Al polymers is depicted in Fig. 13.5. It is clear from this sequence that as the polymers enlarge, the OH/Al ratio increases and the average charge per Al decreases.

Several workers have observed *d*(001) reflections of 18 to 19 Å for smectites intercalated with hydroxy-Al species (Brindley and Sempels, 1977; Lahav and

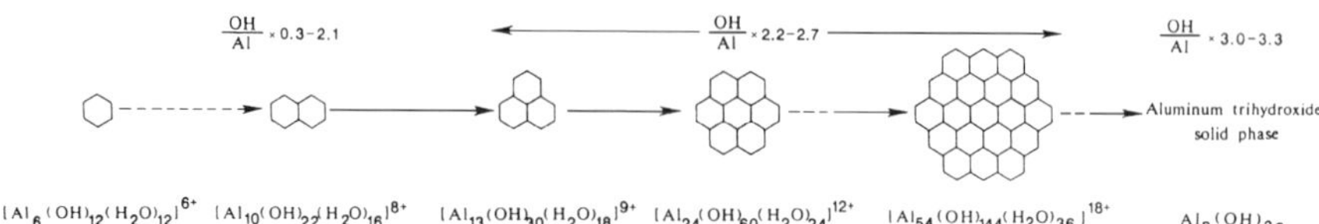


FIG. 13.5 Depiction of aluminium polymerization as individual hexamers coalesce (from Bertsch, 1989).

Shani, 1978). Basal spacings of ~19 Å would arise if two hexameric Al polymers were stacked along the *c*-axis within the interlayer space. Alternatively, the hydroxy-Al species may be of the form depicted in Fig. 13.6. This hydroxy polymer consists of a central tetrahedrally coordinated Al surrounded by four groups of three Al octahedra. The most commonly reported composition for this '$Al_{13}$' polynuclear species is $[AlO_4Al_{12}(OH)_{24}(H_2O)_{12}]^{7+}$ (Bertsch, 1989). Intercalation of smectite with these $Al_{13}$ polymers would also produce the observed 19 Å spacing. In fact, recent solid state $^{27}Al$ and $^{29}Si$ NMR studies have identified $Al_{13}$ as the interlayer component of several clays (Plee *et al.*, 1985; Schutz *et al.*, 1987).

There are significant differences in the nature of hydroxy interlayers formed in natural environments and those prepared in the laboratory. The most common basal spacing of all interlayered clays is ~14 Å. However, laboratory-synthesized clays can display basal spacings which deviate significantly from this value (Barnhisel and Bertsch, 1989). The wide range of basal spacings in the laboratory-prepared clays reflects the diverse structures of the interlayer components, which may include the $Al_{13}$ molecule. The conditions of formation may also govern the distribution of hydroxy polymers within the interlayer space. Two main arrangements have been proposed: (*a*) a uniform distribution; and (*b*) an 'atoll' arrangement (Fig. 13.7) (Dixon and Jackson,

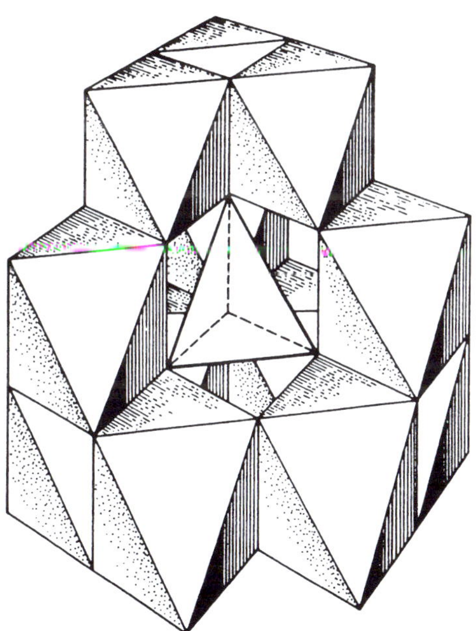

FIG. 13.6 Representation of the $[AlO_4Al_{12}(OH)_{24}(H_2O)_{12}]^{7+}$ polynuclear species showing the central tetrahedrally coordinated Al surrounded by four sets of three Al octahedra (from Baes and Mesmer, 1976).

1962). A uniform distribution of polymers within the interlayer would provide greater support than an atoll arrangement against collapse upon heating. As indicated in Fig. 13.7, a uniform distribution would also allow free access to the exchange sites within the interior of the interlayer space. Given these differences, therefore, one must be cautious when assigning the properties of laboratory-prepared interlayers to those formed in natural environments.

### *13.3.3 Chromium intercalation*

Hydroxy-Cr interlayers are not nearly as well understood as those of Al. The earliest investigations attempted to simply determine the molecular weight of the sorbed hydroxy-Cr polymers (Rengasamy and Oades, 1978). In all cases,

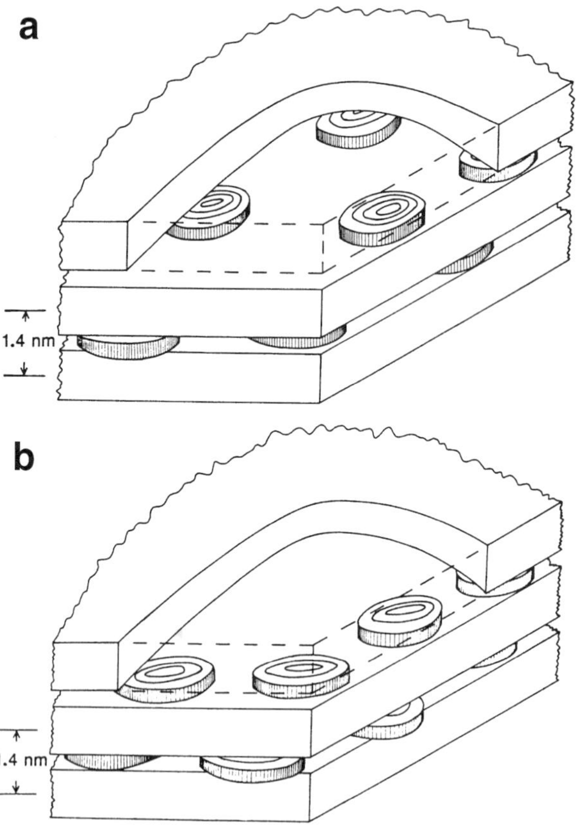


FIG. 13.7 Illustration showing the distribution of hydroxy polymers in the interlayer space of 2:1 layer silicates: (*a*) uniform distribution; (*b*) an "atoll" arrangement (from Barnhisel and Bertsch, 1989).

the hydroxy-Cr polymers were considerably smaller than their Al counterparts. Later, efforts were directed at determining the composition of the interlayered polymeric species (Brindley and Yamanaka, 1979; Carr, 1985). These workers found that the interlayer material had $(OH+H_2O)/Cr$ molar ratios that were consistent with the polymers ranging in size from dimers to hexamers. The *d*(001) reflections of the interlayered montmorillonite ranged from 15 to 17 Å, considerably larger than for a chlorite. These reflections were also broad, indicating that the basal spacings were variable. The intercalation of montmorillonite with hydroxy-Cr oligomers of known size and structure yielded sharp basal reflections (Drljaca *et al.*, 1992). After calculating the gallery height of the interlayered clay, these workers were then able to propose probable orientations for the various oligomers. Because the calculated gallery heights of several of the clays were somewhat smaller than expected, the authors proposed that the oligomers were sorbed covalently to the montmorillonite. The formation of a covalent bond would require coordinated water to be released, thus reducing the effective size of the oligomers and also decreasing the gallery height.

In light of the potential toxicity of Cr, more recent work has continued to focus on the sorptive capacity of montmorillonite for hydroxy-Cr polymers and the mode of Cr complexation (Dubbin and Goh, 1995). Theoretical considerations have been invoked to indicate that Cr is likely to form inner-sphere complexes with the siloxane oxygen of smectites. The principle of hard and soft Lewis acids and bases (HSAB) predicts that, in general, hard acids complex with hard bases and soft acids complex with soft bases (Jensen, 1978). The siloxane oxygen atoms of smectite are known to be relatively soft Lewis bases (Sposito, 1984). The relative softness of $Cr^{3+}$ determined with the Misono softness parameter (Misono *et al.*, 1967) is known to be 2.80, which is considered borderline soft. According to the HSAB principle, therefore, the inner-sphere complexation of $Cr^{3+}$ with the siloxane oxygen is not unexpected.

Difficulty in displacing Cr from smectite further indicates that Cr is bonded specifically (Dubbin and Goh, 1995). If Cr were held through outer-sphere complexes, the smallest hydroxy polymers would be readily displaced by $Ca^{2+}$. There are no known reports of significant amounts of exchangeable Cr from dried hydroxy-Cr interlayered montmorillonite. Similar observations have been made by Drljaca *et al.* (1992). These workers found that, while the montmorillonite was still wet, the adsorbed Cr could be exchanged with other cations very easily. Upon drying, however, the Cr became virtually non-exchangeable. These authors suggested that, as the interlayer region collapsed due to loss of water, the Cr became more intimately bound with the siloxane surface, allowing inner-sphere complexes to form.

Additional evidence for the inner-sphere complexation of Cr is found in the work of Carr (1985), which revealed the surface-catalysed polymerization of intercalated hydroxy-Cr oligomers. Hydrolytic polymerization of $Cr^{3+}$ requires the replacement of $H_2O$ with $OH^-$ within the coordination sphere of $Cr^{3+}$.

Because transition metals of the $d^3$ type must overcome a large energy barrier in the formation of intermediates, ligand displacement reactions of $Cr^{3+}$ complexes are extremely slow (Cotton and Wilkinson, 1988). However, once the first ligand exchange has occurred, subsequent exchanges are more rapid. The inner-sphere complexation of Cr with the siloxane oxygen of montmorillonite (i.e. exchange of $O^{\delta-}$ for $H_2O$) would provide this first ligand exchange and allow for the enhanced polymerization observed by Carr (1985).

The most direct evidence for the inner-sphere complexation of Cr has been provided by infrared (IR) spectroscopy (Dubbin and Goh, 1995). In the non-interlayered montmorillonite, the perpendicular Si–O vibrations occur freely and result in an absorption band at ~1100 $cm^{-1}$. The formation of a Cr–O–Si linkage in the dried interlayered clay would result in a redistribution of electrons at the clay-water interface. Electron requirements of Cr(III) cause a shift in the electron density from the Si–O bond toward the newly-formed Cr–O bond. Consequently, the Si–O bond weakens, lengthens, and its vibration frequency decreases. This weakened Si–O bond now vibrates at 1015–1020 $cm^{-1}$.

Montmorillonite has been shown to have a high capacity to sorb hydrolysed Cr. A plot of sorbed *vs.* equilibrium solution Cr concentration shows that, although sorption continues to increase at all Cr concentrations, there is an obvious decrease in the affinity of montmorillonite for hydroxy-Cr polymers at concentrations >1200 cmol(+)/kg (note the decreasing slope in Fig. 13.8). Once the montmorillonite had sorbed 1200 cmol(+)/kg, much of the internal surface would be complexed with the hydroxy-Cr species, as determined by charge

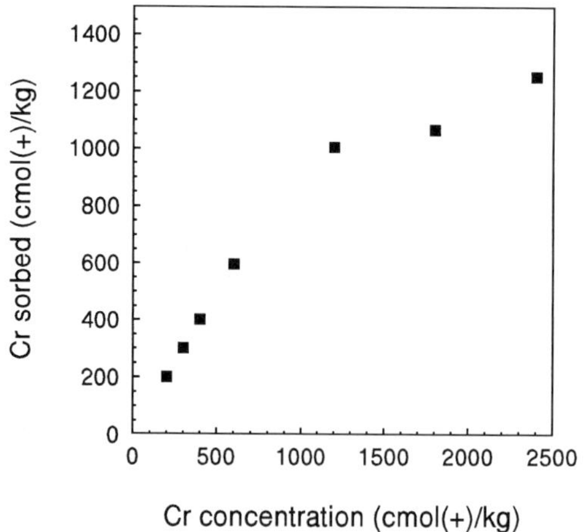


FIG. 13.8 Plot of sorbed *vs.* equilibrium solution Cr concentration. Sorbed Cr is considered to be the difference in supernatant Cr concentration before and after ageing (from Dubbin and Goh, 1995).

balance calculations. Only the smallest and most remote exchange sites would remain available for sorption. At concentrations >1200 cmol(+)/kg, therefore, Cr sorption could occur through only two mechanisms. First, those polymers in solution which are sufficiently small would be able to navigate the porous network of the interlayer and access the remote exchange sites within the interior. In this first mechanism, Cr from the small polymers would complex directly with the siloxane surface. The second mechanism involves the addition of Cr on to those polymers which are already sorbed. If the subsequent polymer growth occurred along the *a*- or *b*-axes, pore space would decrease. Polymer growth along the *c*-axis, however, would increase the gallery height, thereby increasing the pore space and, consequently, also the sorptive capacity of the montmorillonite.

The broad basal reflections in X-ray powder diffraction (XRD) patterns indicate that the interlayered hydroxy-Cr polymers are very poorly ordered (Dubbin and Goh, 1995). Because these polymers are so poorly ordered, they would not be packed efficiently within the interlayer region. In contrast to the well-ordered, and often complete, octahedral sheet of interlayer Al, the hydroxy-Cr interlayer is very fragmented and often characterized by abundant pore spaces. However, this porous network allows for the sorption of large amounts of hydrolysed Cr.

### *13.3.4 Intercalation of other hydroxy-metal polymers*

The intercalation of metals that hydrolyse less extensively than Al and Cr has, for the most part, received only cursory attention. The more limited focus given to the intercalation of metals such as Cu, Fe, Mn and Ni is due not only to their less extensive hydrolysis, but, with the exception of Fe, also to their generally lower abundance in soils and sediments. However, the intercalation of hydrolytic polymers of Cu, Fe, Mn, and Ni could play a significant role in governing the bioavailability of these metals. In environments where toxicity of these elements may be a concern, expanding layer silicates could sorb sufficient quantities of the hydroxy-metal polymers in a non-exchangeable form and thereby reduce aqueous metal concentrations to levels that are non-toxic to plants and animals.

In a comparison of hydroxy-Fe and hydroxy-Al interlayers, Carstea *et al.* (1970) reported much poorer crystallinity of the former. Consequently, the hydroxy-Fe interlayers were found to collapse readily upon heating. Similar observations were made by Indraratne *et al.* (1996). However, both studies observed that large amounts of Fe were sorbed non-exchangeably, as revealed by significant reductions in cation exchange capacity. Intercalation of hydrolytic polymers of Cu (Bassett, 1958), Mn (Dubbin and Goh, 1999), and Ni (Yamanaka and Brindley, 1978; Ghesquiere *et al.*, 1982) in non-exchangeable form further illustrate the potential of expanding layer silicates to ameliorate environments containing toxic levels of these metals. In the case of

Mn, however, intercalation of hydrolysed species was brought about following oxidation of sorbed Cr(III) by Mn(IV). It is not known if such intercalation would occur in the absence of an interlayered reductant.

### *13.3.5 Coprecipitation*

As Al is the cation that has received the most attention in studies of interlayering, experimental studies of coprecipitation invariably involve Al along with one other cation, principally Cr(III), Fe(III) or Si(IV). The coprecipitation of Al and Cr produces interlayer materials that possess similar gibbsite-like structures, giving rise to 14 Å basal spacings (Dubbin *et al.*, 1994). Substitution of Cr for Al in these polymers is supported by both XRD and IR spectroscopic data. These data indicate that Al may serve as a template in the formation of the gibbsite-like polymers. Chromium increases the resistance of the coprecipitated Al-Cr interlayers to displacement. With increasing Cr content, displacement of Cr by both K and Ca becomes more difficult. Aluminium, however, increases the thermal stability of the interlayer. As the Al content increases, the interlayer becomes more resistant to collapse upon heating. The presence of both Cr and Al in the hydroxy polymers results in chemical and physical properties which seem to be contradictory. The most heat-stable interlayers (those with the most Al) are also the least resistant to displacement. Meanwhile, the least heat-stable interlayers (those with the most Cr) are the most resistant to displacement. The explanation proposed by Dubbin *et al.* (1994) to explain this behaviour is that Al increases the thermal stability of the interlayers by increasing the long-range order of the polymers, while Cr increases the resistance to displacement by forming inner-sphere complexes with the siloxane surface.

Coprecipitated hydroxy-(Fe,Al) interlayers were prepared first by Indraratne *et al.* (1999). These workers synthesized the mixed-cation interlayers by adding a 1:1 molar ratio of Fe and Al to a montmorillonite suspension. The resulting hydroxy-(Fe,Al) interlayered clays exhibited greater crystallinity and heat stability than the Fe-only clays. Presumably the Al increases the stability of the hydroxy-Fe polymers in the mixed-cation clays in a fashion similar to that proposed by Dubbin *et al.* (1994) for the hydroxy-(Cr,Al) interlayers.

Several decades ago Dixon and Jackson (1962) speculated that an allophane-like material consisting of both Al and Si may occur in the interlayer of expanding layer silicates. More recently, Lou and Huang (1988) observed the intercalation of hydroxy-aluminosilicate species in montmorillonite and identified the interlayered polymers as possessing an imogolite-like structure that becomes stable with ageing. More recent still, Inoue and Satoh (1992) and Lou and Huang (1995) reported the intercalation of hydroxy-aluminosilicate ions into the interlayer spaces of vermiculite. Coprecipitation of Al and Si within the interlayer spaces of expanding layer silicates plays a major role in governing the fate of not only Al, but also of other pollutants in terrestrial and aquatic ecosystems. Polymerized hydroxy-aluminosilicate ions sorbed to smectite were found by Inoue *et al.* (1988) to contain higher pH dependent charge than hydroxy-Al interlayers alone. The presence of these coprecipitated hydroxy-aluminosilicate interlayers is therefore likely to have a significant influence on the sorption of nutrients and pollutants in these environments.

## 13.4 Intercalation of organic contaminants

The intercalation of organic contaminants by expanding layer silicates is controlled by a number of properties of both the layer silicate and the organic molecule. Most important among these for the organic molecule are its charge, polarity, size and shape, as well as the identity and acidity or basicity of the functional groups attached to the molecule. Despite the diversity of organic contaminants that may occur in polluted environments, they may all be conveniently classed into two broad groups, based on their charge and polarity.

### *13.4.1 Ionic molecules*

Although the intercalation of organic cations has received considerable attention, only a small number of these molecules are used as pesticides or occur in soil as pollutants. Glyphosate, diquat and paraquat will serve as examples to illustrate the intercalation of this group of molecules. Glyphosate [*N*-(phosphonomethyl)glycine, $(HO)_2$-PO-$CH_2$-$H_2N^+$-$CH_2$-COOH] is a broad-spectrum herbicide that is widely used in agriculture, forestry and urban settings. It is one of the most heavily-applied pesticides in the USA, with an annual use >12 million kg (Aspelin, 1997). Upon addition to soil, glyphosate is readily sorbed, losing both its efficacy as a herbicide and its potential for translocation. At low pH, glyphosate possesses a net positive charge and may therefore, readily exchange with native cations on the clay surface. More commonly, however, glyphosate will exist within the interlayer space as a zwitterion, a dipolar ion possessing spatially separated positive and negative charges, even at low pH. Adsorption of the zwitterion glyphosate onto various homoionic montmorillonite suspensions was found by Sprankle *et al.* (1975) to increase as follows:

$$Ca^{2+} < Mn^{2+} < Zn^{2+} < Mg^{2+} < Fe^{3+} < Al^{3+}$$

Subsequent work postulated that glyphosate was held on the interlayer surfaces by complexation of the phosphonate and carboxyl groups with sorbed polyvalent cations, especially $Fe^{3+}$ and $Al^{3+}$ (Shoval and Yariv, 1979).

Adsorption of two organic cation pesticides, diquat (1,1′-ethylene-2,2′-bipyridylium ion) and paraquat (1,1′-dimethyl-4,4′-bipyridinium ion), by layer silicates occurs via cation exchange (Hayes *et al.*, 1978; Zachara *et al.*, 1990). These two pesticides are adsorbed more strongly by layer silicates, particularly smectites, than by organic matter. Adsorption of these molecules by smectite yields an H-type isotherm, whereas L-type isotherms are more common for adsorption by organic matter, indicating stronger adsorption by the former. A thorough discussion of these and other adsorption isotherms is given by Sposito (1984). The rapid inactivation of these pesticides when applied to soils rich in smectites can be explained by their strong adsorption to the large internal surfaces of this clay. However, intercalation of these organic contaminants will often prevent their subsequent microbial degradation, thereby increasing their persistence in the environment.

The selectivity of layer silicates for various ionic organic molecules depends on several factors, one of the most important of which is molecular weight. Generally, larger molecules are sorbed preferentially. The sorption of quaternary ammonium cations by smectite, for example, has revealed the following affinity sequence (McBride, 1994):

$$(CH_3)_4N^+ < (CH_3CH_2)_4N^+ < (CH_3CH_2CH_2)_4N^+$$

The principal reason for this order of preference arises from the greater number of water molecules that are displaced from the clay surface by the larger organic cations. With a larger number of water molecules displaced from the surface and liberated to solution, the entropy term, $\Delta S$, becomes larger and the reaction is more highly favoured.

Intercalation of organic contaminants depends not only on the properties of the organic molecule, but also on those of the layer silicate. The expansion of smectites is greater, and occurs more easily than that of vermiculite, due to the smaller layer charge of the former. This difference in expandability among vermiculite and the smectites gives rise to the phenomenon of molecular sieving. Vermiculites typically expand to give a maximum basal spacing of ~14.5 Å, producing an interlayer spacing of ~5 Å. Smectites, however, may expand sufficiently to give an interlayer distance of nearly 30 Å. This is particularly true of smectites with a small layer charge that originates mainly from the octahedral sheet. Owing to their ease of expansion, smectites may therefore intercalate organic contaminants of a much greater variety of size and shape than vermiculites. This difference in selectivity gives rise to molecular sieving (Fig. 13.9).

### *13.4.2 Non-polar and non-ionic molecules*

There exists a large class of organic molecules which are either non-polar or non-ionic. Layer silicates are generally ineffective sorbents for these poorly water soluble (hydrophobic) organic contaminants because the native hydrated inorganic cations on the exchange sites impart a hydrophilic character to the clay surface. These native inorganic cations must first be replaced with a variety of organic cations to form organo-clays that, unlike native clay, are effective sorbents for removing organic contaminants from water (Jaynes and Boyd, 1991). A class of organic cations that is commonly used to prepare these organo-clays are quaternary ammonium cations of the form $[(CH_3)_2NR_2]^+$

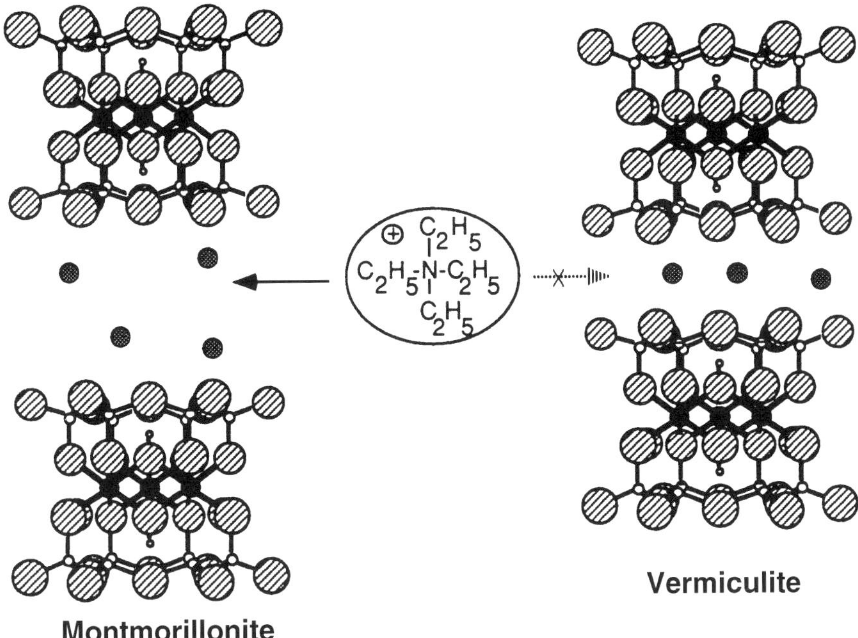


FIG. 13.9 Illustration of the sieving action of expanding layer silicates in the presence of large organic cations (from McBride, 1994).

or $[(CH_3)_3NR]^+$, where R is an aromatic or alkyl hydrocarbon group. Exchanging the native metal cations for these organic cations markedly alters the clay surface and interlayer environment, which both change from hydrophilic to organophilic. Additionally, the organic cations serve to prop open the interlayer spaces and thus allow for easier intercalation of the contaminants.

## 13.5 Summary

Expanding layer silicates are one of the most important mineral groups controlling the dynamics of organic and inorganic contaminants in soils and sediments. The fate and bioavailability of many metals are determined, to a large extent, by the presence of vermiculite and smectite group minerals. Coprecipitation of two or more of these metals may result in intercalated precipitates that have properties unlike either of the single-metal intercalates. Of the two broad classes of organic molecules, those that are ionic may be intercalated through simple cation exchange. Sorption of non-polar organic compounds, however, requires that the clay first be treated with any of a number of organic molecules that give the clay an organophilic character. The use of expanding layer silicates as effective sorbents for a range of organic and inorganic contaminants requires that the interlayer material exhibit stability in both terrestrial and aquatic ecosystems. Future research is therefore likely to focus on the long-term stability and transformations of the intercalated species across a range of environments.

## References


Aspelin, A.L. (1997) *Pesticides industry sales and usage: 1994 and 1995 market estimates*. US EPA Publ. 733-R-87-002, Washington, D.C.

Baes, C.F. and Mesmer, R.E. (1976) *The Hydrolysis of Cations*. John Wiley & Sons, Toronto, ON.

Bailey, S.W. (1980) Summary of recommendations of AIPEA nomenclature committee. *Clay Miner.*, **15**, 85–93.

Bailey, S.W. (1988) *Hydrous phyllosilicates (exclusive of the micas)*. Reviews in Mineralogy, **19**. Mineralogical Society of America, Washington, D.C.

Barnhisel, R.I. and Bertsch, P.M. (1989) Chlorites and hydroxy-interlayered vermiculite and smectite. Pp. 729–88 in: *Minerals in Soil Environments*, 2[nd] edition (J.B. Dixon and S.W. Weed, editors). Soil Science Society of America Book Series No. **1**, Madison, WI.

Bassett, W.A. (1958) Copper vermiculites from northern Rhodesia. *Amer. Mineral.*, **43**, 1112–33.

Bertsch, P.M. (1989) Aqueous polynuclear aluminum species. Pp. 87–115 in: *The Environmental Chemistry of Aluminum* (G. Sposito, editor). CRC Press, Boca Raton, FL.

Brindley, G.W. and Brown, G. (1980) *Crystal Structures of Clay Minerals and their X-ray Identification*. Monograph **5**, Mineralogical Society, London.

Brindley, G.W. and Kao, C.C. (1980) Formation, compositions and properties of hydroxy-Al- and hydroxy-Mg-montmorillonite. *Clays Clay Miner.*, **28**, 435–43.

Brindley, G.W. and Sempels, R.E. (1977) Preparation and properties of some hydroxy-aluminum beidellites. *Clay Miner.*, **12**, 229–37.

Brindley, G.W. and Yamanaka, S. (1979) A study of hydroxy-chromium montmorillonites and the form of the hydroxy-chromium polymers. *Amer. Mineral.*, **64**, 830–35.

Carr, R.M. (1985) Hydration states of interlamellar chromium ions in montmorillonite. *Clays Clay Miner.*, **33**, 357–61.

Carstea, D.D., Harward, M.E. and Knox, E.G. (1970) Comparison of iron and aluminum hydroxy interlayers in montmorillonite and vermiculite: I. Formation. *Soil Sci. Soc. Amer. Proc.*, **34**, 517–21.

Cotton, F.A. and Wilkinson, G. (1988) *Advanced Inorganic Chemistry*, 5[th] edition. Wiley-Interscience, New York, NY.

Dixon, J.B. and Jackson, M.L. (1962) Properties of intergradient chlorite-expansible layer silicates of soils. *Soil Sci. Soc. Amer. Proc.*, **26**, 358–62.

Dixon, J.B. and Weed, S.B. (1989) *Minerals in Soil Environments*, 2[nd] edition. Soil Science Society of America Book Series No. **1**, Madison, WI.

Drljaca, A., Anderson, J.R., Spiccia, L. and Turney, T.W. (1992) Intercalation of montmorillonite with individual chromium(III) hydrolytic oligomers. *Inorg. Chem.*, **31**, 4894–7.

Dubbin, W.E. and Goh, T.B. (1995) Sorptive capacity of montmorillonite for hydroxy-Cr polymers and the mode of Cr complexation. *Clay Miner.*, **30**, 175–85.

Dubbin, W.E. and Goh, T.B. (1999) Permanganate-mediated dissolution of hydroxy-Cr interlayers from montmorillonite. *Appl. Clay Sci.*, **15**, 381–92.

Dubbin, W.E., Goh, T.B., Oscarson, D.W. and Hawthorne, F.C. (1994) Properties of hydroxy-Al and -Cr interlayers in montmorillonite. *Clays Clay Miner.*, **42**, 331–6.

El Swaify, S.A. and Emerson, W.W. (1975) Changes in the physical properties of soil clays due to precipitated aluminum and iron hydroxides: I. Swelling and aggregate stability after drying.

*Soil Sci. Soc. Amer. Proc.*, **39**, 1056–63.

Fong, D.W. and Grunwald, E. (1969) Kinetic study of proton exchange between the $Al(OH_2)_6^{3+}$ ion and water in dilute acid. Participation of water molecules in proton transfer. *J. Amer. Chem. Soc.*, **91**, 2413–22.

Ghesquiere, C., Lemaitre, J. and Herbillon, A.J. (1982) An investigation of the nature and reducibility of Ni-hydroxymontmorillonites using various methods including temperature-programmed reduction (TPR). *Clay Miner.*, **17**, 217–30.

Hayes, M.H.B., Pick, M.E. and Toms, B.A. (1978) The influence of organocation structure on the adsorption of mono- and bipyridinium cations by expanding lattice clay minerals. I. Adsorption by $Na^+$-montmorillonite. *J. Coll. Interf. Sci.*, **65**, 254–65.

Herrera, R. and Peech, M. (1970) Reaction of montmorillonite with iron(III). *Soil Sci. Soc. Amer. Proc.*, **34**, 740–2.

Hsu, P.H. (1989) Aluminum hydroxides and oxyhydroxides. Pp. 331–78 in: *Minerals in Soil Environments*, 2nd edition (J.B. Dixon and S.W. Weed, editors). Soil Science Society of America Book Series No. **1**, Madison, WI.

Hsu, P.H. and Bates, T.F. (1964) Formation of X-ray amorphous and crystalline aluminum hydroxides. *Mineral. Mag.*, **33**, 749–68.

Indraratne, S.P, Goh, T.B., Hawthorne, F.C. and Oscarson, D.W. (1996) Characterization of interlayered clay minerals. Pp. 131–42 in: *Proc. of the 39th Annual Meeting of the Manitoba Soc. of Soil Sci.*

Indraratne, S.P., Goh, T.B., Oscarson, D.W. and Hawthorne, F.C. (1999) Hydroquinone-derived humic polymers formed in the presence of hydroxy-interlayered clays. Pp. 127–39 in: *Proc. 11th Int. Clay Conf.*, Ottawa, Canada.

Inoue, K. and Satoh, C. (1992) Electric charge and surface characteristics of hydroxyalumino-silicate- and hydroxy-aluminum-vermiculite complexes. *Clays Clay Miner.*, **40**, 311–8.

Inoue, K., Pavan, M.A. and Yoshida, M. (1988) Fixation of hydroxy aluminosilicate ions (proto-imogolite) on smectite. *Soil Sci. Plant Nutr.*, **34**, 277–85.

Jaynes, W.F. and Boyd, S.A. (1991) Hydrophobicity of siloxane surfaces in smectites as revealed by aromatic hydrocarbon adsorption from water. *Clays Clay Miner.*, **39**, 428–36.

Jensen, W.B. (1978) The Lewis acid-base definitions: A status report. *Chem. Rev.*, **78**, 1–22.

Keren, R. (1980) Effects of titration rate, pH, and drying process on cation exchange reduction and aggregate size distribution of montmorillonite hydroxy-aluminum complexes. *Soil Sci. Soc. Amer. J.*, **44**, 1209–12.

Lahav, N. and Shani, U. (1978) Cross-linked smectites. II. Flocculation and microfabric characteristics of hydroxy-aluminum-montmorillonite. *Clays Clay Miner.*, **26**, 116–24.

Lou, G. and Huang, P.M. (1988) Hydroxy-aluminosilicate interlayers in montmorillonite: implications for acidic environments. *Nature*, **335**, 625–7.

Lou, G. and Huang, P.M. (1995) Adsorption of hydroxy-aluminosilicate ions by vermiculite. Pp. 187–91 in: *Proc. 10th Int. Clay Conf.*, Adelaide, Australia.

McBride, M.B. (1994) *Environmental Chemistry of Soils*. Oxford University Press, Oxford.

Misono, M., Ochiai, E., Saito, Y. and Yoneda, Y. (1967) A new dual parameter scale for the strength of Lewis acids and bases with the evaluation of their softness. *J. Inorg. Nucl. Chem.*, **29**, 2685–91.

Nordstrom, D.K. and May, H.M. (1989) Aqueous equilibrium data for mononuclear aluminum species. Pp. 29–53 in: *The Environmental Chemistry of Aluminum* (G. Sposito, editor). CRC Press, Boca Raton, FL.

Plee, D., Borg, L., Gatineau, L. and Fripiat, J.J. (1985) High resolution solid state $^{27}Al$ and $^{29}Si$ nuclear magnetic resonance study of pillared clay. *J. Amer. Chem. Soc.*, **107**, 2362–9.

Rai, D., Sass, B.M. and Moore, D.A. (1987) Chromium(III) hydrolysis constants and solubility of chromium(III) hydroxide. *Inorg. Chem.*, **26**, 345–9.

Rausell-Colom, J.A. and Serratosa, J.M. (1987) Reactions of clays with organic substances. Pp. 371–422 in: *Chemistry of Clays and Clay Minerals* (A.C.D. Newman, editor). Monograph, **6**,

Mineralogical Society, London.
Rengasamy, P. and Oades, J.M. (1978) Interaction of monomeric and polymeric species of metal ions with clay surfaces. III. Aluminium(III) and chromium(III). *Aust. J. Soil Res.*, **16**, 53–66.
Rich, L.D., Cole, D.L. and Eyring, E.M. (1969) Hydrolysis kinetics of dilute aqueous chromium(III) perchlorate. *J. Phys. Chem.*, **73**, 713–6.
Sakurai, K. and Huang, P.M. (1998) Intercalation of hydroxy-aluminosilicate and hydroxy-aluminum in montmorillonite and resultant physicochemical properties. *Soil Sci. Soc. Amer. J.*, **62**, 362–8.
Schutz, A., Stone, W.E.E., Poncelet, G. and Fripiat, J.J. (1987) Preparation and characterization of bidimensional zeolitic structures obtained from synthetic beidellite and hydroxy-aluminum solutions. *Clays Clay Miner.*, **35**, 251–61.
Shoval, S. and Yariv, S. (1979) The interaction between roundup (glyphosate) and montmorillonite. Part I. Infrared study of the sorption of glyphosate by montmorillonite. *Clays Clay Miner.*, **27**, 19–28.
Shriver, D.F., Atkins, P.W. and Langford, C.H. (1990) *Inorganic Chemistry*. W.H. Freeman, New York, NY.
Sposito, G. (1984) *The Surface Chemistry of Soils*. Oxford University Press, New York, NY.
Sposito, G. (1994) *Chemical Equilibria and Kinetics in Soils*. Oxford University Press, New York, NY.
Sprankle, P., Meggitt, W.F. and Penner, D. (1975) Adsorption, mobility, and microbial degradation of glyphosate in the soil. *Weed Sci.*, **23**, 229–34.
Sterte, J. and Shabtai, J. (1987) Cross-linked smectites. V. Synthesis and properties of hydroxy-silicoaluminum montmorillonites and fluorhectorites. *Clays Clay Miner.*, **35**, 429–39.
Van Olphen, H. and Fripiat, J.J. (1979) *Data Handbook for Clay Materials and other Non-metallic Minerals*. Pergamon Press, Toronto.
Yamanaka, S. and Brindley, G.W. (1978) Hydroxy-nickel interlayering in montmorillonite by titration method. *Clays Clay Miner.*, **26**, 21–4.
Yamanaka, S. and Brindley, G.W. (1979) High surface area solids obtained by reaction of montmorillonite with zirconyl chloride. *Clays Clay Miner.*, **27**, 119–24.
Yamanaka, S., Yamashita, G. and Hattori, M. (1980) Reaction of hydroxybismuth polycations with montmorillonite. *Clays Clay Miner.*, **28**, 281–4.
Zachara, J.M., Ainsworth, C.C. and Smith, S.C. (1990) The sorption of N-heterocyclic compounds on reference and subsurface smectite clay isolates. *J. Contam. Hydrol.*, **6**, 281–305.

CHAPTER FOURTEEN

# Uranium behaviour in natural environments

K. V. RAGNARSDOTTIR[1,2] AND L. CHARLET[2]

[1] *Department of Earth Sciences, University of Bristol, Wills Memorial Building, Queens Road, Bristol BS8 1RJ, UK (E-mail: Vala.Ragnarsdottir@bris.ac.uk)*

[2] *Environmental Geochemistry Group, LGIT, University of Grenoble-I, BP 53-38041 Grenoble Cedex 09, France (E-mail: Laurent.Charlet@obs.ujf-grenoble.fr)*

**ABSTRACT**

In crustal fluids uranium transport and deposition is dominated by redox conditions. The concentrations of dissolved U(IV) species never exceed $\mu g\ l^{-1}$ levels, whereas the concentration of U(VI) species reaches tens of $mg\ l^{-1}$, depending on pH and ligand concentration(s). As a result, uranium is transported in aqueous fluids of high Eh but deposited when Eh decreases, e.g. in sedimentary environments and engineered barriers that contain sulphide, organic matter, and/or Fe(II)-bearing minerals; bacteria and mineral surfaces are known to catalyse these reduction reactions. Mining activities result in remobilization of uranium as U(VI). In the presence of oxy(hydr)oxides or clay minerals, aqueous uranyl ions are sorbed at intermediate pH. In rivers, uranium transport is dominated by surface-sorbed uranium on colloids and particles. Upon entering the oceans, uranium is deposited in estuarine sediments due to flocculation, or desorbed due to the presence of carbonate ligands in seawater. Uranyl-carbonate complexes can be transported over long distances and thus the uranium residence time in the oceans is greater than that of water. In the presence of appropriate ligands uranium precipitates to form stable minerals and the transport path can be traced by U isotopic ratios of alteration products. The mobility of uranium and its enrichment in the Earth's crust is magnified by subduction zone processes.

## 14.1 Introduction

Uranium is present in all crustal and mantle rocks in trace amounts (e.g. Bailey and Ragnarsdottir, 1994). The continental crust is two orders of magnitude enriched in U relative to the upper mantle. This excess of uranium is reintroduced into the upper mantle by subducting continental crust, redistributing uranium over wide ranges of temperature and pressure conditions.

Fluids carry small concentrations of uranium under reducing conditions where uranium (U(IV)) forms an aqua $U^{4+}$ species but concentrations several

Ragnarsdottir, K.V. and Charlet, L. (2000) Uranium behaviour in natural environments. Pp. 245–289 in: *Environmental Mineralogy: Microbial Interactions, Anthropogenic Influences, Contaminated Land and Waste Management* (J.D. Cotter-Howells, L.S. Campbell, E. Valsami-Jones and M. Batchelder, editors). Mineralogical Society Series, **9**. Mineralogical Society, London. 0 903056 20 8.

orders of magnitude greater in oxidizing fluids where uranium (U(VI)) is present as the oxyanion $UO_2^{2+}$. After a brief review of uranium mineralogy, the distribution of uranium in the earth and the isotopic compositon of uranium, we discuss the health risks associated with human exposure to uranium. Here we separate the risk associated with uranium as a heavy metal and that of radiation exposure due to uranium decay. As for other heavy metals, knowledge of the aqueous geochemistry of uranium demonstrates that the mobility of uranium is a trade-off between solution complexation by inorganic and organic ligands, which increase its mobility, and sorption and (a)biotic reduction reactions which tend to immobilize uranium. The transport of uranium and its immobilization is observed on a large scale on the oceanic floor and on a localized scale in the formation of mineral ore deposits. After reviewing the major uranium ore deposits and their exploitation we outline the risk and methods for confining mine tailings and waste. We conclude by discussing the uranium cycle and residence time in the hydrosphere.

## 14.2 Physical and chemical properties of uranium

### *14.2.1 Uranium distribution in the Earth*

The average crustal abundance of uranium is 2–3 ppm (Plant *et al.*, 1985, 1999). Uranium concentrations in oceanic sediments range from 0.3 to 3.8 mg $kg^{-1}$ (Church, 1973; Ben Othman *et al.*, 1989), with averages of 1.2–1.3 mg $kg^{-1}$ in sedimentary rocks and 2.2–15 mg $kg^{-1}$ in granites (Langmuir, 1978). Arc magmas have concentrations from 0.1 to 3.2 mg $kg^{-1}$ (e.g. McCulloch and Gamble, 1991; Sigmarsson *et al.*, 1990; Hawkesworth *et al.*, 1997*a,b*; Turner *et al.*, 1997) and ophiolites range from 0.1 mg $kg^{-1}$ (pristine) to 0.4 mg $kg^{-1}$ (hydrothermally altered) (Valsami-Jones and Ragnarsdottir, 1997*a*). Higher concentrations are found in manganese nodules which have concentrations as high as 7.7 mg $kg^{-1}$ (Ben Othman *et al.*, 1989) and phosphate rocks range from 20 to 120 mg $kg^{-1}$ (Langmuir, 1978). In comparison, primitive mantle has an average uranium concentration of 0.18 mg $kg^{-1}$ (Gill and Williams, 1990). The total silicate earth has been estimated to have a uranium concentration of 0.019 mg $kg^{-1}$ with the core containing no uranium, the content in the mantle ranging from 0.007 mg $kg^{-1}$ (upper mantle) to 0.017 mg $kg^{-1}$ (lower mantle) and in the crust from 0.373 to 1.67 with an average value of 1.34 mg $kg^{-1}$ for the upper crust and 0.471 mg $kg^{-1}$ for the lower crust (Kramers and Tolstikhin, 1997).

Because the uranium ion has a high charge, it is unlikely to substitute for similarly sized ions in the structures of major early crystallizing minerals and tends to be concentrated in the late, silicic differentiates of igneous melts (e.g. Levinson, 1980). Thus silicic plutons, volcanic rocks and volcaniclastic sediments are relatively uranium-rich, containing ~5 ppm uranium. Circulating groundwater can leach dispersed uranium from such rocks and concentrate it in hydrologically and geochemically favourable areas (Ingebritsen and Sanford, 1998).

### *14.2.2 Uranium oxidation states*

Uranium is known to have several oxidation states. These include U(III), U(IV), U(V), and U(VI). In aqueous fluids U(III) and U(IV) exist as the aqua ions $U^{3+}$ and $U^{4+}$ whereas U(V) and U(VI) form the oxyanions $UO_2^+$ and $UO_2^{2+}$ (uranyl ion). Thermodynamic evaluation of available data led Shock *et al.* (1997*a*) to conclude that the predominant species in weathering systems and hydrothermal solutions are the U(IV) and U(VI) species, linked by

$$U^{4+} + 2H_2O = UO_2^{2+} + 4H^+ + 2e^- \quad \log K^o = -9.038 \tag{14.1}$$

(Grenthe *et al.*, 1992) and the U(III) and U(V) species are not expected to persist or exist at these conditions; hence they will be ignored in the analysis presented below.

### *14.2.3 Uranium solid phases*

In many uranium ore deposits uraninite and pitchblende (poorly crystalline), both nominally $UO_2$, are the principal primary ore minerals (Rich *et al.*, 1977). These minerals are present in high-temperature hydrothermal vein and pegmatite deposits, and sedimentary unconformity type (formed at 150–300°C) and 'roll-front' deposits (formed at ~50°C) (e.g. Galloway, 1978; Nash *et al.*, 1981b; Komninou and Sverjenski, 1996) (see section 14.6.1). Coffinite, $USiO_4$, is also present in some uranium deposits in significant quantities.

Weathering of primary U(IV) minerals results in the formation of secondary uranyl minerals such as schoepite ($\gamma$-$UO_3 \cdot 2H_2O$) as well as silicate, phosphate and vanadite uranyl minerals. These commonly include soddyite ($(UO_2)_2SiO_4 \cdot 2H_2O$), autunite ($Ca(UO_2)_2(PO)_4 \cdot xH_2O$), carnotite ($K_2(UO_2)_2(VO_4)_2$), tyuyamunite ($Ca(UO_2)_2(VO_4)_2$) and Ca-uraninite ($(Ca,U)O_{3-x}$). It is important to note that phases such as $U_4O_9$, $U_3O_7$ and $U_3O_8$ are often included in thermodynamic evaluations of uranium systems. These phases represent solid solutions between $UO_2$ and $UO_3$ (Allen and Tempest, 1986) and thus do not need to be considered as mineral phases.

### *14.2.4 Uranium isotopes*

Uranium has three isotopes with an average natural composition of 0.0057% $^{234}U$, 0.719% $^{235}U$ and 99.275% $^{238}U$ (Rösler and Lange, 1972) (Table 14.1). $^{233}U$ and $^{239}U$ are produced in nuclear reactors and thus are present in spent nuclear fuel (e.g. Notz, 1990). Enriched uranium fuel contains levels of $^{235}U$ which are increased by several percent over natural abundance. The half lives of uranium isotopes ($t_{1/2}$), vary from several days (6.75 d, $^{239}U$), to hundreds of thousands of years ($1.6 \times 10^5$ y, $^{233}U$ and $2.46 \times 10^5$ y, $^{234}U$) and millions to billions of years ($7.038 \times 10^8$ y, $^{235}U$ and $4.468 \times 10^9$ y, $^{238}U$) (Lederer *et al.*, 1967; Steiger and Jäger, 1977; Langmuir, 1997) (Table 14.1). $^{235}U$ and $^{238}U$

TABLE 14.1 Half lives and abundances of uranium isotopes (after Rösler and Lange, 1992).

| Uranium isotope | Natural abundance (%) | Half life $t_{1/2}$ | Final decay product |
|---|---|---|---|
| $^{233}U$* | | $1.6 \times 10^5$ y | |
| $^{234}U$ | 0.0057 | $2.46 \times 10^5$ y | |
| $^{235}U$* | 0.719 | $7.038 \times 10^8$ y | $^{207}Pb$ |
| $^{238}U$ | 99.27 | $4.468 \times 10^9$ y | $^{206}Pb$ |
| $^{239}U$* | | 6.75 d | |

* Produced in nuclear reactors

are the parents of series of radioactive daughters ending with stable isotopes of lead. For $^{238}U$ this chain involves $^{234}$(Th to Pa then U), $^{230}Th$, $^{226}Ra$, $^{222}Rn$, $^{218}$(Po to At then Rn), $^{214}$(Pb to Bi then Po), $^{210}$(Tl to Pb to Bi then Po) and finally $^{206}$(Hg to Tl, terminating with Pb) (Faure, 1986). The overall decay reactions for $^{238}U$ and $^{235}U$ can be written as

$$^{238}U \rightarrow {}^{206}Pb + 8\ {}^{4}He + 6\beta^- + Q_1 \tag{14.2}$$

$$^{235}U \rightarrow {}^{207}Pb + 7\ {}^{4}He + 4\beta^- + Q_2 \tag{14.3}$$

and shows that for each atom of $^{238}U$, one atom of $^{206}Pb$ is produced by emission of eight $\alpha$ particles ($^4He$), six $\beta$ particles and heat ($Q_1 = 47.4$ MeV per atom or $9.4 \times 10^5$ J kg$^{-1}$ s$^{-1}$ (Wetherill, 1966)). $^{235}U$ decays to $^{207}Pb$ with the emission of seven $\alpha$ particles and four $\beta$ particles. The uranium decay series form the basis of the U-Pb dating methods which have been used extensively to determine the age of rocks that have not undergone significant metamorphism or weathering (e.g. Faure, 1986) and cast light on the evolution of the Earth and crustal recycling in the Earth's mantle (Kramers and Tolstikhin, 1997; Elliott *et al.*, 1999). Lead isotopes from the uranium and thorium series are sensitive indicators for crustal contamination of mantle-derived magmas (e.g. Tilton and Barreiro, 1980) as well as for regional or global recycling of continental-derived sediments into the mantle (Hofmann and White, 1982). Recent developments in isotope geochemistry now allow uranium isotopic ratios ($^{234}U/^{238}U$) to identify alteration by fluids and to trace palaeo-fluid flow paths (Sato *et al.*, 1998).

### *14.2.5 U-toxicology*

Uranium is harmful to living organisms if the metal or its decay products enter the body. In the body, uranium has the form of the uranyl ion which has a high affinity for carboxylate functional groups (e.g. Martell and Hancock, 1998). The toxicity of uranium as a heavy metal has been shown to be pH (and hence speciation) dependent (Hyne *et al.*, 1992) and a deliberate overdose of uranium (as uranium acetate) resulted in, amongst other symptoms, acute renal failure, liver disfunction and a paralytic ileum (Pavlakis *et al.*, 1996). Uranium exposure can result from drinking uranium-contaminated water, eating uranium-contaminated food or breathing in uranium-rich dust or decay products of uranium such as radon gas, leading to the production of free radicals and related cell damage in affected body areas (e.g. Prabhavathi *et al.*, 1995; Durakovic, 1999). Uranium toxicity is enhanced by the production of both $\alpha$ and $\beta$ particles due to radioactive decay. The decay of radon also produces solid particles that are toxic, including Po, Pb and Bi. Treatments for uranyl ion intoxication include the ingestion of bicarbonate, catechol derivatives, salicylic acid, gallic acid, quinamic acid, DTPA or EDTA, which form stable complexes or chelates with uranyl and promote excretion (Basinger *et al.*, 1983; Tao *et al.*, 1987; Ortega *et al.*, 1989; Martell and Hancock, 1998). These ligands function as antagonists for a considerable number of metal ions.

Current legal drinking water limits for uranium vary from 20 $\mu g\ l^{-1}$ in the USA to a maximum acceptable value in Canada of 100 $\mu g\ l^{-1}$ (e.g. Langmuir, 1997). The low solubility of uranium phases indicates that uranium concentrations in natural waters are usually not high enough to constitute a health risk (see section 14.3.3). Uraniferous areas in the USA typically have 10 $\mu g\ l^{-1}$ U in streams to 120 $\mu g\ l^{-1}$ in groundwater. Rarely are higher concentrations seen, with the exception of leachates from mill tailings. There U(IV) mineral ores have been converted to U(VI) salts, and concentrations of up to 20 mg $l^{-1}$ U are found (Langmuir, 1997).

### *14.2.6 Radiation exposure*

The SI unit of radioactivity is the becquerel (Bq). It is defined as the amount of a radioisotope which gives one disintegration per second. The radioactivity of a sample is measured by electrical or photographic methods and is expressed as the number of disintegrations observed per unit time. Gray (Gy) is the SI unit for the absorbed dose (J $kg^{-1}$). The SI unit for dose equivalent is Sievert (Sv), measured in grays multiplied by a quality factor for the type of radiation and a weighting factor for the tissue irradiated. Sievert is numerically equivalent to gray for electrons and X-rays irradiating the whole body.

Human radiation exposure in the natural and urban environment is focused upon radon (Rn), the only radioactive gas which is formed by the decay of uranium and thorium (e.g. Ball *et al.*, 1991; Bottrell, 1993). Radon has three isotopes $^{219}$Rn (action; $t_{1/2}$ s), $^{220}$Rn (thoron; $t_{1/2}$ s) and $^{222}$Rn (radon; $t_{1/2}$ = 3.82 days). The gas is ingested respiratorally where it can cause damage as severe as lung cancer (Jones, 1995). $^{222}$Rn decays by the emission of an alpha particle. In the open air, the concentration of radon is generally very low (<1 Bq $m^{-3}$). However, in areas built on or out of rocks containing high quantities of uranium, such as granite, the radon levels can rise well above background levels (Ball and Miles, 1993) which is 1 mSievert $y^{-1}$. In the UK the National Radiological Protection Board has defined areas where radiation of >200 Bq $m^{-3}$ is measured in >1% of houses as 'action areas'. In these areas, which are primarily centred around the Cornubian granites of Devon and Cornwall, special measures must be taken in ventilating houses (e.g. Bell, 1998).

Of note is that Rn levels in groundwater are used to predict earthquakes (e.g. Silver and Wakita, 1996). This is due to the fact that shortly before the rupture of the crust, a large quantity of microcracks open, releasing to the groundwater trapped radon produced by the decay of uranium and thorium in the rocks. For example, the radon content peaked dramatically in a well ~30 km northeast of the epicentre of the disastrous M 7.2 earthquake at Kobe, Japan in 1995 (Igarashi *et al.*, 1995).

## 14.3 Aqueous uranium chemistry

### *14.3.1 Thermodynamics and structure of aqueous uranium*

For natural waters it is necessary to consider all U(IV) and U(VI) species known to exist at varying pH and redox conditions (Garrels and Christ, 1965; Langmuir, 1978, 1997; Langmuir and Herman, 1980; Ryan and Rai, 1983; Shock *et al.*, 1997*a*). Under reducing conditions and low pH, uranium is present as the aqua $U^{4+}$ species. As the pH increases, uranium hydroxides form $UOH_3^+$, $U(OH)_2^{2+}$, $U(OH)_3^+$, $U(OH)_4^0$ and $U(OH)_5^-$. Ryan and Rai (1983), Parks and Pohl (1988) and Rai *et al.* (1990) show that there is no dependence on pH of the solubility of $UO_2(s)$ up to 300°C and at pH values >4 and therefore $U(OH)_5^-$ is only of minor importance. Table 14.2 lists the thermodynamic constants for the dissolution of $UO_2(s)$ under reducing conditions

$$UO_2(s) + 2H_2O \rightleftharpoons U(OH)_4^0 \qquad (14.4)$$

TABLE 14.2 Range of reported solubility of uranium phases in terms of log molal and ppm (or ppb) concentrations, depending on mineral crystallinity.

| $UO_2$ crystallinity | Solubility as log $[U(OH)_4^0]$ | Solubility as w/w | Source |
|---|---|---|---|
| Amorphous | −8.0 | 3.0 ppb | Rai *et al.* (1990) |
| Microcrystalline | −8.7 | 0.6 ppb | Yajima *et al.* (1995) |
| Microcrystalline | −7.3 | 15 ppb | Casas *et al.* (1998) |
| Natural uraninite | −8.5 | 0.9 ppb | Casas *et al.* (1998) |
| Well crystalline | −9.5† | 0.1 ppb | Parks and Pohl (1988) |
| Schoepite (am)** | −4.5 | 9.4 ppm | Bruno *et al.* (1987) |
| Well crystalline | −17.1* | | Grenthe *et al.* (1992) |

* calculated (see text); ** probable phase (Langmuir, 1997)
† extrapolated from 100°C by Shock *et al.* (1997*a*)
am: amorphous

Figure 14.1 gives experimental results plotted as a function of equation 14.4 above. The figure indicates that $U(OH)_2^{2+}$ and $U(OH)_3^+$ are of minor importance with $UOH^{3+}$ dominating below pH 4 and $U(OH)_4^0$ above pH 4. It is to be noted, however, that considerable uncertainty exists in the hydrolysis constants quoted in the literature (e.g. Grenthe *et al.*, 1992; Shock *et al.*, 1997*a*). Thermodynamic speciation calculations show that at low redox conditions ($\log f_{H_2}$ of −5) and at 25°C the dominating species are $U^{4+}$ from pH 0–1, $UOH^{3+}$ from pH 1–2, $U(OH)_4^0$ from pH 2–10 and $U(OH)_5^-$ from pH 12–14 (Shock *et al.*, 1997*a*). The concentration of these species is, however, always extremely low (well below μg $l^{-1}$ levels), as long as the pH is >4.

Under oxidizing conditions, U(VI) is present as the uranyl species. The uranyl ion has a linear $(O{-}U{-}O)^{2+}$ structure with an axial U–O distance of 1.79 Å (e.g. Thompson *et al.*, 1997; Schofield *et al.*, 1999). In addition, the $U^{6+}$ ion is surrounded by five equatorial water or hydroxide molecules. As pH increases, the uranyl species hydrolyses from $UO_2OH^+$, $UO_2(OH)_2^0$, $UO_2(OH)_3^-$, $UO_2(OH)_4^{2-}$, $(UO_2)_3(OH)_5^+$ and $(UO_2)_3(OH)_4^{2+}$. Stability constants for these species have been under debate for decades because the speciation of experimental solutions has been determined from potentiometric studies which are subject to difficulties with computer fittings. Therefore some of the species mentioned above are unconfirmed (see Grenthe *et al.*, 1992 for a detailed discussion). In the absence of ligands other than $OH^-$ speciation distribution calculations for oxidizing conditions of $\log f_{H_2}$ of −20 give $UO_2^{2+}$ as the

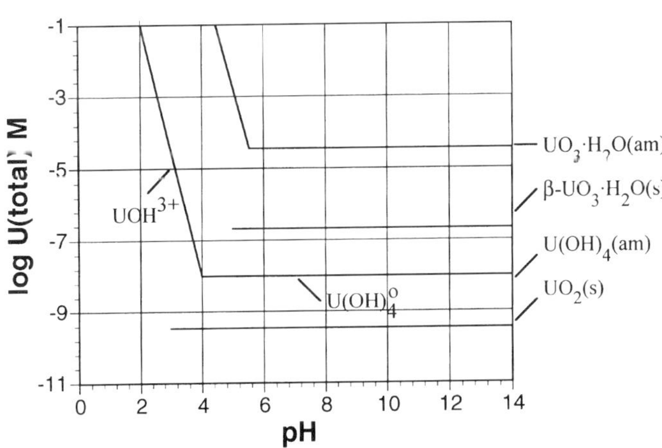


Fig. 14.1 The solubility of uranium solids based on data from Gayer and Leider (1955, 1957) and Bruno *et al.* (1987) for $UO_3{\cdot}H_2O$(am); calculated from thermodynamic data from Langmuir (1997) for β-$UO_3{\cdot}H_2O$; Ryan and Rai (1983) and Rai *et al.* (1990) for $U(OH)_4$(s); and extrapolated to 25°C from 100°C data of Parks and Pohl (1988) for $UO_2$(s). The solid lines are best fit lines for the reactions $U(OH)_4(am) + 3H^+ = UOH^{3+} + 3H_2O$ and $U(OH)_4(s) = U(OH)_4^0$ as reported by Rai *et al.* (1990). (am: amorphous)

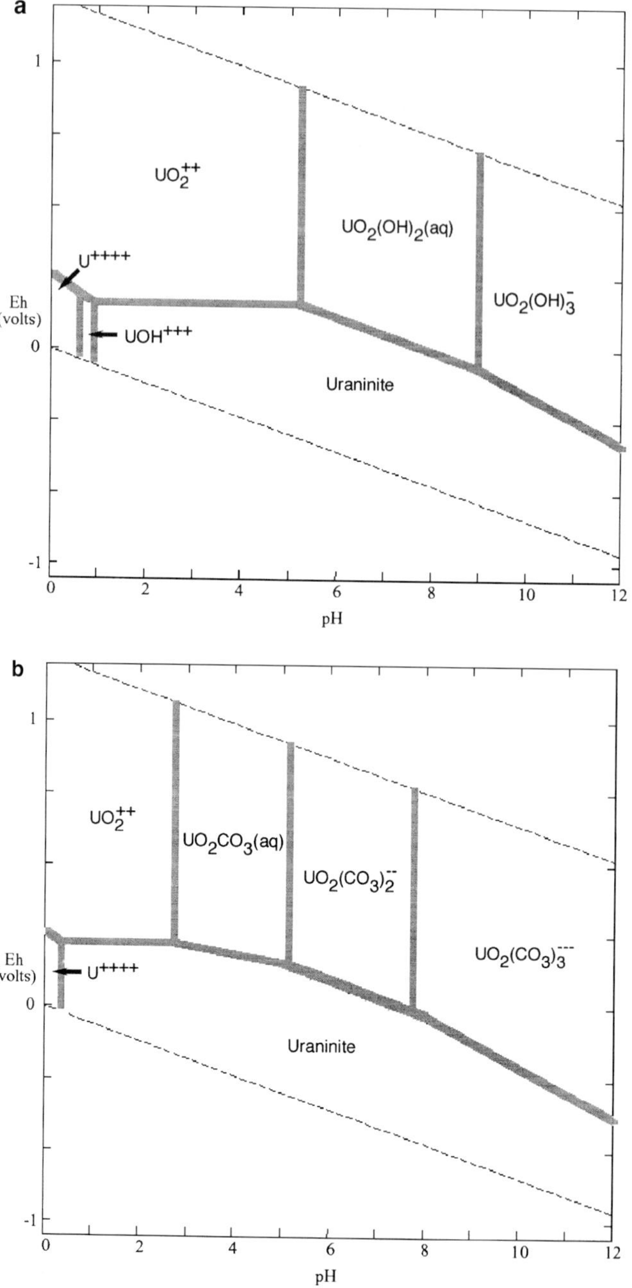


FIG. 14.2 Eh-pH diagram for aqueous species in the U-$O_2$-$CO_2$-$H_2O$ system in pure water at 25°C, 1 bar total pressure and for total U = $10^{-8}$ M: (*a*) $P_{CO_2} = 10^{-3.8}$ and (*b*) $P_{CO_2} = 10^{-2.0}$ bar. The diagram was plotted with the aid of *Geochemists Workbench* (Bethke, 1994, 1996).

dominant species from pH 0–5, $UO_2(OH)_2^0$ for pH 5–9, $UO_2(OH)_3^-$ from pH 9–13 and $UO_2(OH)_4^{2-}$ from pH 13–14 (Shock *et al.*, 1997*a*). This leads to mg $l^{-1}$ level concentrations of uranium in aqueous solutions above pH 5, and up to thousands of mg $l^{-1}$ at pH 2. Figure 14.2*a* shows that the dominating uranyl species are $UO_2^{2+}$ (pH <6), $UO_2(OH)_2^0$ (pH 6–9) and $UO_2(OH)_3^-$ (pH >9). At intermediate redox conditions, both U(IV) and U(VI) species exist.

The most important species for near neutral waters are the uncharged $UO_2(OH)_2^0$ and $U(OH)_4^0$ species. Therefore in the pH range of many natural waters (6<pH<9), the concentration of uranium in a VI valence state is about four orders of magnitude greater than that of uranium in its IV valence state. In the presence of carbonate, carbonate complexes form (Fig. 14.2*b*). Figure 14.2 shows that uranium concentrations are governed by Eh control and that the lowest uranium concentrations are found at low Eh where uraninite is stable. Complexing of uranyl with organic ligands can further increase uranium concentrations (Shock and Koretsky, 1993; 1995; Silva and Nitche, 1995). Caution needs to be exercised when interpreting and modelling with uranium data because some workers use different conventions in representing aqueous species (e.g. Wagman *et al.*, 1982; Shock *et al.*, 1997*b*). For example Shock *et al.* (1997*b*) represent $U(OH)_4^0$ as $UO_2^0$ and $UO_2(OH)_2^0$ as $UO_3^0$.

It is evident from the discussion above that in the absence of additional ligand(s) and under reducing conditions dissolved uranium levels are below μg $l^{-1}$ concentrations but as the Eh increases the concentrations can reach mg $l^{-1}$ level concentrations of uranium in aqueous solutions above pH 5, and up to thousands of mg $l^{-1}$ at pH 2. Uranium concentrations of shallow groundwaters show values of μg $l^{-1}$ to mg $l^{-1}$ uranium, e.g. uranium in groundwaters from the Laurivuoma bedrock feeding the Kalix River in Sweden has measured values from 2 to 122 μg $l^{-1}$ (Porcelli *et al.*, 1997). Similarly, uranium from groundwaters in the Stripa granite, Sweden, has been measured at 0.9 to 90 μg $l^{-1}$ (Andrews *et al.*, 1989) and shallow groundwaters from the Helsinki area in Finland have uranium concentrations that range from 10 μg $l^{-1}$ to 14.8 mg $l^{-1}$ (Asikainen and Kahlos, 1979).

### *14.3.2 Uranium complexing agents*

The most effective complexing agents for uranium in natural waters are hydroxide, carbonate, phosphate, fluoride, chloride and organic ligands. Table 14.3 lists the stability constants of important uranium complexes. A typical groundwater has a $CO_2$ pressure of $10^{-2}$ bar (and rivers $10^{-3.8}$ bar). In groundwater conditions, the uranium-carbonate complexes $UO_2CO_3^0$ (pH 3–5), $UO_2(CO_3)_2^{2-}$ (pH 5–8) and $UO_2(CO_3)_3^{4-}$ (>pH 8) predominate over uranyl-hydroxide complexes (Fig. 14.2*b*). Uranyl hydroxide and uranyl carbonate complexes thus have a neutral or a negative charge in near neutral to alkaline solutions, and this has a profound effect on the adsorption behaviour of uranyl (see section 14.4.3) and thus the residence time in seawater (see section 14.7.1).

In the presence of phosphate, the uranyl ion forms the dominating complexes $UO_2H_2PO_4^+$ (pH 2–3), $UO_2HPO_4^0$ (pH 3–6.6), $UO_2PO_4^-$ (pH 6.9–9.5) (Sandino and Bruno, 1992). While other phosphate species have been suggested by Baes and Mesmer (1956), the three species quoted above are likely to be predominant in natural systems where $UO_2$(s) ore is oxidized and uranium is transported as phosphate complexes and eventually precipitated as uranyl phosphate minerals such as $(UO_2)_3(PO_4)_2{\cdot}4H_2O$ (Sandino and Bruno, 1992). In such systems the uranyl ion forms much stronger complexes with phosphate than carbonate and indeed correlation between phosphate and uranium concentrations has been observed in groundwaters (Duerden, 1990). Phosphate complexing is thought to have been important for U(VI) phosphate ore formation at Koongarra in Australia (Duerden, 1990), Southern Karoo, South Africa and Northwestern province of Zambia. Low concentrations of uranium are associated with sedimentary prosphates such as the Florida Phosphate uranium province which reaches from Florida to Georgia in the US (Cathcart *et al.*, 1984).

Uranyl is also known to be associated with organic matter such as humic and fulvic acids present in natural waters (e.g. Sheppard *et al.*, 1980; Anderson, 1982; Plater *et al.*, 1992). Despite the ubiquitous presence of natural organic matter in aquatic systems, and the well documented ability of uranium to form uranyl-organic complexes (e.g. Moulin and Moulin, 1995; Pompe *et al.*, 1996; Silva and Nitsche, 1996; Martell and Hancock, 1998; Nefedow *et al.*, 1998), the confirmation of stability constants has proven somewhat elusive (e.g. Zeh *et al.*, 1997; Lenhart and Honeyman, 1999). Stability constants for 1:1 and 1:2 (metal:ligand) complexes of uranyl with various humic acids were only recently determined by potentiometry (Lubal *et al.*, 2000). The existence of these complexes has been demonstrated for groundwater around the Gorleben salt domes in eastern Germany (Zeh *et al.*, 1997) where 75% of the natural uranium is humic colloid-borne, and in uranium mines in eastern Germany and the Czech Republic where tree stumps are used to clad the mine walls (Schmidt *et al.*, 1996). Upon flooding of the mines, the waters contain high concentrations of humic acids that cause elevated uranium concentrations. Studies of miners in the area have revealed that they suffer from lung and liver cancer as well as leukaemia (Enderle and Friedrich, 1999).

$UO_2(Cl)_2^0$ and $UO_2Cl^+$ are considered to have been important species for transporting uranium during the uranium ore forming event (200°C, pH 4.5) of the Koongarra unconformity-type ore formation in Australia (Komninou and Sverjenski, 1996). In geological systems that contain fluoride (e.g. granitic systems) U-F complexation would result in much higher concentrations of uranium (Romberger, 1984) than in the presence of chloride, as shown by the high stability constant for the first U-F complex relative to the first U-Cl complex (Table 14.3; Grenthe *et al.*, 1992).

TABLE 14.3 Selected association constants ($I = 0$) for the formation of some U complexes.

| Association reaction | Log $K$ |
|---|---|
| $U^{4+} + Cl^- = UCl^{3+}$ | 1.72[a] |
| $U^{4+} + F^- = UF^{3+}$ | 9.28[a] |
| $U^{4+} + H_2O = UOH^{3+} + H^+$ | −0.65[b] |
| $U^{4+} + 4H_2O = U(OH)_4^0 + 4H^+$ | −12.0[b] |
| $U^{4+} + SO_4^{2-} = USO_4^{2+}$ | 6.58[a] |
| $UO_2^{2+} + CO_3^{2-} = UO_2CO_3^0$ | 9.68[a] |
| $UO_2^{2+} + 2CO_3^{2-} = UO_2(CO)_3^{2-}$ | 16.94[a] |
| $UO_2^{2+} + 3CO_3^{2-} = UO_2(CO)_3^{4-}$ | 21.60[a] |
| $UO_2^{2+} + H_2O = UO_2OH^+ + H^+$ | −5.2[b] |
| $UO_2^{2+} + 2H_2O = UO_2(OH)_2^0 + 2H^+$ | −12.0[b] |
| $UO_2^{2+} + 3H_2O = UO_2(OH)_3^- + 3H^+$ | −19.2[b] |
| $UO_2^{2+} + PO_4^{3-} = UO_2PO_4^-$ | 13.23[a] |
| $UO_2^{2+} + SO_4^{2-} = UO_2SO_4^0$ | 3.15[a] |

[a] Grenthe *et al.* (1992)
[b] Langmuir (1997)

### *14.3.3 Solubility of uranium minerals*

The solubility of uraninite under reducing conditions is very low (e.g. Langmuir, 1997). The ability to obtain good solubility data has been hampered by a number of factors: uraninite is often poorly crystalline; the difficulty in keeping the experimental solutions reduced; the presence of oxidized surface layers on the $UO_2$ starting material; and the problem of keeping $CO_2$ out of the experimental solutions. Analytical detection limits and the reliability of instruments pose further difficulty. As discussed in the speciation sections 14.3.1 and 14.3.2, the most important species under near neutral conditions and in the absence of ligands are the neutral hydroxide species ($U(OH)_4^0$ and $UO_2(OH)_2^0$). Also, the solubility of uranium solids has been shown to be independent of pH above pH 4 from 25 to 300°C (e.g. Parks and Pohl, 1988; Rai *et al.*, 1990; Torrero *et al.*, 1998; Yajima *et al.*, 1995) (Fig. 14.1)). Similar results have been obtained for Ca-uranate ($Ca,U_{1.6}O_{5.8}{\cdot}2H_2O$) (Valsami-Jones and Ragnarsdottir, 1997*b*). By assuming that the activity of pure end-member uraninite is one and that the activity of water is also one, the solubility of uraninite can be expressed as the activity of the uranium hydroxide species:

$$UO_2(s) + 2H_2O \rightleftharpoons U(OH)_4^0 \qquad K = a_{U(OH)_4^0} = 10^{-9.47} \tag{14.5}$$

In a summary of the results of experimental studies, Langmuir (1997) presented preferred values for the log $[U(OH)_4^0]$ as −8.0 for amorphous $UO_2$, −8.7 for microcrystalline $UO_2(s)$, −9.47 for crystalline $UO_2(s)$ and −17.1 (0.002 pg $l^{-1}$) for a well crystallized $UO_2(s)$ (Table 14.2). This reaction is very important for the mobility of uranium from intermediate and high level radioactive waste repositories many of which are designed to have low Eh, and primarily have waste forms of $UO_2$ (e.g. Savage, 1995). The lowest uraninite solubility values ($[U(OH)_4^0] = 10^{-17.1}$) is a calculated value and has not been measured in the laboratory due to the difficulties outlined above (Grenthe *et al.*, 1992). The data summarized here shows that the solubility of uraninite varies over nine orders of magnitude, as a function of the crystallinity of the uraninite solid. There is a discrepancy in the literature as to which solubility value for uraninite should be used to represent the crystalline phase for geochemical modelling purposes. Most workers in the uranium geochemistry field use the solubility value of $10^{-9.47}$ from Parks and Pohl (1988) for the solubility of crystalline $UO_2$ under reducing conditions (e.g. Raffensberger and Garven, 1995; Komninou and Sverjenski, 1996; Langmuir, 1997; Shock *et al.*, 1997*a*; Valsami-Jones and Ragnarsdottir, 1997*b*) and these are compiled in several geochemical databases.

By assuming that the activity of water is 1, the solubility of schoepite can be described as the activity of the neutral species, giving the log $[UO_2(OH)_2^0]$ of $-6.8$ ($mg\ l^{-1}$). Schoepite ($\beta$-$UO_3 \cdot 2H_2O$) thus has a three orders of magnitude greater solubility than uraninite

$$\beta\text{-}UO_3 \cdot 2H_2O\ (s) \rightleftharpoons UO_2(OH)_2^0 + H_2O \qquad K = 10^{-6.8} \tag{14.6}$$

(Langmuir, 1997). Figure 14.1 shows the pH independence above pH 4 but the solubility values depicted are greater, representing the amorphous nature of the solids used in the experiments and perhaps also the presence of carbonate contamination in the experimental solutions.

Uranyl-phosphate minerals solubility results show that uranium complexes with phosphate in solution and that the total uranium concentration in solutions is of the order of $10^{-7}$ M uranium at pH 5–7, which corresponds to 32 $\mu g\ l^{-1}$ $UO_2H_2PO_4^+$ (Sandino and Bruno, 1992). Of note is that phosphate fertilizers have a soluble phosphate form ($Ca(H_2PO_4)_2$; e.g. Lee, 1991). Because these fertilizers contain trace quantities of uranium, their addition to the world's soils is ultimately changing the budget of uranium in the oceans. It has, for example, been suggested that Everglade wetlands contain fertilizer-derived uranium based on $^{234}U/^{238}U$ ratio (Zielinski *et al.*, 2000) and that Mediterranean waters have high uranium content, due to phosphate fertilizer inputs (Schmidt and Reyss, 1991).

### *14.3.4 Radiolysis of water*

Radiolysis of water occurs as an $\alpha$ particle, produced by uranium decay, loses its energy by way of ionizing the water. One $\alpha$ particle with the energy of 1 MeV can ionize $10^5$ water molecules (Dubessy *et al.*, 1988), producing the stable secondary products $H_2$ and $O_2$. Other species produced include $H_2O_2$ and $HO_2$ which are powerful oxidizing agents. If organic matter is present, organic free radicals, i.e. organic compounds or moieties possessing one or more unpaired electrons, are also produced (e.g. Rigali and Nagy, 1997). In the natural environment, hydrogen can easily diffuse away, but this is not the case for oxygen and therefore the radiolysis of water may account for the oxidation of uranium ores and hence, the mobilization of uranium. This is supported by the fact that uraninite surfaces never show true $UO_{2.00}$ stoichiometry in natural systems (e.g. Cathelineau *et al.*, 1982). Rather, the surface layer of $UO_2$ is closer to $UO_3$ in composition. Of note is that uranium fuel consists of >95% $UO_2$ and thus the long term risk assessment for future sites for radioactive waste is dominated by the oxidation of $^{238}U$ and $^{235}U$ oxides and by the chemistry of the daughter products.

### *14.3.5 Oxidative dissolution of uraninite*

Uranium is mobilized by the oxidative dissolution of uraninite. This process has recently been outlined by Torrero *et al.* (1998) to occur in the following manner

$$>UO_2 + 0.5O_2 \rightarrow >UO_2{\cdot}0.5O_2 \rightarrow >UO_3 \qquad (14.7)$$

where $>UO_2$ and $>UO_3$ refer to U(IV) and U(VI) surface species. Once $>UO_3$ is formed, the dissolution of uraninite follows the scheme of calcite dissolution (Busenberg and Plummer, 1986; Van Cappellen *et al.*, 1993), i.e. it occurs at low pH by the surface co-ordination of $H^+$

$$>UO_3 + H^+ \rightarrow >UO_3\text{-}H^+ \rightarrow UO_2(OH)^+ \qquad (14.8)$$

and at neutral to alkaline pH, by the surface co-ordination of $H_2O$

$$>UO_3 + H_2O \rightarrow >UO_3\text{-}H_2O \xrightarrow{\text{fast}} UO_2(OH)_2^0 \qquad (14.9)$$

Torrero *et al.* (1998) showed the rate of dissolution to depend on the concentration of available surface sites. At constant solid surface concentration, the rate increases with increasing $H^+$ and $O_2$ activity

$$k\ (\text{mol m}^{-2}\ \text{s}^{-1}) = 3.5\ (\pm 0.8)10^{-8}\ [H^+]^{0.37}[O_2]^{0.31} \qquad (14.10)$$

The dissolution rate is highest at low pH and high $f_{O_2}$ and reaches a minimum above pH 6 (Torrero *et al.*, 1998). The presence of carbonate ions also enhances the dissolution rate of uraninite (de Pablo *et al.*, 1999).

## 14.4 Transport of uranium

### *14.4.1 Uranium colloids*

Colloid-sized particles can remain in stable suspension and be transported by natural waters, including river and groundwaters, moving at any velocity (e.g. McCarthy and Degueldre, 1993). Colloids have a diameter of <10 μm by definition (Stumm and Morgan, 1981), but membrane filters for taking out colloids commonly have a diameter of 0.45 μm. Colloids usually have significant unsatisfied surface charge that is related to their colloidal behaviour but also makes them potential sorbents for dissolved species in water. Colloidal particles remain in suspension because of mutual charge repulsion and Brownian motion (Stumm and Morgan, 1996). Rivers, particularly turbulent ones, can transport colloids over long distances. Once the rivers enter estuaries, the suspensions of the colloids are destabilized by increases in ionic strength (e.g. Mayer, 1982). Colloids can vary in composition from iron (oxy)hydroxide (FeOOH) to organic matter such as humic substances and bacteria. Clays, silica ($SiO_2(s)$) and calcium carbonate ($CaCO_3$) also exist in nature as colloid-sized particles. What all colloids have in common is a huge surface area (commonly 10–1000 $m^2\ g^{-1}$) and high surface-site density of charged sites per unit area. These sites are available for sorption of dissolved ions in water (e.g. Langmuir, 1997).

The uranium concentrations of river waters in large drainage basins are largely determined by dissolution of limestones. This is due to the weathering of uraniferous black shales which contain sulphide minerals which produce sulphuric acid and reduce the pH of water upon oxidation. This leads to dissolution of limestone and the associated aqueous input of carbonate ions, leading to high U levels in some rivers (Palmer and Edmond, 1993) due to the formation of uranyl-carbonate complexes. A significant proportion of U in rivers is transported as suspended particles (Martin and Whitfield, 1983). By definition, suspended particles are 1 μm to 5 mm in size (Stumm and Morgan, 1981). The nature of these particles has been debated in the literature (e.g. Ames *et al.*, 1983*a,b*; Plater *et al.*, 1992) but recent studies have shown that a large proportion of uranium transported in rivers is sorbed to particles (Andersson *et al.*, 1998) and colloids (Porcelli *et al.*, 1997). These studies found that uranium transport by the Kalix River in northern Sweden into the Baltic Sea is dominated by U sorbed to Fe-oxyhydroxide and biogenic phases which are >0.45 μm in size. The colloid fraction, which was defined in the study to be the size fraction from 10 kD (kilo Daltons) to 0.45 μm, carries 30–90% of the transported uranium and the authors conclude that uranium is being transported by humic acids. This is supported by the observation that ~50% of the <0.45 μm filtered water fraction of discharged Kalix River water uranium is removed at high salinities and neutral pH, as expected if uranium is associated with colloidal material which flocculates as the ionic strength increases (Boyle *et al.*, 1977; Sholkovitz *et al.*, 1978; Fox, 1983). Uranium is thus primarily transported in rivers by colloidal particles and is deposited in estuaries (Porcelli *et al.*, 1997). This behaviour has also been reported for the following rivers: Ganghes-Brahmaputra (Carroll and Moore, 1993), the Forth, UK (Toole *et al.*, 1987) and the Glatt, Switzerland (Lienert *et al.*, 1994). In contrast, uranium has been found to behave conservatively in the estuaries of the rivers Saire (Martin *et al.*, 1978) and the Clyde and Tamar, UK (Toole *et al.*, 1987), probably due to lack of colloid material in the river water. Porcelli *et al.* (1997) conclude that uranium removal in estuaries is not likely to be due to adsorption onto inorganic colloids, as adsorption is diminished in higher ionic strength waters. Rather they suggest that uranium is more likely to be removed as U-humate complexes.

There is a large amount of literature devoted to colloid-mediated transport as a means for the long distance transportation of strongly adsorbing species. Little evidence exists in the literature, although many studies have been undertaken on former military sites, where depleted uranium shields litter the landscape (Guy *et al.*, 1998). A recent NATO report shows that ~10 tons of uranium was dispersed in Kosovo during the 1999 war (UNEP, 1999*a,b*) and hundreds of tons were spread over Iraq during the Gulf war, calling for urgent studies of U behaviour in these areas. Uranium has been reported to be transported in porous media or fractures either as U(VI), sorbed on organic colloids (Read *et al.*, 1998) or as U(IV) oxide colloids (Van der Lee *et al.*, 1992). There is further evidence that particle and colloid transport in groundwaters does occur and may affect contaminant migration (e.g. Dearlove *et al.*, 1991; Mills *et al.*, 1991). In contrast to the previous statement, Lienert *et al.* (1994) conclude that for a groundwater recharge from a river, colloids have no role to play in determining uranium mobility. It has also been suggested that porous rocks which contain micropores (<2 nm) and mesopores (2–50 nm) can filter out most of the colloidal materials carried by soil water and groundwater (Davis and Kent, 1990). It is thus likely that many rock formations with low hydraulic conductivity will act as active filters for colloid particles. The role of colloid-related uranium transport in groundwaters is thus unclear and needs to be addressed further to cast light on uranium transport from future repositories of radioactive waste and U behaviour in former war zones.

### *14.4.2 Aqueous transport of uranium*

When considering geochemical reactions which may occur along a fluid flow path in groundwater systems the law of conservation of mass can be applied (e.g. Freeze and Cherry, 1979). The overall relation that takes account of fluid flow and mass flux is

| Net change of mass of solute in volume | = | Flux of solute out of the volume | − | Flux of solute into the volume | ± | Loss or gain of solute mass due to reactions or adsorption | (14.11) |
|---|---|---|---|---|---|---|---|

The mathematical formulation of this relationship is the advection-dispersion equation which describes the transport of metals such as uranium in saturated porous media (Bear, 1979). The assumptions made are that the formation is homogeneous and that fluid flow is at steady state and follows Darcy's law

$$\left(1 + \frac{\rho_b K_d}{n}\right) \frac{\partial C}{\partial t} = D_e \frac{\partial^2 C}{\partial x^2} - V \frac{\partial C}{\partial x} - \lambda C - \sum_i k_i C \qquad (14.12)$$

where $C$ is the uranium (VI) concentration in water (mol $l^{-1}$), $t$, the time (days), $x$ is length or depth (m), n is the porosity of the medium (dimensionless), $\rho_b$ the bulk density of the medium (g $m^{-3}$), and $V$ the pore water velocity (m $day^{-1}$). The uranium solute dependent parameters are: $D_e$, the effective dispersion coefficient ($m^2$ $day^{-1}$) in porous media, $\lambda$ is the uranium decay rate ($day^{-1}$) and $k_i$ is the first order rate ($day^{-1}$) of the transformation reaction $i$ (e.g. $i$ = reductive precipitation of uranium), and $K_d$ is the distribution coefficient between particles and solution ($cm^3$ $g^{-1}$). The parenthesis on the left hand side of equation 14.12 is referred to as the retardation factor and is equal to the velocity ratio between uranium and water. By far the most intensively studied solute-dependent parameter has been the empirical $K_d$ value because it is relatively easy to determine. As shown below, this empirical coefficient can be obtained from thermodynamic cation exchange coefficients or surface complexation constants.

### *14.4.3 Adsorption phenomena*

According to equation 14.12, sorption processes dictate the retardation of uranium during groundwater movement through porous or fractured media. Three major sorbents can be distinguished in groundwater systems, depending on the mineralogical and chemical composition: clay minerals, oxy(hydr)oxides and bio/organic matter. Information on sorption phenomena can be obtained from macroscopic studies and spectroscopic measurements and are reviewed together in the following sections.

*14.4.3.1 Clay minerals.* Clay minerals have a layered structure. Clay platelets are composed of octahedral layers (with metals such as Fe(II) or Mg surrounded by six oxygens) and tetrahedral layers (with Si surrounded by four oxygens) (e.g. Sposito, 1984). Substitution of $Al^{3+}$ for $Si^{4+}$ results in a charge imbalance which is taken care of by interlayer cations such as $Na^+$ or $Ca^{2+}$. Sorption on clay particles (e.g. smectite or montmorillonite) occurs by two distinct mechanisms: cation exchange between interlayer cations and solution cations (e.g. $2Na^+$ for $UO_2^{2+}$), and surface complexation on the particle edges (Morris *et al.*, 1994; Geipel *et al.*, 1996; Boult *et al.*, 1998). The $Na^+/UO_2^{2+}$ cation exchange reaction can be represented as

$$2NaX + UO_2^{2+} \rightleftharpoons X_2UO_2 + 2Na^+ \qquad (14.13)$$

where $X^-$ represents one mole of the clay exchanger. In this reaction the uranyl ions remain fully hydrated, i.e. the uranyl ion, with its two axial O atoms ($d_{U-O}$ = 1.79–1.81 Å) is surrounded by five water molecules or hydroxyl ions in the equatorial positions ($d_{U-O}$ = 2.44 Å) (Dent *et al.*, 1992; Chisholm-Brause *et al.*, 1994; Muskett *et al.*, 1998). Very few measurements for the cation exchange constant of equation 14.13 have been reported in the literature (Tsunashima *et al.*, 1981; Giaquinta *et al.*, 1997; Arnold *et al.*, 1998; Boult *et al.*, 1998). The affinity of the clay interlayer sorption sites for the uranyl ion is of the order of that observed for the fully hydrated divalent cations such as $Ca^{2+}$.

Complexation of uranyl ions with the clay edge sites starts to dominate the sorption phenomena over cation exchange mechanism at low uranium concentration (i.e. uranium concentration lower than the available number of edge sites) and at pH >4 (Boult *et al.*, 1998) (Fig. 14.3*a*). The general equation describing the formation of an $(x,y,z)$ complex in waters containing carbonate ions is

$$z(>SOH) + UO_2^{2+} + xOH^- + yL^{e-} \rightleftharpoons (>SO)_zUO_2(OH)_{x-z}(L)_y^{2-x-ey} + zH_2O \quad \beta_{(x,y,z)} \qquad (14.14)$$

with >S representing surface Si or Al, and $L^{e-}$ standing for ligands ($CO_3^{2-}$, $SO_4^{2-}$, etc.), and $\beta_{(x,y,z)}$ representing the sorption constant. Sorption of uranyl ions on clay edges has been described as (2,0,1) and (4,0,1) surface complex formations, in which a monodentate surface complex is assumed to be formed (Boult *et al.*, 1998). Surface complexes formed in interlayer spacing and on clay edges are shown in Fig. 14.3*b*. However, a close inspection of EXAFS spectroscopic data indicates that the uranyl ion may share equatorial oxygens with the edges of the clay octahedral layer, in a mechanism which is described in detail below for adsorption on oxide surfaces. This mechanism leads to a sharp reduction in the number of equatorial oxygen atoms, from 5 in solution to 2.9, and to a subsequent decrease of the $d_{U-Oeq}$ from 2.44 to 2.39 Å (Chisholm-Brause *et al.*, 1994). The tendency for these complexes to form increases slightly with temperature from 20 to 60°C in alkaline conditions, at pH 10.

*14.4.3.2 Oxyhydroxides.* Adsorption on oxy(hydr)oxides (ferrihydrite, goethite, hematite, silica and gibbsite) occurs only by surface complexation,

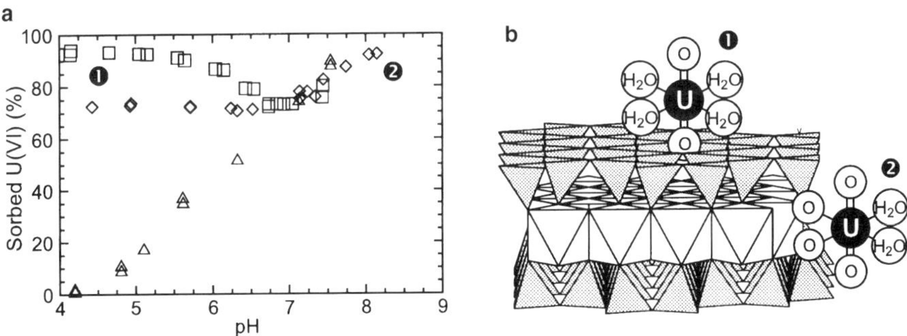


FIG. 14.3 Sorption of uranium (VI) on clay particles. Sorption isotherm (*a*) after Boult *et al.* (1998). (*b*) Structure of the uranyl cation on the surface of a clay mineral: (1) shows the cation involved in a 'cation exchange' reaction i.e. a solvated cation in interlayer spacing; (2) gives the inner sphere structure and co-ordination of the uranyl ion sorbed onto the clay particle 'edge'.

i.e. by the mechanism described in equation 14.14 with S representing Fe, Si, Al. To date the majority of uranium sorption studies that have been undertaken have been dedicated to iron oxy(hydr)oxides and to silica gel. Spectroscopic studies for uranium complex characterization have been performed using two methods: extended X-ray absorption fine structure (EXAFS) spectroscopy for iron oxy(hydr)oxides and time resolved laser induced fluorescence (TRLIF) spectroscopy for silica.

In the absence of $CO_2$ and at near neutral pH values, adsorption occurs mainly via the formation of (2,0,1) and (3,0,1) surface complexes. A large literature is devoted to the corresponding formation constants for Fe-oxy(hydr)oxides and to the impact of competing complexing ligands (e.g. phosphate and sulphate ions) on uranium adsorption (Ho and Doern, 1985; Hsi and Langmuir, 1985; Ho and Miller, 1986; Sagert *et al.*, 1989; Pabalan and Turner, 1993; Waite *et al.*, 1994; Bruno *et al.*, 1995; Morrison *et al.*, 1995; Gabriel *et al.*, 1998). Uranyl complexes on $SiO_2$ were identified by Gabriel (1998) using TRLIF spectroscopy. Two surface complexes formed with distinct lifetimes and luminescence emission spectra. In these complexes the uranyl ion is bound to silica surfaces via two equatorial O atoms, according to EXAFS spectroscopy (Dent *et al.*, 1992). This results in a splitting of the equatorial shell, with four oxygen atoms at 2.29 Å not participating in the surface complex, and two O atoms at 2.50 Å from U being shared with the mineral surface (Reich *et al.*, 1996). Note that co-ordination numbers can only be determined with a $\pm 1$ accuracy by EXAFS spectroscopy (e.g. Randall *et al.*, 1999) and therefore the six equatorial oxygens reported by Reich *et al.* are not at odds with five equatorial O determined in many other solution and surface studies. On ferrihydrite, uranyl is also found to form a bidentate surface complex. Similarly to what has been reported for silica, two equatorial oxygen atoms (with $d_{U-O}$ = 2.52 Å) are shared with one surface iron octahedron (edge-sharing bond), while the remaining three equatorial oxygen atoms ($d_{U-O}$ = 2.44 Å) do not participate in the complex (Manceau *et al.*, 1992; Waite *et al.*, 1994; Muskett *et al.*, 1998). Ageing ferrihydrite transforms to goethite or hematite, and results in a limited possible desorption of the adsorbed uranium as the uranium becomes trapped in the neoformed mineral structure (Payne *et al.*, 1994).

While many dissolved ligands compete with surface reactive groups for U complexation, some ligands, namely citrate and carbonate, may enhance U sorption on mineral particles via the formation of ternary complexes (equation 14.14). This has been reported for citrate at pH 5 (Bencheikh-Latmani *et al.*, 2000) and for carbonate at pH 8.5 and $P_{CO_2}$ ranging from atmospheric pressure to higher ($P_{CO_2} = 10^{-2}$). The formation of a ternary (2,1,1) uranium surface complex, possibly with additional (3,1,1), (2,2,1) and (2,3,1), have been invoked to account for the sharp decrease in uranyl adsorption above pH 7 for an overestimated competition between surface ferranol sites and dissolved bicarbonate ligands (Hsi and Langmuir, 1985;

Kohler *et al.*, 1992; Waite *et al.*, 1994; Morrison *et al.* 1995, Gabriel *et al.*, 1998). This is shown in Fig. 14.4 where uranyl sorption has a maximum at intermediate pH. The FTIR and EXAFS investigations showed uranyl ions to form ternary complexes with hematite surface sites at atmospheric $P_{CO_2}$ and for pH values as low as 4.75. The presence of carbonate ion co-adsorbed with uranyl is confirmed, by a pair of C–O stretching frequencies which are not observed in the absence of carbonate or uranyl ions.

*14.4.3.3 Bacteria, particulate organic carbon and lichen.* Uranium may be transported in groundwater or from the ocean to sediments either complexed with organic matter, or by sorption and uptake of dissolved U by plankton and bacteria (Anderson *et al.*, 1989a). Microorganisms are able to concentrate metals from dilute aqueous solutions using mechanisms that immobilize, complex or otherwise remove metals from solution such as volatilization, extracellular complexing and subsequent accumulation, binding to the cell surface and intracellular accumulation (Brierley, 1990; see also chapters 1–5). These processes may be operative individually or in concert to sequester uranium. Indeed, U and $C_{org}$ contents in oceanic sediments have been shown to be well correlated (Anderson *et al.*, 1989*b*). In surface waters, the U concentrations are larger in the summer than in winter as a consequence of: (1) (photo)reductive dissolution of U-bearing Fe oxide particles (Barnes and Cochran, 1991; Lienert *et al.*, 1994); (2) formation of stable uranyl-humate and uranyl-saccharate complexes when biological activities are higher (Ho and Miller, 1985; Choppin, 1992; Read *et al.*, 1998); and (3) sorption of uranium on bacteria. Uranium (VI) forms strong complexes with organic matter (Nash

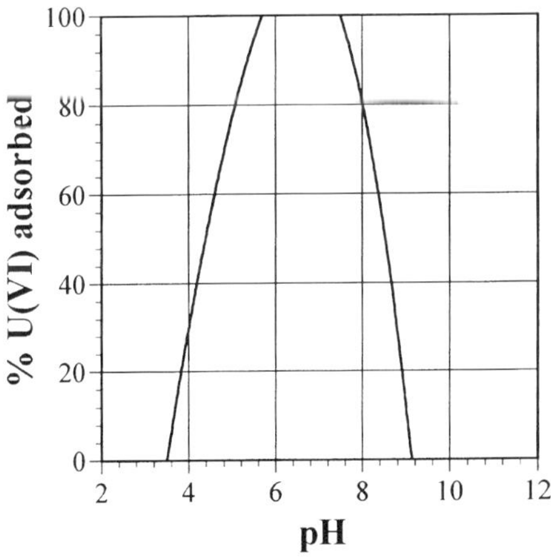


FIG. 14.4 Shape of adsorption isotherm determined by diffuse layer (DL) modelling of U(VI) adsorption by hydrous ferric oxide (FeOOH) at $10^{-3}$ M as Fe, as a function of pH, for total U = $10^{-6}$ M. Model fit to adsorption data at I = 0.1 M, 0.00032 bar air between pH 3 and 10. After Waite *et al.* (1994).

*et al.*, 1981*a*) and represents a likely mechanism for U loss in algae and microbial systems. Indeed, a recent study done with *Pseudomonas fluoresens* in the presence of goethite indicates that U is removed equally well by *P. fluoresens* and by Fe-oxy(hydr)oxide particles (Benkleich-Latmani *et al.*, 2000). While organic ligands such as citrate form strong complexes with uranium, U is observed to be primarily chelated by cell membrane. Strong biosorption of U(VI) on alga (*Chlorella*) (Duff *et al.*, 1999) and lichen (*Peltigera membranacea*) has also been reported (Haas *et al.*, 1998; McLean *et al.*, 1998). For the lichen, EDS analysis show a correlation between P and U, suggesting the involvement of biomass surface functional groups in the uptake process.

### *14.4.4 Reactive transport*

The effective diffusion coefficient of uranium in granitic rocks and bentonite has been determined recently (Yamaguchi and Nakayama, 1998). At pH 4, the effective diffusion coefficient $D_e$ of the free uranyl ion is $(3.6 \pm 1.2)\ 10^{-14}\ m^2\ s^{-1}$ in granitic rock matrix, while at pH 9.3 and in a 0.1 mol $l^{-1}$ $NaHCO_3$ solution, the $D_e$ value for the anionic uranyl carbonate species is four times larger ($1.4 \pm 0.2 \times 10^{-13}\ m^2\ s^{-1}$; Yamaguchi and Nakayama, 1998). This difference could be due both to sorption of free uranyl ions on mineral surfaces at pH 4, and at high pH due to repulsion of the $(UO_2)(CO_3)_3^{4-}$ ion by the negatively charged mineral surfaces. In compacted bentonite the apparent diffusivity is one order of magnitude faster or $10^{-12}\ m^2\ s^{-1}$ for pure bentonite and $2 \times 10^{-12}\ m^2\ s^{-1}$ for a mixture of 10% bentonite and 90% quartz sand (Albinsson and Engvist, 1991). Diffusion in solid matrices is thus a slow process and in general dispersion due to fluid flow is a more important transport process than diffusion.

Studies on the advective-diffusive transport of uranium in porous media have been conducted on columns filled with quartz grains, sandstone and goethite-coated sand (Kohler *et al.*, 1996; Gabriel *et al.*, 1998; Read *et al.*, 1998). Experiments run at pH ranging from 4 to 5 demonstrated that small differences in input pH lead to very large differences in retardation factors. This could be modelled adequately by surface complexation models since no adsorption occurs on silica at pH lower than 3.5, whereas the adsorption is nearly complete at pH 5 (Kohler *et al.*, 1996; Read *et al.*, 1998). In the absence of additional ligands, uranyl sorption on silica is largely reversible. Addition of complexing ligands (such as fluoride, saccharic acid or EDTA) to the input solutions reduces the uranyl retardation factor by an order of magnitude. However, flushing with these ligands had a limited effect in removing pre-adsorbed uranyl ions from a sandstone column, demonstrating the non reversibility of sorption reaction on sandstone (Read *et al.*, 1998).

An elegant field study of uranium retardation in groundwaters was performed in Switzerland (Lienert *et al.*, 1994). Uranium concentrations fluctuate in the river Glatt, from high values in the summer, when bloom production of organic acids leads to the (photo)reductive dissolution of U-rich Fe-oxyhydroxides, to low values in the winter, when Fe and U settle as coprecipitates in an upstream lake. The river Glatt recharges an aquifer; in a well which is located only 5 m away from the river bank, the dissolved uranium concentration oscillation is found to occur six months after that observed in the river (Lienert *et al.*, 1994) (Fig. 14.5)! Since the water flow in groundwater is 5 m day$^{-1}$, the retardation factor is equal to 180, and the corresponding $K_d$ value was found to match that predicted for surface complexation on iron oxide particles present in the aquifer (Lienert *et al.*, 1994).

Bicarbonate ions appear to be very efficient in removing uranyl ions adsorbed on Fe-oxyhydroxides (Fig. 14.4) and silicate ions compete for surface sites available to U adsorption (Liger *et al.*, 1999). A shift of pH from 8 to 9 in the bicarbonate rich input water led to the total removal of pre-adsorbed

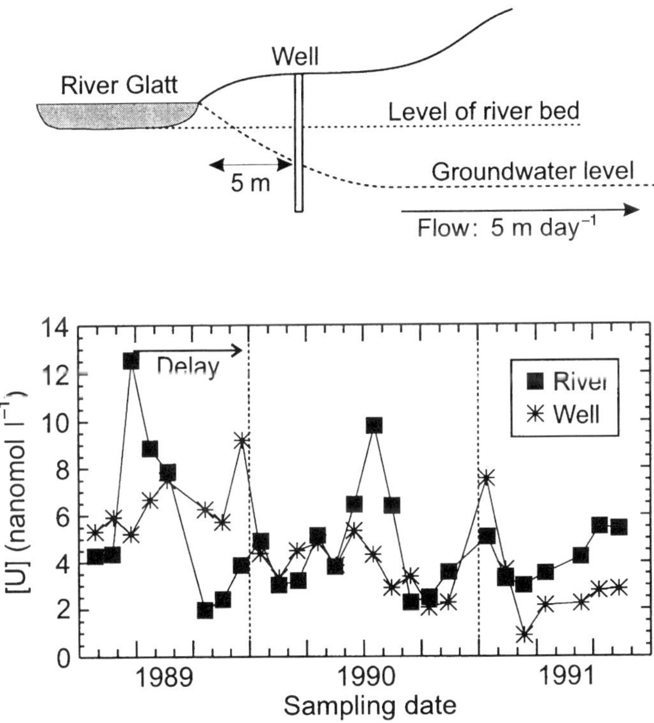


Fig. 14.5 Groundwater that is recharged by a river and uranium(VI) retardation (after Short, 1992; Lienert *et al.*, 1994). (*a*) Cross section of the Glatt river field site in Switzerland. (*b*) Fluctuation of the monthly U(VI) concentration in the River (■) and in a well located 5 m from the river (✳).

uranyl ions from a goethite-coated sand column, and thus to a significant uranium mobilization pulse (Gabriel, 1998). Therefore, incorrect management of low-level contaminated sites, such as surface treatment with chalk, may lead to enormous problems in abstracted local groundwaters. Experiments run at constant pH and in the presence of bicarbonate ions led to $K_d$ values which were about three times larger in batch experiments than in column experiments. The retardation factor value obtained in static and dynamic studies are indeed often very different. In Gabriel's study, this was shown to be due to a difference in water residence time, and thus in silica dissolution. Silicic acid competes with uranyl for sorption sites onto goethite and could account for the low $K_d$ value obtained in static batch experiments (Gabriel *et al.*, 1998).

Uranyl fixation by saleeite ($Mg(UO_2)_2(PO_4)_2 \cdot 10H_2O$) has been observed under oxidizing conditions in the vicinity of the secondary uranium ore deposit at Koongarra, Australia (Murakami *et al.*, 1997). There, uranium released from the deposit is precipitated as saleeite and replaces apatite. This mechanism has also been observed in the laboratory where the presence of uranyl ions causes apatite to dissolve and a uranyl phosphate to precipitate, causing further apatite dissolution (Valsami-Jones *et al.*, 1996). This process continues until all of the apatite in the system has dissolved, provided there is enough aqueous uranium to drive the reaction. Similarly, the diffusion of U into buried archaeological bones and adsorption/precipitation phenomena with apatite led to U-shaped profiles in bones, which together with $^{234}U/^{238}U$ ratio may be used to reconstruct former groundwater composition (Hedges and Millard, 1995; Pike, 2000).

## 14.5 Sinks of uranium

As outlined above, the solubility of uranium(VI) oxide is many orders of magnitude greater than that of uranium(IV) oxide. Sinks of uranium are thus locations where U-reductive precipitation occurs. These reactions have been shown to involve a variety of inorganic reducants and microorganisms.

### *14.5.1 Reduction processes*

Potentially important reducants of U(VI) in low-temperature geochemical systems include sulphides (e.g. $HS^-$ or $FeS_2$), Fe(II) (e.g. in magnetite or biotite) and organic matter (e.g. organic acids, peat). Sulphide minerals are frequently found in close association with supergene uranium deposits (Langmuir and Chatam, 1980; Nash *et al.*, 1981*b*; Brookins, 1988). Furthermore X-ray photoelectron (XPS) and Auger spectroscopic analyses by Wersin *et al.* (1994) have demonstrated conclusively that galena and pyrite are capable of reducing U(VI) under anoxic conditions. Reaction of U(VI) with hydrogen sulphide, produced by bacterial sulphate reduction, has been proposed as a mechanism for uranium retention in organic rich marine sediments (Anderson *et al.*, 1989*a*) and in waste water disposal ponds (Duff *et al.*, 1999) where the U(IV) to U(VI) ratio increases with depth of burial (Duff *et al.*, 1997), suggesting that a slow reduction transformation is occurring. This process probably follows the reaction suggested by Klinkhammer and Palmer (1991).

$$4(UO_2(CO_3)_3)^{4-} + HS^- + 3H^+ + 4H_2O \rightleftharpoons 4UO_2(s) + SO_4^{2-} + 12HCO_3^- \quad (14.15)$$

Magnetite is also frequently found in close association with supergene uranium deposits, such as on the Poccos de Caldas site (Lichtner, 1996). In organic- (and hence bicarbonate-) rich sediments, Klinkhammer and Palmer (1991) found U removal to occur after the onset of Fe oxidation. Therefore U reduction may occur according to the reaction

$$UO_2(CO_3)_3^{4-} + 2Fe_3O_4(s) + H^+ + 10H_2O \rightleftharpoons UO_2(s) + 6Fe(OH)_3(s) + 3HCO_3^- \quad (14.16)$$

Uranium deposition often involves the complexation of uranyl ions with dissolved organic matter in groundwaters, leading to tabular uranium deposits (Sanford 1990, 1992; Hansley and Spirakis, 1992), and in the ocean when sedimentation of organic U-rich particles occurs. This is indicated by the strong correlation observed between U and $C_{org}$. In the latter case, however, the $U/C_{org}$ ratio has been observed to be four times greater in the sediment than in the settling particles (Anderson *et al.*, 1989*a*). This may be due to adsorption of U on the solid phase, after particulate matter is deposited on the basin floor, or to the reduction of U(VI) by organic matter (Barnes and Cochran, 1993), according to the reaction

$$2(UO_2(CO_3)_3)^{4-} + CH_2O + H^+ + 2H_2O \rightleftharpoons 2UO_2(s) + 7HCO_3^- \quad (14.17)$$

However, this process has been reported not to proceed at a measurable rate at temperatures below 45°C (Nakashima *et al.*, 1984; Disnar and Sureau, 1990). Most of these reactions are kinetically inhibited in homogeneous aqueous solutions and like many redox reactions (Wehrli *et al.*, 1989), they appear to be catalysed by mineral surfaces and bacteria.

### *14.5.2 Catalysis by bacteria and mineral surfaces*

*14.5.2.1 Bacteria.* Microorganisms and mineral surfaces affect the kinetics rather than the thermodynamics of the uranium reduction. Bacteria are able to reduce uranium either directly, or indirectly. In the former case, the bacteria catalyse the reduction reaction of U(VI) to U(IV), coupled with the oxidation of an organic substrate. In the latter case, the bacteria reduce Fe(III)-oxyhydroxides, and then U(VI) is reduced by the Fe(II) (e.g. that present in magnetite), according to the overall reaction 14.19.

Direct enzymatic U(VI) reduction is coupled to the anaerobic oxidation of acetate to $CO_2$ for the *GS-15* strain of bacteria (Loveley *et al.*, 1991), i.e. the bacteria are respiring uranium! *Shewanella putrefaciens*, *Alteromonas putrefaciens* and *Desulfibrio desulfuricans*, which are well known Fe-reducing bacteria, can reduce U(VI) indirectly, i.e. bacteria reduce Fe(III) to Fe(II) which in turn reduces uranium. However, *Shewanella putrefaciens* can grow in the laboratory from the oxidation of $H_2$ coupled with the reduction of uranyl chloride (Lovley *et al.*, 1991, 1992*a,b*), and has been observed to reduce U(VI) in groundwaters associated with uranium mill tailings. *S. putrefaciens* can also couple U(VI) reduction to organic carbon (lactate) oxidation. This bacterium has been identified in a groundwater associated with uranium mill tailings; complete reduction of U(VI) was observed in three weeks, leading to the precipitation of a calcium uraninite $(U,Ca)O_2(s)$ solid solution (Abdelouas *et al.*, 1998). The reduction reaction was accompanied by the reduction of sulphate to sulphide. Thus, the indirect reduction of U(VI) by S(-II) cannot be excluded. However, uranium reduction by sulphide ions is extremely slow, particularly when uranyl ions are complexed with carbonate ligands. The enzyme responsible for the U reduction is the same as that for sulphate reduction, cytochrome $c_3$ (Lovley *et al.*, 1993). The overall kinetics of U(VI) reduction has recently been determined for *D. desulfuricans* and a mixed culture of wild-type sulphate-reducing bacteria and the former is slightly faster in reducing uranium (Spear *et al.*, 1999). Using XPS and TEM, Spear *et al.* (1999) demonstrated that the uranium was present as an amorphous extracellurar mass of U(IV). See Suzuki and Banfield (1999) for further detail about uranium geomicrobiology.

*14.5.2.2 Mineral surfaces.* The impact of mineral surfaces on U(VI) reduction kinetics has been reported for the reaction of U(VI) with hydrogen sulphide, and has been demonstrated for U(VI) reaction with Fe(II) (Liger *et al.*, 1999) at pH 7.5 and in a $CO_2$-free environment. Colloidal hematite ($\alpha$-$Fe_2O_3$) was used as a model solid. At pH 7.5, U(VI) and Fe(II) specific adsorptions are nearly complete. In the presence of excess Fe(II), the initial U(VI) reduction rate exhibits a first order dependence on the total concentration of the uranyl ion. Analysis of kinetic data in terms of Fe(II) surface speciation, reveals the following kinetic law (Liger *et al.*, 1999)

$$d[U(VI)]/dt = -k\ [{>}Fe^{III}OFe^{II}OH^0][U(VI)] \qquad (14.18)$$

where $>Fe^{III}OFe^{II}OH^0$ is a hydrolysed ternary Fe(II) surface complex. The rate coefficient k has a value of 399 $min^{-1}$. The apparent first order reaction rate is of the same order of magnitude as those reported for enzymatic bacterial reduction by organic acids (Lovely *et al.*, 1991; Liger *et al.*, 1999).

### *14.5.3 Anoxic sediments as sink*

Reductive precipitation of uranium in sediments may occur in lakes, estuaries and in the ocean (Anderson *et al.*, 1989*a*; Barnes and Cochran, 1990, 1993; Klinkhammer and Palmer, 1991; Liger *et al.*, 1999) as well as in groundwaters

(Galloway, 1978; Lichtner and Waber, 1992). Pore water studies conducted in coastal and pelagic sediments have shown evidence for a substantial flux of dissolved uranium into suboxic or anoxic sediments (Anderson *et al.*, 1989*b*; Barnes and Cochran, 1990; Liger *et al.*, 1999) and particularly into organic-rich sediments (Klinkhammer and Palmer, 1991). As shown below (section 14.6.1), the continuous flow of groundwater through a fixed oxic/anoxic boundary has led to the formation of many uranium ore deposits.

## 14.6 Uranium mines and tailings

### *14.6.1 Uranium deposit formation*

Several classifications of uranium deposits have been proposed (e.g. OECD/NEA, 1998; Plant *et al.*, 1999). The OECD/NEA 'Red Book' (1998) is updated every two years and presents a description and a map with the location of 650 deposits. On the basis of geological setting and in order of economic importance the classes of uranium deposits are: (1) unconformity related; (2) sandstone; (3) quartz-pebble conglomerate; (4) veins; (5) breccia complex; (6) intrusive; (7) phosphorite; (8) collapse breccia; (9) volcanic; (10) surficial; (11) metasomatite; (12) metamorphic; (13) lignite; and (14) black shale. Plant *et al.* (1999) simplified this scheme on the basis of the geological setting and the overall uranium enrichment of the Earth. They recognize two main groups of deposits, those of igneous plutonic or volcanic association (including metamorphic and hydrothermal deposits) and those of sediment/sedimentary basin association.

Uranium deposits are known to have formed from the Proterozoic to the Tertiary. Uranium deposit formation involves two major steps: (1) weathering of rocks under oxic conditions and transport of U(VI) ions; and (2) reductive precipitation of U(IV) in reducing carbon or iron-rich environments. The non-existence of Archaean uranium deposits has been attributed to the lack of oxygen in an early atmosphere, relating to the strong redox control on uranium transport and deposition (Rich *et al.*, 1977). A recent study by Kramers and Tolstikhin (1997) of uranium series isotopes indicates that the Th/U ratio of Archaean (~2.7 Ga) mantle-derived volcanic rocks and xenoliths was 3.32. This is much higher than the present day value of 2.6 for the upper mantle. The lowering of the Th/U ratio has been suggested to have occurred due to enhanced U transport into the upper mantle by subduction zone processes (Zartman and Haines, 1988; Bailey and Ragnarsdottir, 1994). For this model to work, enhanced U transport requires environmental conditions to be oxidizing enough to produce soluble U(VI) and erosion itself must be significant. Kramers and Tolstikhin (1997) suggest that soluble U(VI) may have been produced as early as 2 Ga ago, but their model suggests that erosion only became important at ~1.6 Ga. Uranium mobilization in the Earth thus coincides with the assembly of the first supercontinent between 2.6 and 2.4 Ga ago by processes comparable to modern plate tectonics (Plant *et al.*, 1999).

These processes culminated in continent-continent collision followed locally by extensional collapse and erosion of overthickened crust in a regime of intense tectono-magmatic activity. Over the same time granitoids became enriched in uranium and other incompatible elements.

As stated above, uranium deposits have been classified to have several types of origin (c.f. Guilbert and Park, 1986), ranging from magmatic, to metamorphic, hydrothermal through to sedimentary. Some of the largest and most productive deposits are of sedimentary origin; these include the conglomerates of Witwatersrand, South Africa, and Blind River, Canada. In this review chapter only the major processes resulting in the aqueous transport and deposition of uranium ore are outlined. For further detail see Plant *et al.* (1999).

*14.6.1.1 Sedimentary deposits.* Many sedimentary deposits are stratabound in arkosic continental sedimentary rocks (Guilbert and Park, 1986). These include the Grants deposit of New Mexico and associated belts on the Colorado Plateau which encircles the four corners of the US states Utah (San Rafael Swell, Lisbon Valley), Colorado (Uravan), Arizona (Cameron, Monument Valley) and New Mexico (Grants, Chruch Rock, Ambrosia Lake) and the Wyoming (Pumkin Buttes, Southern Powder River Basin, Shirley Basin, Crooks Gap). These areas contain at least 1.7 Mt of uranium metal reserves and are primarily of Triassic and Jurassic age, although deposits of Palaeozoic to Cenozoic age have been identified. The deposits listed above consist of uraninite, pitchblende and coffinite and several uranium-vanadium minerals. The deposits are formed by typical 'roll-front' deposition (Fig. 14.6) where oxidized water flows into sandstone containing ferrous clay minerals (e.g. Wyoming), in palaeo-riverine channels which were rich in organic debris such as tree trunks (Colorado). They are referred to as 'roll-fronts' because over time they 'roll' downgradient as the upgradient edge of the deposit dissolves and is redeposited near the down-gradient edge. The deposits consist of yellow carnotite ($K(UO_2)_2(VO_4)_2{\cdot}3H_2O$), tyuyamunite ($Ca(UO_2)(VO_4)_2{\cdot}5-8H_2O$), and yellow-orange-red gummite (hydrous oxides of U, Th and Pb) in addition to pitchblende and coffinite (Guilbert and Park, 1986). Also identified in these areas of the US are hydrothermal pitchblende deposits (Marysvale, Utah) and humate uranium deposits (Grants mineral belt, New Mexico). The latter are of Jurassic age and are formed in lake sediments which were rich in humate. The Witwatersrand conglomerate deposits in South Africa contain uranite, and uraniferous carbon in association with gold (c.f. Edwards and Atkinson, 1986) and are most likely of secondary origin.

Two thirds of the world's uranium deposits have formed below 50°C in near-surface sedimentary environments (Nash *et al.*, 1981*b*). Generally, the uranium is leached from acidic rocks such as granite or rhyolite that are somewhat enriched in U. The circulating fluids are thought to have been oxic, $UO_2^{2+}$ rich, and near neutral in pH. The uranium is transported as the uranyl ion until a redox front is encountered such as zones rich in sulphide minerals, organic matter or graphite

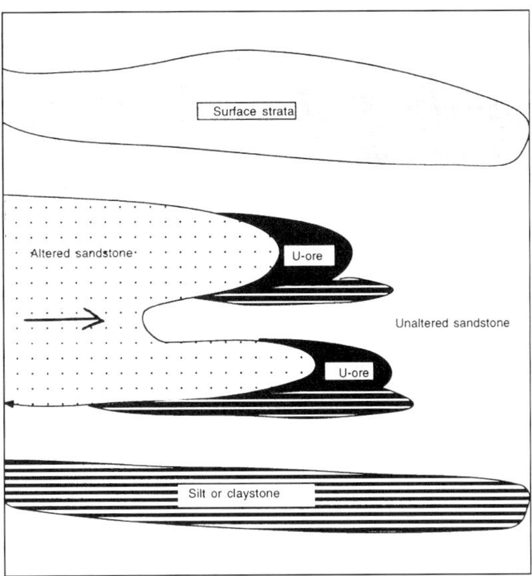


FIG. 14.6 Schematic cross-section on an idealized uranium 'roll-front' ore body showing the zonation of elements and primary hydrologic and geochemical features. Oxidized groundwaters flow from left to right. The 'roll front' and associated redox interface moves in the same direction. After Larson (1978) and Langmuir (1997).

(see section 14.5.1 for details of this mechanism). Within these zones the uranium is reduced from the VI to IV oxidation state, inducing $UO_2$ precipitation. Such deposits are found in the United States (Colorado, Texas, Shirley Basin, Wyoming), Czech Republic (e.g. Straz), France and eastern Germany. Other elements such as As, Mo, Se and V, which are also redox sensitive, are found in connection with 'roll-front' uranium deposits (Wanty *et al.*, 1987). There is still only limited information available about eastern European U deposits due to their past perceived strategic importance.

*14.6.1.2 Unconformity deposits.* Unconformity-type uranium deposits are known from at least two large districts in Australia and Canada. Most of these deposits are located within the Athabasca Basin, Saskatchewan, Canada (Cluff Lake, Rabbit Lake, Key Lake) and the McArthur Basin, Northern Territory, Australia (Ranger-Jabiluka and Alligator River) (Raffensberger and Garven, 1995). Redox control of uranium transport and deposition is demonstrated by recent geochemical modelling of unconformity-type uranium deposits which account for >25% of the world's proven uranium reserves (Raffensberger and Garven, 1995). Thermodynamic modelling of secondary minerals observed in the Precambrian (Proterozoic) Koongarra unconformity-type ore deposit in northern Australia suggests that at 200°C the pre-ore fluids had a log $f_{O_2}$ of −35 and contained 0.1 μg $l^{-1}$ uranium whereas the syn-ore fluids had an oxygen fugacity ten orders of magnitude greater, $f_{O_2}$ of −25, and concentrations of 17 mg $l^{-1}$ (Komninou and Sverjenski, 1996). Such high fluid concentrations are considered to be necessary for the formation of uranium deposits on a geologically realistic time scale (~1 Ma) (Raffensberger and Garven, 1995). The high oxygen fugacity during the Koongarra ore formation is thought to have resulted from the flow of aqueous fluids through evaporite sequences which contained sulphate minerals; these evaporites also being the source of the high salinity in the basinal brines (Sverjenski, 1987). Chloride complexes such as $UO_2Cl_2^0$ further enhance its mobility (Komminou and Sverjensky, 1996). Minerals rich in reduced iron (garnet, biotite, hornblende or chlorite), present in the host rocks were the major cause of uranium reduction. The deposition of uraninite occurs simultaneously with the precipitation of hematite (Komninou and Sverjenski, 1996)

$$2Fe^{2+}(\text{Fe-silicates}) + UO_2Cl_2^0 + 3H_2O \rightarrow Fe_2O_3 + UO_2(s) + 2Cl^- + 6H^+ \quad (14.19)$$

Graphite is also an important chemical reductant for uranium deposition in the sandstone cover of unconformity-type ore deposits such as the Cigar Lake deposit (Raffensberger and Garven, 1995).

### *14.6.2 Conventional uranium mining*

Conventional mining of uranium is still undertaken in several localities around the world, including the Rozna metamorphic deposit in southern Moravia in the Czech Republic. Here uraninite is mined from veins that cross cut Precambrian gneiss and the ore contains 0.1−0.5 wt.% U (1 to 5 kg $t^{-1}$) (Zeman, pers. comm.). The mine has been in operation since 1956 and the depth of present ore extraction is 1200 m. Uranium is leached out of the crushed ore with sodium-carbonate and precipitated as 'yellow cake' using ammonia. Mining will cease in 2002. Mine tailings have been covered with soil and vegetation and leaching from the tailings is less than leaching from the country rocks.

### *14.6.3 Acid uranium leaching*

Uranium can also be leached out of uraniferous sediments by the use of sulphuric acid. *In situ* acid leaching is still undertaken at the sedimantary U-deposit in the Straz area of Northern Bohemia in the Czech Republic (Sulovski, pers. comm.). The uranium is in a typical 'roll-front' deposit located in a Cretaceous sedimentary basin. The uranium ore is found in association with pyrite and organic matter (bitumen). The leaching operation consists of a network of wells, with acid injection wells (60 g $l^{-1}$ $H_2SO_4$) and extraction wells which recover solutions containing ~60 mg $l^{-1}$ uranium (Sulovski, pers. comm). The uranium is then extracted by the use of ion exchange gels, precipitated as 'yellow cake' and the acids are re-injected for further leaching.

The Straz mining operation will cease in 2030 due to extreme acid contamination of the local shallow aquifer. Similar uranium leaching operations as well as *ex situ* acid or bicarbonate leaching have been used widely and are currently being wound up in the Erzgebirge mountains of eastern Germany (Hambeck *et al.*, 1996).

### *14.6.4 Uranium mine tailings*

Uranium mine tailings provide a rich source of uranium for the surrounding hydrosphere. A study of six mining sites in the USA has shown uranium concentrations at pH ~7 of up to 10 mg $l^{-1}$ in leachate waters below a clay dam holding the tailings at the Midnite Mine in eastern Washington, USA (Landa and Gray, 1995). This mine was in operation from 1956 to 1982. A pale yellow precipitate containing aluminous salts and gypsum with 6 wt.% $U_3O_8$ was observed in 1978 for a distance of 0.8 km downstream from the site. The Canon City ore processing site in Colorado operated from 1958 to 1989 and both acid leaching and sodium carbonate leaching were used at the mills to extract uranium from the ore. The tailings and raffinate were discharged into large lined ponds. Water in the adjoining shallow aquifers has been measured to have up to 5.7 mg $l^{-1}$ uranium. Concentrations of up to 20 mg $l^{-1}$ values have also been reported (Langmuir, 1997) where the pH is low. Similar tailings and raffinate are spread out in southern former East Germany, covering 37 $km^2$ (Hambeck *et al.*, 1996). The uranium tailing inventory has been estimated to accumulate at the rate of 25 to 75 million tonnes per year at mine and mill sites throughout the world (NEA, 1984).

### *14.6.5 Bioleaching of uraninite ores*

Bioleaching of metals out of rock formations is used to speed up natural weathering in oxic environments. In the 1940s *Thiobaccillus ferrooxidans* was isolated from acid mine drainage and it became clear that the bacteria oxidizes pyrite to form ferric sulphate and sulphuric acid (Colmer and Hinkle, 1947). Uranium minerals are often found associated with, or locked inside pyrite and since ferric iron is a excellent oxidant for U(IV), the dissolution of uranium oxides can be represented with these two reactions (Bruynesteyn and Hackl, 1990)

$$UO_2(s) + Fe_2(SO_4)_3 + 2H_2SO_4 \rightarrow UO_2(SO_4)_3^{4-} + 2FeSO_4 + 4H^+ \qquad (14.20)$$

$$UO_3(s) + 3H_2SO_4 \rightarrow UO_2(SO_4)_3^{4-} + H^+ + H_2O \qquad (14.21)$$

The amount of pyrite oxidized is far greater than the amount of uranium solubilized and therefore the process becomes self-sustaining in terms of oxidant and acid requirements. This process has been used at the Elliot Lake ore deposit in Ontario since the 1980s. The deposit is mined both conventionally (high grade ore) and by underground bacterial leaching (low grade ore, 0.02 wt.% U mixed with 5–10 wt.% pyrite). Mining of the higher grade ores produces caverns that act as leach vessels. The low grade ore is then sprayed with a solution containing the bacteria and their action extracts ~75% of the available uranium. Uranium is separated from impurities in the solution and precipitated as $U_3O_8$. In 1986 360,000 kg of $U_3O_8$ was produced by the underground bacterial leaching of low grade ores (Bruynesteyn and Hackl, 1990).

In soil and in granitic subsurface environments, natural weathering occurs in two steps: oxidative dissolution of U-bearing minerals, and complexation of U(VI) with organic chelators such as acetic acid. This process is mediated by *Pseudomonas fluorescens, Pseudomonas putida* and *Achromobacter* (Zajic, 1969; Magne *et al.*, 1974). A single bacterial strain, e.g. *Thiobacillus ferroxidans*, can both oxidize and reduce uranium (DiSpirito and Tuovinen, 1981, 1982*a,b*).

For the indirect pathway, U(IV) is oxidized under acidic conditions by Fe(III) according to equation 14.20. *Thiobacillus ferrooxidans* then regenerates $Fe^{3+}$ in a reaction where protons are consumed:

$$2Fe^{2+} + 0.5O_2 + 2H^+ \rightleftharpoons 2Fe^{3+} + 2H_2O \qquad (14.22)$$

Thus the pH increases and ferric iron precipitates as pure ferric hydroxide, or jarosite, which is a ferric sulphate mineral. The continuous production of free ferric ions by *T. ferrooxidans* keeps the process going, without the need for continual supply of external ferric iron to the system. The direct oxidation of U(IV) to U(VI) can also be performed by *Thiobacillus ferrooxidans*, which can use some of the energy from the reaction to assimilate $CO_2$ (DiSpirito and Tuovinen, 1981, 1982*a,b*). However, due to the low solubility of uraninite, bioleaching is believed chiefly to involve the indirect oxidation mechanism.

### *14.6.6 Treatment of U-contaminated sites*

Uranium-contaminated sites are typically former military fields, mine and mill sites (Brown *et al.*, 1998). Discrete U solids found on military sites can easily be removed. However, smaller particles as well as the ground up U mineral residue, found in mine and mill tailings, are usually too voluminous to be extracted. The main remediation procedures used in this case include: (1) minimizing the contact of water and oxygen with the solid residue by placing an impermeable barrier above the contaminated site; (2) confining and collecting the water percolating from these sites with a funnel and gate design; and (3) treating the effluent waters which may contain as much as 1.2 to 20 mg $l^{-1}$ of dissolved uranium (Brown *et al.*, 1998). The USGS is currently testing the mechanisms involved and the commercial viability of the method (J.A. Davis, pers. comm.).

Water treatment can be undertaken *in situ* (permeable reactive barriers) or *ex situ* (pump-and-treat method). The removal can be based on either biological or abiotic processes. Abiotic removal of U from waters is achieved by bringing the water into contact with a reactive material, which can be: (1) granular zero-valent iron; (2) bone char phosphates; or (3) amorphous ferric oxyhydroxide. As outlined above, uranyl is adsorbed on ferric oxyhydroxide (Brown *et al.*, 1998), coprecipitated with phosphate minerals and reductively precipitated by zero-valent iron (Fiedor *et al.*, 1998; Gu *et al.*, 1998). The removal efficiency for input concentrations as high as 20 mg $l^{-1}$, ranges from 88.1% (method 1) to 99.9% (method 3) (Brown *et al.*, 1998; Fiedor *et al.*, 1998).

The biological removal of U by *Shewanella putrefaciens* in oxidizing carbonated waters is an efficient removal method (Abdelouas *et al.*, 1998). In acidic (pH 4) conditions, *Desulfovibrio desulfuricans* may remove uranium from mine drainage waters and contaminated site groundwaters (Lovley and Phillips, 1992*b*). *Saccharomyces cerevisiae* and *Pseudomonas aeruginosa* accumulate 10–15% of their dry weight of uranium (Strandberg *et al.*, 1981). However, the mechanism of accumulation differs for the two microorganisms. Cells of *S. cerevisiae* accumulate the uranium on the cell surface in a layer that was measured to be 0.2 μm thick. In contrast *P. aeruginosa* accumulates uranium internally, but metabolism was not required for uranium to cross the cell membrane. Biosorption is often limited due to solution complexation of uranyl with carbonate and hydroxyl ions.

*Rhizopus arrhizus*, a fungal by-product of industrial fermentation has also been tested for use as biosorbent of uranium (Tzesos and Volesky, 1981; Tzesos, 1990). These fungal cells have uranium uptake capacity of >180 mg $g^{-1}$ dry weight. Three mechanisms were identified for the biosorption: (1) coordination with the amine nitrogen of the chitin component of the cells walls; (2) complexed uranium acted as a nucleation site for accumulation of additional uranium; and (3) the hydrolysis and subsequent precipitation of uranyl hydroxide ($UO_2(OH)_2$) on the cell wall. The uranium can subsequently be eluted using bicarbonate solution and the fungal biomass is reusable through multiple cycles of accumulation and elution (Tsezos, 1984).

### *14.6.7 Analogue deposits*

The behaviour of uranium in natural settings can be assessed from geochemical modelling and by the careful study of uranium deposits. This is due to the fact that spent fuel is chiefly $UO_2$ (95 wt.%). Several ore deposits have been studied in detail in recent years in order to validate experimental and modelling work for future geological waste repositories. These sites include Pena Blance, Mexico; Cigar Lake in Saskatchewan, Canada; Pocos de Caldas, Brazil and Oklo, Gabon (e.g. Cramer and Smellie, 1994; Gauthier-Lafaye *et al.*, 1996; Lichther, 1996). The Oklo deposit has the added bonus of having reached criticality in nature due to the accumulation of critical mass of fissile uranium that can undergo continuous chain reactions. Therefore the site not only contains natural uranium and its decay products, but also fission products such as neptunium and plutonium and their derivatives.

*14.6.7.1 Oklo and Bangombe.* The Oklo and the nearby Bangombe natural fission reactors are located in the oldest (Palaeoproterozoic) non-metamorphic sedimentary basin known in the world. The sites consist of hydrothermally altered clastic sedimentary rocks that contain abundant uraninite and authigenic clay minerals. Studies of these reactors have identified which factors are most important in retaining the uranium and their fission products (Gauthier-Lafaye *et al.*, 1996; Lichtner, 1996). First, the presence of organic matter in the sediments helped to reduce uranium so that $UO_2$ precipitated, thereby concentrating sufficient quantities of uranium to initiate the nuclear chain reactions (Nagy *et al.*, 1993). This organic matter, along with hydrothermally-produced graphite, also kept the uranium reduced through geologic time and has largely prevented uranium migration from the deposits. The high stability of U(IV) minerals in reducing environments means that the fission products are retained within the structure of the uraninite. Other important factors that kept the uranium and radionuclides preserved on these sites include the geological stability of the Frenceville Basin and the low permeability of the rocks enclosing the reactors. The latter is due mainly to the presence of clay layers enclosing the core of the reactors and in part to the closure of the porosity and permeability of the surrounding sandstones during diagenesis of the basin. Finally, the presence of low solubility phosphate minerals such as hydroxy-apatite is also an important factor for the retention of uranium and a variety of fission products, including Pu, Nd, Sm, Sr and Rb within the formation.

*14.6.7.2 Pocos de Caldas.* Another analogue site for uranium redox fronts is the Pocos de Caldas formation in Brazil which has been studied petrologically and modelled by reactive transport modelling (Waber *et al.*, 1990; Lichtner and Waber, 1992; Lichtner, 1996). Deposition of uraninite and ferrihydrite is modelled to occur in response to the changing redox potential as it varies from oxidizing at the water inlet to reducing further into the rock where the fluid has equilibrated with pyrite. Ferrihydrite formed as iron released by oxidation of pyrite diffused towards the fluid flow inlet and came in contact with oxidizing fluid. Uraninite precipitated in a narrow zone separated by a clearly defined gap from the ferrihydrite zone. This site was studied as an example of a uranium deposit where the uranium transport over time is very limited even though the site is not sealed by clay.

*14.6.7.3 Cigar Lake.* The Cigar Lake uranium deposit is located in northern Saskatchewan, Canada. The mid-Proterozoic ($1.3 \times 10^9$ y) deposit is located at a depth of ~450 m below the surface in a water-saturated sandstone at the unconformity with the high-grade metamorphic basement rocks of the Canadian Shield. The uranium mineralization consists primarily of uraninite and remains extremely well preserved, making it an ideal analogue deposit for radioactive waste repositories (Cramer and Smellie, 1994). The ore is surrounded by a clay-rich halo in both sandstone and basement rocks. This clay halo appears to have been a very effective natural barrier for uranium transport, because at the surface there is no indication of the deposit at depth. The groundwaters around the uranium deposit are characterized by neutral to near-neutral pH (6–8), by low contents of total dissolved solids (60–240 mg $l^{-1}$), low Eh (buffered by the $Fe^{3+}/Fe^{2+}$ and $S^{2-}/SO_4^{2-}$ redox couples), and by overall low U contents (~30 μg $l^{-1}$) (Cramer, 1994). Radiocolloid transport is not considered to be of importance for uranium transport at the Cigar Lake site (Vilks, 1994). The detailed study of the Cigar Lake site shows that accumulations of U can be stable and practically immobile in low-Eh groundwaters for geologically significant time periods. Similarly, the clay content of the surrounding rocks has been used to demonstrate that clay minerals are effective sorbents for any U that is mobilized, preventing large-scale migration of U from the ore body (Cramer and Smellie, 1994).

## 14.7 Conclusion – the uranium cycle

### *14.7.1 Uranium residence time in the ocean*

The residence time of U in the ocean depends on both input and output fluxes. After the dramatic increase in oxygen fugacity of the atmosphere at the end of the Archaean, U has been weathered from the continents in its soluble oxidized state. The flux of uranium into the oceans from rivers is poorly known, but has been estimated by Palmer and Edmond (1993) to be ~$4.5 \pm 1.5 \times 10^7$ mol $yr^{-1}$. The natural present-day riverine U flux cannot be measured more accurately in view of the important anthropogenic input linked with the widespread use of U-rich phosphate amendments. Uranium shows variable behaviour during estuarine mixing, ranging from its release from shelf sediments in the Amazon delta (Barnes and Cochran, 1993) where mixing takes place well out at sea above shelf sediments, to U removal in the reducing environment of the Ganges-Brahmaputra mangrove swamps (Palmer and Edmond, 1993). The measured $^{238}$U concentration in the ocean is $3.3 \times 10^{-6}$ g $l^{-1}$ (Santschi and Honeyman, 1989).

The uranium deposition from the ocean is two fold. First, uranium diffuses into the Fe(II)-rich, organic-rich continental shelf sediment, where it is fixed by reduction (Anderson *et al.*, 1989*a*). This flux is estimated to be equal to $2.8 \times 10^7$ mol $y^{-1}$, i.e. ~75% of the uranium removal from the oceans (Klinkhammer and Palmer, 1991). Second, U is removed from the ocean by deep circulating seawater which recharges hydrothermal cells at mid-ocean ridges. In these cells, U(VI) reacts with Fe(II)-bearing basaltic rocks and with $CH_4$(g) and/or $H_2S$ diffusing out from the magma chambers. Exhalations of high-temperature hydrothermal waters at black smokers and studies of altered ophiolites indicate that uranium is quantitatively stripped from these deep circulating seawaters (Michard *et al.*, 1983; Valsami-Jones and Ragnarsdottir, 1997*a*). This high temperature uranium uptake in the hydrothermal cells is estimated to be $1.6 \pm 0.4 \times 10^7$ mol $y^{-1}$ (Palmer and Edmond, 1989; Elliott *et al.*, 1999). The uranium present in the oceanic crust is transported, by subduction, back into the depleted MORB mantle (Elliott *et al.*, 1999).

The resulting residence time of U in the oceans is $5 \times 10^5$ y (Santschi and Honeyman, 1989), i.e. the residence time is an order of magnitude larger than the mean residence time of water in the oceans (35,000 y). This is due to the formation of hydroxide and carbonate U(VI) complexes in the ocean. Indeed, in the physicochemical conditions prevailing in the oceans, uranium forms stable hydroxo- and carbonate uranyl complexes, i.e. $UO_2(OH)_2^0$, $UO_2(CO_3)_2^{2-}$, $(UO_2)_2CO_3(OH)_3^-$ and $UO_2(CO_3)_3^{4-}$, which are not adsorbed on negatively-charged particles. This has been demonstrated on ferrihydrite and hematite particles (Waite *et al.*, 1994), but should be even more pronounced for negatively-charged organic particles.

### *14.7.2 Uranium cycle in the anthroposphere*

The material highlighted in this chapter demonstrates that uranium once introduced by man into the hydrosphere is mobile when the redox conditions are high. Under such conditions weathering of uranium-rich formations releases U(VI) which forms complexes with hydroxide, carbonate or chloride ions, resulting in long distance transport of uranium. Retardation of uranium occurs as: (1) sorption on biological or mineral surfaces strongly retard its migration; and/or (2) abiotic (e.g. on $Fe^0$ or $Fe^{2+}$-rich particles) or biologically-mediated reduction of uranium. While the inorganic reactions appear to be fairly well understood, biological reactions still need to be better quantified in terms of reaction rates and long-term efficiency. Sorption and reduction mechanisms, which appear to have successfully confined uranium in analogue deposits, are presently used in the design of engineered barriers to confine nuclear waste, military fields, mine tailings and other anthropogenically-contaminated uranium sites.

### *14.7.3 Future research needs*

Future research needs to quantify the anthropogenic influx of uranium into the hydrosphere, atmosphere and biosphere due to U-bearing coal combustion, use of U-bearing fertilizers and use of U-bearing bombs in recent wars. Researchers also need to focus on the toxicology of depleted (99.8% $^{238}U$) uranium as a heavy metal due to the massive surface distribution of uranium in recent military operations. In general, uranium is known: (1) to be taken up by ion channels; (2) to interact with enzymes by complexation with thio-sites; and/or (3) to replace $PO_4$ in the DNA backbone – similar to other oxyanions such as chromate and arsenate. Other mechanisms of toxicity may also be of importance for uranium. The effect of depleted uranium on the health of populations affected by military operations has not been studied but preliminary reports indicate 'hot zones' in the Balkans (UNEP/UNCHS, 1999*a,b*) and that uranium may be the cause of a large number of birth defects and a high incidence of leukaemia and other cancers in the Gulf region (Durakovic, 1999).

## Acknowledgements

KVR is indebted to Geoff Allen, Liz Bailey, Jordi Bruno, Chris Hawkesworth, Chris Muskett, Heino Nitsche, Jane Plant, Don Porcelli, David Savage, Everett Shock, Éva Valsami-Jones and Bernie Wood for many enlightening deliberations about uranium in a variety of environments, from nuclear reactors to hydrothermal systems and from rivers to subduction zones. LC thanks Emmanuelle Liger and Uta Gabriel for their fruitful laboratory research and for exciting discussions on the behaviour of uranium in dynamic environments. Pat Brady, Chris Hawkesworth, Philippe van Cappellen, two anonymous reviewers and the editor, Éva Valsami-Jones provided detailed comments which greatly improved the clarity and completeness of this review chapter. Michel Schlegel and James Wilson are thanked for drawing some of the figures. KVR acknowledges support from Centre National de la Recherche Scientifique (CNRS) during her visit to Grenoble in 1999.

## References

Abdelouas, A., Lu, Y., Lutze, W. and Nuttall, H.E. (1998) Reduction of U(VI) to U(IV) by indigenous bacteria in contaminated groundwater. *J. Contam. Hydrol.*, **35**, 217–33.

Albinsson, Y. and Engkvist, I. (1991) Diffusion of Am, Pu, U, Np, Cs, I and Tc in compacted sand-bentonite mixture. *Radioactive Waste Management,* **15**, 221–39.

Allen, G.C. and Tempest, P.A. (1986) Ordered defects in the oxides of uranium. *R. Soc. Lond. A*, **406**, 325–44.

Ames, L.L., McGarrh, J.E., Walker, B.A., and Salter, P.F. (1983*a*) Uranium and radium sorption on amorphous ferric oxyhydroxide. *Chem. Geol.,* **40**, 135–48.

Ames, L.L. McGarrh, J.E. and Walker, B.A. (1983*b*) Sorption of trace constituents from aqueous solutions onto secondary minerals. 1. Uranium. *Clays Clay Miner.,* **31**, 321–34.

Anderson, R.F. (1982) Concentration, vertical flux, and remineralization of particulate uranium in sea-water. *Geochim. Cosmochim. Acta*, **46**, 1293–9.

Anderson, R.F., Fleisher, M.Q. and LeHuray, A.P., (1989b) Concentration, oxidation state, and particulate flux of uranium in the Black Sea. *Geochim. Cosmochim. Acta*, **53**, 2215–24.

Anderson, R.F., LeHuray, A.P., Fleisher, M.Q. and Murray, J.W. (1989a) Concentration, oxidation-state, and particulate flux of uranium in the Black-Sea. *Geochim. Cosmochim. Acta* **53**, 2205–24.

Andersson, P.S., Porcelli, D., Wasserburg, G.J. and Ingri, J. (1998) Particle transport of $^{234}U$-$^{238}U$ in the Kalix River and the Baltic Sea. *Geochim. Cosmochim. Acta,* **62,** 385–92.

Andrews, J.N., Ford, D.J., Hussain, N., Trivedi, D., and Youngman, M.J. (1989) Natural radioelement solution by circulating groundwaters in the Stripa granite. *Geochim. Cosmochim. Acta,* **53**, 1791–802.

Arnold, T., Zorn, T., Bernhard, G. and Nitsche, H. (1998) Sorption of uranium(VI) onto phyllite. *Chem. Geol.*, **151**, 129–41.

Asikainen, M. and Kahlos, H. (1979) Anomalously high concentrations of uranium, radium and radon in water from drilled wells in the Helsinki area. *Geochim. Cosmochim. Acta,* **43**, 1681–6.

Bailey, E.H. and Ragnarsdottir, K.V. (1994) Uranium and thorium solubilities in subduction zone fluids. *Earth. Planet. Sci. Let.,* **124**, 119–29.

Ball, T.K. and Miles, J.C.H. (1993) Geological and geochemical factors affecting the radon concentration in homes in Cornwall and Devon, UK. *Environ. Geochem. Health*, **15**, 27–36.

Ball, T.K., Cameron, D.G., Colman, T.B. and Roberts, P.D. (1991) Behaviour of radon in the geological environment – A review. *Q. J. Eng. Geol.*, **24**, 169–82.

Baes, C.F. and Mesmer R.E. (1976) *The Hydrolysis of Cations*. Wiley-Interscience, New York.

Barnes, C.E. and Cochran J.K. (1990) Uranium removal in oceanic sediments and the oceanic-U balance. *Earth Planet. Sci. Lett.*, **97**, 94–101.

Barnes, C.E. and Cochran J.K. (1991) Geochemistry of uranium in Black-Sea sediments. *Deep-Sea Res*., **38**, S1237–54.

Barnes, C.E. and Cochran J.K. (1993) Uranium geochemistry in estuarine sediments – controls on removal and release preocesses. *Geochim. Cosmochim. Acta,* **57**, 555–69.

Basinger, M.A., Forti, R.L., Burka, L.T., Jones, M.M., Mitchell, W.M., Johnson, J.E. and Gibbs, S.J. (1983) Phenolic chelating-agents as antidotes for acute uranyl acetate intoxication in mice. *J. Toxicol. Environ. Health*, **11**, 237–42.

Bear, J. (1979) *Dynamics of Fluids in Porous Media.* Elsevier, New York.

Bell, F.G. (1998) *Environmental Geology: Principles and Practice*. Blackwell, Oxford.

Bencheikh-Latmani, R., Leckie, J.O. and Sporman, A. (2000) Solution interactions between *Pseudomonas fluorescens* and uranyl: bioavailability of chelator, inhibition and fate of uranyl. *Env. Sci. Techol.* (submitted).

Ben Othman, D., White, W.M. and Patchett, J. (1989) The geochemistry and age of Timiskaming alkali volcanics and the Otto syenite stock, Abitibi, Ontario. *Earth Planet. Sci. Lett.,* **94**, 1–21.

Bethke, C.M. (1994) *The Geochemists Workbench, Version 2.0, A Users guide to Rxn, Act2, Tact, React and Gtplot.* Hydrogeology Program, University of Illinois, Urbana, IL.

Bethke, C.M. (1996) *Geochemical Reaction Modelling*. Oxford University Press, Oxford.

Bottrell, S.H. (1993) Redistribution of uranium by physical processes during weathering and implications for radon production. *Environ. Geochem. Health*, **15**, 21–5.

Boult, K.A., Cowper, M.M., Heath, T.G., Sato, H., Shibutani, T. and Yui, M. (1998) Towards and understanding of the sorption of U(V) and Se(IV) on sodium bentonite. *J. Contam. Hydrol.*, **35**, 141–50.

Boyle, E.A., Edmonds, J.M. and Sholkovitz, E.R. (1977) The mechanism of iron removal in estuaries. *Geochim. Cosmochim. Acta*, **41**, 1313–24.

Brierley, J.A. (1990) Microbial processes for recovery of metals. Pp. 286–95 in: *Biohydrometallurgy* (G.I. Karavaiko, G. Rossi and Z.A. Avakyan, editors). International Seminar on Dump and Underground Bacterial Leaching Metals from Ores. Centre for International Projects, USSR State Committee for Environment Protection, Moscow.

Brookins, D.G. (1988) *Eh-pH Diagrams for Geochemistry*. Springer Verlag, Berlin.

Brown, P.L., Guerin, M., Hankin, S.I. and Lowson, R.T. (1998) Uranium and other contaminant migration in groundwater at a tropical Australian uranium mine. *J. Contam. Hydrol.*, **35**, 295–303.

Bruno, J., Casas, I., Lagerman, B. and Munoz, M. (1987) The determination of the solubility of amorphous $UO_2$(s) and the mononuclear hydrolysis constants of uranium (IV) at 25°C. *Mat. Res. Soc. Symp. Proc.,* **84**, 153–60.

Bruno, J., de Pablo, J., Duro, L. and Figuerola, E. (1995) Experimental-study and modeling of the U(VI)-$Fe(OH)_3$ surface precipitation coprecipitation equilibria. *Geochim. Cosmochim. Acta,* **59**, 4113–23.

Bruynesteyn, A. and Hackl, R.P. (1990) Mineral biotechnology: Present and future applications. Pp. 10–20 in: *Biohydrometallurgy* (G.I. Karavaiko, G. Rossi and Z.A. Avakyan, editors). International Seminar on Dump and Underground Bacterial Leaching Metals from Ores. Centre for International Projects, USSR State Committee for Environment Protection, Moscow.

Busenberg, E. and Plummer, L.N. (1986) A comparative study of dissolution and crystal growth kinetics of calcite and aragonite. Pp. 139–68 in: *Studies in Diagenesis* (F.A. Mumpton, editor). US Geological Survey Bulletin, **1578**.

Carroll, J. and Moore, W.S. (1993) Uranium removal during low discharge in the Ganges-Brahmaputra mixing zone. *Geochim. Cosmochim. Acta.*, **16**, 155–71.

Casas, I., de Pablo, J., Gimenez, J., Torrero, M.E., Bruno, J., Cera, E., Finch, R.J. and Ewing, R.C. (1998) The role of pe, pH, and carbonate on the solubility of $UO_2$ and uraninite under nominally reducing conditions. *Geochim. Cosmochim. Acta,* **62**, 2223–31.

Cathcart, J.B., Sheldon, R.P. and Gulbrandsen, R.A. (1984) Phosphate rock resources of the United States. *US Geol. Surv. Circ.,* **888**.

Cathelineau, M., Cuney, M., Leroy, J., Lhote, F., Nguyen Trung, C., Pagel, M. and Poty, B. (1982) Caracteres mineralogiques des pechblendes de la province hercynienne d'Europe. Comparison avec les oxydes d'uranium du Proterozoique de differents gisements d'Amerique du Nord, d'Afrique et d'Australie. Pp. 159–77 in: *Vein Type and Similar uranium Deposits in Rocks Younger than Proterozoic.* AIEA, Vienna.

Chisholm-Brause, C., Conradson, S.D., Buscher, C.T., Eller, P.E. and Morris, D.E. (1994) Speciation of uranyl sorbed at multiple binding-sites on montmorillonite. *Geochim. Cosmochim. Acta*, **58**, 3625–31.

Choppin, G.R. (1992) The role of natural organics in radionuclide migration in natural aquifer systems. *Radiochim. Acta*, **58-9**, 113–20.

Church, S.E. (1973) Limits of sediment involvement in the genesis of orogenetic volcanic rocks. *Contrib. Mineral. Petrol.*, **39**, 17–32.

Colmer, A.R. and Hinkle, M.E. (1947) The role of microorganisms in acid mine drainage: a preliminary report. *Science*, **106**, 253–5.

Cramer, J. (1994) Hydrogeochemistry. Pp. 65–6 in: *Final report of the AECL/SKB Cigar Lake analogue study* (J. Cramer and J. Smellie, editors). SKB Technical Report 94-04, Stockholm, Sweden.

Cramer, J. and Smellie, J., editors (1994) *Final report of the AECL/SKB Cigar Lake analogue study.* SKB Technical Report 94-04, Stockholm, Sweden.

Davis, J.A. and Kent, D.B. (1990) Surface complexation modeling in aqueous geochemistry. Pp. 177–260 in: *Mineral-Water Interface Geochemistry* (M.F. Hochella and A.F. White, editors). Reviews in Mineralogy, **23**. Mineralogical Society of America, Washington D.C.

de Pablo, J., Casas, I., Gimenez, J., Molera, M., Rovira, M., Duro, L. and Bruno, J. (1999) The oxidative dissolution mechanism of uranium dioxide. 1. The effect of temperature in hydrogen carbonate medium. *Geochim. Cosmochim. Acta*, **63**, 3097–104.

Dearlove, J.P.L., Longworth, G., Ivanovich, M., Kim, J.I., Delakowitz, B. and Zeh, P. (1991) A study of groundwater-colloids and their geochemical interactions with natural radionuclides in Gorleben aquifer systems. *Radiochim. Acta*, **52/53**, 83–9.

Dent, A.J., Ramsay, J.D.F. and Swanton, S.W. (1992) An EXAFS study of uranyl-ion in solution and sorbed onto silica and montmorillonite clay colloids. *J. Coll. Interf. Sci.*, **150**, 45–60.

Disnar, J.R. and Sureau, J.F. (1990) Organic-matter in ore genesis – progress and perspectives. *Org. Geochem.*, **16**, 577–99.

DiSpirito, A.A. and Tuovinen, O.H. (1981) Oxygen-uptake coupled with uranous sulfate oxidation by *thiobacillus-ferrooxidans* and *thiobacillus-acidophilus. Geomicrobiol. J.,* **2**, 275–91.

DiSpirito, A.A. and Tuovinen O.H. (1982*a*) Uranous ion oxidation and carbon-dioxide fixation by *thiobacillus-ferrooxidans. Arch. Microbiol.*, **133**, 28–32.

DiSpirito, A.A. and Tuovinen O.H. (1982*b*) Kinetics of uranous ion and ferrous iron oxidation by *thiobacillus-ferrooxidans*. *Arch. Microbiol.*, **133**, 33–7.

Draganic, I.G. and Draganic, Z.D. (1971) *The Radiation Chemistry of Water. Physical Chemistry.* A series of monographs. Academic Press, New York.

Dubessy, J., Pagel, M., Beny, J.-M., Christenson, H., Hickel, B., Kosztolanyi, C. and Poty, B. (1988) Radiolysis evidenced by $H_2O_2$ and $H_2$-bearing fluid inclusions in three uranium deposits. *Geochim. Cosmochim. Acta,* **52**, 1155–67.

Duerden, P. (1990) *Alligator River Analogue Project.* Australian Nuclear Science and Technology organisation (ANSTO).

Duff, M.C., Amrhein, C., Bertsch, P.M. and Hunter, D.B. (1997) The chemistry of uranium in evaporation ponds sediment in the San Joaquin Valley, California, USA, using X-ray fluorescence and XANES techniques. *Geochim. Cosmochim. Acta*, **61**, 73–81.

Duff, M.C., Hunter, D.B., Bertsch, P.M. and Amrhein, C. (1999) Factors influencing uranium reduction and solubility in evaporation pond sediments. *Biogeochem.*, **45**, 95–114.

Durakovic, A. (1999) Medical effects of internal contamination with uranium. *Croat. Med. J.,* **40**, 49–66.

Edwards, R. and Atkinson, K. (1986) *Ore Deposit Geology and its Influence on Mineral Exploration.* Chapman & Hall, New York.

Elliott, T., Zindler, A. and Bourdon, B. (1999) Exploring the kappa conundrum: the role of recycling in the lead isotope evolution of the mantle. *Earth Planet Sci. Lett.,* **169**, 129–45.

Enderle, G.J. and Friedrich, K. (1999) Uranium mining in East Germany ("Wismut"): health consequences, occupational medical care and workers' compensation. *Int. Arch. Occup. Environ. Health*, **72**, M42–9.

Faure, G. (1986) *Principles of Isotope Geology*. John Wiley, New York.

Fiedor, J.N., Bostick, W.D., Jarabek, R.J. and Farrell, J. (1998) Understanding the mechanism of uranium removal from groundwater by zero-valent iron using X-ray photoelectron spectroscopy. *Environ. Sci. Technol.*, **32**, 1466–73.

Fox, L.E. (1983) Geochemistry of humic acid during estuarine mixing. Pp. 407–26 in: *Aquatic and Terrestrial Humic Materials* (R.F. Christman and E.T. Gjessing, editors). Ann Arbor Science, Ann Arbor, Michigan.

Freeze, R.A. and Cherry, J.A. (1979) *Groundwater*. Prentice-Hall, Englewood Cliffs, NJ.

Gabriel, U. (1998) *Transport réactif de l'uranyl: Mode de fixation sur la silice et la goethite; expériences en colonnes et réacteurs fermés, simulations*. PhD thesis. Univ. Grenoble, France.

Gabriel, U., Gaudet, J.P., Spadini, L. and Charlet, L. (1998) Reactive transport of uranyl in goethite column: an experimental and modelling study. *Chem. Geol.*, **151,** 107–28.

Galloway, W.E. (1978) Uranium mineralization in a coastal-plain fluvial aquifer system: Catahoula formation, Texas. *Econ. Geol.*, **73**, 1655–76.

Garrels, R.M. and Christ, C.L. (1965) *Solutions, Minerals and Equilibria.* Harper Row, New York.

Gauthier-Lafaye, F., Holliger, P. and Blanc, P.-L. (1996) Natural fission reactors in the Franceville basin, Gabon: A review of the conditions and results of a "critical event" in a geologic system. *Geochim. Cosmochim. Acta*, **60**, 4831–52.

Gayer, K.H. and Leider, H. (1955) The solubility of uranium trioxide, $UO_3{\cdot}H_2O$ in solutions of sodium hydroxide and perchloric acid at 25°C. *J. Amer. Chem. Soc.*, **77**, 1148–50.

Gayer, K.H. and Leider, H. (1957) The solubility of U(VI) hydroxide in solutions of sodium hydroxide and perchloric acid at 25°C. *Canad. J. Chem.,* **35**, 5–7.

Geipel, G., Bernhard, G., Brendler, V. and Nitsche, H. (1996) Sorption of uranium(VI) on rock material of a mine tailing pile: Solution speciation by fluorescence spectroscopy. *Radiochim. Acta*, **74**, 235–8.

Giaquinta, D.M., Soderholm, L., Yuchs, S.E. and Wasserman, S.R. (1997) The speciation of uranium in a smectite clay: Evidence for catalysed uranyl reduction. *Radiochim. Acta*, **76**, 113–21.

Gill, J.B. and Williams, R.W. (1990) Th-isotope and U-series of subduction-related volcanic rocks. *Geochim. Cosmochim. Acta*, **54**, 1427–42.

Grenthe, I., Fuger, J., Konings, R.J.M., Lemire, R.J., Muller, A.B., Nguyen-Trung, Ch. and Wanner, H. (1992) *Chemical Thermodynamics of Uranium.* North Holland, Amsterdam.

Gu, B., Liang, L., Dickey, M.J., Yin, X. and Dai, S. (1998) Reductive precipitation of uranium(VI) by zero-valent iron. *Environ. Sci. Technol.*, **32**, 3366–73.

Guilbert, J.M. and Park, C.F., Jr. (1986) *The Geology of Ore Deposits.* W.H. Freem and Co., New York.

Guy, C., Crancon, P., Guinois, G., Marchand, D., Meyer, J., Sornein, J.F., Bechu, J., Esteban, R. and Fresquet, D. (1998) *Bilan environmental du T.E.E. - Situation acuelle et evaluation du devenir des depots d'uranium et de metaux lourds.* Commissariat de l'Energie Atomique (CEA) report no CEA/DIF/DASE/SRCE/B3 DO 125 19/05/98. Directin des Applications Militaires, Service Radioanalyse, Chimie et Environnement, 91680 Bruyhres le Chatel, France.

Haas, J.R., Bailey, E.H. and Purvis, O.W. (1998) Bioaccumulation of metals by lichens: Uptake of aqueous uranium by *Peltigera membranacea* as a function of time and pH. *Amer. Mineral.*, **83**, 1494–502.

Hambeck, L., Meyer, J., Thie, F.W. and Willie, F. (1996) Cleaning up Wismut's waste dumps. *Atw-Int. Zeitschr. Kernenergie*, **41**, 103–7.

Hansley, P.L. and Spirakis, C.S. (1992) Organic-matter diagenesis as the key to a unifying theory for the genesis of tabular uranium-vanadium deposits in the Morrison Formation, Colorado Plateau. *Econ. Geol Bull. Soc.*, **87**, 352–65.

Hawkesworth, C.J., Turner, S., Peate, D., McDermott, F. and van Calsteren, P. (1997*a*) Elemental U and Th variations in island arc rocks: Implications for U-series isotopes. *Chem. Geol.*, **139**, 207–21.

Hawkesworth, C.J., Turner, S., Peate, D., McDermott, F. and van Calsteren, P. (1997*b*) U-Th isotopes in arc magmas: Implications for element transfer from the subducted crust. *Science*, **276**, 551–5.

Hedges, R.E.M. and Millard, A.R. (1995) Bones and groundwater – towards the modeling of diagenetic processes. *J. Archaeol. Sci.*, **22**, 155–64.

Ho, C.J. and Doern, D.C. (1985) The sorption of uranys species on a magnetite sol. *Canad. J. Chem.*, **63**, 1100–4.

Ho, C.J. and Miller, N.H. (1985) Effect of humic acid on uranium uptake by hematite particles. *J. Coll. Interf. Sci.*, **106**, 281–8.

Ho, C.J. and Miller, N.H. (1986) Adsorption of uranyl species from bicarbonate solution onto hematite particles. *J. Coll. Interf. Sci.*, **110**, 165–71.

Hofmann, A.W. and White, W.M. (1982) Mantle plumes from ancient oceanic-crust. *Earth Planet. Sci. Lett.*, **57**, 421–36.

Hsi, C.K.D. and Langmuir, D. (1985) Adsorption of uranyl onto ferric oxyhydroxides – application of the surface complexation site-binding. *Geochim. Cosmochim. Acta*, **49**, 1931–41.

Hyne, R.V., Rippon, G.D. and Ellender, G. (1992) PH-dependent uranium toxicity of fresh-water hydra. *Sci. Total Environ.*, **125**, 159–73.

Igarashi, G., Saeki, S., Takaharta, N., Sumikawa, K., Tasaka, S., Sasaki, Y., Takahashi, M. and Sano, Y. (1995) Groundwater radon anomaly before the Kobe earthquake in Japan. *Science*, **269**, 60–1.

Ingebritsen, S.E. and Sanford, W.E. (1998) *Groundwater in Geologic Processes.* Cambridge University Press, Cambridge, UK.

Jones, R.L. (1995) Soil uranium, basement radon and lung-cancer in Illinois, USA. *Environ. Geochem. Health*, **17**, 21–4.

Klinkhammer, G.P. and Palmer, M.R. (1991) Uranium in the oceans – where it goes and why. *Geochim. Cosmochim. Acta*, **55**, 1799–806.

Kohler, M., Wieland, E. and Leckie, J.O. (1992) Metal-ligand-surface interactions during sorption of uranyl and neptunyl on oxides and silicates. Pp. 51–4 in: *Water-Rock Interaction* (Y.K. Kharaka and A.S. Maest, editors). Balkema, Rotterdam.

Kohler, M., Curtis, G.P., Kent, D.B. and Davis, J.A. (1996) Experimental investigation and modeling of uranium(VI) transport under variable chemical conditions. *Water Resour. Res.*, **32**, 3539–51.

Komninou, A. and Sverjenski, D.A. (1996) Geochemical modeling of the formation of an unconformity-type uranium deposit. *Econ. Geol.*, **91**, 590–606.

Kramers, J.D. and Tolstikhin, I.N. (1997) Two terrestrial lead isotope paradoxes, forward transport modelling, core formation and the history of the continental crust. *Chem. Geol.*, **139**, 75–110.

Landa, E.R. and Gray, J.R. (1995) US Geological Survey research on the environmental fate of uranium mining and milling wastes. *Environ. Geol.*, **26**, 19–31.

Langmuir, D. (1978) Uranium solution-mineral equilibria at low temperatures with applications to sedimentary ore deposits. *Geochim. Cosmochim. Acta*, **42**, 547–96.

Langmuir, D. (1997) *Aqueous Environmental Geochemistry*. Prentice-Hall, New Jersey.

Langmuir, D. and Chatham, J.R. (1980) Groundwater prospecting for sandstone-type uranium deposits: A preliminary comparison of the merits of mineral-solution equilibria, and single-element tracer methods. *J. Geochem. Explor.*, **13**, 201–19.

Langmuir, D. and Herman, J.S. (1980) The mobility of thorium in natural water at low temperatures. *Geochim. Cosmochim. Acta*, **44**, 1753–66.

Larson, W.C. (1978) *Uranium in situ Leach Mining in the United States*. Info. Circ. 8777, U.S. Dept. of Interior, Bureau of Mines.

Lederer, C.M., Hollander, J.M. and Perlman, I. (1967) *Table of Isotopes*, 6th ed. John Wiley, New York.

Lee, J.D. (1991) *Concise Inorganic Chemistry*. Chapman & Hall, London.

Lenhart, J.J. and Honeyman, B.D. (1999) Uranium(VI) sorption to hematite in the presence of humic acid. *Geochim. Cosmochim. Acta*, **63**, 2891–901.

Levinson, A.A. (1980) *Introduction to Exploration Geochemistry*. Applied Publishing, Wilmette, IL.

Lichtner, P.C. (1996) Continuum formaulation of multicomponent-multiphase reactive transport. Pp. 1–82 in: *Reactive Transport in Porous Media.* (P.C. Lichtner, C.I. Steefel and E.H. Oelkers, editors). Reviews in Mineralogy, **34**. Mineralogical Society of America, Washington, D.C.

Lichtner, P.C. and Waber, N. (1992) Redox front geochemistry and weathering – theory with application to the Osamu Utsumi uranium-mine, Pocos-de-Caldas, Brazil. *J. Geochem. Explor.*, **45**, 521–64.

Lienert, C., Short, S.A. and Von Gunten, H.R. (1994) Uranium infiltration from a river to shallow groundwater. *Geochim. Cosmochim. Acta,* **58**, 5455–63.

Liger, E., Charlet, L. and van Cappellen, P. (1999) Surface catalysis of uranium(VI) reduction by iron(II). *Geochim. Cosmochim. Acta*, **63**, 2939–55.

Lovley, D.R. and Philips, E.J.P. (1992*a*) Reduction of uranium by *Desulfovibrio-desulfuricans. Appl. Environ. Microb.,* **58**, 850–6.

Lovley, D.R. and Philips, E.J.P. (1992*b*) Bioremediation of uranium contamination with enzymatic uranium reduction. *Environ. Sci. Technol.,* **26**, 2228–34.

Lovley, D.R., Philips, E.J.P., Gorby, Y.A. and Landa, E.R. (1991) Microbial reduction of uranium. *Nature*, **350**, 413–5.

Lovley, D.R., Roden, E.E., Philips, E.J.P. and Woodward, J.C. (1993) Enzymatic iron and uranium reduction by sulfate-reducing bacteria. *Marine Geol.*, **113**, 41–53.

Lubal, P., Fetsch, D., Sirokoy, D., Lubalova, M., Senkyr, J. and Havel, J. (2000) Potentiometric and spectroscopic study of uranyl complexation with humic acids. *Talanta*, **51**, 977–91.

Magne, R., Berthelin, J. and Dommergues, Y. (1974) Solubilisation et insolubilisation de l'uranium des granites par des bactéries hétérotrophes. Pp. 73–8 in: *Formation of Uranium Ore Deposit*. International Atomic Energy Commission, Vienna, Austria.

Manceau, A., Charlet, L., Boisset, M.C., Didier, B. and Spadini, L. (1992) Sorption and speciation of heavy metals on Fe and Mn hydrous oxides. From microscopic to macroscopic. *Appl. Clay Sci.*, **7**, 201–23.

Martell, A.E. and Hancock, R.D. (1998) *Metal Complexes in Aqueous Solutions*. Plenum Press, New York.

Martin, J.M. and Whitfield, M. (1983) The significance of the river input of chemical elements to the ocean. Pp. 265–96 in: *Trace Metals in Sea Water* (C.S. Wong, E. Boyle, K.W. Bruland, J.D. Burton and E.D. Goldbert, editors). Plenum Press, New York.

Martin, J.M., Meybeck, M. and Pusset, M. (1978) Uranium behaviour in Zaïre estuary. *Netherlands J. Sea Res.*, **12**, 338–44.

Mayer, L.M. (1982) Aggregation of colloidal iron during estuarine mixing – kinetics, mechanism, and seasonality. *Geochim. Cosmochim. Acta*, **46**, 2527–35.

McCarthy, J.F. and Degueldre, C. (1993) Sampling and characterizing of colloids and particles in groundwater for studying their role in contaminant transport. Pp. 247–316 in: *Environmental Particles* Vol. 2 (J. Buffle and H.P. van Leeuwen, editors). IUPAC Envir. Analytical and Physical Chem. Ser. Lewis, Boca Raton, FL.

McCulloch, M.T. and Gamble, J.A. (1991) Geochemical and geodynamical constraints on subduction zone magmatism. *Earth Planet. Sci. Lett.*, **102**, 358–74.

McLean, J., Purvis, O.W., Williamson, B.J. and Bailey, E.H. (1998) Role of lichen melanins in uranium remediation. *Nature*, **391**, 649–50.

Michard, A., Albarède, F., Michard, G., Minster, J.F. and Charlou, J.L. (1983) Rare-earth elements and uranium in high-temperature solutions from East Pacific Rise hydrothermal vent field (13-degrees-N). *Nature*, **303**, 795–7.

Mills, W.B., Liu, S. and Fong, F.K. (1991) Literature-review and model (COMET) for colloid metals transport in porous-media. *Ground Water*, **29**, 199–208.

Morris, D.E., Chisholm-Brause, C.J., Barr, M.E., Conradson, S.D. and Eller, P.G. (1994) Optical spectroscopic studies of the sorption of $UO_2^{2+}$ species on a reference smectite. *Geochim. Cosmochim. Acta*, **58**, 3613–23.

Morrison, S.J., Spangler, R.R. and Tripathi, V.S. (1995) Adsorption of uranium(VI) on amorphous ferric oxyhydroxide at high concentrations of dissolved carbon(IV) and sulfur(VI). *J. Contam. Hydrol.*, **17**, 333–46.

Moulin, V. and Moulin, C. (1995) Fate of actinides in the presence of humic substances under conditions relevant to nuclear waste disposal. *Appl. Geochem.*, **10**, 573–80.

Murakami, T., Ohnuki, T., Isobe, H. and Sato, T. (1997) Mobility of uranium during weathering. *Amer. Mineral.*, **82**, 888–99.

Muskett, C., Sherman, D.M. and Ragnarsdottir, K.V. (1998) The adsorption of uranium onto goethite and chinochlore. *Mineral. Mag.*, **62A**, 1048–9.

Nagy, B., Gauthier-Lafaye, F., Hollinger, P., Mossman, D.J., Leventhal, J.S. and Rigali, M.J. (1993) Role of organic matter in the Proterozoic Oklo natural fission reactors, Gabon, Africa. *Geology*, **21**, 655–8.

Nakashima, S., Disnar, J.R., Perruchot, A. and Trichet, J. (1984) Experimental study of mechanisms of fixation and reduction of uranium by sedimentary organic matter under diagenetic or hydrothermal conditions. *Geochim. Cosmochim. Acta*, **48**, 2321–9.

Nash, K., Fried, S., Friedman, A.M. and Sullivan, J.C. (1981*a*) Redox behaviour, complexing, and adsorption of hexavalent actinides by humic acid and selected clays. *Environ. Sci. Technol.*, **15**, 834–7.

Nash, J.T., Granger, H.C. and Adams, S.S. (1981*b*) Geology and concepts of genesis of important types of uranium deposits. *Econ. Geol. 75th Anniv. Vol.*, 32–116.

NEA (1984) *Long term Radiological Aspects of Management of Wastes from Uranium Mining and Milling*. OECD, Paris.

Nefedov, V.I., Teterin, Y.A., Lebedev, A.M., Teterin, A.Y., Dementjev, A.P., Bubner, M., Reich, T., Pompe, S., Heise, K.H. and Nitsche, H. (1998) Electron spectroscopy for chemical analysis investigation of the interaction of uranyl and calcium ions with humic acids. *Inorg. Chim. Acta*, **273**, 234–7.

Notz, K.J. (1990) *Characteristic Data Base, Systems Instruction Program.* Oak Ridge National Laboratory, Oak Ridge, Tennessee.

OECD Nuclear Energy Agency (1998) *Uranium, Resources, Production and Demand 1997.* OECD, Paris.

Ortega, A., Domingo, J.L., Gomez, M. and Corbella, J. (1989) Treatment of experimental acute uranium poisoning by chelating-agents. *Pharmacol. Toxicol.*, **64**, 247–51.

Pabalan, R. and Turner, D.R. (1993) Sorption modeling for high-level waste performance assessment. Pp. 67–80 in: *High level Radioactive Waste Research at CNWRA*. CNWRA Report 93-01S.

Palmer, M.R. and Edmond, J.M. (1993) Uranium in river water. *Geochim. Cosmochim. Acta,* **57,** 4947–55.

Parks, G.A. and Pohl, D.C. (1988) Hydrothermal solubility of uraninite. *Geochim. Cosmochim. Acta,* **52**, 863–75.

Pavlakis, N., Pollock, C.A., McLean, G. and Bartrop, R. (1996) Deliberate overdose of uranium: Toxicity and treatment. *Nephron*, **72**, 313–7.

Payne, T.E., Davis, J.A. and Waite, T.D. (1994) Uranium retention by weathered schists – the role of iron minerals. *Radiochim. Acta*, **66/67**, 297–303.

Pike, A.W.G. (2000) *U-series dating of archeological bone by T.J.M.D.* PhD thesis, Univ. Oxford, UK.

Plant, J.A., O'Brien, C., Tarney, J. and Hurdley, J. (1985) Geochemical criteria for the racognition of high heat production granites. Pp. 263–85 in: *High Heat Production (HHP) Granites, Hydrothermal Circulation and Ore Genesis*. Institution of Mining and Metallurgy, London.

Plant, J.A., Simpson, P.R., Smith, B., and Windley, B.F. (1999) Uranium ore deposits – products of the radioactive earth. Pp. 321–432 in: *Uranium: Mineralogy, Geochemistry and the Environment* (P. C. Burns and R. Finch, editors). Reviews in Mineralogy, **38**. Mineralogical Society of America, Washington, D.C.

Plater, A.J., Ivanovich, M. and Dugdale, R.E. (1992) Uranium series disequilibrium in river sediments and waters – the significance of anomalous activity ratios. *Appl. Geochem.*, **7,** 101–10.

Pompe, S., Bubner, M., Denecke, M.A., Reich, T., Brachmann, A., Geipel, G., Nicolai, R., Heise, K.H. and Nitsche, H. (1996) A comparison of natural humic acids with synthetic humic acid model substances: Characterization and interaction with uranium(VI). *Radiochim. Acta*, **74**, 135–40.

Porcelli, D., Andersson, P.S., Wasserburgh, G.J., Ingri, J. and Baskaran, M. (1997) The importance of colloids and mires for the transport of uranium isotopes through the Kalix River watershed and Baltic Sea. *Geochim. Cosmochim. Acta,* **61**, 4095–113.

Prabhavathi, P.A., Fatima, S.K., Padmavathi, P., Kumari, C.K. and Reddy, P.P. (1995) Sister-chromatid exchanges in nuclear fuel workers. *Mutat. Res. Lett.*, **347**, 31–5.

Raffensberger, J.P. and Garven, G. (1995) The formation of unconformity-type uranium ore deposits. 1. Coupled hydrogeochemical modeling. *Amer. J. Sci.,* **295**, 639–96.

Rai, D., Felmy, A.R. and Ryan, J.L. (1990) Uranium(VI) hydrolysis constants and solubility product of $UO_2 \cdot xH_2O$(am). *Inorg. Chem.,* **29**, 260–4.

Randall, S.R., Sherman, D.M. and Ragnarsdottir, K.V. (1999) The mechanism of cadmium surface complexation on iron oxyhydroxide minerals. *Geochim. Cosmochim. Acta*, **63**, 2971–88.

Read, D., Ross, D. and Sims, R.J. (1998) The migration of uranium through Clashach Sandstone: the role of low molecular weight organics in enhancing radionuclide transport. *J. Contam. Hydrol.*, **35**, 235–48.

Reich, T., Moll, H., Denecke, M.A., Geipel, G., Bernhard, G., and Nitsche, H. (1996) Characterization of hydrous uranyl silicate by EXAFS. *Radiochim. Acta*, **74**, 219–23.

Rich, R.A., Holland, H.D. and Peterson, U. (1977) *Hydrothermal Uranium Deposits. Developments in Economic Geology*, **6**, Elsevier, Amsterdam.

Rigali, M.J. and Nagy, B. (1997) Organic free radicals and micropores in solid graphitic carbonaceous matter at the Oklo natural fission reactors, Gabon. *Geochim. Cosmochim. Acta*, **61**, 357–68.

Romberger, S.B. (1984) Transport and deposition of uranium in hydrothermal systems at temperatures up to 300°C: geological implications. Pp. 12–7 in: *Uranium Geochemistry, Mineralogy, Geology, Exploration and Resources* (B. De Vivo, F. Ippolito, G. Capaldi and P.R. Simpson, editors). Institution of Mining and Metallurgy, London.

Rösler, H. and Lange, H. (1972) *Geochemical Tables*. Elsevier, Amsterdam.

Ryan, J.L. and Rai, D. (1983) The solubility of uranium(VI) hydrous oxide in sodium-hydroxide solutions under reducing conditions. *Polyhedron*, **2**, 947–52.

Sagert, N.H., Ho, C.H. and Miller, N.H. (1989) The adsorption of uranium(VI) onto a magnetite sol. *J. Coll. Interf. Sci.*, **130**, 283–7.

Sandino, A. and Bruno, J. (1992) The solubility of $(UO_2)_3(PO_4)_2{\cdot}4H_2O(s)$ and the formation of U(VI) phosphate complexes: Their influence in uranium speciation in natural waters. *Geochim. Cosmochim. Acta*, **56**, 4135–45.

Sanford, R.F. (1990) Hydrogeology of an ancient arid closed basin - implications for tabular sandstone-hosted uranium deposits. *Geology*, **18**, 1099–102.

Sanford, R.F. (1992) A new model for tabular-type uranium deposits. *Econ. Geol. Bull. Soc.*, **87**, 2041–55.

Santschi, P.H. and Honeyman, B.D. (1989) Radionuclides in aquatic environments. *Radiat. Phys. Chem.*, **34**, 213–40.

Sato, T., Yanase, N., Williams, I.S., Compston, W., Zaw, M., Payne, T.E. and Airey, P.L. (1998) Uranium micro-isotopic analysis of weathered rock by Sensitive High Resolution Ion Microprobe (SHRIMP II). *Radiochim. Acta*, **82**, 335–40.

Savage, D. (editor) (1995) *The Scientific and Regulatory Basis for the Geological Disposal of Radioactive Waste*. Wiley, New York.

Schmidt, M., Baraniak, L., Bernhard, G. and Nitsche, H. (1996) *Institute of Radiochemistry Annual Report*, 30–1. Rossendorf, Germany.

Schmidt, S. and Reyss, J.L. (1991) Uranium concentrations of Mediterranean seawater with high salinities. *Comptes Rendue Acad. Sci. Serie II*, **312**, 479–84.

Schofield, P.F., Bailey, E.H. and Mosselmans, J.F.W. (1999) Structure and stability of the solvated $(UO_2)^{2+}$ uranyl ion: An *in situ* XAS study. Pp. 465–68 in: *Geochemistry of the Earth's Surface* (H. Armansson, editor). Balkema, Rotterdam.

Sheppard, J.C., Campbell, M.J., Cheng, T. and Kittrick, J.A. (1980) Retention of radionuclides by mobile humic compounds and soil particles. *Environ. Sci. Tech.*, **14**, 1349–53.

Shock, E.L. and Koretsky, C.M. (1993) Metal organic complexes in geochemical processes – calculation of standard partial molal thermodynamic properties of aqueous acetate complexes at high-pressures and temperatures. *Geochim. Cosmochim. Acta*, **57**, 4899–922.

Shock, E.L. and Koretsky, C.M. (1995) Metal organic complexes in geochemical processes – calculation of standard partial molal thermodynamic properties of aqueous complexes between metal-cations and monovalent organic acid ligands at high-pressures and temperatures. *Geochim. Cosmochim. Acta*, **59**, 1497–532.

Shock, E.L., Sassani, D.C. and Betz, H. (1997*a*) Uranium in geologic fluids: Estimates of standard partial molal properties, oxidation potentials, and hydrolysis constants at high temperatures and pressures. *Geochim. Cosmochim. Acta*, **61**, 4245–66.

Shock, E.L., Sassani, D.C., Wills, M. and Sverjensky, D.A. (1997*b*) Inorganic species in geologic fluids: Correlations among standard molal thermodynamic properties of aqueous ions and hydroxide complexes. *Geochim. Cosmochim. Acta*, **61**, 907–50.

Sholkovitz, E.R., Boyle, E.A. and Price, N.B. (1978) The removal of dissolved humic acids and iron during estuarine mixing. *Earth Planet. Sci. Lett.*, **40**, 130–6.

Short, S.A. (1992) *Studies of Radionuclide Geochemistry in a Shallow Glacio-fluvial Aquifer Infiltrated in a River: Glattfelden, Northeastern Switzerland.* Final Report to AC-Laboratory, GRD, Spiez, Switzerland.

Sigmarsson, O., Condomines, M., Morris, J.D. and Harmon, R.S. (1990) Uranium and Be-10 enrichments by fluids in Andean arc magmas. *Nature*, **346**, 163–5.

Silva, R.J. and Nitsche, H. (1995) Actinide environmental chemistry. *Radiochim. Acta*, **70/71**, 377–96.

Silver, P.G. and Wakita, H. (1996) A search for earthquake precursors. *Science*, **273**, 77–8.

Spear, J.R., Figueroa, L.A. and Honeyman, B.D. (1999) Modeling the removal of uranium U(VI) from aqueous solutions in the presence of sulfate reducing bacteria. *Environ. Sci. Technol.*, **33**, 2667–75.

Sposito, G. (1984) *Surface Chemistry of Soils*. Oxford University Press, Oxford.

Steiger, R.H. and Jäger, E. (1977) Subcommission on geochronology: Convention on the use of decay constants in geo- and cosmochronology. *Earth Planet. Sci. Lett.*, **36**, 359–62.

Strandberg, G.W., Shumate, S.E. and Parrott, J.R. (1981) Microbial cells as biosorbents for heavy-metals – accumulation of uranium by *Saccharomyces cerevisiae* and *Pseudomonas aeruginosa*. *Appl. Environ. Microbiol.*, **41**, 237–45.

Stumm, W. and Morgan, J.J. (1981) *Aquatic Chemistry*, 2nd edition. Wiley, New York.

Stumm, W. and Morgan, J.J. (1996) *Aquatic Chemistry*, 3rd edition. Wiley, New York.

Suzuki, Y. and Banfield, J.F. (1999) Geomicrobiology of uranium. Pp. 1130–41 in: *Uranium: Mineralogy, Geochemistry and the Environment* (P.C. Burns and R. Finch, editors). Reviews in Mineralogy, **38**. Mineralogical Society of America, Washington, D.C.

Sverjenski, D.A. (1987) The role of migrating oil-filed brines in the formation of sediment-hosted Cu-rich deposits. *Econ. Geol.*, **82**, 1130–41.

Tao, Z.Q., Xu, X.H., Yan, X.M., Chen, Z.J., Zhang, J.S., Liang, Y.Y. and Xie, Y.Y. (1987) Detoxication and mobilization of uranium by catecholamic acid. *Acta Pharm. Sinica,* **8**, 284–8.

Thompson, H.A., Brown, G.E. and Parks, G.A. (1997) XAFS spectroscopic study of uranyl coordination in solids and aqueous solution. *Amer. Mineral.*, **82**, 483–96.

Tilton, G.R. and Barreiro, B.A. (1980) Origin of lead in Andean calc-alkaline lavas, Southern Peru. *Science*, **210**, 1245–7.

Toole, J., Baxter, M.S. and Thompson, J. (1987) The behaviour of uranium isotopes with salinity change in 3 UK estuaries. *Estuar. Coast. Shelf Sci.*, **25**, 283–97.

Torrero, M.E., Casas, I., de Pablo, J., Duro, L. and Bruno, J. (1998) Oxidative dissolution mechanism of uranium dioxide at 25°C. *Mineral. Mag.,* **62A**, 1429–530.

Tsezos, M. (1984) Recovery of uranium from biological adsorbents – desorption equilibrium. *Biotechnol Bioeng.*, **26**, 973–81.

Tsezos, M. (1990) The mechanism of uranium biosorption. Pp. 325–36 in: *Microbial Mineral Recovery* (H.L. Ehrlich and C.I Brierley, editors). McGraw-Hill, New York.

Tsezos, M. and Volesky, B. (1981) Biosorption of uranium and thorium. *Biotechnol. Bioeng.*, **23**, 583-604.

Tsunashima, A., Brindley, G.W. and Bastovanov, M. (1981) Adsorption of uranium from solutions by montmorillonite – compositions and properties of uranyl montmorillonites. *Clays Clay Miner.*, **29**, 10–6.

Turner, S., Hawkesworth, C., Rogers, N. and King, P. (1997) U-Th isotope disequilibria and ocean island basalt generation in the Azores. *Chem. Geol.*, **139**, 145–64.

UNEP/UNCHS (1999*a*) *The Kosovo Conflict. Consequences for the Environment and Human Settlements*. United Nation's Environment Programme/United Nations Centre for Human Settlements (Habitat), Nairobi, Kenya.

UNEP/UNCHS (1999*b*) *The Potential Effects on Human Health and the Environment Arising* *from Possible use of Depleted Uranium during the 1999 Kosovo Conflict. A Preliminary Assessment.* United Nation's Environment Programme/United Nations Centre for Human Settlements Balkans Task Force (BTF), Geneva, Switzerland.

Valsami-Jones, E. and Ragnarsdottir, K.V. (1997*a*) Controls on uranium and thorium behaviour in ocean-floor hydrothermal systems: examples from the Pindos ophiolite, Greece. *Chem. Geol.,* **135**, 263–74.

Valsami-Jones, E. and Ragnarsdottir, K.V. (1997*b*) Solubility of uranium oxide and calcium uranate in water, and $Ca(OH)_2$-bearing solutions. *Radiochim. Acta,* **79**, 249–57.

Valsami-Jones, E., Ragnarsdottir, K.V., Mann, T. and Kemp, A.J. (1996) An experimental investigation of the potential of apatite as radionuclide and heavy metal scavenger. Pp. 686–9 in: *Geochemistry of the Earth's Surface* (S.H. Bottrell, editor). University of Leeds, UK.

Van Cappellen, P., Charlet, L., Stumm, W. and Wersin, P. (1993) A surface complexation model of the carbonate mineral-aqueous solution interface. *Geochim. Cosmochim. Acta*, **57**, 3505–18.

Van der Lee, J., Ledoux, E. and de Marsily, G. (1992) Modeling of colloidal uranium transport in a fractured medium. *J. Hydrol.*, **139**, 135–8.

Vilks, P. (1994) Colloids. Pp. 219–41 in: *Final report of the AECL/SKB Cigar Lake Analogue Study* (J. Cramer and J. Smellie, editors). SKB Technical Report 94-04. Stockholm, Sweden.

Waber, N., Schorscher, H.D. and Peters, T.J. (1990) *Mineralogy, Petrology and Geochemistry of the Pocos de Caldas Analogue Study Sites, Minas Gerais, Brazil, I. Osamu Utsumi Uranium Mine.* Nagra NTB 90-20. Wettingen, Switzerland.

Wagman, D.D., Evans, W.H., Parker, V.B, Schumm, R.H., Halow, I., Bailey, S.M., Churney, K.L., and Nuttall, R.L. (1982) The NBS tables of chemical thermodynamic properties. *J. Phys. Chem. Ref. Data*, **11** (Suppl. 2).

Waite, T.D., Davis, J.A., Payne, T.E., Waychunas, G.A. and Xu, N. (1994) Uranium(VI) adsorption to ferrihydrite – application of surface complexation model. *Geochim. Cosmochim. Acta*, **58**, 5465–78.

Wanty, R.B., Chathan, J.R. and Langmuir, D. (1987) The solubilities of some major and minor element minerals in ground waters associated with a sandstone-hosted uranium deposit. *Bull. Mineral.*, **110**, 209–26.

Wehrli, B., Sulzberger, B. and Stumm, W. (1989) Redox processes catalyzed by hydrous oxide surfaces. *Chem. Geol.*, **78**, 167–79.

Wersin, P., Hochella, M.F., Persson, P., Redden, G., Leckie, J.O. and Harris, D.W. (1994) Interaction between aqueous uranium(VI) and sulfide minerals – spectroscopic evidence for sorption and reduction. *Geochim. Cosmochim. Acta*, **58**, 2829–43.

Wetherill, G.W. (1966) Radioactive decay constants and energies. Pp. 514–9 in: *Handbook of Physical Constants* (S.E. Clarke, Jr., editor). Geolological Society of America Memoir, **97**.

Yajima, T., Kawamura, Y. and Ueta, S. (1995) Uranium(VI) solubility and hydrolysis constants under reduced conditons. *Mat. Res. Soc. Symp. Proc.*, **353**, 1137–42.

Yamaguchi, T. and Nakayama, S. (1998) Diffusion of U, Pu and Am carbonate complexes in a granite from Inada, Ibaraki, Japan studied by diffusion. *J. Cont. Hydrol.*, **35**, 55–65.

Zajic, J.E. (1969) *Microbial Biogeochemistry.* Academic Press, New York.

Zartman, R.E. and Haines, S. (1988) The plumbotectonic model for Pb isotopic systematics among major terrestrial reservoirs - a case for bi-directional transport. *Geochim. Cosmochim. Acta*, **52**, 1327–39.

Zeh, P., Czerwinski, K.R. and Kim, J.I. (1997) Speciation of uranium in Gorleben groundwaters. *Radiochim. Acta*, **76**, 37–44.

Zielinski, R.A., Simmons, K.R. and Orem, W.H. (2000) Use of U-234 and U-238 isotopes to identify fertilizer-driven uranium in the Florida Everglades. *Appl. Geochem.*, **15**, 369–83.

CHAPTER FIFTEEN

# Metal phosphates and remediation of contaminated land

M. E. Hodson[1,*], E. Valsami-Jones[2] and J. D. Cotter-Howells[3,†]

[1] *Department of Mineralogy, The Natural History Museum, Cromwell Road, London SW7 5BD, UK*

[2] *Department of Mineralogy, The Natural History Museum, Cromwell Road, London SW7 5BD, UK (E-mail: evj@nhm.ac.uk)*

[3] *Department of Plant and Soil Science, University of Aberdeen, St. Machar Drive, Aberdeen AB24 3UU, UK*

**ABSTRACT**

The remediation of metal-contaminated soils is a growing environmental problem. Current practice often involves either capping contaminated sites or removing contaminated material and storing it elsewhere. However, given that metal phosphates are highly insoluble under a wide range of Eh and pH conditions, it has been suggested that the conversion of metals into metal phosphates may represent a sustainable, *in situ*, remediation technique for metal-contaminated soils. In this chapter the solubility of metal phosphates and experimental work investigating the potential for metal phosphate formation as a remediation technique are reviewed. It is concluded that current work indicates that metal phosphate formation may well be a viable treatment. However, further field trials and investigations into the solubility of some less well known metal phosphates, particularly in the presence of organic ligands and microorganisms are required.

## 15.1 Introduction

The presence of potentially toxic metals in soils and waters associated with industrial, mining and waste disposal activities is a common environmental problem (e.g. Iskandar and Adriano, 1997; Young *et al.*, 1997). Many metals can be harmful to biota when present above threshold concentrations in bioavailable forms (Alloway and Ayres, 1997). The remediation of such contaminated soils is therefore required to reduce the associated risks. A

* Current address: Department of Soil Science, University of Reading, PO Box 233, Whiteknights, Reading RG6 6DW, UK. E-mail: m.e.hodson@reading.ac.uk

† Current address: Greenpeace Research Laboratories, Department of Biological Sciences, University of Exeter, Prince of Wales Road, Exeter EX4 4PS, UK (E-mail: j.cotter-howells@exeter.ac.uk)

Hodson, M.E., Valsami-Jones, E. and Cotter-Howells, J.D. (2000) Metal phosphates and remediation of contaminated land. Pp. 291–311 in: *Environmental Mineralogy: Microbial Interactions, Anthropogenic Influences, Contaminated Land and Waste Management* (J.D. Cotter-Howells, L.S. Campbell, E. Valsami-Jones and M. Batchelder, editors). Mineralogical Society Series, **9**. Mineralogical Society, London. ISBN 0 903056 20 8.

variety of bioremediation methods involving plants (phytoremediation) and bacteria are available for the treatment of organic contaminants in soils (Wood, 1997; Fry *et al.*, 1992). Bacterial treatments have been used to successfully remove metals from solutions (Gadd, 1992). However, further development of bioremediation methods is required before such treatments become economically feasible for the decontamination of metal-bearing soils (Salt *et al.*, 1998; Brooks, 1997). Thus the main methods currently in use to clean up metal-contaminated land involve capping contaminated sites or excavating the site and dumping the contaminated soil elsewhere (Wood, 1997). These techniques are not only expensive but are unsustainable. An alternative, cost-efficient, sustainable remediation technique is urgently required and it has been suggested that conversion of the metals present at contaminated sites into metal phosphates may represent such a technique.

Phosphates of many metals (e.g. Pb, Zn, Cu, Cd, Ni, Al and U) have exceptionally low solubilities (Nriagu, 1984). In addition, they are stable over almost the entire range of Eh and pH conditions encountered in the natural surface environment. Therefore, the formation of these stable phosphates in contaminated soils would immobilize metals resulting in reduced leaching of metals into groundwater and reduced bioavailability to plant and animal life (Nriagu, 1984). In some industries (e.g. manufacturers of colour tubes for televisions), the insoluble nature of metal phosphates has been used for some time as a means of cleaning metals from waste waters (e.g. Lewicke, 1972). A successful phosphate treatment of metal contaminated soil would be one that would involve a mechanism of slow release of phosphate ions which would react with metal ions to precipitate metal phosphate compounds. In theory, phosphate treatments of contaminated land could be carried out in isolation or in parallel with the bioremediation of organic pollutants.

A growing body of work has been published which examines the efficacy of various phosphate treatments in reacting with metals to form metal phosphates. The objectives of this chapter are therefore: (1) to discuss the natural occurrence of metal phosphates in the environment; (2) to summarize information on the stability and solubility of metal phosphates published in the scientific literature; and (3) to review the various research projects which have investigated the potential for metal phosphate formation to be a useful remediation treatment.

## 15.2 Metal phosphates in the natural environment

The majority of metal phosphates that are the subject of this chapter occur naturally in the environment as secondary minerals, forming by the alteration of metal ore deposits in the presence of suitable amounts of P. Lead-phosphates such as pyromorphite [$Pb_5(PO_4)_3Cl$], plumbogummite [$PbAl_3(PO_4)_2(OH)_5{\cdot}H_2O$] and corkite [$PbFe_3PO_4SO_4(OH)_6$] occur widely as secondary minerals in the oxidation zone of Pb-bearing ore deposits provided that enough P is available (Nriagu, 1984). It should be noted that the composition of pyromorphite is $Pb_5(PO_4)_3Cl$ whereas the related phases $Pb_5(PO_4)_3F$ and $Pb_5(PO_4)_3OH$ are known as 'fluorpyromorphite' and 'lead hydroxyapatite' (though in the environmental literature, and for convenience in this chapter, this phase is more commonly referred to as "hydroxypyromorphite") respectively.

Although various workers (e.g. Keeton, 1937; MacLean *et al.*, 1969; Cox and Rains, 1972) have suggested that relatively insoluble lead phosphates should form in soils and sediments, it was first suggested by Nriagu (1973, 1974, 1984) that 'hydroxypyromorphite' and pyromorphite should be the most stable form of Pb in a wide range of environments. He suggested that members of the apatite–pyromorphite and plumbogummite–crandallite [$(Pb,Ca)Al_3(PO_4)_2(OH)_5{\cdot}H_2O$] solid solution series should form by roadsides due to the presence of Pb in car exhaust fumes and identified plumbogummite (unpublished data) as being present.

The definitive identification of lead phosphates forming in soils as a result of anthropogenic impacts is difficult due to the potential for substitution of divalent cations for $Pb^{2+}$ and the effects that this can have on unit-cell dimensions and X-ray diffraction (XRD) patterns. For example Bigi *et al.* (1989) and Brückner *et al.* (1995) show how XRD patterns and unit-cell dimensions vary between $Ca_{10}(PO_4)_6(OH)_2$ and $Pb_{10}(PO_4)_6(OH)_2$ end-members (unit-cell dimensions increase towards the Pb end-member). Various workers have identified Pb-phosphates occurring in the environment using analytical transmission electron microscopy (ATEM), extended X-ray absorption fine structure spectroscopy (EXAFS) and XRD. Using ATEM and EXAFS, Cotter-Howells *et al.* (1994) identified a Ca-substituted pyromorphite which had formed in mine-waste soils from the South Pennine Ore Field, UK. The XRD pattern of this phase was similar to that for the pure end-member pyromorphite but was sufficiently different, due to a reduction in the pyromorphite unit-cell dimensions, to prevent identification by conventional XRD. Cotter-Howells (1996) reported the presence of Pb phosphate in urban and roadside soils. Electron microprobe analysis (EMPA) showed that the particles had a similar composition to pyromorphite and XRD patterns obtained for the phase were similar to those determined by Cotter-Howells *et al.* (1994). Morin *et al.* (1999) investigated three soil horizons from a soil developed on a Pb-mineralized sandstone from Ardeché, France and identified plumbogummite on the basis of EXAFS and XRD. Using XRD, Ruby *et al.* (1994) identified Pb-phosphate as well as other Pb minerals in soils from a port area where lead ore used to be off-loaded. The port area was adjacent to a phosphoric acid plant. Davis *et al.* (1993) reported the presence of pyromorphite, plumbogummite and corkite on the basis of particle chemistry in soils from around the Pb ore body near Butte, Montana, USA.

Other metal phosphates, such as cornetite [$Cu_3PO_4(OH)_3$], libethenite [$Cu_2PO_4OH$], hopeite [$Zn_3(PO_4)_2{\cdot}4H_2O$] and spencerite

$[Zn_4(PO_4)_2(OH)_2 \cdot 3H_2O]$ occur as crusts, crystals and alteration products in and around ore bodies (Nriagu, 1984). The turquoise group $[CuAl_6(PO_4)_4(OH)_8 \cdot 4H_2O]$ occurs as dense, cryptocrystalline to fine granular masses as well as in concretions, stalactites and veinlets. More rarely it occurs as small prismatic crystals. Nriagu (1984) attributed the sparsity of Cd- and Ni-phosphates in the geological record to the low concentrations of cadmium and nickel in the earth's crust but pointed out that many Ca- and Al-phosphates contain significant concentrations of Cd and Ni.

## 15.3 Metal phosphate solubility

In inorganic systems, metal phosphates are highly insoluble and are commonly the most stable form in which a metal can exist over a wide range of pH provided that a small amount of phosphorus is present. Phosphate solubility is increased and stability is reduced in organic systems. However, the occurrence of metal phosphates in nature as the result of the alteration of metal ores indicates that they must be relatively stable in the natural environment, even in the presence of organic ligands.

### *15.3.1 Phosphate solubility in simple inorganic systems*

Table 15.1 is a list of the solubility product (log $K_{sp}$) values calculated for some metal phosphates and other metal bearing phases on the basis of their thermochemical constants as published in Wagman *et al.* (1968). Since $K_{sp}$ values are a measure of the solubility of a mineral, the smaller the value of log $K_{sp}$ the less soluble this mineral is. Thus it is clear that a variety of metal phosphates are less soluble than other metal phases.

Nriagu (1972, 1973, 1974) calculated solubilities of 'hydroxypyromorphite' and pyromorphite and showed that they represented the most stable form of $Pb^{2+}$ under a wide range of environmental conditions. Using solubility data, Nriagu (1984) constructed diagrams to show the relative stability of various phases including metal phosphates. Example diagrams for Pb and Zn are shown in Figs 15.1 and 15.2. The precise geometry of such diagrams depends on the chemical reactions considered and the assumed activities of chemical species relevant to the diagrams. However, the diagrams do show that metal phosphates are more stable over a wider range of conditions of pH than other metal phases and that, if phosphate concentrations are high enough, metal phosphates should form at the expense of other metal phases.

Jurinak and Inouye (1962) titrated $ZnCl_2$ and $Cu(NO_3)_2$ solutions with NaOH at different pHs in the presence of $KH_2PO_4$, and observed that crystalline $Zn_3(PO_4)_2 \cdot 4H_2O$ and amorphous $Cu_5(H_2PO_4)_2(OH)_8$ formed. The solubility products for these phases were calculated, and Cu-phosphate was shown to be less soluble than Zn-phosphate. Valsami-Jones *et al.* (1998) carried out experiments investigating the solubilities of naturally-occurring

TABLE 15.1 Metal phosphate solubilities.

| Mineral name | Reaction | Log $K_{sp}$ |
|---|---|---|
| Pyromorphite | $Pb_5(PO_4)_3Cl \rightleftharpoons 5Pb^{2+} + 3PO_4^{3-} + Cl^-$ | −84.4 |
| Plumbogummite | $PbAl_3(PO_4)_2(OH)_5{\cdot}H_2O \rightleftharpoons Pb^{2+} + 3Al^{3+} + 2PO_4^{3-} + 5OH^- + H_2O$ | −99.3 |
| Anglesite | $PbSO_4 \rightleftharpoons Pb^{2+} + SO_4^{2-}$ | −7.7 |
| Cerussite | $PbCO_3 \rightleftharpoons Pb^{2+} + CO_3^{2-}$ | −12.8 |
| Copper pyromorphite | $Cu_5(PO_4)_3OH \rightleftharpoons 5Cu^{2+} + 3PO_4^{3-} + OH^-$ | −65.6 |
| Turquoise | $CuAl_6(PO_4)_4(OH)_8{\cdot}4H_2O \rightleftharpoons Cu^{2+} + 6Al^{3+} + 4PO_4^{3-} + 8OH^- + 4H_2O$ | −179.0 |
| Chalcanthite | $CuSO_4{\cdot}5H_2O \rightleftharpoons Cu^{2+} + SO_4^{2-} + 5H_2O$ | −2.64 |
| Copper carbonate | $CuCO_3 \rightleftharpoons Cu^{2+} + CO_3^{2-}$ | −9.63 |
| Hopeite | $Zn_3(PO_4)_2{\cdot}4H_2O \rightleftharpoons 3Zn^{2+} + 2PO_4^{3-} + 4H_2O$ | −35.3 |
| Zinc pyromorphite | $Zn_5(PO_4)_3OH \rightleftharpoons 5Zn^{2+} + 3PO_4^{3-} + OH^-$ | −63.1 |
| Faustite | $ZnAl_6(PO_4)_4(OH)_8{\cdot}4H_2O \rightleftharpoons Zn^{2+} + 6Al^{3+} + 4PO_4^{3-} + 8OH^- + 4H_2O$ | −177.7 |
| Zinkosite | $ZnSO_4 \rightleftharpoons Zn^{2+} + SO_4^{2-}$ | 3.4 |
| Smithsonite | $ZnCO_3 \rightleftharpoons Zn^{2+} + CO_3^{2-}$ | −9.9 |
| Cadmium phosphate | $Cd_3(PO_4)_2 \rightleftharpoons 3Cd^{2+} + 2PO_4^{3-}$ | −38.1 |
| Cadmium sulphate | $CdSO_4 \rightleftharpoons Cd^{2+} + SO_4^{2-}$ | −2.46 |
| Otavite | $CdCO_3 \rightleftharpoons Cd^{2+} + CO_3^{2-}$ | −11.6 |

Data from Nriagu (1984) and Appelo and Postma (1993)

fluorapatite and synthetic hydroxylapatite and 'hydroxypyromorphite' synthesized after the method of Napper and Smythe (1966). The synthetic hydroxylapatite was 10 log units more soluble than the fluorapatite but dissolved more slowly. It was suggested that the difference in dissolution rates could be due to the higher surface roughness of the natural apatite or the presence of physical and chemical defects in the natural apatite. The solubility of the 'hydroxypyromorphite' was far lower than that of the apatites. Ma (1996) investigated the stability of 'hydroxypyromorphite' with respect to synthetic hydroxylapatite by reacting mixtures of synthetic hydroxylapatite and 'hydroxypyromorphite'. In P-limited solutions (maintained by the presence of a P-adsorbing resin) 'hydroxypyromorphite' dissolved only slightly when P levels were very low (~1 μmol $l^{-1}$), whereas synthetic hydroxylapatite dissolved significantly even when no resin was present and the solution contained appreciable phosphorus (~47 μmol $l^{-1}$). In experiments run with an excess of $Ca^{2+}$, a small amount of $Pb^{2+}$ was released into solution but phosphorus levels remained low, decreasing with increased $Ca^{2+}$, indicating that some 'hydroxypyromorphite' dissolved and some synthetic hydroxylapatite precipitated.

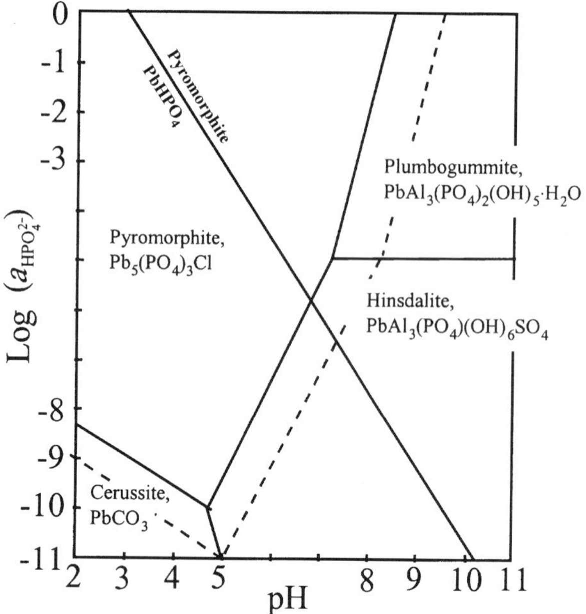


FIG. 15.1 Stability relationships for secondary Pb phosphate minerals, assuming $a_{SO_4^{2-}} = a_{HCO_3^-} = 10^{-3}$; $a_{Pb^{2+}} = 10^{-6}$ (solid lines) and $a_{Pb^{2+}} = 10^{-5}$ (broken lines). The broken lines show the direction of change of 'solubility' with increasing Pb concentration. Minerals with closely related compositions and stabilities are assigned the same stability fields. Adapted from Nriagu (1984).

### *15.3.2 Metal phosphate stability in the presence of organics*

The above work investigated relatively simple, inorganic systems. Little work has been carried out on the stability of metal phosphates in organic systems of the type that would be encountered in the natural environment.

Santillan-Medrano and Jurinak (1975) performed experiments on both Pb and Cd solubility in soils. Calculation and experiment indicated that, at pH 5–9 and concentrations of $10^{-6.6}$ M $PO_4^{3-}$, Pb-phosphates were the least soluble Pb phase. At the same $PO_4^{3-}$ concentration, Cd-phosphate was the least soluble phase at pH 6 to ~7.5 and Cd-carbonate was the least soluble phase at pH 7.5–9. Cadmium phases were more soluble than those of Pb. Ma (1996) reacted a mixture of synthetic hydroxylapatite and 'hydroxypyromorphite' with ethylenediaminetetraacetic acid (EDTA) of varying concentrations (molar concentrations of EDTA to $Pb^{2+}$ ranged from 7 to 26). Most of the lead (85–100%) in the 'hydroxypyromorphite' was released but not all the synthetic hydroxylapatite dissolved. This is thought to be a result of the chelation of $Pb^{2+}$ by EDTA. Sauvé *et al.* (1998) also suggested that at neutral to high pH the solubility of Pb-phosphates, in the presence of organic matter, may be far greater than in inorganic systems due to the Pb-complexing capacity of organic matter which is

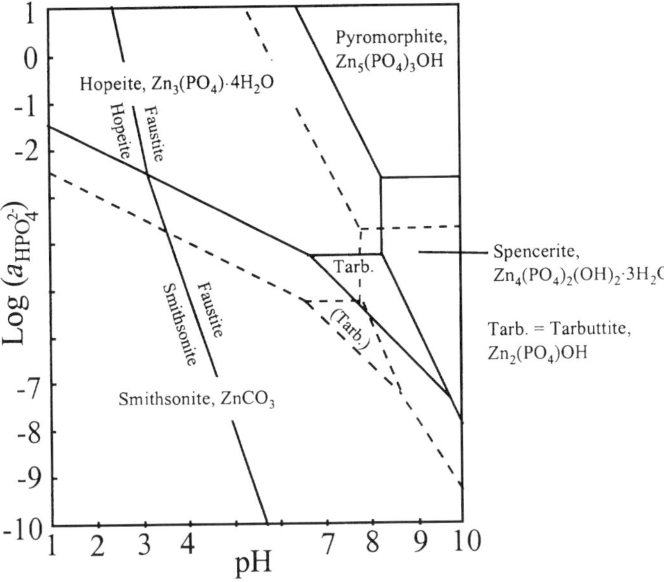


FIG. 15.2 Stability of secondary Zn phosphate minerals, assuming $a_{\mathrm{zinc}^{2+}} = a_{\mathrm{Fe}^{2+}} = 10^{-6}$; $a_{\mathrm{Ca}^{2+}} = a_{\mathrm{HCO}_3^-} = 10^{-3}$ for solid lines. For the broken lines the activity of Zn ions has been assigned a value of $10^{-5}$. Minerals with similar compositions occupy the same domain in the diagram. Adapted from Nriagu (1984).

solubilized at neutral pH and above. Organic complexing of Pb was demonstrated by reacting a suspension of synthetic Pb-orthophosphate and soil (19.5–151 μg $Pb^{2+}$ $g^{-1}$), and comparing the total dissolved Pb in solution to its concentration as determined by a selective electrode which measured labile Pb (assumed to be free $Pb^{2+}$) but not non-labile Pb (assumed to be equivalent to organically complexed $Pb^{2+}$). As the pH rose from 4 to 6, the total dissolved Pb in solution decreased whilst organically-bound lead increased in the high $Pb^{2+}$ experiment and decreased in the low $Pb^{2+}$ experiment. As pH rose from 6 to 8, total dissolved Pb and organically-bound lead increased, whereas estimated free $Pb^{2+}$ decreased or remained constant. The ratio of organically-bound to total Pb in solution varied between ~0.001–1 at pH 4 to ~1 at pH 8.

Ruby *et al.* (1993) investigated the solubility of Pb phases in conditions similar to those encountered in human digestive tracts. Mine wastes containing 5820 μg $Pb^{2+}$ $g^{-1}$ and 40 ml of a fluid designed to mimic the contents of a rabbit's stomach were reacted for 4 h. Lead was always released, but samples which contained more lead oxide, silicate and anglesite and less Pb phosphate released the most Pb.

Sayer *et al.* (1997, 1999) showed that various microorganisms have the ability to dissolve metal phosphates. The fungi *Coriolus versicolor,*

*Penicillium bilaii, Rhizopus arrhizus* and *Aspergillus niger* were grown on agar which contained 0.2 wt.% pyromorphite. All of the microorganisms grew despite the presence of the pyromorphite and all the microorganisms contained Pb in their biomass indicating that pyromorphite was dissolved. The XRD studies on the agar after the experiments had ended showed that pyromorphite below *A. niger* had dissolved and that Pb oxalate dihydrate had formed. Pyromorphite was still present in the agar below the other organisms. Metal oxalates are themselves highly insoluble, but calculations indicated that under the majority of pH and P concentrations likely to be encountered in soils, Pb oxalate would be converted back into pyromorphite over time (Sayer *et al.*, 1999).

## 15.4 Metal phosphate formation in experimental systems using aqueous metal ions

The simplest experiments investigating the potential for metal phosphate formation to remediate contaminated soil examine the formation of metal phosphates as a result of the reaction between a phosphate source and an aqueous metal ion in a monometallic system. Metal ions are usually introduced in the form of nitrates due to the high solubility of these phases. Experiments have also examined the effects of mixed cocktails of elements (both anions and cations) on metal immobilization. The experiments usually take the form of shaking a P source, either P solution or some form of apatite, in a metal ion solution of either fixed or unbuffered pH. Solution composition and solid reaction products are then examined.

Selected experiments are summarized in Table 15.2. Pyromorphites form readily when Pb and P are brought together in solution (e.g. Xu and Schwarz, 1994; Ma *et al.*, 1993). In addition, U- and Zn-phosphates can form and Cd-phosphates may form (e.g. Gauglitz *et al.*, 1992; Valsami-Jones *et al.*, 1998; Chen *et al.*, 1997). The presence of other cations and anions can inhibit metal phosphate formation by competitively reacting with phosphate ions or inhibiting the dissolution of the phosphate source (Ma *et al.* 1994*a,b*). The formation of pyromorphite coatings on the phosphate source can also limit further metal phosphate formation (Xu and Schwarz, 1994). The presence of organic compounds may inhibit phosphate formation by forming highly soluble metal-organic complexes (Ma, 1996; Sauvé *et al.*, 1998).

### *15.4.1 Experiments concerned with Pb*

Experiments in which a P source, either synthetic hydroxylapatite, ground up natural rock apatite or a soluble P salt such as $NaH_2PO_4{\cdot}H_2O$ or $CaHPO_4$, is reacted with $Pb^{2+}$ in solution have been carried out by Ma *et al.* (1993, 1995), Valsami-Jones *et al.* (1998), Xu and Schwarz (1994) and Chen *et al.* (1997). The dominant anion in the $Pb^{2+}$ solution is usually $NO_3^-$ but Xu and Schwarz

TABLE 15.2 Summary of experiments examining aqueous metal immobilization by phosphate formation.

| | Xu and Schwarz (1994) | Ma *et al.* (1993) | Valsami-Jones *et al.* (1998) | Chen *et al.* (1997) | Ma *et al.* (1994*a,b*) | Gauglitz *et al.* (1992) | Ma *et al.* (1995) |
|---|---|---|---|---|---|---|---|
| Mass of P source | 0.16–0.5 g HA* | 0.2–0.07 g HA, $NaH_2PO_4{\cdot}H_2O$, $CaHPO_4$ | 0.02–0.04 g HA | 1.2 g NA | 0.1 g HA | ~0.6–2.5 g HA | 0.1–0.2 g NA† |
| Metal concentration | 0.5–12 mM $Pb^{2+}$ | 0.005–2 M $Pb^{2+}$ | 0.5–4 mM $Pb^{2+}$, $Cd^{2+}$ | $2.5–7.5 \times 10^{-2}$ M $Pb^{2+}$, $Zn^{2+}$, $Cd^{2+}$ | 24.1–482 μM $Pb^{2+}$ | $10^{-2}$ M $U^{6+}$, $Th^{2+}$ | 4.8–48.2 μM $Pb^{2+}$ |
| Additional cations | – | – | – | – | 24.1–3395 μmol $l^{-1}$ $Al^{3+}$, $Cd^{2+}$, $Cu^{2+}$, $Fe^{2+}$, $Ni^{2+}$, $Zn^{2+}$, $Na^{+}$, $K^{+}$, $Ca^{2+}$ | $Na^{+}$, $Ca^{2+}$, $Mg^{2+}$ $K^{+}$ | – |
| Anions | $NO_3^-$, $Cl^-$ | $NO_3^-$ | $NO_3^-$ | $NO_3^-$ | $F^-$, $SO_4^{2-}$, $CO_3^-$, $Cl^-$, $NO_3^-$ | $SO_4^{2-}$, $Cl^-$, $NO_3^-$ | $NO_3^-$ |
| Solution volume (ml) | 100–260 | 150–200 | 50–100 | 35 | 200 | 50 | 200 |
| Initial pH | 3.5–5.8 | 3–7 | 2–7 | 1–12 | 6 | – | 7.6–8.2 |
| Phosphate formation? | Yes | Yes | Yes (Zn), probably Cd | Yes (Pb, Zn) | Yes | Yes (U) | Yes |
| Duration of experiments | Up to 9 days | Up to 24 h | Up to 9 months | 24 h | 2 h | 7 days–5 months | Up to 2 days |

* HA = synthetic hydroxylapatite; † NA = natural apatite

(1994) carried out some experiments with $Cl^-$. In all of the above studies 'hydroxypyromorphite' (or pyromorphite in the presence of $Cl^-$) was identified in reaction products using XRD. The formation of pyromorphites, together with changes in solution chemistry, was taken to indicate that the majority of Pb immobilization was due to dissolution of the P source and precipitation of pyromorphites when P and $Pb^{2+}$ concentrations were high enough. However, several of the studies show that some adsorption of $Pb^{2+}$ by the solid P source may also have occurred.

Using atomic force microscopy (AFM) in conjunction with a fluid cell, Valsami-Jones *et al.* (1998) showed that crystallites, presumed to be 'hydroxypyromorphite', nucleated on growth surfaces of apatite, imaged in their experiments. Xu and Schwarz (1994) found that aqueous Pb removal was more rapid in the presence of $NO_3^-$ than $Cl^-$. Scanning electron microscopy (SEM) indicated that whilst discrete 'hydroxypyromorphite' crystals formed in the presence of $NO_3^-$, crystals of pyromorphite which coated the synthetic hydroxylapatite thereby inhibiting further dissolution and P supply, formed in the presence of $Cl^-$ ions. Chen *et al.* (1997) found that when using natural apatite as a P source, a variety of Pb-phosphates could form. As pH increased in their experiments, XRD indicated that the dominant Pb-phosphate varied from 'fluorpyromorphite' to 'hydroxypyromorphite' to 'carboxy hydrofluro-pyromorphite' [$Pb_{10}(PO_4,CO_3)_6(F,OH)_2$] to cerussite ($PbCO_3$) to 'hydroxy-pyromorphite' and lead oxide fluoride ($Pb_2OF_2$). The pyromorphites became less crystalline with increasing pH and the shift in their composition was due to the varying pH-dependent dissolution rates of the P source and pH-dependent solubilities of the Pb phases. In a slightly different experiment, Ma *et al.* (1993) reacted synthetic hydroxylapatite with an exchange resin that had adsorbed $Pb^{2+}$. 'Hydroxypyromorphite' formation occurred on the resin surface. In a similar experiment, Zhang *et al.* (1997) added goethite, to which $Pb^{2+}$ had adsorbed, to a solution containing $NO_3^-$, $Cl^-$ and phosphate ions. When the goethite was held in a bag made of dialysis tubing, pyromorphite formed on the surface of the tubing. When the goethite was not in dialysis tubing, a coating of pyromorphite formed on the goethite surface. No further pyromorphite formation occurred when the remaining adsorbed $Pb^{2+}$ was isolated from the P ions.

### *15.4.2 Experiments concerned with other metal elements*

Compared to the work done on Pb, relatively few experiments have been carried out on the potential for other single metal phosphates to form.

Valsami-Jones *et al.* (1998) found that removal of $Cd^{2+}$ from solutions in the presence of synthetic hydroxylapatite increased with increasing pH, within the pH range 2.6–5.3, and showed a correlation with $Ca^{2+}$ release into solution (though not as strongly as did $Pb^{2+}$). Less $Cd^{2+}$ was removed from solution than in similar experiments using $Pb^{2+}$ and no Cd-phosphate was detected by

XRD. However, in unpublished work, a mixed Ca-Cd phosphate was identified using ATEM and it seems likely that this formed by dissolution/precipitation mechanisms. Chen *et al.* (1997) found that although aqueous Cd concentrations were reduced in the presence of apatite, this appeared to be due to adsorption of $Cd^{2+}$ and the precipitation of otavite ($CdCO_3$). Middleberg and Comans (1991) obtained similar results. Chen *et al.* (1997) also reacted ground carbonate fluorapatite rock with solutions of $Zn^{2+}$. The behaviour of $Zn^{2+}$ was similar to that of $Cd^{2+}$ except that $Zn^{2+}$ removal increased as pH decreased whereas $Cd^{2+}$ removal had decreased. In addition, hopeite [$(Zn_3(PO_4)_2{\cdot}4H_2O$] and zincite [ZnO] (when pH >8) were identified as reaction products by XRD. The concentration of hopeite decreased as the pH of the reacting solution increased.

### *15.4.3 Multi-element experiments*

In experiments run in the same way as for individual elements, Chen *et al.* (1997) reacted ground carbonate fluorapatite rock with a mixed solution containing $Pb^{2+}$, $Cd^{2+}$ and $Zn^{2+}$. The behaviour of the different ions was very similar to when they were present singularly. The same Pb-phosphates and carbonate formed as did otavite, hopeite and zincite.

Gauglitz *et al.* (1992) reacted synthetic hydroxylapatite with mixed solutions containing either $UO_2^{2+}$ or $Th^{4+}$ and varying concentrations of $Mg^{2+}$, $K^+$, $Na^+$, $Cl^-$, $SO_4^{2-}$ and $NO_3^-$. Their experiments were designed to investigate the influence of synthetic hydroxylapatite added to the backfill material of a radioactive waste repository located in a salt dome. Concentrations of U and Th were initially at the $10^{-2}$ M level but fell by a factor of $10^5$ within 1 h of reaction in the presence of 0.6–2.5 g synthetic hydroxylapatite in 50 ml of solution. In the $UO_2^{2+}$ experiments, saléeite [$Mg(UO_2)_2(PO_4)_2{\cdot}9H_2O$) was identified, using XRD, as a reaction product in experiments conducted at high $Mg^{2+}$ concentrations and meta-autunite [$(Ca(UO_2)_2(PO_4)_2{\cdot}6H_2O$] was identified in experiments with lower $Mg^{2+}$ concentrations. Additional $UO_2^{2+}$ removal from solution may have occurred by adsorption of the U at the surface of the phosphates. In the experiments carried out using $Th^{2+}$ the reaction products were X-ray amorphous and consequently were not identified. Ma and co-workers (Ma *et al.*, 1994*a,b*; Ma, 1996) carried out an extensive survey into the effects that the presence of other cations and anions had on the ability of synthetic hydroxylapatite to immobilize $Pb^{2+}$. Ma *et al.* (1994*a*) found that $NO_3^-$, $F^-$ and $Cl^-$ had no effect on $Pb^{2+}$ immobilization and 'hydroxypyromorphite', 'fluorpyromorphite' and pyromorphite formed respectively. Sulphate did not inhibit $Pb^{2+}$ immobilization and $Na_2Ca(SO_4)_2$ formed as well as 'hydroxypyromorphite'. Lead immobilization was inhibited by $CO_3^{2-}$ though 'hydroxypyromorphite' still formed. It is thought that the high pH associated with the high $CO_3^{2-}$ levels inhibited synthetic hydroxylapatite dissolution thereby limiting the concentration of P ions available to react with the $Pb^{2+}$.

In a second set of experiments, Ma *et al.* (1994*b*) found that the presence of other metals ($Al^{3+}$, $Cu^{2+}$, $Fe^{2+}$, $Cd^{2+}$, $Zn^{2+}$ and $Ni^{2+}$) in solution inhibited, but did not stop, $Pb^{2+}$ immobilization. 'Hydroxypyromorphite' still formed in the experiments and was detected using XRD, though the intensity of the peaks on the patterns decreased as the concentration of non-Pb ions in solution increased. The concentration of the metal ions in solution, other than $Pb^{2+}$, decreased in the presence of synthetic hydroxylapatite. However, no XRD patterns for metal-bearing phases other than 'hydroxypyromorphite' were obtained, though new precipitates were identified using SEM and coloured precipitates were observed in the $Cd^{2+}$, $Cu^{2+}$ and $Fe^{2+}$ experiments. In addition, the ranking of the metals in terms of increasing effectiveness at inhibiting $Pb^{2+}$ immobilization did not correspond to the ranking of metals in terms of the degree of metal immobilization by synthetic hydroxylapatite. This indicates that the inhibition of $Pb^{2+}$ immobilization was not simply a case of other metals competitively scavenging $PO_4^{3-}$ from solution. Metal immobilization, other than for Pb, could therefore have been due to both the formation of new phases and also adsorption.

In another set of experiments, Ma (1996) showed that $Na^+$ and $K^+$ had no effect on $Pb^{2+}$ immobilization by synthetic hydroxylapatite. However, greater $Ca^{2+}$ concentrations resulted in less $Pb^{2+}$ immobilization and calcite inhibited $Pb^{2+}$ removal, though this was independent of calcite concentration. For both the $Ca^{2+}$ and calcite experiments, the concentration of P in solution was reduced and it is thought that $Pb^{2+}$ immobilization was reduced due to a reduction in synthetic hydroxylapatite dissolution. In the case of calcite, this was probably due to the higher pH values associated with the calcite loading. Ma (1996) also investigated the effects of EDTA on $Pb^{2+}$ immobilization by synthetic hydroxylapatite. The EDTA had a far greater inhibitory effect on $Pb^{2+}$ immobilization than the inorganic ions. Solutions containing low concentrations of EDTA (EDTA/$Pb^{2+}$ molar ratio of 1) immobilized only 24–31% of the $Pb^{2+}$ whilst more concentrated EDTA solutions resulted in no immobilization of the $Pb^{2+}$. This was thought to be due to chelation of the $Pb^{2+}$ by EDTA and the formation of aqueous $Pb_2$-EDTA and PbH-EDTA complexes.

## 15.5 Metal phosphate formation in batch experiments using metal-bearing solids

As in the experiments with aqueous metal ions, the majority of these experiments are concerned with $Pb^{2+}$. Some of the experiments use Pb-contaminated soils rather than Pb compounds and these may contain other metal contaminants. One study looks at the potential for apatite to immobilize U in contaminated sediments (Arey *et al.*, 1999). The experiments are summarized in Table 15.3. As with the aqueous metal ion experiments, these studies are usually carried out by reacting the P source, the metal source and

TABLE 15.3 Summary of experiments examining phosphate formation by reaction between P and a metal-bearing material.

| | Rabinowitz (1993) | Laperche *et al.* (1996) | Ma *et al.* (1996) | Ma (1996) | Zhang *et al.** | Arey *et al.* (1999) | Hodson and Valsami-Jones (2000) |
|---|---|---|---|---|---|---|---|
| Mass/concentration of P source | 0.03 M $Na_2HPO_4$ | 0.02–1.0 g HA[†] | 0–2 g HA | 0–1 g HA | 0.004–12 g HA | 0.025–0.25 g HA | 0.025–0.5 g Bonemeal |
| Metal source and concentration | Soil (1233–600 μg Pb $g^{-1}$ | PbO, $PbCO_3$, Soil (829000 μg Pb $g^{-1}$) | Soil (2100 μg Pb $g^{-1}$) | Soil (2100 μg Pb $g^{-1}$) | Soil (34592 μg Pb $g^{-1}$), PbS, $PbSO_4$, $PbCO_3$, goethite | Sediment (1703–2100 μg U $g^{-1}$) | Soil (94, 36, 86 and 233 μg Zn, Ni, Cu and Pb $g^{-1}$) |
| Mass of metal source | 1g | 0.1–1.3 g | 6 g | 5 g | 0.005–1 g | 5 g | 5 g |
| Solution volume (ml) | 10 | 100 | 30 | 30 | 1000 | 25 | 25 |
| Initial pH | Ambient | 5–7 | 4.3 | 4.3 | 2–7 | 3.7–4 | 3.4 |
| Phosphate formation? | Yes | Yes | Probably | Probably | Yes | Probably | Probably |
| Metal immobilization? | Yes | Yes | Yes | Yes | Yes | Yes | Yes |
| Duration of experiments | 1–7 days | 2–14 days | 5–48 h | 2 h | 2 h | 2 h | 3.5–28 days |

* Zhang and Ryan (1998, 1999*a,b*), Zhang *et al.* (1997, 1998)
[†] HA = synthetic hydroxylapatite

solution for a given length of time and then measuring element concentrations in the solution and analysing reaction products by XRD. Metal phosphates form as both the phosphate and metal sources dissolve. The formation of metal phosphates may be limited by the dissolution rate of either the phosphate or metal source, or the lack of saturation of the solution with respect to a metal phosphate phase. Whilst data are usually consistent with metal immobilization by dissolution of the phosphate source and precipitation of a metal phosphate, and EMPA shows phases containing both metals and phosphates, the identification of the metal phosphates by XRD can be difficult due to the lack of either chemical or physical ideality of the metal bearing phase.

### *15.5.1 Experiments concerned with Pb*

Rabinowitz (1993) reacted $Na_2HPO_4$ and NaCl with Pb-contaminated soils. Suspensions were then centrifuged and the solids were reacted with either 10% $HNO_3$, 0.1 N HCl or 10% citric acid. The HCl and citric acid extracts removed more Pb from the untreated than the treated soils, but the additions of phosphate had no effect on the amount of $Pb^{2+}$ removed by the nitric acid extractions. No attempt was made to identify reaction products or determine reasons for changes in extractability of $Pb^{2+}$. Laperche *et al.* (1996) carried out experiments in which synthetic hydroxylapatite was reacted with a suspension of PbO or $PbCO_3$. 'Hydroxypyromorphite' formed in both experiments. Laperche *et al.* (1996) also reacted Pb-contaminated soil with synthetic hydroxylapatite at pH 5 and 7 $HNO_3$ acid. After 1 week the release of $Pb^{2+}$ into solution, compared to controls, had been reduced by 13% in the pH 7 experiment and by 81–99% in the pH 5 experiments. It is thought that 'hydroxypyromorphite' formation was limited by the relatively slow dissolution of synthetic hydroxylapatite at pH 7. 'Hydroxypyromorphite' was not detected by XRD in the residues from the pH 7 experiments. However, for the pH 5 experiments XRD patterns indicated that cerrusite ($PbCO_3$) dissolved, whilst a phase similar to 'hydroxypyromorphite' (but with slightly different peak shapes possibly due to the formation of mixed metal phosphates rather than 'hydroxypyromorphite') precipitated. Ma *et al.* (1993) and Ma (1996) reacted natural apatite and synthetic hydroxylapatite with suspensions of contaminated soil. $Pb^{2+}$ release into solution was reduced by the presence of apatite. The reduction in $Pb^{2+}$ release was most marked in the experiments using synthetic hydroxylapatite. No pyromorphites were identified in the experimental residues, possibly due to lack of ideal mineral chemistries and structures.

Zhang and co-workers (Zhang and Ryan, 1998; Zhang *et al.*, 1998; Zhang and Ryan, 1999*a,b*) carried out a series of experiments investigating the reaction between synthetic hydroxylapatite and various solid Pb phases (anglesite [$PbSO_4$], cerrusite [$PbCO_3$], galena [PbS] and soil particles) in the presence of 0.1 M $NaNO_3$ or 0.01 M NaCl solution. Two styles of experiment were carried out; constant pH experiments at pH 2–7, and variable pH experiments where pH started at 2 and was then increased incrementally to 7. In the constant pH experiments, synthetic hydroxylapatite dissolution increased with decreasing pH. At pH 2 pyromorphite did not form, probably because the solutions were not saturated with respect to pyromorphite. At higher pHs, synthetic hydroxylapatite dissolution was incomplete and, for reactions involving anglesite and cerrusite, this limited pyromorphite formation. Galena dissolution rates increased with increasing pH and so did pyromorphite formation. For anglesite and cerrusite more pyromorphite formed when more synthetic hydroxylapatite was added but with galena, pyromorphite formation was limited by the rate of galena dissolution rather than P supply. For reactions with the soil particles, synthetic hydroxylapatite amendments reduced the amount of $Pb^{2+}$ passing into solution at pH 4 and above but no pyromorphite was identified as having formed. Below pH 4, solutions were not saturated with respect to pyromorphite. It is thought that release of $Ca^{2+}$ from the soil inhibited synthetic hydroxylapatite dissolution. Above pH 4, $Pb^{2+}$ immobilization increased with pH due to both (assumed) pyromorphite formation (none was identified by XRD) and adsorption of $Pb^{2+}$ onto soil particles. In the dynamic pH experiments synthetic hydroxylapatite and the $Pb^{2+}$ bearing solids dissolved at low pH and then, as pH increased, pyromorphite precipitated out of solution. In all the dynamic experiments, XRD was used to identify pyromorphite as a reaction product.

### *15.5.2 Experiments concerned with other elements*

Arey *et al.* (1999) carried out experiments in which U-contaminated sediments were reacted with synthetic hydroxylapatite. The U concentration in solution decreased substantially, P concentration increased and pH increased. Uranium, Fe and Al phosphates were identified by EMPA but not by XRD. Uranium was not observed to be associated with residual apatite, suggesting that U precipitation, rather than adsorption, was responsible for U immobilization. Sequential extractions indicated that apatite addition resulted in U being transferred from the labile fraction to a less soluble fraction.

### *15.5.3 Multi-element experiments*

Hodson and Valsami-Jones (2000) mixed varying concentrations of bone meal (a microcrystalline apatite derived from the grinding of bones, similar in structure to synthetic hydroxylapatite) with metal-contaminated soil. Increased concentrations of bone meal reduced $Zn^{2+}$, $Ni^{2+}$ and $Cu^{2+}$ release (no $Pb^{2+}$ was released in any of the experiments) but no metal phosphates were identified using XRD. In addition, the bone meal amendments led to reduced bioavailability of the metals as assessed by DTPA (diethylenetriaminepentaacetic acid) and $CaCl_2$ extractions.

## 15.6 Metal phosphate formation in soil column and field experiments

Few studies have actually been carried out on additions of P to soils where the soil was maintained in its natural state and water content (rather than reacting the soil and P source in suspension). Some of the experiments summarized here investigated not only metal phosphate formation but also plant availability of metals after P treatment. The addition of phosphates to soils results in the immobilization of metals and reductions in the bioavailability of those metals. Although most of the experimental evidence is consistent with the formation of metal phosphates and EMPA analysis indicates the presence of metal phosphate compounds, it is difficult to identify these phases by XRD.

### *15.6.1 Pb-contaminated soils*

Laperche *et al.* (1997) grew Sudax (*Sorghum bicolor* L. Moench) in the presence of various sand, sand + phosphate source, and Pb-contaminated soil (37,026 μg $Pb^{2+}$ $g^{-1}$) + phosphate source combinations. The plants extracted $Pb^{2+}$ and P from the 'hydroxypyromorphite' when it was the only P source, but in the presence of synthetic hydroxylapatite or natural apatite the $Pb^{2+}$ content of the plants usually decreased by a factor of 10–100. The exception was when concentrations of synthetic hydroxylapatite were high (>3.2 μg P $g^{-1}$) when pyromorphite formed in the plant roots, possibly due to the high levels of P present. The EMPA and XRD indicated that either pyromorphite or a metal phosphate, thought to be a mixed Ca-Pb phosphate on the basis of the similarity of the XRD peak positions to those for 'hydroxypyromorphite', formed in the soil depending on the source of the P (synthetic hydroxylapatite or natural apatite respectively). Brown *et al.* (1999) investigated the efficacy of a variety of *in situ* treatments to reduce the bioavailability of $Pb^{2+}$ in soils. Soils were amended with various P, P + Fe, compost, or P + compost treatments. After three months Tall Fescue seeds were sown on the treated soils and were harvested three months later. Amendments of 3.2% P reduced plant $Pb^{2+}$ and *in vitro* accessible $Pb^{2+}$ most effectively. The rock phosphate amendments, some of the compost amendments and Fe-rich amendments also reduced $Pb^{2+}$ availability. The 1% P amendment did not significantly alter $Pb^{2+}$ availability although in previous, carefully controlled tests, it had done so. It was speculated that the amendment might have caused a decrease of pH leading to solubilization of $Pb^{2+}$ phases and reprecipitation of $Pb^{2+}$ as relatively soluble orthophosphates rather than as insoluble pyromorphite.

Ma *et al.* (1995) investigated the effects that different mixing methods between ground phosphate rock and 2,560 μg $Pb^{2+}$ $g^{-1}$ of contaminated soil had on $Pb^{2+}$ immobilization. In a series of soil column experiments, phosphate rock was either mixed uniformly with the soil, placed in a sandy layer at the base of the column, or placed in a separate container through which soil leachate was poured. In all cases $Pb^{2+}$ concentrations in leachates were reduced. In a separate set of experiments soil + synthetic hydroxylapatite mixtures were kept moist but not continuously drained. Soil water was extracted after different periods of incubation and it was shown that incubation time had little effect on the amount of $Pb^{2+}$ immobilization. Solid products were not identified but the results were consistent with metal immobilization by 'hydroxypyromorphite' formation.

In the only reported literature study where P amendments had little or no effect on $Pb^{2+}$ immobilization, Davenport and Peryea (1991) carried out leaching experiments using monoammonium phosphate and monocalcium phosphate on Pb-As-contaminated land. They found that the fertilizer amendments retarded but did not prevent soil $Pb^{2+}$ movement. In addition, the fertilizer applications caused a flushing of As from the system.

### *15.6.2 Mixed metal contaminated soils*

Cotter-Howells and Caporn (1996) took two soils from former mining areas containing 129,000 $\mu g\ g^{-1}$ $Pb^{2+}$ and 4,700 $\mu g\ g^{-1}$ $Zn^{2+}$ and 9,670 $\mu g\ g^{-1}$ $Pb^{2+}$ and 10,900 $\mu g\ g^{-1}$ $Zn^{2+}$ respectively and added either 10 wt.% $Na_2HPO_4$ to saturate the soil solution with P, or peat to give the soils 10% organic matter. *Agrostis capillaris* L. (a grass) was grown on the soil. After three months, EMPA indicated that Pb- and Zn-phosphate minerals formed in the phosphate-amended bulk soil and on plant roots. A Ca-Pb pyromorphite similar to that identified in previous studies by Cotter-Howells *et al.* (1994) was detected by XRD. Hettiarachchi and Pierzynski (1999) incubated soils with total metal concentrations ranging from 1200–9100 $\mu g\ Pb^{2+}\ g^{-1}$, 30–190 $\mu g\ Cd^{2+}\ g^{-1}$ and 4500–42600 $\mu g\ Zn^{2+}\ g^{-1}$ with seven different phosphate treatments (no P, rock phosphate, fertilizers) for up to 48 weeks at 20% gravimetric moisture content and 25°C. Air-dried samples were then analysed for pH and extracted by the *in vitro* bioaccessibility test (Ruby *et al.*, 1996). All the P treatments reduced $Pb^{2+}$ bioavailability under both stomach and intestinal conditions. Phosphate rock was the most efficient treatment at reducing bioavailability of $Zn^{2+}$ and $Cd^{2+}$, and for the majority of soils it was also the most effective at reducing the bioavailability of $Pb^{2+}$. Pyromorphite was present in the soils prior to treatment but XRD indicated that additional pyromorphite formed during treatment. Hodson and co-workers (Hodson and Valsami-Jones, 2000; Hodson *et al.*, 2000*a,b*) carried out leaching column experiments using bone meal (ground bone apatite) as a source of P on a range of soils containing up to 14,000 $\mu g\ Zn^{2+}\ g^{-1}$, 2,300 $\mu g\ Cu^{2+}\ g^{-1}$, 136,300 $\mu g\ Pb^{2+}\ g^{-1}$, 56 $\mu g\ Cd^{2+}\ g^{-1}$ and 50 $\mu g\ Ni^{2+}\ g^{-1}$ and with pH ranging from 2.7 to 7.1. The bone meal amendments reduced metal release from the soils. The results were consistent with metal phosphate formation and $Zn^{2+}$ and $Pb^{2+}$ phosphates were identified tentatively by XRD in two of the four soils investigated. $CaCl_2$ and DTPA extractions indicated that the bone meal amendments reduced the bioavailability of the metals. In some experiments release of $Cu^{2+}$ and the bioavailability of $Cu^{2+}$ increased with bonemeal amendments. This was thought to be due to formation of highly soluble Cu-organic complexes at circum-neutral (6–7) pH.

## 15.7 Concluding remarks

Much work has been done to show that the formation of metal phosphates, particularly, Pb phosphates, has the potential to be a sustainable, economic method for the remediation of metal-contaminated land.

In the case of Pb, sufficient work has been carried out in the laboratory that progress with this technique will probably only be made through careful field trials. For other metal phosphates perhaps more laboratory work could be carried out to show that metal phosphates can form and are sufficiently stable to immobilize pollutant metals. In particular the stability of metal phosphates in the presence of organic materials and microorganisms requires more attention. In addition, experimental work on the characterization of mixed-metal phosphates is necessary to aid identification of such phases in phosphate-treated soils in the future.

## Acknowledgements

The authors acknowledge funding from The BOC Foundation, the Environment Agency and the Leverhulme Trust. Thoughtful reviews by Gary Pierzynski (Kansas State University) and an anonymous referee are also gratefully acknowledged.

## References

Alloway, B.J. and Ayres, D.C. (1997) *Chemical Principles of Environmental Pollution*. Blackie, London.

Appelo, C.A.J. and Postma, D. (1993) *Geochemistry, Groundwater and Pollution*. Balkema, Rotterdam.

Arey, J.S., Seaman, J.C. and Bertsch, P.M. (1999) Immobilization of uranium in contaminated sediments by hydroxyapatite addition. *Environ. Sci. Technol.*, **33**, 337–42.

Bigi, A., Ripamonti, A., Brückner, S., Gazzano, M., Roveri, N. and Thomas, S.A. (1989) Structure refinements of lead-substituted calcium hydroxyapatite by X-ray powder fitting. *Acta Crystallogr.*, **B45,** 247–51.

Brooks, R.R. (1997) Plant hyperaccumulators of metals and their role in mineral exploration, archaeology, and land remediation. Pp. 123–33 in: *Remediation of Soils Contaminated with Metals* (I.K. Iskandar and D.C. Adriano, editors). Science Reviews, Northwood, UK,

Brown, S.L., Chaney, R. and Berti, B. (1999) Field test of amendments to reduce the in situ availability of soil lead. Pp. 506–7 in: *Proc. 5th Int. Conf. on the Biogeochemistry of Trace Elements, Vienna.*

Brückner, S., Lusvardi, G., Menabue, L. and Saladini, M. (1995) Crystal structure of lead hydroxyapatite from powder X-ray diffraction data. *Inorg. Chim. Acta,* **236**, 209–12.

Chen, X., Wright, J.V., Conca, J.L. and Peurrung, L.M. (1997) Effects of pH on heavy metal sorption on mineral apatite. *Environ. Sci. Technol.*, **31**, 624–31.

Cotter-Howells, J. (1996) Lead phosphate formation in soils. *Environ. Pollut.*, **93**, 9–16.

Cotter-Howells, J. and Caporn, S. (1996) Remediation of contaminated land by formation of heavy metal phosphates. *Appl. Geochem.*, **11**, 335–42.

Cotter-Howells, J.D., Champness, P.E., Charnock, J.M. and Pattrick, R.A.D. (1994) Identification of PM in mine-waste contaminated soils by ATEM and EXAFS. *Eur. J. Soil. Sci.*, **45**, 393–402.

Cox, W.J. and Rains, D.W. (1972) Effects of lime on lead uptake by five plant species. *J. Environ. Qual.*, **1**, 167–9.

Davenport, J.R. and Peryea, F.J. (1991) Phosphate fertilizers influence leaching of lead and arsenic in a soil contaminated with lead arsenate. *Water Air Soil Pollut.*, **57-58**, 101–10.

Davis, D., Drexler, J.W., Ruby, M.V. and Nicholson, A. (1993) Micromineralogy of mine wastes in relation to lead bioavailability, Butte, Montana. *Environ. Sci. Technol.*, **27**, 1415–25.

Fry, J.C., Gadd, G.M., Herbert, R.A., Jones, C.W. and Watson-Craik, I.A. (1992) *Microbial Control of Pollution*. Cambridge University Press, UK.

Gadd, G.M. (1992) Microbial control of heavy metal pollution. Pp. 59–88 in: *Microbial Control of Pollution* (J.C. Fry, G.M. Gadd, R.A. Herbert, C.W. Jones and I.A. Watson-Craik, editors). Cambridge University Press, UK.

Gauglitz, R., Holterdorf, M., Franke, W. and Marx, G. (1992) Immobilization of heavy metals by hydroxylapatite. *Radiochim. Acta.*, **58/59**, 253–7.

Hettiarachchi, G.M. and Pierzynski, G.M. (1999) Effect of phosphorus and other soil amendments on soil lead, cadmium, and zinc bioavailability. Pp. 514–5 in: *Proc. 5$^{th}$ Int. Conf. on Biogeochemistry of Trace Elements, Vienna.*

Hodson, M.E. and Valsami-Jones, E. (2000) *Remediation of toxic metal pollution in soil using bone meal.* R&D Technical Report P234. Environment Agency. ISBN 1-857-05033-9.

Hodson, M.E., Valsami-Jones, E. and Cotter-Howells, J.D. (2000*a*) Bonemeal additions as a remediation treatment for metal contaminated soil. *Environ. Sci. Technol.*, **36**, 3501–7.

Hodson, M.E., Valsami-Jones, E., Cotter-Howells, J.D., Dubbin, W.E., Kemp, A.J., Thornton, I. and Warren. A. (2000*b*) Effects of bone meal (calcium phosphate) amendments on metal release from contaminated soils – a leaching column study. *Environ. Pollut.* (in press).

Iskandar, I.K. and Adriano, D.C. (1997) *Remediation of soils contaminated with metals.* Northwood, UK.

Jurinak, J.J. and Inouye, T.S. (1962) Some aspects of zinc and copper phosphate formation in aqueous systems. *Soil. Sci. Soc. Proc.*, **26,** 144–7.

Keeton, C.M. (1937) Influence of lead compounds on the growth of barley. *Soil Sci.*, **43**, 401–11.

Laperche, V., Traina, S.J., Gaddam, P. and Logan, T.J. (1996) Chemical and mineralogical characterizations of Pb in a contaminated soil: Reactions with synthetic apatite. *Environ. Sci. Technol.*, **30**, 3321–6.

Laperche, V., Logan, T.J., Gaddam, P. and Traina, S.J. (1997) Effect of apatite amendments on plant uptake of lead from contaminated soil. *Environ. Sci. Technol.*, **31**, 2745–53.

Lewicke, C.K. (1972) Treating lead and fluoride wastes. *Environ. Sci. Technol.*, **6,** 321–2.

Ma, L.Q. (1996) Factors influencing the effectiveness and stability of aqueous lead immobilization by hydroxyapatite. *J. Environ. Qual.*, **25**, 1420–9.

Ma, Q.Y., Traina, S.J. and Logan, T.J. (1993) In situ lead immobilization by apatite. *Environ. Sci. Technol.*, **27**, 1803–10.

Ma, Q.Y., Logan, T.J., Traina, S.J. and Ryan, J.A. (1994*a*) Effects of $NO_3^-$, $Cl^-$, $F^-$, $SO_4^{2-}$ and $CO_3^{2-}$ on $Pb^{2+}$ immobilization by hydroxyapatite. *Environ. Sci. Technol,*. **28**, 408–18.

Ma, Q.Y., Logan, T.J., Traina, S.J. and Ryan, J.A. (1994*b*) Effects of aqueous Al, Cd, Cu, Fe(II), Ni and Zn on Pb immobilization by hydroxyapatite. *Environ. Sci Technol.*, **28**, 1219–28.

Ma, Q.Y., Logan, T.J. and Traina, S.J. (1995) Lead immobilization from aqueous solutions and contaminated soils using phosphate rocks. *Environ. Sci. Technol.*, **29**, 1118–26.

MacLean, A.J.R., Halstead, R.L. and Finn, B.J. (1969) Extractability of added lead in soils and its concentration in plants. *Canad. J. Soil Sci.*, **49**, 327–34.

Middleburg, J.J. and Comans, R.N.J. (1991) Sorption of cadmium on hydroxyapatite. *Chem. Geol.*, **90**, 45–53.

Morin, G., Juillot, F., Ildefonse, P., Dumat, C., Benedetti, M., Chevallier, P., Brown Jr., G.E. and Calas, G. (1999) Chemical forms of lead in a soil developed on a Pb-mineralized sandstone (Ardèche, France). Pp. 1094–5 in: *Proc. 5th Int. Conf. on the Biogeochemistry of Trace Elements, Vienna.*

Napper, D.H. and Smythe, B.M. (1966) The dissolution kinetics of hydroxyapatite in the presence of kink poison. *J. Dental Res.*, **45**, 1775–84.

Nriagu, J.O. (1972) Lead orthophosphates – I. Solubility and hydrolysis of secondary lead phosphates. *Inorg. Chem.*, **11**, 2499–503.

Nriagu, J.O. (1973) Lead orthophosphates – II. Stability of chloropyromorphite at 25°C. *Geochim. Cosmochim. Acta,* **37**, 367–77.

Nriagu, J.O. (1974) Lead orthophosphates – IV. Formation and stability in the environment. *Geochim. Cosmochim. Acta,* **38**, 887–98.

Nriagu, J.O. (1984) Formation and stability of base metal phosphates in soils and sediments. Pp. 318–329 in: *Phosphate Minerals* (J.O. Nriagu and P.B. Moore, editors). Springer-Verlag, Berlin.

Rabinowitz, M.B. (1993) Modifying soil lead bioavailability by phosphate addition. *Bull. Environ. Contam. Toxicol.*, **51**, 438–44.

Ruby, M.V., Davis, A., Link, T.E., Schoof, R., Chaney, R.L., Freeman, G.B. and Bergstrom, P. (1993) Development of an *in vitro* screening test to evaluate the *in vivo* bioaccessibility of ingested mine-waste lead. *Environ. Sci. Technol.*, **27**, 2870–7.

Ruby, M.V., Davis, A. and Nicholson, A. (1994) *In situ* formation of lead phosphates in soils as a method to immobilize lead. *Environ. Sci. Technol.*, **28**, 646–54,

Ruby, M.V., Davis, A., Schoof, R., Eberle, S. and Sellstone, C.M. (1996) Estimation of bioavailability using a physiologically based extraction test. *Environ. Sci. Technol.*, **30**, 42–430.

Salt, D., Smith, R.D. and Raskin, I. (1998) Phytoremediation. *Ann. Rev. Plant Physiol. Plant Mol. Biol.*, **49**, 643–68.

Santillan-Medrano, J. and Jurinak, J.J. (1975) The chemistry of lead and cadmium in soil: solid phase formation. *Soil Sci. Soc. Amer. Proc.*, **39**, 851–6.

Sauvé, S., Mcbride, M. and Hendershot, W. (1998) Lead phosphates solubility in water and soil suspensions. *Environ. Sci. Technol.*, **32**, 388–93.

Sayer, J.A., Kierans, M. and Gadd, G.M. (1997) Solubilization of some naturally occurring metal-bearing minerals, limescale and lead phosphate by *Aspergillus niger*. *FEMS Microbiol. Lett.*, **154**, 29–35.

Sayer, J.A., Cotter-Howells, J.D., Watson, C., Hillier, S. and Gadd, G.M. (1999) Lead mineral transformation by fungi. *Curr. Biol.*, **9**, 691–4.

Valsami-Jones, E., Ragnarsdottir, K.V., Putnis, A., Bosbach, D., Kemp, A.J. and Cressey, G. (1998) the dissolution of apatite in the presence of aqueous metal cations at pH 2–7. *Chem. Geol.*, **151**, 215–33.

Wagman, D.D., Evans, W.H., Halow, I., Parker, V.B., Bailey, S.M. and Schumm, R.H. (1968) *Selected values of chemical thermodynamic properties*. NBS Techn. Note 270–3.

Wood, P.A. (1997) Remediation methods for contaminated sites. Pp. 47–71 in: *Contaminated Land and its Reclamation* (R.E. Hester and R.M. Harrison, editors). Thomas Telford Publishing, London.

Xu, Y. and Schwartz, F.W. (1994) Lead immobilization by hydroxyapatite in aqueous solution. *J. Contam. Hydrol.*, **15**, 187–206.

Young, P.J., Pollard, S. and Crowcroft, P. (1997) Overview: Context, calculating risk and using consultants. Pp. 1–24 in: *Contaminated Land and its Reclamation* (R.E. Hester and R.M. Harrison, editors). Thomas Telford Publishing, London.

Zhang, P. and Ryan, J.A. (1998) Formation of pyromorphite in anglesite-hydroxyapatite suspensions under varying pH conditions. *Environ. Sci. Technol.,* **32,** 3318–24.
Zhang, P. and Ryan, J.A. (1999*a*) Formation of chloropyromorphite from galena (PbS) in the presence of hydroxyapatite. *Environ. Sci. Technol.,* **33**, 618–24.
Zhang, P. and Ryan, J.A. (1999*b*) Transformation of Pb(II) from cerrusite to chloropyromorphite in the presence of hydroxyapatite under varying conditions of pH. *Environ. Sci. Technol.,* **33**, 625–30.
Zhang, P., Ryan, J.A. and Bryndzia, L.T. (1997) Pyromorphite formation from goethite adsorbed lead. *Environ. Sci. Technol.,* **31**, 2673–8.
Zhang, P., Ryan, J.A. and Yang, J. (1998) In vitro soil Pb solubility in the presence of hydroxyapatite. *Environ. Sci. Technol.* **32,** 2763–8.

CHAPTER SIXTEEN

# Section 4: Minerals and waste management

L. S. CAMPBELL

*Telford Institute of Environmental Systems, Salford University, Frederick Road, Salford M6 6PU, UK (E-mail: L.S.Campbell@salford.ac.uk)*

## 16.1 Introduction

The purpose of this introduction is to outline briefly why and how minerals are currently used in various waste disposal and waste management situations, and to introduce three papers on specific aspects of minerals and wastes. A further purpose is to highlight the wider potential of minerals in connection with waste management, by reference to examples from the current and recent literature. Many recent papers on wastes and the geosphere, of relevance to the mineralogist, can be found in Metcalfe and Rochelle, (1999).

At the heart of utilization of any type of material, for any purpose, is an appreciation of the interactive characteristics of that material. In the context of minerals and waste disposal, these characteristics include the intrinsic physical properties of minerals themselves (e.g. density), and in addition, the attributes governing the behaviour of a mineral in relation to its environment. Hence, mineral stability is dependent on both the nature of the mineral itself, and the nature of the surrounding environment. For the purpose of this discussion, the term also encompasses many geotechnical properties of minerals or mineral aggregates, such as plasticity.

## 16.2 Interactive characteristics of minerals

Interactive characteristics of prime importance in the field of mineralogy and waste management are identified as: cation exchange capacity; adsorption capacity; physical strength and plasticity; stability (resistance to alteration); water retention; neutralization capability; density and swelling behaviour; diffusion resistance; and permeability. Reference to most of these characteristics will be variously found in the three research contributions that follow in this section. Many requirements for effective physical and chemical barriers to wastes (as outlined by Hermanns Stengele and Plötze, 2000), are satisfied by these characteristics, individually or collectively.

Campbell, L.S. (2000) Section 4: Minerals and waste management. Pp. 313–318 in: *Environmental Mineralogy: Microbial Interactions, Anthropogenic Influences, Contaminated Land and Waste Management* (J.D. Cotter-Howells, L.S. Campbell, E. Valsami-Jones and M. Batchelder, editors). Mineralogical Society Series, **9**. Mineralogical Society, London. ISBN 0 903056 20 8.

Considering the essential requirements of materials used in waste management, there are three mineral groups that eclipse the rest in terms of a summation of their useful characteristics. These groups are the zeolites, the clay minerals and the carbonates. Reasons for their prominence are explained in the paragraphs that follow. Also, examples from recent lines of research are given, with the aim of putting into context each of the papers in this thematic set, and of acknowledging studies as to the potential use of other mineral groups, such as the oxides, in waste management applications.

## 16.3 Zeolites

Outstripping all other minerals in terms of their ion exchange properties, are, of course, the zeolites. As well as their abilities in selective ion exchange, their usefulness in waste applications sometimes relies on their silicate framework stabilities over wide ranges of pH, for example, or when subject to significant doses of $\alpha$, $\beta$ or $\gamma$-radiation (**Dyer, chapter 17**), or to temperature changes (Chipera and Bish, 1997). Their non-swelling behaviour is another advantageous characteristic in their use within solid waste composites such as cements and glasses. Further discussions of waste applications of zeolite minerals can be found in the regularly produced volumes on natural zeolites (e.g. Kirov *et al.*, 1997). However, in the current volume, a full and comprehensive examination of zeolite applications in the treatment of nuclear wastes is given by Dyer (2000). In addition to reviewing zeolite usage for nuclear facilities globally, he outlines studies concerning the effect of zeolites on radionuclides in the geo- and biospheres, including *in vivo* uses in contaminated humans. These wider studies tend to be generated as a result of the consequences of nuclear accidents. With 309 references, 12 tables and numerous specific examples from around the world, the contribution can be considered a definitive work on the subject at this point in time.

## 16.4 Clay minerals

As with the zeolites, it is the ion exchange properties of clays that often constitute the principal reason for their usage in waste disposal. Unlike the zeolites, however, the swelling behaviours, plasticities, diffusion resistance and low permeabilities of clays provide distinct advantages over other materials, and in particular, for waste applications. In this volume, **Donohew *et al.* (chapter 18)** report on the physical behaviour of a variety of clays in response to gas injection. The series of experiments was undertaken with water-saturated clay pastes, and have relevance, for example, to gas migration through landfill liners. The authors identified distinct mechanisms of gas migration, and found systematic relationships between gas entry pressure, water content and clay composition.

Clay liners in landfill applications have been employed for several years now, but important design developments have occurred in recent years. These developments essentially relate to new research on geosynthetic clay liners (GCLs) as opposed to the earlier compacted clay liners (CCLs). **Rowe and Lake (chapter 19)** review the developments in the third contribution to this section. Frequent reference is made to the interactive characteristics of clay components with the synthetic membranes and the leachates. Such intense and specific characterization of the geochemical and engineering performance of GCLs serves to highlight the increasing need for tailor-made designs in modern times. For a comparison of methods and strategies of waste management by landfill that are in use around the world, the reader is referred to Hermanns Stengele and Plötze (2000).

## 16.5 Carbonates

Calcite, in its abundant occurrence as limestone or marble, is widely and cheaply available in most parts of the world. Its characteristic of prime relevance to waste management is its capacity to neutralize acidity. As demonstrated by Jambor (2000), no other mineral group performs in this respect, to the same degree as the carbonates. Very specifically, carbonates (as limestone), are used in mine waste applications, where acidity has been generated through oxidation of pyrite (Keith and Vaughan, 2000). A recent example at the Martha gold mine in New Zealand showed that both limestone and dolomite (and, indeed, fluidized bed boiler ash) were successful in neutralizing acidity in the pyritic pitwall rock (Gurung *et al.,* 2000). But these approaches to waste management are not without problems. Gurung *et al.* (2000) reported the enclosure, and hence, isolation, of a portion of the total carbonate used, by growth of secondary mineral phases (i.e. the oxyhydroxide products of reaction).

Geochemical reactions involving carbonates are also known in landfill situations. Manning (2000) predicted that carbonates, among other phases, should precipitate from landfill leachates as the leachates evolve with anaerobic microbiological activity. Sampling and analysis demonstrated a correct prediction, with the detection of calcite, gypsum and iron minerals. Comparisons were made with similar processes in natural sedimentary systems.

In a different type of application, it is the physical strength of carbonate cements that is important, and it is the reactivity that potentially causes problems. Klich *et al.* (1999) undertook a six-year study on the fate of hazardous wastes solidified with carbonate cements. These authors found that both physical and chemical alteration occurred, similar to that experienced with the degradation of concretes.

## 16.6 Other mineral groups

Mineral groups other than the zeolites, clays and carbonates also deserve some attention, since there continues to be a number of active areas of research being undertaken, with potential application in waste management. Long recognized for their sorption characteristics, the Fe oxides/oxyhydroxides have been the subject of some recent studies in connection with wastes. Collins *et al.* (1998) identified co-ordination spheres of Sr sorbed to goethite surfaces at high pH, and it was found that inner sphere sorption occurred. The study therefore has relevance to the nuclear industry in relation to the geochemical mobility of $^{90}Sr$. In another study, Meima and Comans (1998) found that Sb was sorbed or coprecipitated with Fe and Al hydroxides, when salts of these common metals were added to the bottom ash of municipal solid waste incinerators. The results appear to suggest that Sb leaching can be controlled in this context. Other potentially toxic metals from municipal solid waste processes have been immobilized in experiments using soluble phosphate (Eighmy *et al.*, 1998). The dry phosphate was mixed with vitrification dusts to produce various insoluble phosphates, including those of the apatite family. The stability of phosphates in the context of contaminated land remediation is discussed in Hodson *et al.* (2000).

The stability, or resistance to alteration, of some minerals is an interactive characteristic that has drawn attention to minerals of the perovskite ($ABX_3$) and zirconolite ($A_2B_2O_7$) groups. These are being used for incorporation into solid waste composites such as synroc (Curtis, 2000). For nuclear wastes, radiation damage is a key area of research, as exemplified in the study by Kuramoto *et al.* (1998). Using actinide-doped perovskites, Kuramoto *et al.* (1998) reported that leach rates using acidic solutions increased with increases in accumulated doses of $\alpha$-radiation, suggesting structural damage to perovskite. Gieré *et al.* (1998) surveyed 300 zirconolites to ascertain a natural compositional range, and to determine some of the controls on substitutions involving actinides. It was suggested that redox conditions play a role since $Fe^{2+}$ is involved in charge balancing. The generally high durability of zirconolite was attributed to the relatively limited numbers of important substitutions observed in the natural samples.

## 16.7 Conclusions

From this overview of present research on mineralogy and waste management, it is clear that any evaluation of specific waste strategies must take account of long term mineral-waste-environment interactions. Waste management applications of minerals constitute classic examples of 'geochemical engineering', another relatively new field emerging at the start of the twenty-first century. Mineralogists and geochemists have a central role to play in these rapidly advancing fields, and must continue to value and provide the fundamental research that will underpin novel developments in the future.

## Acknowledgements

The editorial handling of the 'Minerals and Waste Management' section was undertaken by Meryl Batchelder and Kevin Murphy.

## References

Chipera, S.J. and Bish, D.L. (1997) Equilibrium modeling of clinoptilolite-analcime equilibria at Yucca Mountain, Nevada, USA. *Clays Clay Miner.,* **45,** 226–39.

Collins, C.R., Sherman, D.M. and Ragnarsdottir, K.V. (1998) The adsorption mechanism of $Sr^{2+}$ on the surface of goethite. *Radiochim. Acta,* **81,** 201–6.

Curtis, C.D. (2000) Mineralogy in long term nuclear waste management. Pp. 333–50 in: *Environmental Mineralogy* (D.J. Vaughan and R.A. Wogelius, editors). EMU Notes in Mineralogy, **2**. Eötvös University Press, Budapest.

Donohew, A.T., Horseman, S.T. and Harrington, J.F. (2000) Gas entry into unconfined clay pastes at water contents between the liquid and plastic limits. Pp. 369–94 in: *Environmental Mineralogy: Microbial Interactions, Anthropogenic Influences, Contaminated Land and Waste Management* (J.D. Cotter-Howells, L.S. Campbell, E. Valsami-Jones and M. Batchelder, editors). Mineralogical Society Series, **9**. Mineralogical Society, London.

Dyer, A. (2000) Applications of natural zeolites in the treatment of nuclear wastes and fallout. Pp. 319–68 in: *Environmental Mineralogy: Microbial Interactions, Anthropogenic Influences, Contaminated Land and Waste Management* (J.D. Cotter-Howells, L.S. Campbell, E. Valsami-Jones and M. Batchelder, editors). Mineralogical Society Series, **9**. Mineralogical Society, London.

Eighmy, T.T., Crannell, B.S., Krzanowski, J.E., Butler, L.G., Cartledge, F.K., Emery, E.F., Eusden, J.D., Shaw, E.L. and Francis, C.A. (1998) Characterization and phosphate stabilization of dusts from the vitrification of MSW combustion residues. *Waste Management,* **18,** 513–24.

Gieré, R., Williams, C.T. and Lumpkin, G.R. (1998) Chemical characteristics of natural zirconolite. *Schweizerische Mineral. Petrogr. Mitt.,* **78,** 433–59.

Gurung, S.R., Stewart, R.B., Gregg, P.E.H. and Bolan, N.S. (2000) An assessment of requirements of neutralising materials of partially oxidised pyritic mine waste rock. *Aust. J. Soil Res.,* **38**, 329–44.

Hermanns Stengele, R. and Plötze, M. (2000) Suitability of minerals for controlled landfill and containment. Pp. 291–331 in: *Environmental Mineralogy* (D.J. Vaughan and R.A. Wogelius, editors). EMU Notes in Mineralogy, **2**. Eötvös University Press, Budapest.

Hodson, M.E., Valsami-Jones, E. and Cotter-Howells, J.D. (2000) Metal phosphates and remediation of contaminated land. Pp. 291–311 in: *Environmental Mineralogy: Microbial Interactions, Anthropogenic Influences, Contaminated Land and Waste Management* (J.D. Cotter-Howells, L.S. Campbell, E. Valsami-Jones and M. Batchelder, editors). Mineralogical Society Series, **9**. Mineralogical Society, London.

Jambor, J. (2000) The relationship of mineralogy to acid- and neutralization-potential values in ARD. Pp. 141–59 in: *Environmental Mineralogy: Microbial Interactions, Anthropogenic Influences, Contaminated Land and Waste Management* (J.D. Cotter-Howells, L.S. Campbell, E. Valsami-Jones and M. Batchelder, editors). Mineralogical Society Series, **9**. Mineralogical Society, London.

Keith, C. and Vaughan, D.J. (2000) Mechanisms and rates of sulphide oxidation in relation to the problems of acid mine drainage. Pp. 117–39 in: *Environmental Mineralogy: Microbial Interactions, Anthropogenic Influences, Contaminated Land and Waste Management* (J.D. Cotter-Howells, L.S. Campbell, E. Valsami-Jones and M. Batchelder, editors). Mineralogical Society Series, **9**. Mineralogical Society, London.

Kirov, G., Filizova, L., and Petrov, O. (editors) (1997) *Natural Zeolites Sofia '95.*

Klich, I., Wilding, L.P., Drees, L.R., Landa, E.R. (1999) Importance of microscopy in durability studies of solidified and stabilized contaminated soils. *Soil Sci. Soc. Amer. J.,* **63**, 1274–83.

Kuramoto, K., Mitamura, H., Banba, T., Muraoka, S. (1998) Development of ceramic waste forms for actinide-rich waste. Radiation stability of perovskite and phase and chemical stabilities of Zr- and Al-based ceramics. *Progress Nucl. Energy,* **32**, 509–16.

Manning, D.A.C. (2000) Carbonates and oxalates in sediments and landfill: monitors of death and decay in natural and artificial systems. *J. Geol. Soc.,* **157**, 229–38.

Meima, J.A. and Comans, R.N.J. (1998) Reducing Sb-leaching from municipal solid waste incinerator bottom ash by addition of sorbent minerals. *J. Geochem. Expl.,* **62**, 299–304.

Metcalfe, R. and Rochelle, C.A. (editors) (1999) Chemical containment of waste in the geosphere. *Spec. Publ.,* **157**. Geological Society, London.

Rowe, R.K. and Lake, C.B. (2000) Geosynthetic Clay Liners (GCLs) for municipal solid waste landfills. Pp. 395–406 in: *Environmental Mineralogy: Microbial Interactions, Anthropogenic Influences, Contaminated Land and Waste Management* (J.D. Cotter-Howells, L.S. Campbell, E. Valsami-Jones and M. Batchelder, editors). Mineralogical Society Series, **9**. Mineralogical Society, London.

CHAPTER SEVENTEEN

# Applications of natural zeolites in the treatment of nuclear wastes and fall-out

A. DYER

*Chemistry Division, Science Research Institute, University of Salford, Salford M5 4WT, UK (E-mail: a.dyer@salford.ac.uk)*

**ABSTRACT**

Inorganic ion-exchangers are the preferred media for the treatment of nuclear wastes. Because of this, natural zeolites have received considerable attention as potential highly selective cation exchangers. Reasons for this, based upon the properties of zeolites, are described. This article provides an introduction to the nature and sources of wastes produced by the nuclear industry and reviews the earlier work in the 1950s which investigated the potential of natural zeolites as scavengers of specific radioisotopes from aqueous waste streams. Comprehensive coverage is given to the type and sources of the zeolites used as well as the range of radioisotopes investigated. Examples of industrial plants using natural zeolites to treat various wastes are given. Their future potential as barrier and back-fill or overpack materials is cited.

Cases where these materials have been used successfully to clean up after nuclear accidents are documented (including Chernobyl and Three Mile Island) with a discussion of the compatibility of zeolites to ceramic and cementitious waste forms, coupled with their safe disposal.

The scavenging of fall-out products from living hosts, including man, is described, together with the influence of zeolite amendments to soils and growth media.

Finally work in which zeolites have been studied as sorbents for gaseous emissions from nuclear plant is commented on briefly.

## 17.1 Introduction

The world-wide quest for power has resulted in more than 400 working nuclear reactors on the Earth's surface. Combined with the current and past emphasis on nuclear weaponry, these stations have created an enormous volume of undesirable radioactive isotopes. The isotopes need to be processed and contained in such a way as to reduce their potential harmfulness to man to the smallest possible level. Such processing must be carried out so as to satisfy the increasingly more stringent requirements of the governmental regulatory agencies charged with monitoring the processing and storage of radioisotopes.

Dyer, A. (2000) Applications of natural zeolites in the treatment of nuclear wastes and fall-out. Pp. 319–368 in: *Environmental Mineralogy: Microbial Interactions, Anthropogenic Influences, Contaminated Land and Waste Management* (J.D. Cotter-Howells, L.S. Campbell, E. Valsami-Jones and M. Batchelder, editors). Mineralogical Society Series, **9**. Mineralogical Society, London. ISBN 0 903056 20 8.

At the present time, the sources of isotopic waste from nuclear installations include: nuclear reactors; facilities for producing nuclear fuel or weapons; nuclear fuel reprocessing plants; clean-up from historic use of nuclear power/ weaponry; clean-up from nuclear accidents; decommissioning of reactors no longer in use; and commercial production facilities for radioisotopes. The radioisotopes commonly present in nuclear effluents (liquid or gaseous) are $\alpha$ emitters, such as isotopes of Th or U; $\gamma$ emitters, such as $^{60}Co$ and $^{137}Cs$; or $\beta$ emitters, such as $^{3}H$ (tritium) and $^{90}Sr$. Examples of these and other isotopes are listed in Table 17.1, along with their half lives and sources.

Gaseous emissions occur primarily in the coolant gases of power stations or in the off-gas of reprocessing plants. The nature of liquid effluents from the nuclear industry ranges from the comparatively simple waters from nuclear coolants or turbines to the mixed solvent systems used in isotope production

TABLE 17.1 Radioisotopes of environmental concern (not exclusive)[1].

| Isotope | Half-life | Type of radiation emitted | Source |
|---|---|---|---|
| $^{3}H$ | 12.26 y | $\beta^-$ | Reprocessing, fission, neutron activation of B and Li [LMFBR, BWR, HTGR, AGR, Magnox and He(HTGR)][2] |
| $^{41}Ar$ | 110 months | $\beta^-, \gamma$ | Fission product |
| $^{60}Co$ | 5.26 y | $\beta^-, \gamma$ | Neutron activation of reactor materials |
| $^{85}Kr$ | 10.6 y | $\beta^-, \gamma$ | Fission product |
| $^{90}Sr/Y$ | 28 y | $\beta^-$ | Fission product |
| $^{95}Zr$, $^{95}Nb$ | 65, 35 days | $\gamma$ | Fission product, nuclear activation of reactor materials |
| $^{99}Tc$ | $2.1 \times 10^{5}$ y | $\beta^-$ | Fission product, nuclear fuel production |
| $^{106}Ru$ | 1 y | $\gamma$ | Fission product |
| $^{129}I$ | $16 \times 10^{16}$ y | $\beta^-, \gamma$ | Fission product, reprocessing |
| $^{134, 137}Cs$ | 2.30 y, 30 y | $\beta^-, \gamma$ | Fission products |
| $^{144}Ce$ | 285 days | $\beta^-, \gamma$ | Fission product |
| $^{154}Eu$ | 16 y | $\beta^-, \gamma$ | Fission product |
| $^{232}Th$ | $1.4 \times 10^{4}$ y | $\alpha$ | Nuclear fuel product and reprocessing |
| $^{234}Th$ | 24.1 days | $\beta^-, \gamma$ | Nuclear fuel product and reprocessing |
| $^{235}U$ | $7.13 \times 10^{8}$ y | $\alpha, \gamma$ | Nuclear fuel product and reprocessing |
| $^{238}U$ | $4. 51 \times 10^{9}$ y | $\alpha, \gamma$ | Nuclear fuel product and reprocessing |
| $^{237}Np$ | $2.2 \times 10^{6}$ y | $\alpha, \gamma$ | Nuclear fuel product and reprocessing |
| $^{239}Pu$ | $2.44 \times 10^{4}$ y | $\alpha, \gamma$ | Nuclear fuel product and reprocessing |
| $^{241}Am$ | 458 y | $\alpha, \gamma$ | Nuclear fuel product and reprocessing |
| $^{244}Cm$ | 18 y | $\alpha, \gamma$ | Nuclear fuel product and reprocessing |

[1]Decommissioning of nuclear facilities will involve consideration of these isotopes and many more arising from neutron activation of constructional materials
[2]BWR – Boiling Water Reactor; PWR – Pressurized Water Reactor; LMFBR – Liquid Metal Fast Breeder Reactor; HTGR – High Temperature Gas-cooled Reactors; AGR – Advanced Gas-cooled Reactors

and reprocessing. Other common effluents are high-salt evaporator wastes and high- and low-pH solutions from fuel-rod cooling ponds and nitric acid reprocessing streams, respectively. Materials able to scavenge tiny quantities of radioactive species from hugh quantities of bulk gaseous or aqueous effluent are required.

This chapter concentrates on the utility of natural zeolites as selective ion exchangers, with some consideration being given to the lesser-used ability of dehydrated natural zeolites to act as selective molecular sorbents. The removal of radioisotopes from aqueous waste streams is treated in detail and in addition the following applications of natural zeolites are considered: (1) as potential 'backfill' or 'overpack' material to buffer storage facilities from the environment; (2) in the separation of fission products for use in research and industry; (3) as decontamination materials in nuclear accidents; (4) in the removal of radioisotopes from animals and human beings; (5) as adsorbents for treating gaseous effluents from nuclear facilities; and (6) as materials compatible with waste disposal/storage protocols.

## 17.2 Historical perspective

The needs for selective ion exchangers (organic and inorganic) in the nuclear industry were stated succinctly in a 1967 publication of the International Atomic Energy Agency (IAEA, 1967). It reviewed experience in the United States, the United Kingdom and elsewhere dating back to that at Chalk River, Canada, in 1947 (IAEA, 1967, p. 100).

Table 17.2 is a general summary of the advantages and disadvantages of potential ion exchangers relevant to nuclear use, as drawn from this early publication (IAEA, 1967). The specific plants described therein generally used organic resins, but as early as 1959, Ames (1959) reported the ability of clinoptilolite to recover caesium and strontium from low-level waste streams at the Hanford Laboratories, Richland, Washington. Similarly, the National Reactor Testing Station in Idaho Falls, Idaho, reported (Wilding and Rhodes, 1963) the use of clinoptilolite for removing $^{137}$Cs and $^{90}$Sr from pond water (i.e. high-purity water used to shield irradiated fuel elements prior to reprocessing). $^{137}$Cs removal from low-level wastes using 'Filtrolit', a zeolite-bearing tuff from the Lacher See area of Germany, was achieved at the Hahn-Meitner Institute in Berlin (H.W. Levi cited in IAEA, 1967, p. 66). In another IAEA report (1972) the use of various local minerals in the treatment of radioactive wastes was surveyed (Table 17.3). These IAEA reports (1967, 1972) provide excellent descriptions of the more technical aspects of liquid radioactive waste treatment. They also contain information on the types of materials tested in this time period and outline ion-exchange theory and practice. A third report (IAEA, 1987) updated 'radwaste' treatment in the context of the world-wide industry. Early work in the United States was reviewed by Mercer and Ames (1978) and an overview of Eastern European

TABLE 17.2 Comparison of selected properties of resin and natural zeolite cation exchangers.

| Property | Strong acid resin | Zeolite |
|---|---|---|
| Bulk density ($g/cm^{3}$) | 0.38 | 1.92[1] |
| Cation exchange capacity (mEq/g) | ~5 | 2.25(theoretical)[2] |
| Cation selectivity | Low | High |
| Temperature stability (max. $T$ (°C) usable) | ~120 | ~750 |
| Resistance to chemical 'poisoning' e.g. from S and N compounds | Relatively low | Good |
| Resistance to oxidation | Varies | Good |
| Resistance to reduction | Varies | Good |
| Resistance to physical attrition | Good but can leak organic residues | Varies |
| Resistance to acid conditions | <pH 1 | ~ pH 3 |
| Resistance to alkaline conditions | >pH 13 | >pH 10 |
| $\alpha$, $\beta$, $\gamma$ radiation resistance | Moderate | Good |

[1]Clinoptilolite tuff
[2]Based on Si/Al ratio of a typical clinoptilolite

practice was presented by Tsitsishvili *et al.* (1992). Sherman (1978) reported useful data for natural zeolites in his review of the ion-exchange properties of zeolites.

## 17.3 Zeolites as specific ion exchangers for radioisotopes

### *17.3.1 General*

Table 17.3 suggests that the early emphasis was on the removal of $^{137}Cs$ and $^{90}Sr$ using mainly clinoptilolite–heulandite minerals to treat liquid effluents. More recent work relates to aspects of waste disposal and containment (*vide infra*). Particular approaches are necessary to establish for each chemical system the utility of zeolites as selective materials for radioisotopes. Clearly, a major attraction of using natural zeolites to treat such wastes is their relative cheapness; however, their low cost does not relieve suppliers and users, from the responsibility of fully characterizing their samples. Unfortunately, only vague details were provided for many of the materials studied by earlier workers, leaving the reader to wonder about the overall applicability of the results. This probably arose from an ignorance of the ion-exchange process, which is always based upon the relative selectivity that the material displays to the cations present. This means that selectivity can be critically affected by the initial cation compositions of a natural zeolite.

TABLE 17.3 Early references to the use of zeolitic and related minerals in the treatment of radioactive wastes (IAEA, 1972).

| Zeolite | Isotopes studied | References |
|---|---|---|
| Analcime | $^{137}$Cs, $^{90}$Sr, $^{22,\,24}$Na | Tamura (1962); Ames (1966*a*); Branca and Gresson (1972) |
| Wairakite | $^{137}$Cs, $^{90}$Sr, $^{22,\,24}$Na, $^{45}$Ca | Ames (1966*a*); Hawkins (1967) |
| Mordenite | $^{137}$Cs, $^{90}$Sr, $^{42}$Ca | Ames (1961); Berak (1963); Platt (1967); Hawkins (1967) |
| Stilbite | $^{137}$Cs, $^{90}$Sr | Berak (1963); Ames (1966*b*) |
| Gismondine | $^{137}$Cs, $^{90}$Sr | Berak (1963) |
| Heulandite | $^{144}$Ce, $^{137}$Cs, $^{90}$Sr, $^{60}$Co | Ames (1960, 1961, 1962*a,b,c*); Ames and Knoll (1962). |
| Clinoptilolite | $^{51}$Cr, $^{42}$Ca, $^{22}$Na | Frysinger (1962); Mathers and Watson (1962); Berak (1963); Ames (1963); Mercer and Ames (1963); Hawkins (1964); Nelson *et al.* (1964); Hawkins and Short (1965); Rhodes and Wilding (1965); Amberson and Rhodes (1966); Mercer (1966); Ames (1966*a*); Diest and Talibudeen (1967); Hawkins (1967) |
| Faujasite | $^{137}$Cs, $^{90}$Sr | Berak (1963); Ames (1962) |
| Harmotome | $^{137}$Cs | Berak (1963) |
| Brewsterite | $^{137}$Cs | Berak (1963) |
| Phillipsite | $^{137}$Cs | Ames (1962); Berak (1963); Nelson *et al.* (1964) |
| Chabazite | $^{137}$Cs | Ames (1961); Berak (1963) |
| Natrolite | $^{137}$Cs | Berak (1963); Branca and Gresson (1972) |
| Scolecite | $^{137}$Cs | Berak (1963) |
| Thomsonite | $^{137}$Cs | Berak (1963) |
| Erionite | $^{137}$Cs, $^{24}$Na | Ames (1961); Ames (1962*a,d*); Ames (1963); Nelson *et al.* (1964) |
| Ferrierite | $^{85}$Sr, $^{42}$Ca | Hawkins (1967) |
| Yugarwaralite | $^{85}$Sr, $^{42}$Ca | Hawkins (1967) |
| Related minerals and rocks | | |
| Leucite | $^{137}$Cs, $^{90}$Sr | Berak (1963) |
| Pollucite | $^{137}$Cs, $^{90}$Sr | Berak (1963) |
| Sodalite | $^{137}$Cs, $^{90}$Sr | Berak (1963) |
| Cancrinite | $^{137}$Cs, $^{90}$Sr | Berak (1963) |
| Neopolitan Yellow Tuff, Phillipsite-rich | $^{137}$Cs, $^{90}$Sr | Bocola *et al.* (1968) |

Natural zeolitic materials available for study include: (1) those from large tuffaceous deposits in sedimentary environments, usually formed by alteration of rhyolitic volcanic ash; and (2) samples from relatively small occurrences in basalts or other basic igneous rocks, crystallized from hydrothermal solutions, meteoric water, or during diagenesis. Early research centred on the basaltic occurrences, and large crystals were recovered from amygdales, fractures and veins of the rock by appropriate hand picking, screening and simple flotation. Once the widespread occurrence of many natural zeolites in tuffs became known in the 1960s, most of the successful operations have used the tuffaceous materials. Zeolitic tuffs contain 25–90% zeolites; hence not only should the principal zeolites be fully characterized, a knowledge of the 'contaminants' is also necessary. These range in amount and nature, but typical mineral impurities are quartz, cristobalite, feldspars, opals, clay minerals, calcite and unreacted volcanic glasses. An example of a tuff composition is given in Table 17.4.

The tuffs are generally microcrystalline and relatively porous. In practice, the rock is crushed and sized for column use. Tuffs rich in clinoptilolite, mordenite, chabazite, phillipsite and ferrierite have commonly been studied and/or used. Synthetic mordenite and ferrierite, prepared as pelletized microcrystalline materials, have also been used in the nuclear industry both as selective ion-exchangers and as gaseous sorbents. Synthetic mordenites have been available as 'large-port' or 'small-port', a classification based upon their different molecular sieving properties. When details of the mordenites have been recorded it has been the 'large-port' variety that has been used. Clinoptilolite is not yet commercially available as a synthetic material.

### *17.3.2 Ion exchange characterization*

*17.3.2.1 Cation exchange capacity (CEC).* Traditionally, natural zeolites used in the nuclear industry have been ranked on the basis of their ability to take up the ammonium ion ($NH_4^+$) from solution, mainly because of the observed ability of this ion to replace most of the exchangeable cations from the zeolite framework. Another reason for using the ammonium ion CEC is its enormous import in agronomy, sewage treatment and in animal nutrition. The ammonium ion, unfortunately, does not always totally replace exchangeable cations in

TABLE 17.4 Mineralogical content, as determined by XRD, of a clinoptilolite-rich tuff from Vulture Creek, Zululand, South Africa (Thermistocleous, 1990).

| Mineral | Content (%) |
|---|---|
| Clinoptilolite | 80–85 |
| Cristobalite | ~5 |
| Sanidine | ~5 |
| Opal-C | <5 |
| Opal-CT | <5 |

zeolites, so the ammonium CEC should be considered along with the zeolite mineral content of the sample, the nature of the cations initially present, as well as the theoretical CEC calculated from a knowledge of the Si/Al content of the zeolite as determined by chemical analysis. For these reasons CEC values will not be included in this review.

*17.3.2.2 Cation selectivity.* The literature on zeolite ion exchange provides selectivity series for most common cations. The selectivity series are generated from ion-exchange isotherms based on homoionic (or nearly homoionic) forms of natural samples. An isotherm is a plot of the concentration of one ion in solution *vs.* its concentration in the zeolite phase. The concentrations are from equilibrating zeolite samples with solutions containing two cations. One of these cations is that originally present in the homoionic form of the zeolite; the other is the cation initially in solution seeking to replace the cation in the zeolite. The solutions are isonormal, i.e. of constant ionic strength. An example of a typical Na/Cs isotherm, determined in the author's laboratory (Las, 1989) for a clinoptilolite from West Java, Indonesia, purified according to the method of Barrer and Townsend (1976), is shown in Fig. 17.1. The preference (or selectivity) of the zeolite for $Cs^+$ over $Na^+$ can be deduced by the high capacity of the zeolite for Cs at low concentrations of that cation in solution.

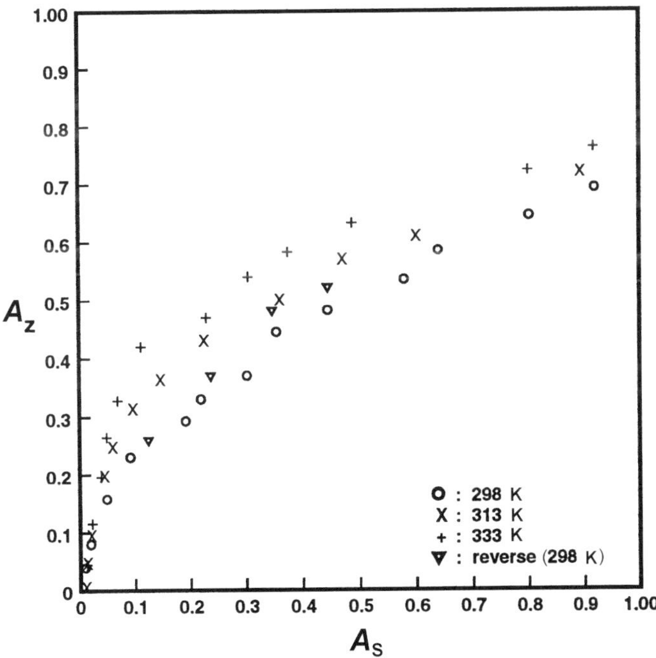


FIG. 17.1 $Na^+ \rightleftharpoons Cs^+$ exchange isotherm on purified clinoptilolite from West Java Indonesia. $A_Z$ = equivalent fraction of $Cs^+$ in zeolite, $A_S$ = equivalent fraction of $Cs^+$ in solution (Las, 1989).

It can be seen that not all the original cations have been replaced. This is a common observation and arises from the phenomenon known as 'ion-sieving' (e.g. Dyer, 1991).

Isotherms can be interpreted to quantify the free energies of exchange, and values of this parameter for different cations can be used to produce a selectivity series. A full account of this procedure and appropriate theory was prepared by Dyer (1991). Preliminary kinetic experiments must be performed to establish the correct equilibrium times for isotherm construction and these can also be used to assess the commercial potential of the envisaged separations.

For scavenging radioisotopes by ion exchange, three points are important: (1) radioisotopes exist in picogram amounts in most aqueous nuclear wastes and selectivities exhibited in the concentrations normally used to prepare isotherm data (i.e. >0.01 N) may well be irrelevant at trace concentrations; (2) the radioisotope in question may not be in true cationic form at picogram concentration levels (i.e. cation speciation may occur); and (3) selectivities determined in binary cation mixtures may not be the same in the presence of other competing cations. Many studies have not gone to the length of plotting ion-exchange isotherms to estimate selectivity. Under most circumstances, determining a distribution coefficient ($K_d$) is sufficient to predict the potential of a natural (or other inorganic ion-exchange substrate) for aqueous nuclear waste treatment. The pragmatism of this approach is obvious when assessing inexpensive, naturally-occurring minerals.

The distribution coefficient is defined as

$$\mathrm{K_d} = (A_o - A_f)V/A_f W$$

where $A_o$ and $A_f$ are the original and final radioactivities in the solution phase of volume $V$ (ml) in contact with a weight $W$ (g) of zeolite. In some instances decontamination factors ($D_F$) are used, where $D_F$ = (initial solution radioactivity)/(final solution radioactivity). Such studies commonly report data as % radioactivity removed ($R_F$).

Before leaving the subject of selectivity, the attention of the reader is drawn to the earlier comments on the importance of proper zeolite characterization and the correct assessment of CECs which are dependent on the selectivity series for the cations and zeolites under study.

### *17.3.3 Other properties*

*17.3.3.1 Radiochemical stability.* The aluminosilicate framework of both natural and synthetic zeolites is resistant to $\gamma$ radiation, even at $8.4 \times 10^9$ rad absorbed dose (Fullerton, 1961) and no authors have reported evidence of structural damage from either $\alpha$ or $\beta$ radiation. The investigations of Lai and Rees (1976) on Szilard-Chalmers processes inside zeolite frameworks show that little permanent damage ensues from quite high neutron fluxes

($1.3 \times 10^{12}$ n cm$^{-2}$ s$^{-1}$). These important properties are significant both in the large-scale use of zeolites for the treatment of aqueous nuclear waste and in their use in the long-term storage of nuclear wastes.

*17.3.3.2 Column use.* For column use, particles of reasonably uniform size and robustness (wet and dry attrition resistance) are needed. Column performance can be assessed by 'breakthrough' curves in which efficiency of removal of an isotope is plotted against the number of bed volumes of solution passing through the column (see Fig. 17.2).

*17.3.3.3 Chemical factors.* Nuclear wastes range widely in composition. A reprocessing waste stream may be 8 M nitric acid, whereas water from fuel-rod storage (pond water) may have a pH of ~11.5, with respect to sodium hydroxide/sodium carbonate (Howden and Pilot, 1984). Wastes can contain organic solvents (e.g. tributyl phosphate and kerosene), high salt concentrations and complexing agents (e.g. EDTA). At low pH, zeolites tend to leach Al from their frameworks, and a broad indication of their likely acid stability would be that those with a higher Si/Al ratio in their framework can withstand a pH as low as 2. At high pH (~10), zeolites may eventually alter to another phase or even dissolve.

## 17.4 Scavenging of radioisotopes from aqueous solution

### *17.4.1 Clinoptilolite*

*17.4.1.1 Selectivity for Cs and Sr.* The relatively high framework Si/Al ratio of clinoptilolite (generally ~5:1) suggests that it has a strong preference for

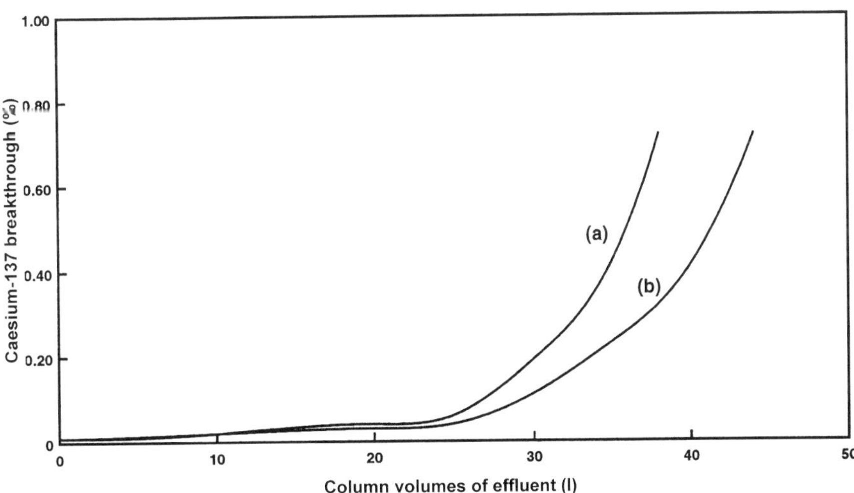


FIG. 17.2 $^{137}Cs$ breakthrough curves for purified clinoptilolite columns. Carrier solution = 0.0001 N in $Cs^+$ in a simulated fuel-storage pond water: (*a*) clinoptilolite from Mud Hills, Barstow, California; (*b*) clinoptilolite from West Java, Indonesia (Las, 1989).

large, monovalent cations of low hydration. This was first noted experimentally by Ames (1964, 1965). The preference can by explained by the electrostatic arguments developed by Eisenman (1962) whereby solid exchange media with lower charges are satisfied by larger ions which are better able to bridge the more disperse negative framework charges. Conversely, those materials with a higher charge prefer smaller ions to neutralize their closer-spaced charge.

Ames (1960, 1964) presented typical monovalent and divalent ion selectivity series for clinoptilolite, which illustrates this: $Li<Na<NH_4<K<Rb<Cs$ and $Mg<Ca<Sr<Ba$. The divalent cation series offers some explanation for the useful capacity that clinoptilolite exhibits for Sr isotopes. Many clinoptilolite-rich tuffs have been tested for Cs and Sr removal.

Early work in the United States (Table 17.3) used samples from Hector (California), Buckhorn (New Mexico) and Sheaville (Oregon). The most significant of this early work was later summarized by Mercer and Ames (1978) who described details of the physical crushing and attrition testing of the Hector and Buckhorn tuffs, with details of the plant at Idaho National Engineering Laboratory, Idaho Falls, which used a local clinoptilolite from Owyhee County, Idaho, to remove Sr and Cs isotopes from low-level waste.

Workers from many other countries have used local clinoptilolites for radwaste treatment. Table 17.5 illustrates this and confirms the utility of clinoptilolite as an inexpensive scavenger for Cs and Sr even with the wide variation in the nature of the exchangeable cations present inevitably present in samples from such diverse localities. The list also contains a sample of heulandite (Chernyavskaya, 1985) which has the same aluminosilicate framework as clinoptilolite, but a lower Si/Al ratio and a higher Ca content and is selective for Na rather than Sr. Chernyavskaya *et al.* (1983) described the details of the elution and regeneration steps of such processes.

From the above, at first sight the initial exchangeable cation composition of clinoptilolite does not seem to be of major importance. Both Nikashina *et al.* (1988) and Senyavin *et al.* (1989), however, demonstrated that a Ca-rich clinoptilolite readily exchanges $Ca^{2+}$ for $Cs^+$, whereas $K^+$ ions are more reluctant to be replaced from the framework. Knowledge of the cations present in bulk concentrations in the waste streams to be treated have prompted several studies in which the natural mineral has first been converted to a homoionic form and then examined for Sr and Cs selectivity. The choice of the homoionic cation form ($H_3O^+$, $Na^+$, $NH_4^+$) arose from the previously mentioned acid nature of some aqueous nuclear wastes coupled with the common use of $Na^+$ or $NH_4^+$ ions to neutralize these solutions as part of their treatment. Sodium ions are also the major cation present in pond storage water. Many authors report that conversion to near homoionic $Na^+$ or $NH_4^+$ ions improves the uptake of both Cs and Sr in comparison to that of the untreated mineral (e.g. Goto *et al.*, 1982; Chernyavskaya *et al.*, 1983; Chales Suarez *et al.*, 1998). This, of course, can be deduced from the selectivity series shown above. Recently Rajec *et al.* (1998) compared clinoptilolite- (and mordenite-)

TABLE 17.5 Examples of the take-up of radioisotopes from aqueous nuclear waste streams by native clinoptilolites (Cp).

| Source of Cp | Isotopes | Comments | References |
|---|---|---|---|
| Zlatokop, Vranje, Serbia | Cs | ~74% Cp in Ca form | Ceranic *et al.* (1987) |
| Russia | Cs, Sr, Na, Ca | Heulandite and Cp | Chernyavskaya *et al.* (1983, 1985) |
| Uzbekistan | Cs, Co | | Vdovina *et al.* (1976, 1977) |
| Nizny Hrobovec Slovakia | Cs, Ba | 40–50 wt.% Cp | Chmielewska-Horvathova and Lesny (1992, 1998) |
| Zaloska Gorica, Slovenia | Cs | | Juznic (1988); Jusznic *et al.* (1990) |
| Georgian Republic, | Cs, Sr | with clay mineral | Kalandiya (1979) |
| Kampendyanskoe, Russia | Sr | 70–95 wt.% Cp | Kravchenko *et al.* (1985) |
| Republic of Korea | Cs | | Lee (1988) |
| Metaxades, Thrace, Greece | Cs | | Misaelides *et al.* (1995) |
| Dzegvi and Tedzami, Georgian Republic | Cs, Sr | various cations present | Nikashina *et al.* (1988); Senyavin *et al.* (1989) |
| Beli Plast and Beli Bair, Bulgaria | Cs, Sr | | Nikashina *et al.* (1988); Senyavin *et al.* (1989); Gradev and Gulabova (1982); Dzhurova *et al.* (1989) |
| Aidag, Azerbaijan | Cs, Sr | | Ozdhagov *et al.* (1986) |
| Polatli/Mulk, Turkey | Cs, Sr | | Senel and Senvar (1981) |
| Cankiri-Coram Basin, Anatolia, Turkey | Cs, Sr | | Aryuez (1996) |
| Itaya, Japan | Cs | | Takagaki (1978); Goto *et al.* (1982) |
| Cuba | Cs, Sr | | Chales Suarez *et al.* (1987, 1998) |

tuffs from Slovakia, Bulgaria, Ukraine and Greece and correlated Cs and Sr isotope uptake to their mineralogical compositions.

Conversion to an acid form ($H_3O^+$) can be achieved because of the relatively high Si/Al ratio (~5) of clinoptilolite. Dyer and Keir (1989) showed that clinoptilolite from Death Valley Junction (California) was resistant to quite high molarities of nitric acid (8 M, pH $\approx$ 0) unlike a phillipsite (Si/Al~1) from Pine Valley, Nevada, which became amorphous in the presence of 1 M nitric acid (pH $\approx$ 0.1). An acid wash (0.5 M) will remove exchangeable cations from clinoptilolite rather than seriously damage the framework by leaching Al from it. Several of the zeolites studied by Ames (1963) were acid-washed commercial zeolites (e.g. AW-500 and AW-400, pelletized natural chabazite and natural erionite respectively). It is interesting to note in passing that the removal of small amounts of Al will, of course, further increase the Si/Al ratio, thereby enhancing Cs and Sr selectivity. Not enough information is available to be able to quantify this. Another route to the acid form comes by $NH_4^+$ exchange followed by calcination (~500°C) and then rehydration. This route would not be expected to significantly alter the Si/Al ratio.

Several authors have reported improved selectivities for Cs and Sr for the acid form of clinoptilolite (e.g. Hlozek *et al.*, 1992*a,b*; Lukac *et al.*, 1992*b*) and this has been corroborated by work in the author's laboratory (Las, 1989). However, Tsitsishvili *et al.* (1992), in reviewing the Russian literature, reported a reduction of selectivity for fission products. Mimura *et al.* (1981, 1984) and Kanno *et al.* (1987) cited limiting concentrations of nitric (pH $\geqslant$ 1), 2 M acetic and $\leqslant$0.1 M hydrofluoric acids, above which a deterioration in Cs and Sr uptake was observed. Suess and Pfepper (1981) included two clinoptilolite samples ('white' and 'green') in their comparative study of Cs isotope scavenging from a nitric acid waste simulant. Both materials performed well in comparison to other natural sorbents tested (bentonite and vermiculite) and the authors concluded that acid pre-treatment would make feasible the possible application of clinoptilolite in acid solution.

Boric acid is another common constituent of nuclear wastes. Lee *et al.* (1979) found it to have a negligible effect on the Cs uptake of clinoptilolite, in agreement with Andreev and Chernyavskaya (1982) who reported the lack of influence of boric acid (pH 2–3) on Cs, Sr scavenging by both clinoptilolite and mordenite from solutions containing uranyl ions. Isotherm studies from the author's laboratory (Princewill, 1994) indicated that these reports may be too optimistic and Figs 17.3 and 17.4 suggest that the presence of boric acid can

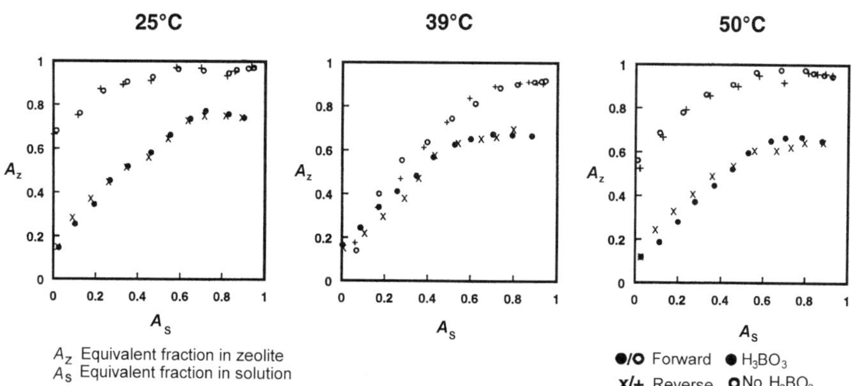


FIG. 17.3 Effect of 5 M boric acid on $Na^+ \rightleftharpoons Cs^+$ exchange for purified clinoptilolite from Death Valley Junction, California (Princewill, 1994). $A_Z/A_S$ is the equivalent fraction of $Cs^+$ in zeolite/in solution. ●: forward, X: reverse, (without boric acid present); ○: forward, +: reverse (with boric acid present).

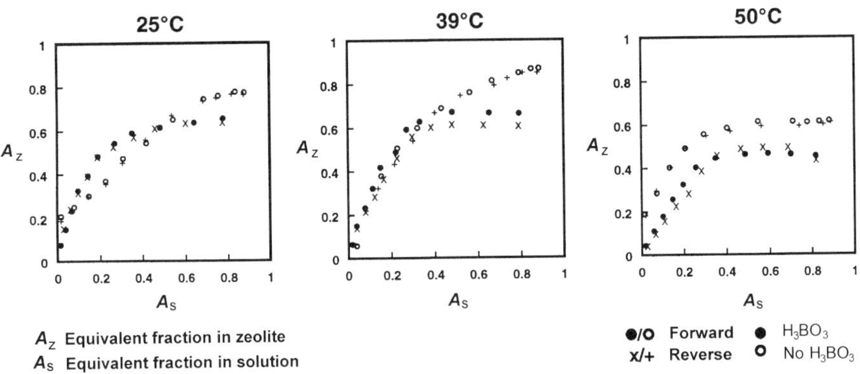


FIG. 17.4 Effect of 5 M boric acid on $Na^+ \rightleftharpoons 1/2Sr^{2+}$ exchange for purified clinoptilolite from Death Valley, Junction, California (Princewill, 1994). $A_Z/A_S$ is the equivalent fraction $Cs^+$ in zeolite/in solution. (Symbols as in Fig. 3).

indeed alter uptake equilibrium values for both Cs and Sr on a homoionic Na-clinoptilolite.

Based on the known high selectivity of hexacyanoferrates (II) for Cs, Harjula and Lehto (1993) and others (see Table 17.6) occluded hexacyanoferrates (II) inside clinoptilolite to form a composite sorbent for Cs. Most authors report that

TABLE 17.6 Other work on Cs and Sr radioisotope uptake on clinoptilolite (Cp).

| Isotope | Comments | Reference |
|---|---|---|
| Cs, Sr | Ferrocyanide-modified Cp and bentonite, gave only modest $K_d$ values, reduced by the presence of other salts | Panasyug *et al.* (1993) |
| $^{137}Cs$ | Higher affinity noted for hexacyanoferrate-modified Cp rather than that for natural Cp | Novosad *et al.* (1992) |
| Cs, Sr | Successful removal from Moscow river water | Milyutin *et al.* (1993) |
| Cs | Rates of removal assessed on Cp as 1st order kinetics | Proko'fev *et al.* (1990) |
| $^{90}Sr$ | Cp and chabazite not as good as other inorganic exchangers for simulated pond-water treatment, but more stable | Dyer and Khor (1980) |
| Sr | Modelling of ion-exchange process on Cp from natural fresh waters | Nikashina *et al.* (1988) |
| $^{137}Cs$ | 80–100% removal in batch process, 40–80% in flow-through system at pHs typical for water purification and wastewater treatment | Zaitsev *et al.* (1995) |

TABLE 17.7 Removal of isotopes other than Cs or Sr from aqueous solution by clinoptilolite (Cp).

| Isotopes | Comments | References |
|---|---|---|
| $^{45}Ca$, $^{22}Na$ | Used to study Sr uptake on Cp and heulandite | Chernyavskaya (1985) |
| $^{133}Ba$ | Kinetic and isotherm data | Chmielewska-Horvathova and Lesny (1992) |
| $^{95}Zr$, $^{95}Nb$<br>$^{181}Hf$ | Comparison to other zeolites<br>Cp not successful | Dyer and Kadhim (1989) |
| $^{60}Co$ | Uzbekistan: Cp not successful | Vdovina *et al.* (1977) |
| $^{232}Th$<br>$^{60}Co$ | Surface precipitation of thoria<br>pH 12.5–13.7; salts ~200 g/dm$^3$ Cp not effective | Dyer and Josefowicz (1992)<br>Gulis and Timulak (1986) |
| U | Preactivation with HCl, kinetics at various pH and concentrations of U | Gumus *et al.* (1992) |
| $^{103}Ru$, $^{57}Co$<br>$^{64}Zn$, $^{54}Mn$ | Cp was effective, as were active charcoal and a resin containing ferrocyanide, in treatment of lab. waste water containing sea water | Imai *et al.* (1990) |
| $^{60}Co$ | Pretreatment with NaOH improved uptake | Lukac and Foldersova (1990, 1994); Lukac *et al.* (1992*a*,*b*) |
| $^{106}Ru$, $^{144}Ce$,<br>$^{95}Nb$, $^{95}Zr$, | Filters made from Cp and Cp/hexacyanoferrate, successful | Luneva *et al.* (1994) |
| $^{60}Co$, $^{51}Cr$<br>$^{59}Fe$, $^{160}Tb$<br>$^{152,\ 154}Eu$ | Comparison with synthetic zeolites and Cp, carrier free, pH 1–3 | Mimura and Kanno (1978*a*,*b*) |
| $^{60}Co$, $^{144}Ce$ | Na enriched Cp, 3 Cuban samples | Chales Suarez *et al.* (1998) |
| Th, U | NaCl pretreatment, 50–20,000 mg/l conc., resistance to low pH | Misealides *et al.* (1995); Constantopoulou *et al.* (1992, 1994) |
| U | $U^{6+}$ sorption high at ~neutral pH | Pabalan *et al.* (1993) |
| U | Comparison with other minerals in the presence of Na | Ames *et al.* (1983*a*) |
| $^{238,\ 233}U$ | Uptake from pH 4 (with Sr and Cs ions) | Andreeva and Chernyavskaya (1982) |
| U | Uptake from 4% $H_3BO_3$ solution on Ca-Cp | Pekov *et al.* (1980) |
| U | Uptake pH 6–9 as U(VI) (also chabazite, erionite, mordenite) | Reddy and Cai (1996) |
| $^{241}Am$ | Uptake from synthetic and natural ground waters | Triay *et al.* (1991) |
| $^{147}Nd$ | Cp not as effective as montmorillonite | Symeopoulos *et al.* (1996) |
| $^{237}Np$ | $Np^{5+}$uptake as $NpO_2(OH)^0$ at pH 7–9.5 | Bertetti *et al.* (1996, 1998) |
| Th, U | Ion-exchange and surface ppt. | Godelitsas *et al.* (1996*a*,*b*)<br>Misaelides *et al.* (1995) |

TABLE 17.7 (*contd.*)

| Isotopes | Comments | References |
|---|---|---|
| Ra | Cp effective for removal from 0.01 M NaCl | Ames *et al.* (1983*b*) |
| $^{226}Ra$ | Wastewaters from U ore mining | Dutu *et al.* (1996) |
| $^{203}Hg$ | Tracer study for industrial waste water treatment | Misaelides and Godelitisas (1995) |
| $^{59}Fe$ | Slovakian and Chinese Cp in Na form | Chmielewska and Lesny (1997) |

the technique is successful (e.g. G'oshev *et al.*, 1987) especially if $K^+$ ions are present in the effluents (10 g/l $K^+$, with ~200 g/l total salt).

Pre-treatment with hydrazine (and $Na^+$, $NH_4^+$) results in a 2–4 uptake in Sr (Chernyavskaya, 1985). Kanno and Hashimoto (1976) used EDTA to elute Cs and Sr from clinoptilolite and other natural zeolites, to provide pure isotopic solutions intended for commercial use. Mimura *et al.* (1995) described a chromatographic separation of Sr and Cs on mixed zeolites columns which include clinoptilolite. Table 17.6 summarizes other work on Cs/Sr isotope uptake on clinoptilolite.

*17.4.1.2 Uptake of radioisotopes other than Cs and Sr.* The wide availability of clinoptilolite-rich tuffs has encouraged workers to investigate their potential to remove many other isotopes from aqueous nuclear wastes (Table 17.7). Unfortunately many of these studies are speculative and do not consider the effects of competing ions or pH. Some reports have described the uptake as surface precipitation rather than ion-exchange. For example, Dyer and Josefowicz (1992) demonstrated that the uptake of Th by clinoptilolite was mainly by surface deposition of thoria. $ThO_2$ precipitates on the highly alkaline surface of the zeolite, much like the oxide/hydroxides of $Zn^{2+}$, $Co^{2+}$, $Cd^{2+}$ and other metals on synthetic zeolites (Wark *et al.*, 1993). Obviously oxide/hydroxide precipitation onto the zeolite may be an acceptable method of isotope scavenging.

The solvents tributyl phosphate (TBP) and odourless kerosene (OK) are used widely to extract Pu (IV) and U (V) in nuclear reprocessing plants. Although these solvents were originally chosen for their low affinity for fission products (e.g. Cs, Sr, Ru), small quantities of these isotopes remain in the solvents. Dyer and Aggrawal (1995) included clinoptilolite in a general study to examine the potential of zeolites to remove fission products from TBP/OK/$H_2O$ mixtures. They observed that Ru uptake exhibited $K_d$ values of the order of $10^2$ ml/g, with only synthetic zeolite Y being more efficient. The same authors (1997) showed that clinoptilolite was less successful in Cs and Sr scavenging from mixed organic solvents, except in the presence of water when it matched, or bettered, the $K_d$ values attained by other zeolites. A similar $K_d$ value for Ru nitrosyl removal by clinoptilolite from a simulated alkaline nuclear waste, was attained but $K_d$ values for acid simulants were very low.

Clinoptilolite was used in a programme of work (Dyer and Shaheen, 1995) designed to show how zeolite cation exchange can be sensitive to the anion species present if the cations are present at the very low concentrations of a radiotracer (so-called 'carrier free', i.e. no concentration of the cation other than that of the radioisotope, normally in the picogram range). Figure 17.5 demonstrates this for Zn, Co and Ni radioisotope uptake onto clinoptilolite. The variations in uptake may be related to the presence of $[MX_m]^{n+}$ species in solution, where $M = Zn^{2+}$, $Co^{2+}$, or $Ni^{2+}$ and $X = SO_4^{2-}$, $NO_3^-$, or $Cl^-$. Variation in their size and/or charge causes changes in the ability of the transition metal to enter the zeolite structure. These effects are not seen at normal cation concentrations when the cations are present as $M^{2+}$ hydrated species.

The problem of the change in selectivity that can occur in a zeolite exchanger at very low concentrations of the in-going cation ($<10^{-4}$ M) is not well understood and the only information available is that for synthetic zeolites (Harjula *et al.*, 1992). Changes in selectivities have been noted in natural zeolite binary and ternary cation systems but not reasons for the changes (Gopala Roa, 1995). This is understandable bearing in mind the complex nature of the systems involved.

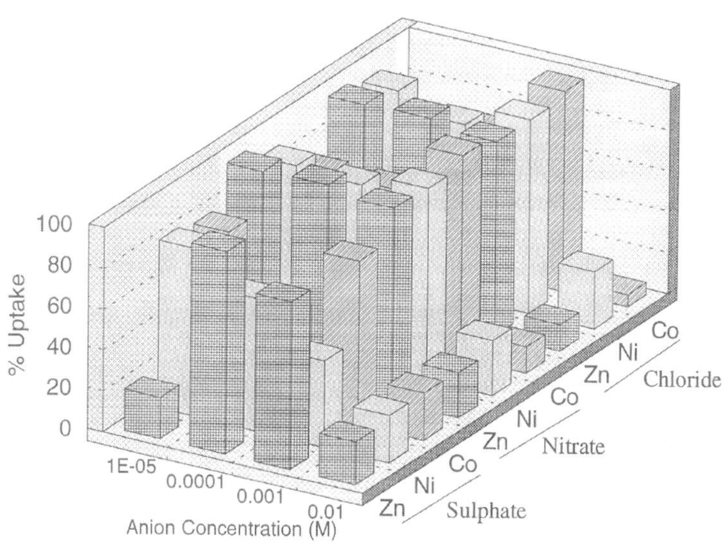


FIG. 17.5 Effect of nature and concentration of anions on $Zn^{2+}$, $Co^{2+}$ and $Ni^{2+}$ uptake on purified clinoptilolite from Mud Hills, Barstow, California (Dyer and Shaheen, 1995).

### *17.4.2 Mordenite*

Mordenite resembles clinoptilolite in its relatively high Si/Al framework composition. Their structures and chemical composition are similar and the similarity extends to the observed selectivity of mordenite for Cs and Sr.

The use of mordenite to treat radioactive waste is included in Table 17.3. One of the references (Hawkins, 1967) deals largely with synthetic zeolites but includes comparison with a natural mordenite from Nova Scotia, Canada and the early work of Ames (1961) made use of a natural mordenite from an unspecified location.

Synthetic mordenites, such as Zeolon 900 (Norton Co.), are available as extrudates and pellets suitable for column use and they have been employed as Cs and Sr scavengers in nuclear facilities (e.g. Ames, 1963). The material AW-300 (Union Carbide) has also been used and Ames (1963) refers to it as a synthetic mordenite. This is also implied by Sherman (1978), although originally it may have been a pelletized form of a natural mordenite from Union Pass, Arizona, mined by the Union Carbide (later UOP) company who marketed AW-300.

Table 17.8 summarizes the more recent work in which natural mordenites have been used in the treatment of aqueous nuclear wastes.

### *17.4.3 Chabazite*

Although its structure differs from that of clinoptilolite and mordenite, Ames (1961) showed a sample of chabazite from the Bay of Fundy, Nova Scotia, to have useful selectivities for Cs and Sr. Its selectivity series were Mg<Ca<Sr<Ba and Li<Na<K<Cs. Berak (1963) also included chabazite in his early studies of Cs and Sr uptake onto minerals. The earlier literature cited a product AW-500 (Union Carbide) as a 'synthetic chabazite' (IAEA, 1967, p. 22). This was used at Hanford (Washington) as part of a process for the purification of Cs, present at $10^{-4}$ M, from highly active wastes containing 4.5 M Na salts but it was actually a pelletized form of a natural chabazite (Mercer and Ames, 1978) from the Bowie deposit (Arizona) (Mondale *et al.*, 1978). Further examples of similar studies on chabazite can be found in Table 17.9.

### *17.4.4 Other natural zeolites*

*17.4.4.1 Ferrierite.* Ferrierite is another zeolite that resembles clinoptilolite and mordenite in structure and Si/Al ratio. It was available in pelletized form, as Zeolon 700 (Norton Co.), from a deposit near Lovelock, Nevada. Dyer and Keir (1984) reported the selectivity for Sr of unpelletized ferrierite from this deposit. Their column work showed that ferrierite breakthrough curves, for both $^{89}$Sr and $^{90}$Sr, were identical to those of a clinoptilolite from Death Valley Junction, California. Mimura *et al.* (1993) achieved similar results using a synthetic ferrierite mixed with synthetic zeolite A.

TABLE 17.8 Studies involving mordenite for the up-take of isotopes.

| Isotopes | Comments | References |
|---|---|---|
| $^{133}Ba$, $^{137}Cs$, $^{60}Co$ | Bartosova Lehotka-Jastraba, Slovakia: kinetics and isotherm data | Chmielewska-Horvathova and Lesny (1992); Lukac *et al.* (1992*a*) |
| $^{95}Zr$, $^{95}Nb$, $^{181}Hf$ | Lovelock, Nevada, kinetics and $K_d$ values | Dyer and Kadhim (1989) |
| $^{232}Th$ | Mordenite/Cp tuff, Eastgate, Nevada: surface precipitation | Dyer and Josefowicz (1992) |
| U | Comparison to other minerals in the presence of Na | Andreeva and Chernyavskaya (1982) |
| $^{134,137}Cs$, $^{60}Co$ | pH 12.3–13.7, ~200 g/dm$^3$ salt concentration | Gulis and Timulak (1986) |
| Cs, Sr | Kinetic data | Liang and Hsu (1993); Liang and Joseph (1995) |
| $^{137}Cs$ | Isotherm and kinetic data, change in pH and Cs conc. | Hsu *et al.* (1994) |
| $^{137}Cs$, $^{90}Sr$ | Chromatographic separations from high-level wastes, mordenite better than chabazite, Cp | Kanno and Hashimoto (1976); Laske and Laube (1989); Mimura *et al.* (1995) |
| $^{137}Cs$, $^{85}Sr$, $^{152,\ 154}Eu$, $^{144}Ce$, $^{160}Tb$ | pH 1–3, carrier free, effect of multivalent cation | Mimura and Kanno (1978*a,b*); Mimura *et al.* (1984) |
| $^{60}Co$, $^{51}Cr$, $^{59}Fe$ | Comparison with synthetic zeolites and Cp | Mimura and Kanno (1978*b*) |
| $^{137}Cs$, $^{90}Sr$ | Equilibrium measurements, fixation studies | Gradev and Gulubova (1982); Dzhurova *et al.* (1989); Goto *et al.* (1982) |

*17.4.4.2 Erionite.* Erionite has attractive selectivity properties for radioisotopes. It also was available as a Union Carbide pelletized product (AW-400) from the Pine Valley deposit (Nevada) and tested as such in early work, as was an erionite from Jersey Valley, Nevada (see references in Table 17.3). More recently Carrera *et al.* (1993) and Pacheco *et al.* (1994) found that $^{60}Co$ could be removed from aqueous solution by natural erionite from Agua Prieta, Sonora, Mexico. Olguin *et al.* (1994) used the same material to study the uptake of $UO_2^+$ (pH = 4) by erionite and synthetic zeolite Y. They pre-treated with NaCl and provided equilibrium curves for zeolites in contact with 0.1 and 0.0005 M solutions of uranyl ions. The capacity of erionite for the uranyl ion was quoted as 0.2037 mEq/g and that of Y as 0.7345 mEq/g. Some unexplained surface modification was noted on erionite in contact with uranyl ions for 8 days. Work on the same zeolite has been extended to Sr (Olguin *et al.*, 1996*a*) and to $^{239}Np$ and $^{235}U$ fission products (Olguin *et al.*, 1996*b*).

TABLE 17.9 Studies involving the use of chabazite (Ch) to treat aqueous nuclear wastes.

| Isotope | Comments | References |
|---|---|---|
| $^{137}Cs$, $^{85}Sr$ | Synthetic zeolite X and Ch both used to decontaminate cooling water, surface diffusion model developed | Kim *et al.* (1995) |
| Cs, Sr | AW-500 best for Cs; synthetic zeolite X best for Sr, in comparison with synthetic zeolite A, AW-300 and clinoptilolite | Lee *et al.* (1993*a,b*) |
| $^{137}Cs$, $^{90}Sr$ | Mixture of synthetic zeolite X and Ch gave best $K_d$ values for both elements, over synthetic zeolites A, Y, mordenite and natural mordenite, Ch, clinoptilolite for treatment of high activity waters | Mimura *et al.* (1988*a,b*; 1989) |
| $^{137}Cs$, $^{90}Sr$ | Ch better than synthetic zeolite A for separation and fixation of both isotopes | Mimura and Kanno (1985) |
| $^{137}Cs$, $^{90}Sr$ | Chromatographic separation from high-level wastes, on columns, natural mordenite, better than Ch, or clinoptilolite | Mimura and Kanno (1995) |
| $^{137}Cs$ | Chinese Ch better than clinoptilolite, studied at various pHs and salt concentrations | Weng and Que (1986) |
| $^{137}Cs$, $^{90}Sr$ | Extensive studies on Mt. Vulture, Italy, Ch-phillipsite tuffs | Lenzi and Cassano (1988) |
| $^{137}Cs$, $^{90}Sr$ | Mixture of AW-500 and synthetic zeolite A effective in removal of both isotopes at pH 3–7 | Kwon *et al.* (1996) |
| $^{232}Th$ | Bowie, Arizona and Skye, Scotland Ch not effective | Dyer and Josefowicz (1992) |

*17.4.4.3 Phillipsite.* Dwairi (1992) showed that a phillipsite-rich tuff from Aritan, Jordan, removed 15.8% of the Cs present in a simulated alkaline nuclear waste, in comparison to the 13.5% taken out by a phillipsite from Shoeshone, California and 23.7% removed by one from Pine Valley, Nevada. Komarneni (1985) described Cs and Sr immobilization on phillipsite, showing the fixation on phillipsite and analcime to be superior to that on chabazite, erionite, mordenite and clinoptilolite. Dyer and Kadhim (1989) indicated that Pine Valley phillipsite and a mordenite from Lovelock, Nevada, were able to scavenge Zr, Hf and Nb radioisotopes from aqueous solutions (pH, 0.5, 8.5 and 12.4) better than analcime (Wikieup, Arizona), clinoptilolite (Death Valley Junction, California), Zeolon 900 or amorphous zirconium phosphate. The Pine Valley phillipsite did not take up $^{232}Th$ (Dyer and Josefowicz, 1992).

Finally the tuffs from Mt. Vulture, Basilicata, Italy, contain phillipsite and can take up both Cs and Sr isotopes (Table 17.8).

*17.4.4.4 Laumontite.* Laumontite is an attractive candidate for radwaste treatment because of its open structure, which can be expected to provide fast exchange kinetics, and its occurrence in deposits of relatively high purity. At least one deposit is mined commercially in California and one near Bor, Serbia has the potential to produce a commercial product (Obradovic, 1988). Dyer *et al.* (1991) showed laumontite, from a vein occurrence in Skye, Scotland, to be potentially useful in the removal of Cs and Sr even in the presence of competing $Na^+$, $K^+$, $Ca^{2+}$ and $NH_4^+$ ions. These results confirmed earlier work on vein laumontite from Finland (Romantschuk *et al.*, 1966) and Usbekistan (Vdovina *et al.*, 1976, 1977).

*17.4.4.5 Harmotome.* A column of harmotome removed 75–99% of the $^{137}Cs$, $^{90}Sr$, $^{90}Y$, $^{144}Ce$, $^{235}U$ and $^{239}Pu$ isotopes from aqueous solution at pH 4.6–7.5. Only Cs and Sr were taken up by ion exchange. The other isotopes were held on the external surface of the zeolite (Hafez *et al.*, 1978).

*17.4.4.6 Natrolite.* Natrolite shows little capacity for exchange but Sultanov *et al.* (1976) have reported its use in removing $^{60}Co$, $^{90}Sr$ and $^{137}Cs$ from mildly radioactive waters. The ability of a low silica natrolite to take up Co and Cs was observed by other workers (Vdovina and Rudyuk, 1992), who also confirmed its stability under $\gamma$ radiation.

*17.4.4.7 Other zeolites.* Vochten *et al.* (1990, 1991) reported external surface sorption of uranyl-hydroxo complexes onto scolecite and stilbite (also chabazite and heulandite). Misaelides *et al.* (1998) recorded high surface concentrations from Rutherford backscattering measurements on scolecite in contact with Cs and Sr solutions. Dyer and Kadhim (1989) found that analcime (from Wikieup, Arizona) had no exchange capacity for Zr, Hf or Nb isotopes. The same analcime and brewsterite from Strontian (Scotland), do not take up $^{232}Th$ (Dyer and Josefowicz, 1992). Analcime from Anatolia (Turkey) was not as effective as a local clinoptilolite for Cs and Sr uptake according to Akyuez (1996).

## 17.5 Treatment of radioactive wastes from nuclear facilities

The pioneering work described above for using natural zeolites to remove $^{137}Cs$ and $^{90}Sr$ from nuclear wastes led to the design of industrial-scale facilities in several countries. The early experience in the United States has been described by Mercer and Ames (1978) and Sherman (1978) and is summarized in Table 17.10. Roddy (1981) evaluated the performance of zeolites for the control of water quality in nuclear storage basins in United States nuclear facilities. The early development of plant work outside the United States was covered in the IAEA publications mentioned above (IAEA, 1967, 1972, 1987).

TABLE 17.10 Processes developed in the United States for the treatment of aqueous nuclear wastes by zeolites (Mercer and Ames, 1978; Sherman, 1978).

| Waste | Isotope removed | Location | Zeolite | Other details |
|---|---|---|---|---|
| High-level waste | $^{137}Cs$ | Hanford, Washington | AW-500[1] | 300 $ft^3$ bed, several millions gal. treated without loss of capacity |
| | $^{137}Cs$ | Hanford, Washington | Zeolon-900[2] | Purification of effluent from above |
| Low-level water from fuel storage basin | $^{137}Cs$, $^{90}Sr$ | National Reactor Testing Station, Idaho Falls, Idaho | Clinoptilolite Castle Creek, Idaho | Non-regenerative treatment, 12,000 gal. waste water /$ft^3$ zeolite |
| Evaporator overheads and misc. wastewater | $^{137}Cs$ | Savannah River Laboratory, Aiken, South Carolina | AW-500 | $\geqslant$76,000 gal. overheads and 12,000 gal. other effluent (9.2 $ft^3$ bed) |
| Process condensate wastewater | $^{137}Cs$ | Hanford Washington | Zeolon-900 | 1.8 m bed including organic resin. Regenerated with sodium nitrate |

[1] AW-500 pelletized natural chabazite from Bowie, Arizona
[2] Zeolon-900 synthetic mordenite, from Norton Company

### *17.5.1 Oak Ridge National Laboratory*

Researchers at the Oak Ridge National Laboratory, at Oak Ridge, Tennessee, have investigated the potential use of chabazite for the removal of radioisotopes from waste streams. Actual waste waters were used to acquire bench and pilot-scale data (Robinson *et al.*, 1987, 1988), which were then used to develop process flow sheets followed by a pilot-scale demonstration (Robinson and Parrott, 1989). Coupled to these studies has been the desire to gain a deeper understanding of the ion-exchange processes relevant to this use of chabazite. These entailed the acquisition of data, first on the effect of the presence of Ca and Mg ions (Robinson *et al.*, 1990) and later for binary and multi-component exchanges involving $Na^+$, $Cs^+$, $Mg^{2+}$, $Ca^{2+}$ and $Sr^{2+}$ and then extending to equilibrium studies involving all five cations. Both batch and column experiments were conducted (Robinson *et al.*, 1991). The same authors (Robinson *et al.*, 1994) measured mass transfer coefficients in Na-Ca-Mg-Cs-Sr ion-exchange on chabazite and concluded that the primary diffusional resistance arose from cation movement through the zeolite crystals, although macropore diffusion made a significant contribution to mass-transfer resistance (De Paoli and Perona, 1996).

Further modelling was performed on multicomponent ion-exchange equilibria in chabazite columns designed to treat Oak Ridge wastes (Perona, 1993*a*,*b*). A successful model for Sr breakthrough allows the prediction of bed depths and cycle times needed to achieve the required decontamination factors at high zeolite utilization efficiencies over a wide range of flow conditions (Perona *et al.*, 1995).

Robinson *et al.* (1995) gave a useful summary of these studies with a comparison of the natural zeolites chabazite, ferrierite and clinoptilolite, to a wide range of organic resins and inorganic ion exchangers (see also Kent *et al.*, 1993; Collins *et al.*, 1993). The chabazites used were; Ionsiv IE-95 which is the pelletized chabazite from Bowie, Arizona (Union Carbide Corporation – now UOP), IE-96 (its Na form) and TSM-300 from Steelhead Industries. The product Zeolon 500 (Norton), which is also from the Bowie deposit and contains chabazite and erionite, was also used. The summary included details of the process waste-water (mainly tap and groundwaters), newly generated low-level waste, legacy-liquid low-level wastes and Cleanex raffinate waste streams (see Table 17.11). Chabazite was effective in removing $10^{-12}$ M amounts of Cs and $10^{-6}$ M amounts of Sr from the process waste-water which contained as much as $1 \times 10^{-3}$ M of non-radioactive salts. Natural zeolites were not especially useful for the removal of Cs and Sr from streams containing 0.1–4 M salts. However, chabazite was selected at Oak Ridge for a new process waste-water plant, designed to treat ~300,000 $m^3$ of waste water annually. It was also selected (Na-form) as a standard to which to other candidate materials can be compared for $^{137}Cs$ and $^{90}Sr$ removal from process water (Bostick *et al.*, 1997).

### *17.5.2 National reactor testing station*

Wilding and Rhodes (1963) described the use of four 5.3 $ft^3$ beds of clinoptilolite at the National Reactor Testing Laboratory, Idaho Falls, Idaho, to remove Cs and Sr isotopes from low-level waste water arising from fuel storage basins. It was a non-regenerative use, able to treat ~12,000 gallons per cubic foot of zeolite. The saturated zeolites were removed and then buried in steel drums.

### *17.5.3 Savannah River*

At the Defense Waste Processing Facility at the Savannah River Laboratory, Aiken, South Carolina, chabazite (AW-500) was used as a 9.2 $ft^3$ bed to take up Cs from evaporator overheads and other waste-water streams (IAEA, 1967, 1972; Mercer and Ames, 1978). A more recent report describes the novel use of IE-96 to take up Cs from a secondary waste stream which was largely benzene (Bibler and Deininger, 1991).

TABLE 17.11 Chemical compositions of simulated waste streams at Oak Ridge National Laboratory, Oak Ridge, Tennessee (Robinson *et al.*, 1995).

| Component | Molarity | | | |
|---|---|---|---|---|
| | LL[1] | NG[2] | CR[3] | PW[4] |
| $NaNO_3$ | 4 | 0.06 | – | – |
| NaOH | 0.2 | 0.3 | – | – |
| $Na_2CO_3$ | 0.1 | 0.6 | – | – |
| NaCl | 0.1 | 0.03 | 0.1 | – |
| $KNO_3$ | 0.2 | – | – | – |
| $Al(NO_3)_3$ | 0.005 | – | – | – |
| $NaAlO_2$ | – | 0.012 | – | – |
| HCl | – | – | 0.03 | – |
| LiCl | – | 0.03 | 0.04 | – |
| $CaCl_2$ | – | – | – | $2 \times 10^{-3}$ |
| $MgCl_2$ | – | – | – | $6 \times 10^{-4}$ |
| $^{90}Sr$ | $3 \times 10^{-5}$ | $2 \times 10^{-6}$ | $2 \times 10^{-3}$ | $2 \times 10^{-6}$ |
| $^{137}Cs$ | $5 \times 10^{-7}$ | $9 \times 10^{-6}$ | $9 \times 10^{-11}$ | $1 \times 10^{-12}$ |
| pH[5] | 13 | 13 | 1.5 | 7 |

[1] LL – Legacy liquid low-level waste
[2] NG – Newly generated liquid low-level waste
[3] CR – Cleanex raffinate waste generated from transuranic isotopes production
[4] PW– Slightly contaminated process water
[5] No pH adjustment

### *17.5.4 Hanford*

Treatment of high-level wastes by AW-500 to remove Cs was reported in the early work (Table 17.10) from the Hanford Site, Washington. The wastes contained large concentrations of Na salts (4.5 M), and a bed 1.83 m in diameter and 3.16 m deep was able to remove ~50% of the 250,000 Curies (Ci) of $^{137}Cs$ from the $3 \times 10^6$ gallons of waste which passed through the bed.

### *17.5.5 Canyon Diablo Power Plant*

A mixture of inorganic ion exchangers, which included clinoptilolite and IE-96, was found to be effective in removing Cs and Co from low concentration, non-recyclable, liquid waste from the Diablo Canyon Power Plant, Avila Beach, California (James and Miller, 1989; Propst *et al.*, 1988).

### *17.5.6 British Nuclear Fuels (BNF) and other UK work*

In the United Kingdom extensive laboratory and pilot plant experiments were commissioned by British Nuclear Fuels at the United Kingdom Atomic Energy

Authority, Harwell, Oxon laboratories (see Howden, 1983; Howden and Pilot, 1984; British Nuclear Fuels, 1987).

The studies encompassed all available inorganic cation exchangers, including a wide range of synthetic and natural zeolites. The intent was to determine the most efficient material to remove Cs and Sr from fuel-storage ponds and other aqueous effluents, at the BNF Sellafield Works, Cumbria (on the Irish Sea). Detailed examination of potential specific geological localities of clinoptilolite, which was the most successful material, resulted in the choice of a high-purity clinoptilolite from Mud Hills, Barstow, California. This is in use in their SIXEP (Site Ion Exchange Effluent Plant) process. The pond water is pre-carbonated, using carbon dioxide, to adjust the working pH to 6.8–8.8. Since commissioning in 1985, the plant has treated 3000 $m^3$ of liquid effluent per day.

The chabazite product AW-500 was used to treat contaminated pond water at the Trawsfynydd Nuclear Power Station at Bala, North Wales (now closed).

### *17.5.7 Lovissa Nuclear Power Plant, Finland*

Treatment of effluents from the Pressurised Water Reactors (PWR) at the Lovissa Nuclear Power Plant, Finland, favours the use of hexacyanoferrate columns over AW-500 to deal with $^{137}Cs$ at pH 6–13 in solutions containing high K (and Na) concentrations (Harjula and Lehto, 1986; Harjula *et al.*, 1994).

### *17.5.8 Central Research Institute of Electrical Power Industry, Japan*

Takagi (1978) describes the development of liquid effluent treatment for the Japanese Nuclear Power Industry that makes use of two Japanese clinoptilolites (from Futatsiu, Askita Prefecture and Itaya, Yamagata Prefecture) to scavenge Cs.

### *17.5.9 Italian Commission for Nuclear and Alternative Energy Sources (ENEA)*

The ENEA has developed a process (SERSE – Separazione Rifuiti Serbatoi EUREX) to treat high-level wastes coming from Moderated Thermal Reactors (MTR), *CAN*ada *D*euterium *U*ranium (CANDU) Reactors and Elk River nuclear fuels reprocessing campaigns. EUREX is the name of a pilot plant. Both IE-95 and IE-96 chabazites were able to achieve capacities of 1 mg Cs/g of zeolite in a fixed bed column operation. The effluents were of high nitrate salt concentrations (Na-80.5, Al-13, g/l, Hg-260, K-13, Sr-0.12, Fe-3, mg/l) at pH 13.5 (Marrocchelli and Pietrelli, 1989; Calle *et al.*, 1990). Other work with CANDU reactor circuit clean-up, in Canada, has compared various exchangers for Cs removal and found that IE-96 gave the highest $K_d$ value of the inorganic materials tested (Dias and Nott, 1990).

## 17.6 Clean-up of radioactive waste from nuclear accidents and fall-out

### *17.6.1 Nuclear plants and fuel-reprocessing facilities*

*17.6.1.1 Three Mile Island.* The 1979 accident at Three Mile Island Nuclear Power Station, Middletown, Pennsylvania, generated ~2800 $m^3$ of high-level radioactive water and severely contaminated two organic resin demineralizers (Collins *et al.*, 1986). The resins degraded both thermally and radiolytically due to the high gamma fields created by $^{137}Cs$ and $^{134}Cs$ isotopes, which were accidentally dumped into the demineralizers (Bond *et al.*, 1986). Early experiments considered clinoptilolite as a means of selectively taking up these fission products (and $^{90}Sr$), but ultimately a 60/40% by volume mixture of IE-96 and A-51, synthetic zeolite A (Union Carbide Corporation), was chosen. The mixture was used in a Submerged Demineralizer System (SDS) to decontaminate the sump water in the containment building of the reactor, water of the reactor coolant system and contaminated resin eluates (rich in sodium borate/boric acid and NaOH). The SDS system was also employed in a Defuelling Water Clean-up System (DWCS) to remove caesium from the refuelling canal, the spent-fuel pond and the reactor vessel during refuelling (see reviews by Hofstetter *et al.*, 1982; Hofstetter and Hitz, 1983; King *et al.*, 1984).

The EPICORE ll system containing mainly resins with some mixed resin/zeolite beds was developed separately to process 565,000 gallons of contaminated water that had collected in the Auxiliary Building of the plant. By 1985 the SDS had immobilized 340,000 Ci of fission products in zeolite beds from >$1.5 \times 10^6$ gallons of water (Hofstetter and Hintz, 1986).

*17.6.1.2 Chernobyl.* In 1986, the world's most publicized nuclear accident took place at the Chernobyl nuclear power station, ~100 km north of Kiev, Ukraine (then part of the Soviet Union). Fallout from the reactor spread to most neighbouring countries to the west and north of the area and was detectable as far as northern Scandinavia and Scotland. Chelishchev (1995*a,b*) reported that 50,000 tons of natural clinoptilolite was processed for use at Chernobyl from the Sokarnitsa (Ukraine), Tedsami (Georgian Republic) and Holinmskoe and Shivirtui (Russia) deposits to treat contaminated water in the reactor and soil in the vicinity, as well as to construct a dyke surrounding a large containment pond. Many thousands of tons of clinoptilolite from the Beli Plast deposit in Bulgaria, were reported to have been dropped by airplane onto the burning reactor to ameliorate the release of radiocaesium to the environment. Secrecy makes it difficult to reach an accurate assessment of the total amount of clinoptilolite that was used at Chernobyl or, in fact, the manner in which it was utilized, although Chelishchev (1995*a*) listed the treatment of low-level activity sewage and laundry waters and the concentration of $^{137}Cs$ and $^{90}Sr$ for later burial as two significant areas of use.

*17.6.1.3 West Valley.* From 1966 to 1972 nuclear fuel was reprocessed on a commercial basis at a Purex plant at the West Valley Nuclear Services site, near West Valley, New York. During this period, the operation produced $\sim 2 \times 10^6$ litres of highly active radioactive waste and stored it in two underground steel tanks shielded in cement vaults (McIntosh *et al.*, 1988). These tanks leaked and zeolites were introduced to contain and store this waste in an environmentally acceptable form (Gockley *et al.*, 1983; Saha, 1985). The alkaline waste in one tank had formed two layers; a supernatant containing $17 \times 10^6$ Ci predominantly as Cs, and an underlying sludge containing approximately the same activity but present mainly as Sr. The laboratory and a 1/12$^{th}$ scale pilot plant used a mixture of synthetic zeolite A(A-51) and a phillipsite-rich tuff from Pine Valley, Nevada, to process simulated waste streams containing high concentrations of $NaNO_3$, NaOH, $Na_2SO_4$ and NaCl (Burn *et al.*, 1987). The success of these tests led to the development of the West Valley Supernatant Treatment System (STS) to treat the supernatant; including the testing of IE-96 (Grant *et al.*, 1987*a,b*). Reports are available that describe the design considerations for STS (Borisch and Leap, 1988; Borisch *et al.*, 1988). Later reports describe the use of IE-96, rather than the phillipsite/A-51 mixture, to decontaminate the supernatant and also solutions from sludge washings, prior to on-site vitrification (Grant *et al.*, 1988; Borisch *et al.*, 1989; Kurath *et al.*, 1990). These authors cite decontamination factors in the range $10^4 - 10^5$ for Cs removal, which means that decontamination of the supernatant should be completed by using ~47,000 kg of IE-96 zeolite. Additionally, Schiffhauser and Thompson (1993) described the use of IE-96 pre-treated with titanium salts in this plant.

### *17.6.2 Radioactive fallout*

The accident at Chernobyl focused the minds of the world on the undesirability of the release of radioisotopes into the environment. In truth, scientific concern and monitoring of radioactivity in the environment started from Hiroshima and continued through the era of nuclear weapons testing. In this respect it is instructional to note the information (see Fig. 17.6) assiduously collected by the University of Helsinki, Finland, on the $^{137}Cs$ whole-body burden of Lapp reindeer herders (Tillander *et al.*, 1989). It can be seen that Chernobyl was responsible for only part of the discharge of this isotope to the environment since 1961. In 1986 the measurements were carried out 3 weeks prior to Chernobyl.

### *17.6.3 Soils and plants*

Before Chernobyl, the ability of soils containing clinoptilolite to take up Cs and Sr radioisotopes was described in a study on the desert soil at Frenchman Flat, a closed drainage basin on the Nevada Test Site (Kautsky, 1984). Earlier Nishita and Haug (1972) had found that addition of 1% clinoptilolite retarded the uptake of $^{90}Sr$, but not $^{137}Cs$, onto beans, clover and barley in pot plant studies.

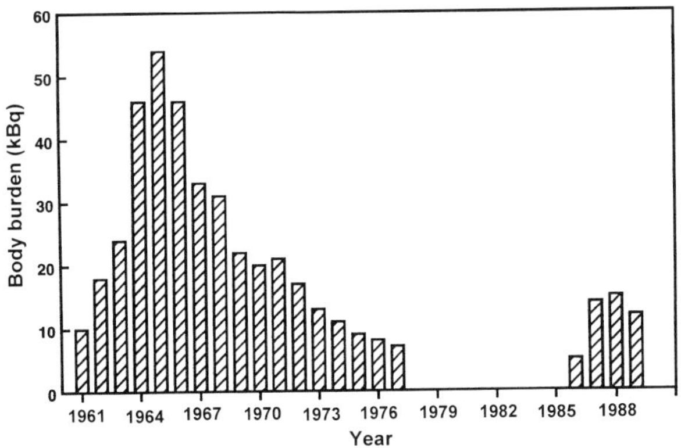


FIG. 17.6 Whole-body measurements of $^{137}Cs$ in male reindeer herders from Finnish Lapland (Tillander *et al.*, 1989).

Soil remediation on samples from Bikini Atoll used a K-rich clinoptilolite from Owyhee County, Idaho. Levels of clinoptilolite additions up to 8 kg/m$^2$ significantly reduced Cs uptake in plants, despite the presence of K (from the zeolite and the fertilizer). Future experiments of this type may aid the return of the Bikinians to their home (Robinson and Stone, 1988).

Since Chernobyl, the work carried out can be divided into that dealing with application of clinoptilolite to the immediate vicinity of the reactor and to work considering a wider geography affected by the event. Ninety per cent of the activity released at Chernobyl was as Cs and Sr isotopes (Chelishchev, 1995*b*) and the local experience is summarized in Table 17.12. Farther afield, successful binding of $^{90}Sr$ and Cs isotopes in various clinoptilolite-amended soils confirms the results listed in Table 17.12. The soils were from Belarus (Davydov, *et al.*, 1990, who also studied $^{144}Ce$ binding), Russia (Velichko *et al.*, 1993), Slovakia (Szabova and Mitro, 1993) and Sweden (Shenber and Erikson, 1989; Shenber and Johanson, 1993). Additions of clinoptilolite to the soils studied were in the range 10–20 tons/ha.

In the United Kingdom the areas most affected by fall-out from Chernobyl were North Wales and Cumbria. Field trials in these areas, carried out over a three-year period, showed that $^{137}Cs$ contents in plant species varied both seasonally and from the effects of the ameliorative agents applied (clinoptilolite, bentonite, hexacyanoferrate). A broad conclusion was that all the agents showed potential for the reduction of radiocaesium soil–plant transfer (Paul and Jones, 1995)

Campbell (1996) suggested that the lack of Cs uptake observed by Nishita and Haug, (1972) may have been due to the smallness of their addition (1%) and to the particle size used. Campbell and Davies (1997) found that clinoptilolite was effective in immobilizing Cs in limed and unlimed soils, when added at 10 wt.% to a lowland loam and an upland peat (both from the

TABLE 17.12 Studies of the use of clinoptilolite at Chernobyl for Cs and Sr radioisotope uptake (taken mainly from Chelishchev, 1995*a*,*b*).

| Use | References |
|---|---|
| Treatment of low-level sewage waters from debalance, reclamation and laundry wastes | Krooglov *et al.* (1990); Nikashina *et al.* (1988) Goncharuk *et al.* (1996) |
| Removal of $^{137}Cs$ and $^{90}Sr$ from potable water | Kravchenko *et al.* (1988) |
| Backfill for engineered barriers and control basins to prevent accidental leakage into areas polluted by radionuclides | Gershevich and Petunkin (1993) |
| Removal and industrial treatment of soils containing $^{137}Cs$ | Gershevitch and Petunkin (1993) |
| Electrodialysis cleaning of radioactive soil by concentrating $^{137}Cs$ on zeolite electrodes | Gershevitch and Petunkin (1993) |
| Zeolite substrates for greenhouse production of uncontaminated vegetables in polluted areas | Chelishchev *et al.* (1986); Chelishchev and Chelishcheva (1980) Firsakova *et al.* (1992) |
| Zeolite addition to agricultural land to decrease $^{137}Cs$ in cultivated crops | Chelishchev *et al.* (1986); Chelishchev and Chelishcheva (1980) Prister *et al.* (1993); Antonova *et al.* (1990) |
| Animal feed supplementation to remove $^{137}Cs$ from farm animals | Chelishchev *et al.* (1985); Prister *et al.* (1993) |
| Zeolite based detergents for laundering radioactively polluted clothes in special laundries | Chelishchev *et al.* (1987) |
| Clinoptilolite used to remove $^{137}Cs$ from the human body and body fluids | Velichkovski *et al.* (1988); Stavitskaya *et al.* (1993) |
| Decontamination of milk and other liquid diary products | Berenshtein *et al.* (1986) |

UK). The efficacy was determined by monitoring the Cs content of rye-grass grown on the soils. A special study on undisturbed coniferous forest soils in Sweden also reports $^{137}Cs$ trapping (Fawaris and Johanson, 1995).

Although not strictly working on soils, Madruga and Cremers (1995) investigated competition between clinoptilolite and freshwater sediments for radiocaesium and Sr uptake to model soil amendments. This theme was developed in a series of recent papers that first established the effect of cations, such as $NH_4^+$, $K^+$, $Mg^{2+}$ and $Ca^{2+}$, on radiocaesium and radiostrontium exchange equilibria on clinoptilolite from Hector, California and a Carpathian mordenite (Valcke *et al.*, 1997*a,b*). A laboratory method was developed to measure the effectiveness of the zeolite amendments made to 15 soils contaminated by Cs and Sr isotopes. The soils were sandy podzols from Belgium, Belarus, Russia and Ukraine; loam-sandy podzols (Ukraine); a forest soil (Belgium); loamy soils from Belgium; and peat soils from Russia and Belarus (Valke *et al*, 1997*c*). Levels of 1–4% zeolite addition created varying reductions in soil isotope levels which could be predicted with reasonable accuracy on the basis of the data from the earlier papers. The final paper (Valcke *et al.*, 1997*d*) studied the growth of spinach on a sandy podzol in the presence of 1–4% zeolite additions. Satisfactory predictions could be made of the observed soil-plant isotope transfer whilst acknowledging that other, as yet undefined, factors also had a role in this transfer.

### *17.6.4 Animals*

Comparisons of *in vivo* uses of clinoptilolite, bentonite and hexacyanoferrates to remove fission products from cattle and sheep have been carried out several times. Additions of hexacyanoferrate (3 g/cow/day), bentonite or clinoptilolite (300 g/cow/day) to feeds reduced the transfer of radiocaesium to milk in lactating cows by 85%, 62% and 35% respectively (Unsworth *et al.*, 1989). Piva *et al.* (1987) reported removals from the same dosage as bentonite $\approx$50% and clinoptilolite $\approx$25%. Clinoptilolite and a hexacyanoferrate were effective in sheep from Scotland, (Phillipo *et al.*, 1988). One study in Prague, (Jandl and Novosad, 1995) using clinoptilolite as a carrier for iron$^{II}$ hexacyanoferrate, showed that a 50 g/sheep/day dose reduced $^{137}$Cs intake into sheep by factors in the range 15–50. Addition of clinoptilolite alone achieved a reduction factor of 10 (see also references in Table 17.11). Forberg *et al.* (1989) found that a synthetic mordenite feed supplement of (10 g/day) for goats and lambs fed with hay contaminated by Chernobyl fall-out reduced radiocaesium body burdens and uptake. Reindeer meat boiled with mordenite lost 8% of its Cs isotope content.

Other related work reports *in vitro* removal of $^{137}$Cs and $^{85}$Sr (but not $^{131}$I) from bovine rumen fluid (Vavrova *et al.*, 1991) by clinoptilolite. De Villiers and De Klerk (1991) recorded that the same zeolite removed 90% radiocaesium and 25% radiostrontium from milk without causing a change in pH. Other materials (titanium phosphate, zirconium phosphate, sodium titanate) gave higher removals but caused unacceptable pH changes.

A dietary supplement of clinoptilolite (2–10%) decreased $^{134}$Cs in the livers, kidneys and femoral muscles of rats with an associated rise of $^{134}$Cs in the faeces (Mizik *et al.*, 1989). Zeolites have been shown to reduce fission products in piglets and broilers (Marinov and Zlatev, 1995).

Japanese work on rats described the reduction in body half-lives for the isotopes $^{54}$Mn (14.4→11.3 days), $^{65}$Zn (16.8→12.7 days) and $^{137}$Cs (5.5→3.9 days) if 10% clinoptilolite was added to their feed (Sato *et al.*, 1993).

### *17.6.5 Man*

Forberg *et al.* (1989), as well as the animal studies cited above, found that synthetic mordenite was not very successful in reducing the body burden of man fed with radiocaesium-contaminated reindeer meat. Stavitskaya *et al.* (1993) examined the use of clinoptilolite for the removal of Cs radionuclides from body fluids and demonstrated that modification of the zeolite with $K^+$ and $Zn^{2+}$ ions produced materials able to balance the salinity of the fluids whilst still removing Cs.

*In vivo* use of a clinoptilolite-based medical drug and as a food additive, can reduce the body burden of radioisotopes in man by factors of 3–5 (Velichkovski *et al.*, 1988). In the Ukraine and Bulgaria, extensive use of clinoptilolite in sweets, pills and other edible forms has helped to eliminate isotopes released by Chernobyl from human beings (L.D. Filizova, pers. comm.). Decontamination of apple and black currant juices by hexacyanoferrates, clays and zeolites (clinoptilolite and mordenite) from $^{134,137}Cs$ arising from Chernobyl, has been demonstrated by Breithaupt *et al.* (1989) with hexacyanoferrates giving the best results, followed by the zeolites.

Thyroid-seeking iodine isotopes can be taken out of water by the use of a silver-exchanged clinoptilolite (Chmielewska-Horvathova and Lesny, 1994). A high capacity (102.7 mg/g) was noted for a sample from Nizny Hrabovec (Slovakia), whilst samples from Wyoming and Mad (Hungary) were less effective (Chmielewska-Horvathova and Lesny, 1996).

## 17.7 Waste containment uses of natural zeolites

The concentration of radioisotopes from large volumes of waste water onto zeolites (or other media) creates radioactive solid waste. Further treatment is then needed to ensure safe containment of these solids into environmentally acceptable forms. If the level of radioactivity is low ($<10^{-4}$ Ci), or intermediate ($1-10^{-4}$ Ci), the containment will probably be in a cement waste form, encapsulated further in a stainless steel container. High-level waste (>1 Ci) is commonly processed as a glass or ceramic prior to encapsulation. (Definitions of types of wastes are those of the IAEA, 1971.) All forms of contained waste are ultimately stored in a repository.

The various waste forms must be well characterized for their thermal, radiation and chemical stabilities. Leaching tests are an important part of this work and involve the elution of waste forms with simulated ground and/or sea waters to determine the effectiveness of the fixation of the radioisotopes in the waste.

### *17.7.1 Natural zeolites in cement waste forms*

That materials, now known to be a zeolite-rich tuff, tend to improve many of the structural properties of cements, has been known from Roman times and pozzolanic concretes containing a much as 25% natural zeolites have been used in many countries (see Aiello, 1995). During the last 25 years, numerous authors have studied the effect of natural zeolites on the compressive strength and resistance to leaching for other non-zeolitic waste forms. For example,

Matsuzoro and Ito (1978) incorporated simulated BWR evaporator concentrates (15 wt.% of Na sulphate adjusted to pH 9–10) into 200 litre drums of Portland and Portland blast-furnace slag cements (Japanese Industrial Standards). The addition of clinoptilolite (25%) reduced leaching of $^{137}$Cs by a factor of 100. Similar results were reported by Pekar and Timul'ak (1985) using a Slovakian clinoptilolite-rich tuff. Elsden *et al.* (1983) described a plant for the immobilization of zeolites in cement and a similar process is being developed in Russia (Bogdanovich *et al.*, 1998).

The past use of organic resins to treat liquid nuclear wastes has left a difficult legacy insofar as the safe containment of these highly radioactive, radiation-unstable materials is concerned. In some cases resin-cement composites have only been a temporary solution and the addition of natural zeolites to the composites has increased their overall stability and resistance to leaching (Nishi *et al.*, 1992; Matsuda *et al.*, 1993; Bagosi and Csetenyi, 1997). Research in Belarus sought to improve the immobilization of low-level wastes from remediation of the Chernobyl area by use of zeolite/polymer/cement composites (Grebenkov *et al.*, 1995) and Xu *et al.* (1986) reported increased compressive strength in grouts and composites created from chemical slurries and cement when 20% additions of clinoptilolite were made. For cemented wastes, consisting of low-level and intermediate-level activity aqueous boric acid and spent resins, Rietmen (1987) found that molten sulphur improved the solidification but produced sulphides which increased the leach rates for both Cs and Co isotopes, as well as decreased the compressive strength. When $H^+$-clinoptilolite was added (~20%), sulphide production was reduced to give a composite with an acceptable leach rate ($9.16 \times 10^{-5}$ g Cs/cm$^2$/day) and a compressive strength of $64.3 \pm 1.2$ N/mm$^2$.

Goto *et al.* (1982) and Mimura and colleagues (e.g. Mimura and Sugano, 1976; Mimura and Kanno, 1985) showed that calcining radioisotope-loaded clinoptilolite, mordenite and chabazite, before incorporation into cement, reduced the leachability of Cs and Sr. Calcination temperatures can be as high as 1300°C, above which pollucite ($CsAlSi_2O_6$) formed. Komarneni and Roy (1981) applied a pressure of 300 bar whilst heating Cs- and Sr-loaded zeolites up to 1000°C (4 h) or 300°C (4 weeks). Leaching with 0.1 M KCl showed that clinoptilolite, chabazite, erionite and mordenites needed the higher temperatures to fix Cs, but 600°C was sufficient to fix Sr. Fixation of both isotopes in analcime and phillipsite was attained at lower temperatures (200–500°C). Pollucite phase formation was given as the major contribution to the fixing of Cs except for analcime, where dehydration alone was successful and phillipsite which formed a Cs feldspar at the lower temperatures. Fixation of Sr may be via a celsian-like phase ($SrAl_2Si_2O_8$).

Glasser and colleagues (e.g. McCulloch *et al.*, 1985) indicated that the pozzolanic nature of clinoptilolite was not wholly advantageous inasmuch as the synergism between zeolite and cement remobilized some Cs from the zeolite, if the zeolite was not pre-calcined to 'fix' the caesium. Aldridge *et al.*

(1994) demonstrated that a Ca-rich clinoptilolite (Werris Creek, NSW, Australia) was less reactive to the calcium hydroxide produced in the hydrating Portland cement than a Na-K clinoptilolite from a Teague-mined deposit in the United States (or synthetic zeolite A). This meant that the Werris Creek zeolite retained Cs more effectively than the other zeolites.

Atkinson *et al.* (1984) reported that the mechanism of leaching from composites of clinoptilolite cement can be considered in terms of the equilibrium and kinetics of ion exchange. A diffusion-based model indicated that Cs leach rates can be described by a $Cs^+$(zeolite) = $\frac{1}{2}Ca^{2+}$ (cement) exchange step, followed by diffusion through the water-filled cement matrix. Strontium leaching is more intricate and involves both chemical interactions and complex ion exchange processes.

### *17.7.2 Natural zeolites in glass and ceramic waste forms*

Ceramic solidification by hot pressing natural zeolites loaded with radioisotopes at 950–1050°C under pressures of 150 kg/cm$^2$ has been shown to be an effective means of mimimizing the leaching of $^{137}$Cs and $^{90}$Sr (e.g. Banba *et al.*, 1980; Briggs *et al.*, 1985; Hash *et al.*, 1996). Vitrification of zeolites results in the formation of glass phases so they are acceptable components in those processes which involve vitrification of wastes (Barner *et al.*, 1984; Bibler *et al.*, 1993). Bogdanova *et al.* (1998) showed that the glassy phases have low leaching properties for Cs isotopes. Mimura and Kanno (1978*c*) reported that some Cs may be volatilized during the fusion process, but Senno *et al.* (1979) demonstrated that zeolite additions at rates as small as 20 g/l were effective in suppressing Cs loss when high-level waste was calcined at 600°C.

Vitrification processes into borosilicate glasses at temperatures in excess of 1100°C have been included in the treatment of waste from Three Mile Island, West Valley and Hanford (Bryan *et al.*, 1987; McIntosh *et al.*, 1988; Bibler *et al.*, 1993). Komarneni *et al.* (1982), based on their work described above, suggested that sintering the loaded zeolite at 1000°C might be an alternative method for treating the Three Mile Island waste.

### *17.7.3 Natural zeolites in nuclear repositories*

The final destination of solid waste containing zeolites loaded with radioisotopes will be in the controlled environments of specially constructed repositories. The potential role of natural zeolites in relation to such repositories can be divided into: (1) improving the retention of radioisotopes within the repository (i.e. the use of crushed zeolite as backfill to enclose waste canisters and to fill tunnels, shafts and rooms) should the steel canisters leak and expose the waste form to groundwaters; and (2) minimizing any external escape of radioisotopes from the repository to the surrounding environment (i.e. groundwaters, host rocks).

*17.7.3.1 Zeolites added to the repository.* Despite earlier negative conclusions (Jacobson, 1977), many reports describe the envisaged use of natural zeolites as backfill materials within sealed repositories. Clinoptilolite or mixtures of montmorillonite (bentonite) and clinoptilolite have been examined for this use (see review by Roy and Burns, 1982). Komarneni and coworkers (e.g. Komarneni and Roy, 1983*a,b*) have tested the stability of materials under expected hydrothermal repository conditions and examined the diffusion of $Cs^+$ and $Cl^-$ through candidate backfill materials in contact with a bittern brine, a likely consequence of a repository constructed in a salt dome (Kumar *et al.*, 1987). Additions of zeolites and silica fume (10–20%) restricted the migration of $Cs^+$.

Chabazite, clinoptilolite and synthetic zeolite A have been tested as potential *in situ* permeable barriers to hinder groundwater Sr migration from the Hanford site to the Columbia River (Cantrell *et al.*, 1994; Cantrell, 1996). All were successful, with clinoptilolite being the most cost-effective (Fuhrmann *et al.*, 1995). Ye *et al.* (1998) concluded that both clinoptilolite and mordenite present in deep geological repositories will remove and strongly retain Sr from groundwaters. In the low-level repository at the Chalk River Laboratories, Ontario, clinoptilolite will be placed on the floor of the repository (Torok *et al.*, 1989) and the same zeolite has been suggested as a reactive barrier to Sr migration (Lee *et al.*, 1998).

Elsewhere, clinoptilolite has been found to retard Sr and Cs radioisotopes (but not Co) leaching from backfill surrounding cemented exchange resins waste (Lee *et al.*, 1992) and Katz *et al.* (1996) have discussed the potential use of engineered soils, which had clinoptilolite additions, for positioning below low-level repositories to restrict Sr and I isotope migration.

*17.7.3.2 Zeolites in bedrock surrounding a repository.* The first high-level nuclear waste repository in the United States is being constructed at Yucca Mountain near the Nevada Test Site in southern Nevada. The repository itself will be within a vitric tuff, a few hundred feet above a thick zeolitic tuff consisting mainly of clinoptilolite (Vaniman and Bish, 1995). These authors described the role of the zeolites in the tuff in retarding radioisotopes that might be released from the waste form in the repository and discussed possible thermal and hydrologic effects on the exchange properties and stability of the zeolite. They concluded that, although the zeolite tuff can be expected to retard the migration of simple fission products (e.g. Sr and Cs), it has no advantage over other common minerals present (e.g. smectites, mica, feldspar, hematite and kaolinite) for complex ions of Np and U. The ability of the clinoptilolite-rich tuff to lose water under thermal loads and to regain water as contaminated solutions pass through it was also foreseen as advantageous in an ageing repository.

Another study (Tsai *et al.*, 1994) recommended that the diffusion of radon through natural mordenite was a good indicator of the attainment of a steady-state in a buffer material associated with a low-level radwaste disposal site.

The take-up of radioisotopes by fracture-filling zeolites in the East Bull Lake Pluton, Massey, Ontario was examined by Ticknor *et al.* (1989) as part of a study of potential nuclear waste repositories in Canada. The sorptive capacity of the fracture-zone minerals for $^{90}Sr$, $^{137}Cs$, $^{144}Ce$ and $^{241}Am$ was greater than that of the primary rock-forming minerals, a significant finding inasmuch as the fracture zones would be likely to play a dominant role in any transport of nuclides from a repository through the rock. Carlos *et al.* (1995) and Chipera *et al.* (1995) drew similar conclusions from studies on clinoptilolite, mordenite, chabazite and analcime as fracture-lining minerals at Yucca Mountain.

## 17.8 Treatment of gaseous emissions from nuclear facilities

Nitrogen oxides can be removed from gaseous effluents produced during the dissolution steps in nuclear fuel reprocessing by using dehydrated clinoptilolite (see Sudo *et al.*, 1989) and $CO_2$ can be sorbed by clinoptilolite from nuclear plant 'blow-offs' (Kefer *et al.*, 1987). The calcination of waste from nuclear fuel reprocessing induces the volatilization ruthenium radioisotopes as $RuO_4$, which can be sorbed from the gaseous state by a mixture of clinoptilolite tuff and synthetic mordenite (Kepak *et al.*, 1990).

The radioisotopes of Ar, Kr and Xe, part of the gaseous emissions from nuclear facilities, can be sorbed in certain natural and synthetic zeolites and committed to long-term storage after hydrothermal vitrification at 340–650°C at pressures >100 bar. Mordenite and chabazite are able to sorb these gases but synthetic zeolite A is the current choice for the process developed at the Kernforschungszen, Karlsruhe, Germany, because of its ability to fix the gases in the amorphous substrate created under the conditions for hydrothermal vitrifcation cited above (Penzhorn *et al.*, 1982).

Finally, it is convenient to mention here that stilbite, natrolite (Dyer and Faghihian, 1998*a,b*) and mesolite (Faghihian and Kazemian, 1998) have been investigated as potential storage materials for tritiated water (3-$H_2O$).

## 17.9 Summary and recommendations for future research

The wide application of natural zeolites to the problems of containing radioisotopes is based principally on their cation exchange properties. Early studies, particularly those of Ames and coworkers (Table 17.1), reported the selectivity of clinoptilolite, mordenite and chabazite for Cs and Sr, and later works have produced accurate isotherms for specific clinoptilolite samples from Yucca Mountain, Nevada (Pabalan, 1991; Pabalan and Bertelli, 1994).

The bulk of data available relates to Cs and Sr and, as yet, the attention given to the other isotopes of environmental concern (Table 17.1), has been relatively sparse (Tables 17.7, 17.8). Similarly clinoptilolite, mordenite and chabazite have been the substrates of choice for the majority of exchange investigations. The requirement for accurate work on multi-cation systems and exchange modelling will continue to grow. It is imperative that cation-exchange selectivities of all natural zeolites be fully understood at the very low concentration levels of cations (radioactive and non-radioctive) present in waste streams to which the zeolites would be exposed in practical operations.

## Acknowledgements

Sincere thanks to Sylvia Thomson for her sterling work in typing the draft of this review, to Laurie Cunliffe for producing the figures and to Dr P. Stone, British Nuclear Fuels, Sellafield, UK, for furnishing the information on the SIXEP process.

## References

Aiello, R. (1995) Zeolitic tuffs as building materials in Italy: A review. Pp. 589–602 in: *Natural Zeolites '93: Occurrence, Properties, Use* (D.W. Ming and F.A. Mumpton, editors). International Committee for Natural Zeolites, Brockport, NY.

Akyuez, T. (1996) Strontium and caesium sorption on some Anatolian Zeolites. *J. Inclusion Phenom. Mol. Recognit. Chem.*, **26**, 89–91.

Aldridge, L.P., Ray, A.S., Stevens, M.G., Day, R.A., Leung, S.H.F., Morassut, P. and Roukis, G. (1994) Immobilization of Cs-contaminated zeolites by Portland cement. *Int. Ceram. Monogr.*, **1**, 1340–5.

Amberson, C.B. and Rhodes, D.W. (1966) Treatment of intermediate- and low-level waste at the National Reactor Testing Station (NTRS). *Proc. Symp. Practices in the Treatment of Low and Intermediate Level Waste, Vienna, 1965*, International Atomic Energy Agency, Vienna, Austria, pp. 419–37.

Ames, L.L., Jr. (1959) *Zeolitic extraction of caesium from aqueous solutions*. United States Atomic Energy Commission Rep., Hanford, Washington, HW-66207, 25 pp.

Ames, L.L., Jr. (1960) The cation sieve properties of clinoptilolite. *Amer. Mineral.*, **45**, 689–700.

Ames, L.L., Jr. (1961) Cation sieve properties of open zeolites. *Amer. Mineral.* **46**, 1120–31.

Ames, L.L., Jr. (1962*a*) Kinetics of caesium reactions with some inorganic cation exchangers. *Amer. Mineral.*, **47**, 1067–78.

Ames, L.L., Jr. (1962*b*) Effect of base cation on the caesium kinetics of clinoptilolite. *Amer. Mineral.*, **47**, 1310–16.

Ames, L.L., Jr. (1962*c*) Characterization of a strontium selective zeolite. *Amer. Mineral.*, **47**, 1317–26.

Ames, L.L., Jr. (1963) Mass action relations of some zeolites in the region of high competing cation concentrations. *Amer. Mineral.*, **48**, 868–82.

Ames, L.L., Jr. (1964) Zeolite equilibria with alkaline earth metal cations. *Amer. Mineral.*, **49**, 1099–110.

Ames, L.L., Jr. (1965) Zeolite cation selectivity. *Canad. Mineral.*, **8**, 325–33.

Ames, L.L., Jr. (1966*a)* Cation exchange properties of wairakite and analcime. *Amer. Mineral.*, **51**, 903–9.

Ames, L.L., Jr. (1966*b*) Exchange of alkali metal cations on a natural stilbite. *Canad. Mineral.*, **8**, 582–592.

Ames, L.L., Jr. and Knoll, KC. (1962) *Loading and elution characteristics of some natural and synthetic zeolites*. Rep. HW–74609, United States Atomic Energy Commission, GEC, Hanford Atomic Products Operation, Richland, Washington, 38 pp.

Ames, L.L., Jr., McGarrah, J.E. and Walker, B.A. (1983*a)* Sorption of trace constituents from aqueous solutions onto minerals. I. Uranium. *Clays Clay Miner.*, **31**, 321–34.

Ames, L.L., Jr., McGarrah, J.E. and Walker, B.A. (1983*b)* Sorption of trace constituents from aqueous solutions onto minerals. II. Radium. *Clays Clay Miner.*, **31**, 335–42.

Andreeva, N.R. and Chernyavskaya, N.B. (1982) Uranyl ion sorption by mordenite and clinoptilolite. *Radiokhimiya*, **24**, 9–13.

Antonova, V.A., Pavlov, I.Y. and Prokofev, O.N. (1990) Possibilities for using zeolites for removing cesium radionuclides from liquid biological samples. *Perspectivy Primeniya Tseditsodorzh Tufov Zabaikal'ya Chita*, 84–9.

Atkinson, A., Nickerson, A.K. and Valentine, T.M. (1984) The mechanisms of leaching from some cement-based nuclear waste forms. *Radioact. Waste Manag. Nucl. Fuel Cycle*, **4**, 357–78.

Bagosi, S. and Csetenyi, T.J. (1997) Immobilization of Cs-loaded ion exchange resins in zeolite-cement blends. *Cement Concrete Res.*, **29**, 479–85.

Banba, T., Tashiro, S., Senno, M. and Araki, K. (1980) Ceramic solidification tests of high level wastes with natural zeolites. I. Effects of treatment procedures and additives. *Japanese Atomic Energy Research Institute*, Tokia, Japan, Rep., JAERI–M-9193, 22 pp.

Barner, J.O., Daniel, J.L. and Marshall, R.K. (1984) *Zeolite vitrification demonstration program. Characterization of radioactive materials*. Rep. GEND-INF-0430, Pacific North West Laboratory, Richland, Washington, 89 pp.

Barrer, R.M. and Townsend, R.P. (1976) Transition metal ion exchange in zeolites. Part 2. Ammines of $Co^{3+}$, $Cu^{2+}$ and $Zn^{2+}$ in clinoptilolite, mordenite and phillipsite. *J. Chem. Soc., Faraday Trans. I,* **72**, 2650–60.

Berak, L. (1963) *The sorption of microstrontium and microcaesium on silicate minerals and rocks*. Ceskoslovniko Akad. Ved. Ustar. Jadereho Vyzkuma, Rez, Rep. UJV-528/63, 40 pp.

Berenshtein, B.G., Donskaya, Ga., Marina, N.B., Marin, V.A. and Samoilova, U.S. (1986) Deactivation of milk on industrial ion exchange plant. Pp. 105–14 in: *Methods of Research of Techological Properties of Fine-grained Minerals and Ores* (N.F. Chelishchev, editor). IMGRE, Moscow (in Russian).

Bertetti, F.P., Pabalan, R.T., Turner, D.R. and Almendarez, M.G. (1996) Neptunium sorption behavior on clinoptilolite, quartz and montmorillonite. *Mater. Res. Soc. Symp. Proc.*, **412**, 631–8.

Bertetti, F.P., Pabalan, R.T., Turner, D.R. and Almendarez, M.G. (1998) Studies of neptunium sorption on quartz, clinoptilolite, montmorillonite and α-alumina. Pp. 131–48 in: *Adsorption of Metals on Geomedia* (E.A. Jenne, editor). Academic, San Diego, California.

Bibler, J.P. and Deininger, J.P. (1991) Decontamination of a benzene waste stream containing caesium and diphenyl mercury. *Waste Manag.*, **1**, 763–6.

Bibler, N.E., Bibler, J.P., Andrews, M.K. and Jantzen, C.M. (1993) Initial demonstration of the vitrification of high-level nuclear waste sludge containing an organic caesium-loaded ion-exchange resin. *Mater. Res. Soc. Symp. Proc.*, **294**, 81–6.

Bocola, W., Boenzi, D., Branca, G. and Lenzi, G. (1968) *Il tuffo gialo Napoletano nel trattamento du effluenti liquidi radioacttivo*. Contratto Euratom Rep. No. 005-65-8-WASI, EUR-3922i.

Bogdanova, V., Fursenko, B.A., Galai, G.I., Belitsky, I.A., Predeinah, M., Pavlyuchenko, VS. and Drobot. IV. (1998) Caesium immobilization using zeolite-containing rocks and high temperature treatment. NATO ASI Series 1, (Defence Nuclear Waste Disposal in Russia. International Perspective), **18**, 69–84.

Bogdanovich, N.G., Konovalov, E.E., Starkov, O.V., Kochetkova, E.A., Grushecheva, E.A., Shumskaya, V.D., Emel'yanov, V.P. and Myshkovskii, M.P. (1998) Caesium and strontium separation from liquid radioactive wastes and immoblization in geocements. *Atomic Energy (N.Y.)*, **84**, 14–8.

Bond, W.D., King, L.J., Knauer, J.B., Hofstetter, K.J. and Thompson, J.D. (1986) Clean-up of demineralizer resins. *ACS Symp. Ser.*, **293**, 250–66. American Chemical Society, Washington, D.C.

Borisch, R.R. and Leap, D.R. (1988) Design and construction of the West Valley supernatant treatment system. *Nucl. Mater. Manag.*, **17**, 54–60.

Borisch, R.R. and Leap, D.R. (1989) The design and construction of the West Valley demonstration project vitrification facility. *Proc. Symp. Waste Manag.*, **1**, 273–7.

Borisch, R.R., Carl, D.E. and Leap, D. (1988) Construction and cold testing of the West Valley supernatant treatment system. *Proc. Symp. Waste Manag.*, **2**, 831–6.

Bostick, D.T., Arnold, W.D., Guo, B. and Burgess, M.W. (1997) The evaluation of sodium-modified chabazite zeolite and resorcinol-formaldehyde resin for the treatment of contaminated waste water. *Sep. Sci. Technol.*, **32**, 793–811.

Branca, G. and Gresson, G. (1972) P. 88 in: *The Use of Local Minerals in the Treatment of Radioactive Waste*. Technical Report No. **136**, International Atomic Energy Agency, Vienna, Austria.

British Nuclear Fuels (1987) *SIXEP and EARP-reduction of radioactive discharges at Sellafield.* British Nuclear Technology Paper 9, Risley, Warrington, UK.

Breithaupt, E., Gahlmann, M., Buehler, K.D. and Grieschner, K. (1989) Possibilities of decontamination of fruit juice containing caesium-134 and caesium-137. *Fluess. Obst.*, **56**, 454–7.

Briggs, A., Jones, D.V.C., Cole, G.B., Valentine, T.H., Preston, R.E., Hawes, R.W.M. and Fones, M.D. (1985) *Immobilization of caesium-137 and strontium-90 in hot pressed clinoptilolite.* Rep. AERE-R-11365, Atomic Energy Research Establishment, Harwell, Oxon, UK, 39 pp.

Bryan, G.H., Gotes, R.W., Knox, C.A. and Knowlton, D.F. (1987) Vitrification of TMI decontaminated zeolite. *Nucl. Chem. Waste Manag.*, **7**, 141–8.

Burn, P., Ploetz, D.K., Saha, A.K., Grant, D.C. and Skriba, M.C. (1987) Design and testing of natural/blended zeolite ion exchange columns at West Valley. *AIChE Symp. Ser.*, **83**, 66–72.

Calle, C., Gili, M., Luce, A., Marrocchelli, A., Pietrelli, L. and Troiani, F. (1990) Chemical treatment of high-level radioactive wastes of ENEA. *Ener. Nucl. (Rome)*, **7**, 69–76.

Campbell, LS. (1996) Radioactive pollution – a mineralogical solution. *Miner. Soc. Bull.*, **110**, 3–5.

Campbell, L.S. and Davies, B.E. (1997) Experimental investigation of plant uptake of caesium from soils amended with clinoptilolite and calcium carbonate. *Plant Soil*, **189**, 65–74.

Cantrell, K.J. (1996) A permeable wall composed of clinoptilolite for containment of Sr-90 in Hanford groundwater. *Proc. 6th Int. Top. Meet. Nucl. Hazard Manag., SPECTRUM'96,* **2**, 1358–65.

Cantrell, K.J., Martin, P.F. and Szecsody, J.E. (1994) Clinoptilolite as an in-situ permeable barrier to strontium migration in ground water. Pp. 839–50 in: *Proc. 33$^{rd}$ in situ Rem. Sci. Basis Curr. Future Technol., Hanford Symp. Health Environ.* (G.W. Gee and R.N. Wing, editors). Battelle Press, Columbus, Ohio, **2**.

Carlos, B.A., Chipera, S.J., Bish, D.L. and Raymond, R. (1995) Distribution and chemistry of fracture-lining zeolites atYucca Mountain, Nevada. Pp. 547–563 in: *Natural Zeolites'93. Occurrence, Properties, Use.* (D.W. Ming and F.A. Mumpton, editors). International Committee for Natural Zeolites, Brockport, New York.

Carrera, L.M., Gomez, S., Bosch, P. and Bulbulian, S. (1993) Removal of cobalt-60 by clays and zeolites. *Zeolites*, **13**, 622–5.

Ceranic, T., Lukic, T. and Jovovic, I. (1987) The ion-exchange capacity for caesium on natural clinoptilolite from the Zlatokep (Vranje) deposit. *J. Serb. Chem. Soc.*, **52**, 553–4.

Chales Suarez, G., Castillo Gomez, R., Jova Sed, L. and De La Cruz Rodriguez, O. (1987) Decontamination of liquid radioactive wastes through chemical treatment and sorption with natural zeolites. *Nucleus Havana*, **2**, 22–7.

Chales Suarez, G., Castillo Gomez, R., Moreno Alvarez, D. and Dominguez Catasus, J. (1998) Study of orption of radionuclides in Cuban natural zeolites by a radioactive indicator method. *Rev. CENIC, Cienc. Quim.*, **29**, 7–11.

Chelishchev, N.F. (1995*a*) Use of natural zeolites at Chernobyl. Pp. 525–532 in: *Natural Zeolites*

*'93: Occurrence, Properties, Use* (D.W. Ming and F.A. Mumpton, editors). International Committee for Natural Zeolites, Brockport, New York.

Chelishchev, N.F. (1995*b*) Use of natural zeolites for liquid radioactive waste treatment (Russian experience). *Stud. Surf. Sci. Catal.*, **98**, 208–10.

Chelishchev, N.F. and Chelishcheva, R.V. (1980) Importance of ion-exchange properties of natural zeolites for the removal of toxic metals from the digestive canal. Pp. 216–26 in: *Natural Zeolites in Agriculture* (A.J. Krupennikova, editor). Metsniereba Press, Tiblisi, Republic of Georgia.

Chelishchev, N.F., Berenshtein, B.G., Marina, N.A., Donskaya, G.A., Marin, B.A., Genkin, M.M., Maslov, S.F. and Chelishcheva, R.V. (1985) Method of cleaning milk and liquid dairy products from radionuclides. *USSR Patent* 1, 366, 399, 4 pp.

Chelishchev, N.F., Berenshtein, B.G., Chelishcheva, R.V., Maslov, S.F. and Genkin, M.M. (1986) Method of cleaning soil from radionuclides. *USSR Patent* 1, 450, 645, 3 pp.

Chelishchev, N.F., Berenshtein, B.G. and Volodkin, V.F. (1987) *Zeolites – New Type of Mineral Raw Materials*. Nedra, Moscow.

Chernyavskaya, N.B., Konstantinovitch, A.A., Andreeva, N.R., Vorob'eva, G.E. and Skirpak, I.Y. (1983) Use of Soviet clinoptilolite for removing caesium and strontium radioisotopes from waste water. *Radiokhimiya*, **25**, 411–4.

Chernyavskaya, N.B. (1985) Sorption of strontium on clinoptilolite and heulandite. *Radiokhimiya*, **27**, 618–21.

Chipera, S.J., Bish, D.L. and Carlos, B. (1995) Equilibrium modelling of the formation of zeolites in fractures at Yucca Mountain, Nevada. Pp. 565–77 in: *Natural Zeolites '93: Occurrence, Properties, Use* (D.W. Ming and F.A. Mumpton, editors). International Committee for Natural Zeolites, Brockport, New York.

Chmielewska, E. and Lesny. J. (1997) A natural ion exchange medium for $^{55}$Fe waste abatement. *J. Radioanal. Nucl. Chem.*, **223**, 243–5.

Chmielewska-Horvathova, E. and Lesny, J. (1992) Adsorption of caesium and barium by zeolite tuffs of Slovakia. *Geol. Carpathica – Clays*, **1**, 47–50.

Chmielewska-Horvathova, E. and Lesny, J. (1994) Sorption of nuclear fission species on natural zeolite. *Miner. Slovakia*, **26**, 290–5.

Chmielewska-Horvathova, E. and Lesny, J. (1996) Sorption of iodine on the surface of modified clinoptilolite. *Pol. J. Environ. Stud.*, **5**, 13–9.

Chmielewska-Horvathova, E. and Lesny, J. (1998) The interaction mechanisms between aqueous solutions of $^{137}$Cs and $^{134}$ Ba radionuclides and local natural zeolites for a deactivation scenario. *J. Radioanal. Nucl. Chem.*, **227**, 151–5.

Collins, E.D., Campbell, D.O., King, L.J., Knauer, J.B. and Wallace, R.M. (1986) Development of the flow sheet for decontaminating high activity level water. *ACS Symp. Series,* **293**, 212–27. American Chemical Society, Washington, D.C..

Collins, J.L., Davidson, D.J., Chase, C.W., Egan, B.Z., Ensor, D.D., Bright, R.M. and Glasgow, D.C. (1993) *Development and testing of ion-exchangers for the treatment of liquid wastes at Oak Ridge National Laboratory*. In: Rep. 1993, Oak Ridge National Laboratory, Oak Ridge, Tennessee, Tm–12315.

Constantopoulou, C., Loisou, Z., Loizidou, M. and Spyrellis, N. (1992) Thorium uptake by natural and synthetic zeolites. Pp. 170–6 in: *Proc. 2nd Eur. Conf. Adv. Mater. Process. 1991* (T.W. Clyne and P.J. Withers, editors). Institute of Materials, London.

Constantopoulou, C.L., Loizidou, M., Loisou, Z. and Spyrellis, N. (1994) Thorium equilibrated with the sodium form of clinoptilolite. *J. Radioanal. Nucl. Chem.*, **178**, 143–51.

Davydov, Y.P., Voronik, N.E., Shatilo, N.N., Vasilevskaya, T.V. and Krupiyanova, I.V. (1990) Problem of caesium-137 and cerium-144 forms determination in the soils of Byelorussia. *Vesti. Akad. Navuk. BSSR. Ser. Fiz. -Ener. Navuk,* **4**, 60–5.

De Paoli, S.M. and Perona, J.J. (1996) Model for Sr-Cs-Ca-Mg-Na ion-exchange uptake kinetics on chabazite. *AIChE J.*, **42**, 3434–41.

De Villiers, W. van Z. and De Klerk, S. (1991) The use of inorganic ion exchangers for the removal of caesium and strontium from milk. *Radiochim. Acta*, **54**, 205–10.

Dias, S.A. and Nott, B.R. (1990) Selective sorbents for water purification in nuclear systems. *Off. Proc. Int. Water. Conf.*, **51**, 330–37.

Diest, J. and Talibudeen, O. (1967) Rb-86 as a tracer for exchangeable potassium in soils. *Soil Sci.*, **104**, 119–22.

Dutu, I.C., Sandor, G.N., Peic, T. and Dinca, G. (1996) Removal of $^{226}$Ra from wastewater resulting from mining and processing of uranium ores with natural zeolites. Pp. 308–10 in: *Proc. 9$^{th}$ Int. Congr. Radiat. Prot. Assoc.* IRPA, Seibersdorf, Austria, **3**.

Dwairi, J. (1992) Evaluation of Jordanian tuffs for decontamination and immobilization of caesium: A comparative study. *Dirasat-Univ. Jordan. Series B*, **19B**, 61–73.

Dyer, A. (1991) Modern theories of ion-exchange and ion-exchange selectivity with particular reference to zeolites. Pp. 33–56 in: *Inorganic Exchangers in Chemical Analysis* (M. Qureshi and K.G. Varshney, editors). CRC Press, Boca Raton, FL.

Dyer, A. and Aggrawal, S. (1995) Removal of fission products from mixed solvents using zeolites. I. Ruthenium removal. *J. Radioanal. Nucl. Chem.*, **198**, 467–74.

Dyer, A. and Aggrawal, S. (1997) Removal of fission products from mixed solvents using zeolites. II. Caesium and strontium removal. *J. Radioanal. Nucl. Chem.*, **221**, 235–48.

Dyer, A. and Faghihian, H. (1998*a*) Diffusion in heteroionic zeolites. I. Diffusion of water in heterionic natrolites. *Microporous Mesoporous Mater.*, **21**, 27–38.

Dyer, A. and Faghihian, H. (1998*b*) Diffusion in heteroionic zeolites. II. Diffusion of water in heteroionic stilbites. *Microporous Mesoporous Mater.*, **21**, 39–44.

Dyer, A. and Josefowicz, L.C. (1992) The removal of thorium from aqueous solutions using zeolites. *J. Radioanal. Nucl. Chem.*, **159**, 47–62.

Dyer, A. and Kadhim, F.H. (1989) Inorganic ion-exchangers for the removal of zirconium, hafnium and niobium radioisotopes from aqueous solutions. *J. Radioanal. Nucl. Chem.*, **131**, 161–9.

Dyer, A. and Keir, D. (1984) Nuclear waste treatment by zeolites. *Zeolites*, **4**, 215–6.

Dyer, A. and Keir, D. (1989) Ion-exchange and pH tolerance in molecular sieve zeolites. *J. Radioanal. Nucl. Chem.*, **132**, 423–41.

Dyer, A. and Khor, A.T. (1980) Inorganic ion-exchangers for the removal of 90-Sr from simulated pond water. Pp. 213–6 in: *Water Chemistry of Nuclear Reactor Systems, 2*. British Nuclear Energy Society, London.

Dyer, A. and Shaheen, T. (1995) Speciation observed by cation exchange. *Sci. Total Environ.* **173–174**, 301–11.

Dyer, A., Abdel Gawad, A.S., Mikhail, M., Enamy, H. and Afshang, M. (1991) The natural zeolite, laumontite, as a potential material for the treatment of aqueous nuclear wastes. *J. Radioanal. Nucl. Chem.*, **154**, 265–75.

Dzhurova, E., Stefanova, I. and Gradev, G. (1989) Geological, mineralogical and ion-exchange characteristics of zeolite rocks from Bulgaria. *J. Radioanal. Nucl. Chem.*, **130**, 425–32.

Eisenman, G. (1962) Cation selective glass electrodes and their mode of operation. *Biophys. J.*, **2**, 259–323.

Elsden, A.D., Sims, J. and Harding, K. (1983) A process for the immobilization of a radioactive waste (inorganic ion-exchange material) in a cement matrix. Pp. 167–76 in: *Proc. Int. Symp. 1982: Conditions for Radioactive Waste Storage and Disposal*. International Atomic Energy Agency, Vienna, Austria.

Faghihian, H. and Kazemian, H. (1998) The use of natural mesolite as a storage material for tritiated water. *J. Radioanal. Nucl. Chem.*, **231**, 153–5.

Fawaris, B.H. and Johanson, K.J. (1995) Sorption of 137-Cs from undisturbed forest soil in a zeolite trap. *Sci. Total Environ.*, **113**, 251–6.

Firsakova, S.K., Grebenchchikova, N.V., Timofeev, S.F. and Novik, A.A. (1992) Effectiveness of agronomic procedures in minimizing the uptake of caesium-137 by pasture lands in the region affected by the accident at the Chenobyl nuclear power reactor. *Dokl. Vses. Akad. S-kh. Nauk imV. I. Lenina*, **3**, 25–7.

Forberg, S., Jones, B. and Westermark, T. (1989) Can zeolites decrease the uptake and acclerate the excretion of radio-caesium in ruminants? *Sci. Total Environ.*, **79**, 37–41.

Frysinger, G.R. (1962) Caesium-sodium exchange on clinoptilolite. *Nature,* **194**, 351–3.

Fuhrmann, M., Aloysius, D. and Zhou, H. (1995) Permeable, subsurface sorbent barrier for 90-Sr. Laboratory studies of natural and synthetic materials. *Waste Manag. (N.Y.)*, **15**, 485–93.

Fullerton, R. (1961) *The effect of $\gamma$ radiation on clinoptilolite.* Rep. HW-69256, United States Atomic Energy Commission, Hanford Atomic Products Operation, Richland, Washington.

Gershevitch, E.G. and Petunkin, N.I. (editors) (1993) *The Use of Natural Zeolites for Reducing Radioactive Contamination in Soils and Agricultural Products.* Republic Techonololgy Centre, Moscow, Russia, 171 pp.

Gockley, G.B., Lahoda, E.J. and Pope, J.M. (1983) Supernatant treatment considerations for the neutralization of West Valley waste. *Mater. Res. Soc. Proc.*, **15**, 663–8.

Godelitsas, A., Misaelides, P., Filippidis, A., Charistos, D. and Anousi, I. (1996*a*) Uranium sorption from aqueous solutions on sodium-form of HEU-type zeolite crystals. *J. Radioanal. Nucl. Chem.*, **208**, 393–402.

Godelitsas, A., Misealides, P., Charistos, D., Fillipidis, A. and Anousis, L. (1996*b*) Interaction of HEU-type zeolite crystals with thorium aqueous solutions. *Chem. Erde*, **56**, 143–56.

Goncharuk, V.V., Kornilovich, B.Y. and Lukachina, V.V. (1996) Removal of radioactive contaminants from water by means of natural sorbents. *Khim. Tekhnol. Vody*, **18**, 131–9.

Gopala Rao, M. (1995) Binary and ternary ion-exchange equilibria in zeolitized tuff: Cs-Sr-Co system. *Sep. Sci. Technol.*, **30**, 1385–96.

G'oshev, G., Darev, K. and Aleksiev, A. (1987) Selective sorption of caesium-137 on a clinoptilolite modified by hexacyanoferrates (II). *Yad. Ener.*, **24**, 74–81.

Goto, Y., Marsuzawa, J. and Matsuda, S. (1982) Removal of caesium and strontium by natural zeolites and the improved zeolite from aqueous solution and their fixation in the zeolites. *Devel. Sedimentol.*, **35**, 789–98.

Gradev, G. and Gulabova, I. (1982) Caesium and strontium ion exchange on clinoptilolite and mordenite. I. Ion-exchangers. Equilibrium characteristics of the ion-exchange process. *Yad. Energ.*, **16**, 64–72.

Grant, D.C., Skriba, M.C., Saha, A.K. and Ploetz, D.K. (1987*a*) Effect of process variables on the removal of contaminants from radioactive waste streams using zeolites. *Proc. Symp. Waste Manag.*, **3**, 609–12.

Grant, D.C., Skriba, M.C. and Saha, A.K. (1987*b*) Removal of radioactive contaminants from West Valley waste streams using natural zeolites. *Environ. Prog.*, **6**, 104–9.

Grant, D.C., Skriba, M.C., Saha, A.K. and Ploetz, D.K. (1988) Design and optimization of zeolite ion exchange system for the treatment of radioactive wastes. *AIChE Symp. Ser.*, **84**, 13–22.

Grebenkov, A.J., Trubinkov, V.P., Verzhynskaja, A.B., Nikolajeva, V.B. and Kapustina, I.B. (1995) Improvement of characteristics of solidified radioactive waste form with polymer-concrete layer. *Mater. Res. Soc. Symp. Proc.*, **353**, 929–34.

Gulis, G. and Timulak, J. (1986) Sorption of radionuclides on natural minerals. *J. Radioanal. Nucl. Chem.*, **159**, 21–5.

Gumus, A., Baksi, G., Taner, M. and Olmez, S. (1992) Adsorption of uranium from aqueous solution by natural zeolite. Pp. 79–83 in: *$8^{th}$ Kim. Kim. Muhenndisligi Semp.* (A. Aydin, editor). Marmara University Faculty of Science and Letters, **3**, Istanbul, Turkey.

Hafez, M.B., Salem, F. and Eldesoki, M. (1978) Fixation mechanism between zeolite and some radioactive elements. *J. Radioanal. Nucl. Chem.*, **47**, 115–9.

Harjula, R. and Lehto, J. (1986) Effect of sodium and potassium ions on caesium absorption from nuclear power plant waste solutions on synthetic zeolites. *Nucl. Chem. Waste Manag.*, **6**, 133–7.

Harjula, R. and Lehto, J. (1993) An industrial-scale process for the removal of caesium-137 utilising hexacyanoferrate columns – development and first run. Pp. 73–80 in: *Ion Exchange Processes. Advances and Applications* (A. Dyer, M.J. Hudson and P.A. Williams, editors). Royal Society of Chemistry, Cambridge, UK.

Harjula, R., Dyer, A., Pearson, S.D. and Townsend, R.P. (1992) Ion-exchange in zeolites. Part 1. Prediction of $Ca^{2+}$- $Na^{+}$ equilibria in zeolites X and Y. *J. Chem. Soc. Faraday Trans. I*, **88**, 1591–7.

Harjula, R., Lehto, J., Tusa, E.H. and Paavolo, A. (1994) Industrial scale removal of caesium with hexacyanoferrate exchanger-process development. *Nucl. Technol.*, **107**, 272–8.

Hash, M.C., Pereira, C., Lewis, M.A., Blaskovitz, R.J., Zyryanov, V.N. and Ackerman, J.P. (1996) Hot pressing of glass-zeolite composites. *Ceram. Trans.* **72** (Environmental Issues and Waste Management Technologies in the Ceramic and Nulcear Industries II), 135–43.

Hawkins, D.B. (1964) *Removal of cobalt and chromium by precipitation and ion-exchange on soil, lignite and clinoptilolite.* Rep. IDO-12036, United States Atomic Energy Commission, Idaho Operations Office, Idaho Falls, Idaho, 24 pp.

Hawkins, D.B. (1967) Zeolite studies II. Ion-exchange properties of some synthetic zeolites. *Mater. Res. Bull.*, **2**, 1021–8.

Hawkins, D.B. and Short, M.L. (1965) *Equation for the sorption of strontium and caesium on soil and clinoptilolite.* Rep. IDO-12046, United States Atomic Energy Commission, Idaho Operations Office, Idaho Falls, Idaho, 40 pp.

Hlozek, P., Lukac, P. and Foldesova, M. (1992*a*) Influence of some cations on the sorption ability of modified clinoptilolites. *J. Radioanal. Nucl. Chem.*, **164**, 109–13.

Hlozek, P., Foldesova, M. and Lukac, P. (1992*b*) Study of sodium hydroxide treated clinoptilolites and their physical and ion-exchange characteristics with regard to cesium(1+) and cobalt(2+). *J. Radioanal. Nucl. Chem.*, **165**, 175–83.

Hofstetter, K.J. and Hitz, C. (1983) The use of the Submerged Demineralizer System (SDS) at Three Mile Island. *Sep. Sci. Technol.*, **18**, 1747–64.

Hofstetter, K.J. and Hitz, C. (1986) Water clean-up systems. *ACS Symp. Series*, **293**, 228–38. American Chemical Society, Washington, D.C.

Hofstetter, K.J., Hitz, C., Harner, K.L., Stoner, P.S., Chevalier, G., Collins, H.E., Grahn, P. and Pitka, W.F. (1982) Chemistry support for submerged demineralizer systems operation at Three Mile Island. Pp. 301–12 in: *Proc. 25$^{th}$ Conf. Anal. Chem. Energy Technol., 1981* (W.S. Lyons, editor). Ann Arbor Science, Ann Arbor, Michigan.

Howden, M. (1983) British Nuclear Fuel's site ion-exchange effluent plant. *Inst. Chem. Eng. Symp. Ser.*, **77**, 253–78.

Howden, M. and Pilot, J. (1984) The choice of ion-exchanger for British Nuclear Fuel's site ion-exchange effluent plant. Pp. 66–73 in: *Ion Exchange Technology* (D. Naden and M. Streat, editors). Horwood, Chichester, UK.

Hsu, C.-N., Liu, D.-C. and Cheng, H.-P. (1994) Evaluation of caesium sorption on natural mordenite. *J. Radioanal. Nucl. Chem.*, **185**, 319–29.

Imai, K., Watari, K., Koyanagi, T., Kitao, K. and Kawamura, S. (1990) Decontamination of radionuclides from laboratory waste water containing seawater with special reference to radioruthenium. *Hoken Butsuri*, **25**, 161–4.

International Atomic Energy Agency (IAEA) (1967) *Operation and control of ion-exchange processes for the treatment of radioactive wastes.* Tech. Rept. **78**, International Atomic Energy Agency, Vienna, Austria, 147 pp.

International Atomic Energy Agency (IAEA) (1971) *Standardization of radioactive waste categories*. Tech. Rept. **101**, International Atomic Energy Agency, Vienna, Austria, 18 pp.

International Atomic Energy Agency (IAEA) (1972) *Use of local minerals in the treatment of radioactive waste*. Tech. Rept. **136**, International Atomic Energy Agency, Vienna, Austria, 113 pp.

International Atomic Energy Agency (IAEA) (1987) *Techniques and practices for the pretreatment of low and intermediate level solid and liquid radioactive wastes*. Tech. Rept.

**272**, International Atomic Energy Agency, Vienna, Austria, 107 pp.

Jacobson, A. (1977) *A short review of the formation, stability and cementing properties of natural zeolites*. Rept. KBS -TR-27, Tek. Hoeg. Luleaa Sweden, 29 pp.

James, K.L. and Miller, C.C. (1989) Ion-exchange media testing for recyclable and nonrecyclable liquids at Diabolo Canyon. *Power Plant Proc. Symp. Waste Manag*., **2**, 431–6.

Jandl, J. and Novosad, J. (1995) *In-vivo* reduction of radiocaesium by modified clinoptilolite in sheep. *Vet. Med. (Prague)*, **40**, 237–41.

Juznic, K. (1988) Migration of radio-caesium in a zeolite-solution system. *J. Radioanal. Nucl. Chem*., **126**, 315–22.

Juznic, K., Golkiewicz, P. and Fajt, P. (1990) Sorption and migration of radiocaesium in natural zeolite-water systems. *Vest. Slov. Kem. Drus*., **37**, 1–8.

Kalandiya, A.A. (1979) Decontamination of drinking water from137-Cs and 90-Sr using Askanite clay activated with acid, and clinoptilolite. *Zh. Prikl. Khim. (Leningrad)*, **52**, 1732–6.

Kanno, T. and Hashimoto, H. (1976) Elution of natural zeolites and the separation of Cs and Sr using EDTA. *Daigaku Senko Seiren Kenkyusho Ito*, **32**, 14–21.

Kanno, T., Mimura, H. and Akiba, K. (1987) Dinitration of high level liquid waste and removal of caesium and stontium with zeolites. Pp. 91–100 in: *Chemical Aspects of a Down Steam Thorium Fuel Cycle* (S. Susuki, editor). Atomic Energy Society of Japan, Tokyo, Japan.

Katz, L.E., Humphrey, D.N., Jankauskas, P.T. and DeMascio, F.A. (1996) Engineered soils for low-level radioactive waste disposal facilities; effects of additives on the adsorptive behaviour and hydraulic conductivity of natural soils. *Hazard. Waste Hazard. Mater.*, **13**, 283–306.

Kautsky, M. (1984) *Sorption of caesium and strontium by arid region desert soil*. Rept. DOE/NV/10162-16, Department of Energy, USA, 111 pp.

Kefer, R.G., Puzanov, I.S. and Starobinets, S.E. (1987) Purification of technological power-plant blow-offs with dissociating coolant from carbon dioxides. *Vesti. Akad. Navuk BSSRR, Ser. Fiz. -Energ. Navuk*, **2**, 117–8.

Kent, T.E., Arnold, W.D., Bond, W.D., Collins, D.L., Davidson, D.J., Taylor, P.A. and Williams, D.F. (1993) Overview of laboratory treatment studies for decontamination of liquid low-level waste at Oak Ridge National Laboratory. *Technol. Programs. Radioact. Waste Manag. Environ. Restor*., **2**, 1701–6.

Kepak, F., Koutova, S. and Kanka, J. (1990) Sorption of ruthenium-106 tetroxide vapours on natural and synthetic zeolites. *Isotopenpraxis*, **26**, 73–8.

Kim, B.T., Lee, H.K., Moon, H. and Lee, K.J. (1995) Adsorption of radionuclides from aqueous solutions by inorganic adsorbents. *Sep. Sci. Technol*., **30**, 3165–82.

King, L.J., Campbell, D.O., Collins, E.D., Knauer, J.B. and Wallace, R.M. (1984) Evaluation of zeolite mixtures for decontamination of high-activity level water in the Submerged Demineralizer System (SDS) at the Three Mile Island Nuclear Power Station, Unit 2. Pp. 660–8 in: *Proc. 6th Int. Zeolite Conf. Reno, Nevada, 1983* (D. Olson and A. Bisio, editors). Butterworth, Guildford, UK.

Komarneni, S. (1985) Phillipsite in caesium decontamination and immobilization. *Clays Clay Miner*., **33**, 145–51.

Komarneni, S. and Roy, R. (1981) Zeolites for fixation of caesium and strontium from radwaste by thermal and hydrothermal treatments. *Nucl. Chem. Waste Manag*., **2**, 259–64.

Komarneni, S. and Roy, R. (1982) Alternative radwaste solidification route for Three Mile Island waste. *J. Amer. Ceram. Soc*., **65**, C198.

Komarneni, S. and Roy, D.M. (1983*a*) Hydrothermal interactions of cement or mortar with zeolites or montmorillonites. *Mater. Res. Soc. Proc*., **15**, 55–62.

Komarneni, S. and Roy, D.M. (1983*b*) Alteration of clay minerals and zeolites in hydrothermal brines. *Clays Clay Miner.*, **31**, 383–91.

Kravchenko, V.I., Lonkunova, A.Y. and Naumova, O.P. (1985) Strontium sorption from process water of concentrating plants. *Khim. Tekhnol. Vody*., **7**, 80–1.

Kravchenko, V.A., Tarasevich, Y.I., Rudenko, G.G., Kujushko, S.G., Koroctishevsky, A.S. and

Kravchenko, N.D. (1988) Purifying drinking water with the use of clinoptilolite filters. *Chem. Technol. Water*, **3**, 255–9.

Krooglov, S.V., Aleksahin, R.M. and Vasileva, N.A. (1990) Forming of radionuclide composition of soils on Chernobyl area. *Pochvovedenie*, **19**, 26–34.

Kumar, A., Komarneni, S. and Roy, D.M. (1987) Diffusion of $Cs^+$ and $Cl^-$ through sealing materials. *Cem. Conc. Res.*, **17**, 153–60.

Kurath, D.E., Bray, L.A., Ross, W.A. and Ploetz, D.K. (1990) Correlation of laboratory testing and actual operations for the West Valley supernatant treatment system. *Ceram. Trans.*, **9**, 529–38.

Kwon, S.G., Kim, E.H., Lee, E.H., Lee, K.I. and Yoo, J.H. (1996) Co-separation of Cs and Sr from simulated liquid wastes by mixed zeolites. *Hwahak Konghak*, **34**, 570–5.

Lai, P.P. and Rees, L.V.C. (1976) Szilard-Chalmers cation recoil studies in zeolites X and Y. Part 2. Recoils from open to locked sites. *J. Chem. Soc. Faraday Trans. I*, **72**, 1818–26.

Las, T. (1989) *Use of natural zeolites for radioactive waste treatment*. PhD thesis, Univ. Salford, UK.

Laske, D. and Laube, K. (1989) Zeolites as ion-exchangers in radioactive waste reprocessing and conditioning. *Swiss. Chem.*, **11**, 27–8, 31.

Lee, D.R., Smyth, D.J.A. Shikase, S.G., Jowett, R., Hartwig, D.S. and Milloy, C. (1998) Wall-and -curtain for passive collection/treatment of contaminant fumes. Descriptions of applied treatment technology. Pp. 77–84 in: *1st Int. Conf. Remediation of Chlorinated and Recalcitrant Compounds* (G.B. Wickramamayake and R.E. Hinchee, editors). Batelle Press, Columbus, Ohio, U.S.A.

Lee, E.H., Lee, W.K. Yoo, J.H. and Park, H.S. (1993*a*) Separation behaviour of Cs and Sr on the various zeolites. *Kongop Hwahak*, **4**, 731–8.

Lee, E.H., Yoo, J.H. and Park, H.S. (1993*b*) Ion exchange kinetics for Cs and Sr in a batch zeolite system. *Kongop Hwahak*, **4**, 739–45 (in Korean).

Lee, J.O., Han, K.W. and Buckley, L.P. (1992) Short-term leaching behaviour of cobalt-60, strontium-85 and caesium-137 from cemented ion-exchange resin waste forms. *ASTM Spec. Techn. Publ.* **STP 1123**, 204–16.

Lee, S.H., Sung, N.J. and Park, W.J. (1979) *Studies on the radioactive waste treatment of caesium. Adsorption by clinoptilolite.* Rept. KAERI-220/RR-89/79, Korean Atomic Energy Research Instititute, Seoul, Korea, 39 pp (in Korean).

Lee, W.K. (1988) A study on caesium adsorption on domestic zeolite. *Kumsok Pyomyon Choli.*, **21**, 47–52.

Lenzi, G. and Cassano, G. (1988) The zeolitized tuffs of Mount Vulture (Basilicata, Italy). Mineralogical and physico-chemical study for use as decontaminanants for radioactive solutions. Pp. 803–40 in: *Occurence, Properties and Ultization of Natural Zeolites* (D. Kallo and H.S. Sherry, editors). Akademai Kiado, Budapest, Hungary.

Liang, T.J. and Hsu, C.N. (1993) Sorption of caesium and strontium on natural mordenite. *Radiochim. Acta*, **61**, 105–8.

Liang, T.J. and Joseph, Y.C. (1995) Sorption kinetics on natural mordenite. *Appl. Radiat. Isotop.*, **46**, 7–12.

Lukac, P. and Foldersova, M. (1990) Zeolite ability in model cobalt-60, caesium-137 labeled solutions. *Jad. Energ.*, **36**, 186–8.

Lukac, P. and Foldersova, M. (1994) Sorption properties of chemically treated clinoptilolites with respect to Cs and Co. *J. Radioanal. Nucl. Chem.*, **188**, 427–37.

Lukac, P., Hlozek, P. and Foldersova, M. (1992*a*) Sorption of caesium-137 and cobalt-60 by chemically treated clinoptilolite and mordenite. *Geol. Carpathica – Clays*, **2**, 125–8.

Lukac, P., Hlozek, P. and Foldesova, M. (1992*b*) Sorption ability of chemically treated clinoptilolites with regard to Cs and Co. *J. Radioanal. Nucl. Chem.*, **164**, 141–6.

Luneva, N.K., Ratko, A.I. and Petushak, I.A. (1994) Mineral-fibre sorbents for concentrating radionuclides. *Radiokhimiya*, **36**, 337–9.

Madruga, M.J. and Cremers, A. (1995) Use of clinoptilolite amendment as a counter-measure for contaminated soils. Pp. 503–11 in: *Proc. Int. Symp., Environmental Impact of Radioactive Releases.* International Atomic Energy Agency, Vienna, Austria.

Marinov, V.M. and Zlatev, A.D. (1995) Use of natural and modified sorbents to reduce the uptake of man-made radionuclides by farm animals. 644–6: *Proc. Int. Symp., Environmental Impact of Radioactive Releases.* International Atomic Energy Agency, Vienna, Austria.

Marrocchelli, A. and Pietrelli, L. (1989) Caesium adsorption with zeolites from nuclear high salt content alkaline wastes. *Solvent Extr. Ion Exch.* **7**, 159–72.

Mathers, W.G. and Watson, L.C. (1962) *A waste experiment using mineral exchange on clinoptilolite.* Rept. AECL-1521, Atomic Energy Commission of Canada, Chalk River, Ontario, Canada, 35 pp.

Matsuda, M., Kikuchi, M., Nishi, T., Tamata S. and Izumida, T. (1993) Conditioning of spent ion exchange resin using high performance cement. *Technol. Programs Radioact. Waste Manag. Environ. Restor.*, **2**, 1569–72.

Matsuzoro, H. and Ito, A. (1978) Immobilization of caesium-137 in cement-waste composites by addition of natural zeolites. *Health Phys.*, **34**, 128–30.

McCulloch, C.E., Angus, M.J., Crawford, R.W., Rahman, A.A. and Glasser, F.P. (1985) Cements in radioactive waste disposal: some mineralogical considerations. *Mineral. Mag.*, **49**, 211–21.

McIntosh, T.W., Bixby, W.W., Krauss, J.E. and Leap, D.R. (1988) An overview of waste management systems at the West Valley Demonstration Project. *Proc. Symp. Waste Manag.*, **2**, 785–90.

Mercer, BW. (1966) *Adsorption of trace ions from intermediate level wastes by ion exchange.* Rept. BNWL-180, Battelle-Northwest, Pacific Northwest Lab., Richland, Washington, 33 pp.

Mercer, B.W. and Ames, L.L., Jr. (1963) *The adsorption of caesium, strontium and cerium on zeolites from multi-cation systems.* Rept. HW-78461, US Atomic Energy Commission. GEC, Hanford Atomic Products Operation, Richland, Washington, 65 pp.

Mercer, BW. and Ames, L.L., Jr. (1978) Zeolite ion-exchange in radioactive and municipal wastewater treatment. Pp. 451–62 in: *Natural Zeolites: Occurrence, Properties, Use* (L.B. Sand and F.A. Mumpton, editors). Pergamon Press, Elmsford, New York.

Milyutin, V.V., Gelis, V.M. and Penzin, R.A. (1993) Sorption-selective characteristics of inorganic sorbents and ion-exchange resins towards caesium and strontium. *Radiokhimiya*, **35**, 76–82.

Mimura, H. and Kanno, T. (1978*a*) Distribution of Cs, Sr, Ce, Eu and Tb into Na- and H-type zeolites. *Tohoku Daigaku Senko Seiren Kenkyusho Iho*, **34**, 85–90.

Mimura, H. and Kanno, T. (1978*b*) Distribution of Co, Cr and Fe into Na and H-type zeolites. *Tohoku Daigaku Senko Seiren Kenkyusho Iho*, **34**, 58–66.

Mimura, H. and Kanno, T. (1978*c*) Processing of radioactive waste with zeolites IV. Volatilization of caesium from zeolites at high temperatures. *Nippon Genshiryoko Gakkaishi*, **20**, 282–6.

Mimura, H. and Kanno, T. (1985) Distribution and fixation of caesium and strontium in zeolite A and chabazite. *J. Nucl. Sci. Technol.,* **22** 284–91.

Mimura, H. and Sugano, T. (1976) Acid resistance and extraction of caesium and strontium from calcined natural zeolites. *Kakuriken Kenkyu Hokoku (Tohuko Daigaku)*, **9**, 157–60.

Mimura, H., Kimura, T. and Kanno, T. (1981) Effect of diverse ions, column temperature and flow rate on the dynamic exchange properties of caesium in various types of zeolites.*Tohoku Daigaku Senko Seiren Kenkyusho Iho*, **37**, 145–52.

Mimura, H., Numano, K. and Kanno, T. (1984) Effect of hydrofluoric acid on distribution of caesium and strontium between zeolites and aqueous solutions. *Tohoku Daigaku Senko Seiren Kenkyusho Iho*, **40**, 173–80.

Mimura, H., Yamagashi, I. and Akiba, K. (1988*a*) Removal of caesium and strontium from high activity level water by zeolites. *Tokoku Daigaku Senko Seiren Kenkyusho Iho,* **44**, 1–7.

Mimura, H., Yamagishi, I. and Akiba, K. (1988*b*) Selective elimination of caesium and strontium in highly contaminated waters by adsorption by zeolites. *Karuriken Kenkyu Hokoku (Tohoku Daigaku)*, **21**, 64–70.

Mimura, H., Yamagashi, I. and Akiba, K. (1989) Removal of radioactive caesium and strontium by zeolites. *Nippon Kagaku Kaishi*, **3**, 621–7.

Mimura, H., Akiba, A. and Igarashi, H. (1993) Removal of heat-generating nuclides from high-level liquid waste through mixed zeolite columns. *J. Nucl. Sci. Technol.*, **30**, 239–47.

Mimura, H., Kobayashi, T. and Akiba, K. (1995) Chromatographic separation of strontium and caesium with mixed zeolite columns. *J. Nucl. Sci. Technol.*, **32**, 60–7.

Misaelides, P. and Godelitsas, A. (1995) Removal of heavy metals from aqueous solutions using pretreated natural zeolitic materials: the case of mercury(II). *Toxicol. Environ. Chem.*, **51**, 21–9.

Misaelides, P., Godelitsas, A. and Filippidis, A. (1995) The use of zeoliferous rocks from Metaxades, Thrace, Greece for the removal of cesium from aqueous solutions. *Fresnius Environ. Bull.*, **4**, 227–31.

Misaelides, P., Godelitsas, A., Filippidis, A., Charistos, D. and Anousis, I. (1995) Thorium and uranium uptake by natural zeolitic materials. *Sci. Total Environ.*, **173**, 237–46.

Misaelides, P., Godelitsas, A., Noli, F. and Kossionidis, S. (1998) $^{12}$C-RBS investigation of scolecite crystals interacted with Cs- and Sr- aqueous solutions. *Nucl. Instrum. Methods Phys. Res., Sect. B*, **139**, 249–52.

Mizik, P., Hrusovsky, J. and Tokosova, M. (1989) Influence of natural zeolite on the excretion and distribution of radio-active cesium in rats. *Vet. Med. (Prague)*, **40**, 467–74.

Mondale, K.D., Mumpton, F.A. and Aplan, F.F. (1978) Benificiation of natural zeolites from Bowie, Arizona. A preliminary report. Pp. 527–37 in: *Natural Zeolites, Occurrence, Properties, Use* (L.B. Sand and F.A. Mumpton, editors). Pergamon Press, Oxford, UK.

Nelson, J.L., Ames, L.L., Jr. and Mercer, B.W. (1964) *Characterization and applications of zeolites for radioactive waste treatment*. Rept. HW–SA–3333, United States Atomic Energy Commission, GEC, Hanford Atomic Products Operation, Richland, Washington, 30 pp.

Nikashina, V.A., Tyurina, V.A., Senyavin, M.M., Stefanov, G.L., Gradev, G.D., Stefanov, I.G. and Avramova, A.F. (1988) Comparative characteristics of the ion-exchange properties of natural zeolites from Bulgaria and the USSR for the purpose of purification of liquid sewage from atomic power stations. Part I. Study of the equilibrium sorption of caesium and strontium ions from solutions of different compositions. *J. Radioanal. Nucl. Chem.*, **105**, 293–8.

Nishi, R., Matsuda, M., Chino, K. and Kikuchi, M. (1992) Reduction of caesium leachablity from cementitious resin forms using natural acid clay and zeolites. *Cem. Concr. Res.*, **22**, 387–92.

Nishita, H. and Haug, R.M. (1972) Influence of clinoptilolite on Sr-90 and Cs-137 uptakes by plants. *Soil Sci.*, **114**, 149–57.

Novosad, J., Jandl, J. and Woollins, J.D. (1992) Characteristics of modified clinoptilolite. *Radioanal. Nucl. Chem.*, **165**, 287–94.

Obradovic, J. (1988) Occurences and genesis of sedimentary zeolites in Serbia. Pp. 59–69 in: *Occurrences, Properties and Utilization of Natural Zeolites* (D. Kallo and H.S. Sherry, editors). Akademai Kiado, Budapest, Hungary.

Odzhagov, G.O., Babaev, I.S., Kasumova, S.M., Abdurakhamanov, F.Y. and Mukhtarova, F.M. (1986) Ion-exchange sorption of strontium and caesium by clinoptilolite from the Aidag deposit of the Azerbaizhan SSR. *Azerb. Khim. Zh.* **5**, 138–42.

Olguin, M.T., Solache, M., Asomoza, M., Acosta, D., Bosch, P. and Bulbulian, S. (1994) $UO_2^{2+}$ sorption in natural Mexican erionite and Y zeolite. *Sep. Sci. Technol.*, **29**, 2161–78.

Olguin, M.T., Garcia-Sosa, I. and Solache-Rios, M. (1996*a*) Sorption of strontium by Mexican erionite. *J. Radioanal. Nucl. Chem.*, **290**, 415–22.

Olguin, M.T., Solache, M., Iturbe, J.L., Bosch, P. and Bulbulian, S. (1996*b*) Sorption of Np-239 and U-235 fission products by zeolite Y, Mexican natural erionite and bentonite. *Sep. Sci. Technol.*, **31**, 2021–44.

Pabalan, R.T. (1991) Nonideality effects on the ion-exchange behaviour of the zeolite mineral clinptilolite. *Mater. Res. Soc. Symp. Proc.*, **212**, 559–67.

Pabalan, R.T. and Bertetti, F.P. (1994) Thermodynamics of ion-exchange between $Na^+/Sr^{2+}$ solutions and the mineral clinoptilolite. *Mater. Res. Soc. Symp. Proc.*, **333**, 731–8.

Pabalan, R.T., Prikryl, J.D., Muller, P.M. and Diefrich, T.B. (1993) Experimental study of uranium(6+) sorption onto the zeolite mineral clinoptilolite. *Mater. Res. Symp. Proc.*, **294**, 777–82.

Pacheco, G., Nova, G., Bosch, P. and Bulbulian, S. (1994) Comparative structural stability of natural clays and zeolites in contact with 60-$Co^{2+}$ aqueous solutions. *J. Radioanal. Nucl. Chem.*, **187**, 431–4.

Panasyug, A.S., Trofimenko, N.E., Masherova, N.P., Rat'ko, A.L. and Golikova, N.I. (1993) Caesium and strontium sorption from mineralized aqueous solutions on natural aluminosilicates modified by heavy-metal ferrocyanides. *Zh. Prikl. Khim. (Saint Petersburg)*, **66**, 2119–22.

Paul, L. and Jones, D.R. (1995) Effects of ameliorative treatment applied to radiocesium contaminated upland vegetation in the United Kingdom. Pp. 513–22 in: *Proc. Int. Symp., Environmental Impact of Radioactive Releases*. International Atomic Energy Agency, Vienna, Austria.

Pekar, A. and Timul'ak, J. (1985) Application of natural zeolite tuff in cementation of radioactive concentrates. *Jad. Energ.*, **31**, 128–30.

Pekov, G., Daiev, K. and Stefanova, I. (1980) Evaporation method for regenerating boric acid from radioactive wastewaters after their preliminary deactivation using natural clinoptilolite. *God Sofii. Univ., Khim. Fak.*, **71**, 59–66.

Penzhorn, R.D., Schuster, P., Lietzig, H. and Noppel, H.E. (1982) Noble gas immobilization in zeolites. *Ber. Bunsenges. Phys. Chem.*, **86**, 1077–82.

Perona, J.J. (1993*a*) Model for strontium-caesium-calcium-sodium ion-exchange equilibria on chabazite. *AIChE J.*, **39**, 1716–20.

Perona, J.J. (1993*b*) *A multi-component ion-exchange equilibrium model for chabazite columns treating ORNL wastewaters*. Rept. ORN4/JM-12272, Oak Ridge National Laboratory, Oak Ridge, Tennessee, 27 pp.

Perona, J.J., Coroneos, A.C., Kent, T.E. and Richardson, SA. (1995) A simple model for strontium break-through on zeolite columns. *Sep. Sci. Technol.*, **30**, 1259–68.

Phillipo, M., Gvozdanovic, S., Gvozdanovic, D., Chesters, J.K., Paterson, E. and Mills, C.F. (1988) Reduction of radiocaesium absorption by sheep contaminated with fallout from Chernobyl. *Vet. Rec.*, **122**, 560–3.

Piva, G., Fusconi, G., Fabri, S., Lusardi, E., Stefani, L., Modenesi, R. and Pacifici, L. (1987) Effect of alimentary addition of sequestering agents on caesium-137 and caesium-134 activity in dairy cow milk. *Ann. Fac. Agrar. (Univ. Cattol. Sacro Cuore)*, **27**, 219–29.

Platt, A.M. (1967) *Fixation of radioactive residues*. Rept. BNWL-434, Battelle-Northwest, Pacific Northwest Lab. Richland, Washington, 29 pp.

Princewill, F.F. (1994) *Studies on the influence of borate species on the uptake of caesium onto natural zeolites used in nuclear waste treatment*. PhD thesis, Univ. Salford, UK.

Prister, B.S., Perepelyatnikov, G.P. and Perepelyatnikov, L.V. (1993) Countermeasures used in the Ukraine to produce forage and animal food products with radionuclides below intervention limits after the Chernobyl accident. *Sci. Total Environ.*, **137**, 183–98.

Proko'fev, O.N., Antonova, V.A. and Pavlov, I.Y. (1990) Evaluation of parameters characterizing the removal of caesium-137 by zeolite. Pp. 58–67 in: *Fiz. -Khim. Med. -Biol-Svoistva Prir. Tseolitov* (N.V. Sobolov, editor). Akad. Nauk. SSSR, Sib. Otd., Inst. Geol. Geofiz., Novosibirsk, USSR.

Propst, R.M., Ekechukwu, O.E., Dameron, H.J., Ward, G.L. and Atherton, N.G. (1988) *Pretreatments and selective materials for improved processing of PWR (pressurized water reactor) liquid radioactive wastes.* Rept. EPRI-NP-5786, Electrical Power Research. Institute, Palo Alto, California, 585 pp.

Rajec, P., Macasek, F., Feder, F., Misaelides, P. and Samajova, E. (1998) Sorption of caesium and strontium on clinoptilolite- and mordenite- containing sedimentary rocks. *J. Radioanal. Nucl. Chem.*, **229**, 49–55.

Reddy, R.G. and Cai, Z. (1996) Removal of radionuclides using zeolites. *Light Metals,* 1173–80.

Rhodes, D.W. and Wilding, M.W. (1965) *Decontamination of radioactive effluent with clinoptilolite*. Rept. IDO–14567, United States Atomic Energy Commission, National Reactor Testing Station, Idaho Falls, Idaho, 16 pp.

Rietmen, K.R. (1987) *Solidification of radioactive wastes with sulphur. Investigation of leaching mechanism*. Rept. EIR-Ber., 630, Eidg. Inst. Reaktorforsch., Wuerenlingen, Switzerland, 140 pp.

Robinson, S.M. and Parrott, J.R., Jr. (1989) *Pilot scale demonstration of process wastewater decontamination using chabazite zeolites*. Rept. ORNL/TM-1O826, Oak Ridge National Laboratory, Oak Ridge, Tennessee, 70 pp.

Robinson, S.M., Begovich., J.M., Brown, C.H., Jr., Campbell, D.O. and Emory, D. (1987) Treatment of radioactive wastewaters by chemical precipitation and ion-exchange. *AIChE Symp. Ser.*, **83**, 52–65.

Robinson, S.M., Begovitch, J.M. and Scott, C. (1988) Low-activity-level process wastewater.Treatment by chemical precipitation and ion-exchange. *J. Water Poll. Control Fed.*, **60**, 2120–27.

Robinson, S.M., Arnold, W.D. and Byers, C.H. (1990) Design of fixed-bed ion exchange columns for wastewater treatment. *Proc. Symp. Waste Manag.*, **2**, 517–25.

Robinson, S.M., Arnold, W.D. and Byers, C.H. (1991) Multicomponent ion-exchange equilibria in chabazite zeolite. *ACS Symp. Ser.*, **468**, 133–52. American Chemical Society, Washington, D.C.

Robinson, S.M., Arnold, W.D. and Byers, C.H. (1994) Mass-transfer mechanisms for zeolite ion-exchange in wastewater treatment. *AIChE J.*, **40**, 2045–54.

Robinson, S.M., Kent, T.E. and Arnold, W.D. (1995) Treatment of contaminated wastewater at Oak Ridge National Laboratory by zeolites and other ion-exchangers. Pp. 579–86 in: *Natural Zeolites'93. Occurrence, Properties, Use* (D.W. Ming and F.A. Mumpton editors). International Committee for Natural Zeolites, Brockport, New York.

Robinson, W.L. and Stone, B. (1988) *Contamination control and revegetation*. Rept. No. 6, Bikini Atoll Rehabilitation Committee (BARC), Berkeley, California, pp. 1–48.

Roddy, J.W. (1981) *Survey. Utilization of zeolites for the removal of radioactivity from liquid waste streams*. Rept. ORNL/TM-7782, Oak Ridge National Laboratories, Oak Ridge, Tennessee, 39 pp.

Romantschuk, H., Mietten, J.K. and Kivalo, P. (1966) Laumonitite – a natural zeolite ion exchange mineral with high selectivity for caesium. *Suomen Kemistilehti B*, **39**, 193–5.

Roy, D.M. and Burns, F.L. (1982) Recent advances in repository seal materials. *Mater. Res. Soc. Symp. Ser.*, **11**, 631–9.

Saha, A.K. (1985) Natural zeolites and clays-the promising media for selective radioisotope partition/immobilization. *Proc. Symp. Waste Manag.*, **3**, 197–201.

Sato, I., Matsusaka, N., Nishimura, Y., Shinagawa, K. and Kobayashi, H. (1993) Availability of zeolite as an eliminator for incorporated radionuclides. Effect on the biological half-life of manganese-54 and zinc-65. *Radioisotopes*, **42**, 298–92.

Schiffhauser, M.A. and Thompson, S.C. (1993) Installation, start-up and operation of a PUREX HLW (high-level waste) sludge mobilization and wash system for the West Valley Demonstration Project. *Technol. Programs Radioact. Waste Manag. Environ. Restor.*, **2**, 957–62.

Senel, S. and Senvar, C. (1981) Decontamination of radioactive waste by ion-exchange. *Chim. Acta Turc.*, **9**, 149–61.

Senno, M., Tashiro, S., Banba, T., Mitamura, H. and Arika, K. (1979) *The effectiveness of zeolites against volatilization of caesium during solidification of high-level wastes*. Rept. JAERI-M-

8572, Japanese Atomic Energy Research Institute, Tokai, Japan, 11 pp.

Senyavin, M.M., Nikasina, V.A., Novikova, V.A., Gradev, G.D., Stefanova, I.G. and Avramova, A.F. (1989) Comparative study of ion-exchange properties in natural clinoptilolites of the USSR and Bulgaria with the aim of purifying liquid sewage from atomic power stations. Part II. Kinetics of strontium sorption by clinoptilolites of different cationic forms. *J. Radioanal. Nucl. Chem.*, **130**, 293–8.

Shenber, M.A. and Erikson, A. (1989) Caesium sorption capacity of zeolite and soil. *Fachverb. Strahlenschutz. [Ber.] FS*, FS-89-48-T, 250–5.

Shenber, M.A. and Johanson, K.J. (1993) Influence of zeolite on the availability of radiocaesium in soil to plants. *Sci. Total Environ.*, **113**, 287–95.

Sherman, J.D. (1978) Ion exchange separations with molecular sieve zeolites. *AIChE Symp. Ser.*, **179**, 98–116.

Stavitskaya, S.S., Gersimenko, N.V., Petrenko, T.P. and Bortun, L.N. (1993) Ion-exchange properties in sorption of trace elements and radionuclides from solutions imitating composition of biological media (body fluids) or the organism. *Ukr. Khim. Zh. (Russian edition)*, **59**, 1268–73.

Sudo, Y., Saito, S., Nakano, S., Uchida, N. and Ohshima, E. (1989) Adsorption of nitrogen dioxide gas on mineral zeolites. *Tokyo Kogyo Koto Senmon Gakku Kenkyu Hokokusho*, **21**, 87–92.

Suess, M. and Pfepper, G. (1981) Investigations of the sorption of caesium from acid solutions by various inorganic sorbents. *Radiochim. Acta*, **29**, 33–40.

Sultanov, A.S., Radyuk, R.I., Tashpulatov, D., Vdovina, E.D., Popova, G.L. and Aripov, E.A. (1976) Separation of long lived isotopes from low activity wastes by natural sorbents. *Radiokhimiya*, **18**, 672–5.

Symeopoulos, B., Soupioni, M., Misaelides, P., Godelitsas, A. and Barbayiannis, N. (1996) Neodymium sorption by clay minerals and zeoliferous rocks. *J. Radioanal. Nucl. Chem.*, **212**, 421–9.

Szabova, T. and Mitro, A. (1993) Effect of zeolite on the sorption and desorption and distribution coefficients of radiostrontium and radiocaesium in different soils. *Pol'nohospodarstvo*, **39**, 1–6.

Takagi, S. (1978) Preliminary study on the use of zeolite for the isolation and removal of long-lived caesium in liquid waste from nuclear power stations. *Nucl. Sci. Technol.*, **15**, 213–21.

Tamura, T. (1962) *Strontium reactions with minerals*. Rept. TID-2028, Oak Ridge National Laboratory, Oak Ridge, Tennnessee, pp. 187–97.

Thermistocleus, T. (1990) *Zeolites and modified zeolite catalysts in the production of hydrocarbons*. PhD, Univ. Witwatersrand, Johannesburg, South Africa.

Ticknor, K.V., Vandergraff, T.T. and Kamineni, D.C. (1989) Radionuclide sorption on primary and fracture-filling minerals from the East Bull Lake Pluton Massif, Ontario. *Appl. Geochem.*, **4**, 163–76.

Tillander, M., Rahola, T., Jaakola, T. and Suemola, M. (1989) Radiocaesium in Finnish Lapps after Chernobyl. *Fachverb. Strahlenschutz, [Ber. ] FS*, FS-89-48-T, 421–6.

Torok, J., Buckley, G., Barton, N.B., Tosello, N.B., Maves, S.R., Blinkie, M.E. and Donaldson, J.R. (1989) *The stability of candidate buffer materials for a low-level radioactive waste repository*. Rept. AECL–10069, Atomic Energy Canada Ltd., Chalk River, Ontario, Canada, 9 pp.

Triay, I.R., Meijer, A., Cisneros, M.R., Miller, G.G., Mitchell, A.J., Ott, M.A., Hobart, D.E., Palmer, P.D., Perrin, R.E. and Aguilar, R.D. (1991) Sorption of americium in tuff and pure minerals using synthetic and natural groundwater. *Radiochim. Acta*, **52–53**, 141–5.

Tsai, S.C., Liang, S.M. and Hsu, C.N. (1994) A numerical method for determining radon diffusion coefficient through buffer materials from low-level radwaste site. *Technol. Programs Radioact. Waste Manag. Environ. Restor.*, **3**, 1771–4.

Tsitsishivili, G.V., Andronikashvili, T.G., Kirov, G.N. and Filizova, L.D. (1992) *Natural Zeolites*.

Ellis Horwood, Chichester, UK.

Unsworth, E.F., Pearce, J., McMurray, C.H., Moss, B.W., Gordon, F.J. and Rice, D. (1989) Investigations of the use of clay minerals and Prussian Blue in reducing the transfer of dietary radiocaesium to milk. *Sci. Total Environ.*, **85**, 339–47.

Valcke, E., Engels, B. and Cremers, A. (1997*a*) The use of zeolites as amendments in radiocaesium- and radiostrontium-contaminated soils. A soil-chemical approach. Part I. Cs-K exchange in clinoptilolite and mordenite. *Zeolites*, **18**, 205–11.

Valcke, E., Engels, B. and Cremers, A. (1997*b*) The use of zeolites as amendments in radiocaesium- and radiostrontium-contaminated soils. A soil-chemical approach. Part II. Sr-Ca exchange in clinoptilolite, mordenite and zeolite A. *Zeolites*, **18**, 212–7.

Valcke, E., Vidal, M., Cremers, A., Ivanov, J. and Perepelyatnikov, G. (1997*c*) The use of zeolites as amendments in radiocaesium- and radiostrontium-contaminated soils. A soil-chemical approach. Part III. A soil-chemical test to predict the potential effectiveness of zeolite amendments. *Zeolites*, **18**, 218–24.

Valcke, E., Elsen, A. and Cremers, A. (1997*d*) The use of zeolites as amendments in radiocaesium- and radiostrontium-contaminated soils. A soil-chemical approach. Part IV. A potted soil experiment to verify laboratory-based predictions. *Zeolites*, **18**, 225–31.

Vaniman, D.T. and Bish, D.L. (1995) The importance of zeolites in the potential high-level radioactive waste repository at Yucca Mountain, Nevada. Pp. 533–46 in: *Natural Zeolites '93: Occurrence, Properties, Use* (D.W. Ming and F.A. Mumpton, editors). International Committee for Natural Zeolites, Brockport, New York.

Vavrova, M., Mustovova, O. and Bartha, S. (1991) *In vitro* studies on the ability of natural zeolites to sorb some radionuclides. *Isotopenpraxis*, **27**, 325–7.

Vdovina, E.D. and Rudyuk, R.I. (1992) Sorption of caesium and cobalt by natrolite. *Radiokhimiya*, **34**, 113–9.

Vdovina, E.D., Rudyuk, R.I. and Sultanov, A.S. (1976) Use of Uzbekistan zeolites for the purification of low-level waste water I. Sorption of radiocaesium. *Radiokhimiya*, **18**, 422–3 (in Russian).

Vdovina, E.D., Rudyuk, R.I., Sultanov, A.S. and Aripova, E.A. (1977) Use of Uzbekistan zeolites for the purification low level wastewater. II Sorption of radioactive cobalt. *Radiokhimiya*, **19**, 185–7.

Velichko, V.A., Okonskii, A.I., Shestakov, E.O. and Panov, N.P. (1993) Regulation of strontium migration and translocation while using local chemical ameliorants. *Dokl. Russ. Akad. S-kh, Nauk*, **5**, 19–22.

Velichkovski, B.T., Korkin, L.G. and Suslova, T.B. (1988) Investigations of medico-biological properties of natural zeolites. Pp. 90–6 in: *Theoretical Industrial Problems of Introduction of Natural Zeolites into the National Economy of Russia*. Kemerovo, Moscow.

Vochten, R.F.C., Van Haverbeke, L. and Goovaerts, F. (1990) External surface adsorption of uranyl-hyroxo complexes in relation to the double layer potential. *J. Chem. Soc., Faraday Trans.*, **86**, 4095–9.

Vochten, R., Van Haverbeke, L. and Goovaerts, F. (1991) External surface adsorption of uranyl-hydroxo complexes on zeolite particles in relation to the double layer potential. *Int. Congr. Appl. Mineral.* **2**, Paper 62, 11 pp.

Wark, M., Lutz, W., Schultz-Ekloff, G. and Dyer, A. (1993) Quantitative monitoring of side products during high loading of zeolites by heavy metals via pH measurements. *Zeolites*, **13**, 658–62.

Weng, H. and Que, Q. (1986) Removal of caesium-137 with natural chabazite. *Beijing Shifan Daxue Xuebao, Ziran Kexueban*, **4**, 51–5.

Wilding, D.W. and Rhodes, M.W. (1963) *Removal of radioisotopes from solution by earth minerals from Eastern Idaho*. Rept. IDO-14624, US Atomic Energy Commission, National Reactor Testing Station, Idaho Falls, Idaho, 58 pp.

Xu, S., Chen, Z., Xu, G., Lu, G. and Zhang, J. (1986) Cement solidification of radioactive chemical species. *Fushe Fanghu*, **6**, 96–100.

Ye, M.L., Lu, S.J., Quin, C.K., Xu, L.H., He, A., Tang, Z.H., Xu, G.Q., Fan, X.L., Gu, J.F. and Du, Z.C. (1998) Investigation of sorption and migration of $^{90}Sr$ on clinoptilolite and mordenite. *Radiochim. Acta*, **81**, 103–6.

Zaitsev, V.N., Kadenko, I.N., Vasilik, L.S. and Oleinik, V.D. (1995) The natural zeolite clinoptilolite as an adsorbent for the removal of radionuclides and heavy metal salts. *Izv. Vyssh. Uchebn. Zaved., Khim. Khim. Tekhnol.*, **38**, 40–5.

CHAPTER EIGHTEEN

# Gas entry into unconfined clay pastes at water contents between the liquid and plastic limits

A. T. DONOHEW[1], S. T. HORSEMAN* AND J. F. HARRINGTON

*British Geological Survey, Keyworth, Nottingham NG12 5GG, UK*
*(*E-mail: stho@bgs.ac.uk)*

**ABSTRACT**

A programme of 143 simple gas injection experiments was performed on unconfined and initially water-saturated clay pastes at water contents between the plastic and liquid limits. The aim was to investigate the relationships between gas entry pressure, water content and plasticity for a range of clay types, to define the principal mechanisms of gas entry and flow by simple visual observations and to determine the effects of previous gas injection and residual gas content on entry pressure. Gas movement was found to be entirely through pressure-induced pathways, including highly-dilated tension fractures, flattened ellipsoidal cavities and bubbles. By examining entry mechanisms across the range of water contents, it was possible to delineate three zones of behaviour. Gas entry pressures in the region of the plastic limit were surprisingly large, particularly for clay types with high total specific surface and plasticity index. The highest individual entry pressure recorded in the study was 1810 kPa for Wyoming bentonite. There was no evidence in any test that gas actually penetrated, or flowed through, the intergranular porosity of the clay matrix. In all cases, gas made its own volume by pushing back the paste and lifting the free surface of the sample. After gas injection, remnant gas-filled voids and cracks remained within the clay. These were re-opened during repeated injections at pressures which were only a fraction of the entry pressures of the gas-free pastes. Gas entry at high pressures was audible and occasionally violent. The significance of these findings to gas migration modelling and the quantitative prediction of gas fluxes in clay formations is briefly examined.

## 18.1 Introduction

Quantitative information on the mechanisms of gas movement in clays and mudrocks has important applications in environmental engineering, soil

---

* Corresponding author
[1] Present address: Campbell Reith Hill, Consulting Engineers, 21 Dartmouth Street, St James's Park, London SW1H 9BP, UK

---

Donohew, A.T., Horseman, S.T., and Harrington, J.F. (2000) Gas entry into unconfined clay pastes at water contents between the liquid and plastic limits. Pp. 369–394 in: *Environmental Mineralogy: Microbial Interactions, Anthropogenic Influences, Contaminated Land and Waste Management* (J.D. Cotter-Howells, L.S. Campbell, E. Valsami-Jones and M. Batchelder, editors). Mineralogical Society Series, **9**. Mineralogical Society, London. ISBN 0 903056 20 8.

mechanics, marine science and petroleum geology. Specific environmental problems which involve gas movement in clay-rich materials include the design of mineral liners and caps for domestic waste landfills (Rowe *et al.*, 1995), calculation of gas migration fluxes from underground waste repositories (Lineham, 1989; Volckaert *et al.*, 1995; Ortiz *et al.*, 1997), analysis of the movement of gases in the bentonite buffer clays which will enclose certain categories of radioactive waste (Pusch *et al.*, 1985; Gray *et al.*, 1996; Horseman *et al.*, 1997, 1999; Tanai *et al.*, 1997; Gallé, 1998; Gallé and Tannai, 1998; Hume, 1999), and assessment of the environmental impacts of natural subsea hydrocarbon seepages (Hovland and Judd, 1988). Gas production is by a diverse range of mechanisms, including biological, chemical and thermal degradation of organic compounds and waste constituents, anoxic corrosion of ferrous metals and the radiolysis of pore-water. This paper examines gas entry into clays and muddy sediments with water contents which are generally much higher than the materials used in waste containment barriers. Specific issues relating to clay fabric such as bedding plane anisotropy, interparticle cementation, role of soil aggregates (e.g. clods), and the effects of mechanical compaction on the gas migration process are not considered. The primary aims of the paper are to reveal some important trends in clay behaviour and to demonstrate that the mineralogy of a clay, together with a number of mineralogy-related geotechnical index properties, are of fundamental importance in determining the critical threshold pressure for gas entry. A number of soil mechanics concepts and parameters will be introduced, the most important of these being soil suction, effective stress, desiccation and shrinkage, plasticity and undrained shear strength. For additional information on these topics, the reader is referred to standard texts such as Scott (1965), Fredlund and Rahardjo (1993) and Mitchell (1993).

Gas migration by the diffusion of gas molecules in pore-water is a slow background process in all clays (Ohsume and Horibe, 1984). The diffusion coefficient and the solubility coefficient are two important parameters in the quantitative treatment of diffusion. The diffusion coefficient is essentially a property of the water-saturated medium. If the concentration of gas in water at the source remains constant with time, the magnitude of the steady-state diffusive flux away from the source is regulated solely by the clay. The steady-state diffusive flux of gas can never exceed the upper bound set by the diffusion coefficient and solubility in water.

In many practical situations, the actual gas flux through a clay is determined by the rate of production of the gas and not by the rate at which a particular transport mechanism can accommodate the flux. We can think of the gas flux as being imposed on the material, rather than being determined by it. When the imposed flux exceeds the upper bound for diffusion, the gas must be transported through the clay as a separate gas phase.

The textbook treatment of gas entry into water-saturated porous media indicates that gas will enter a pore and displace the water from this pore when the capillary pressure (excess gas pressure, $p_g$, relative to pore-water pressure, $p_w$) exceeds the capillary displacement pressure, $p_D$, or

$$(p_g - p_w)_{entry} > p_D = \frac{2\gamma}{a}\cos\theta \qquad (18.1)$$

where $\gamma$ is the interfacial tension coefficient between gas and water, $\theta$ is the contact angle, and $a$ is the pore-throat radius.

The material is assumed to have a distribution of pore-throat radii, so that invasion of the intergranular pores by gas occurs progressively as the gas pressure increases relative to pore-water pressure. The assumptions of pore-water displacement and the progressive invasion of original pore space by the non-wetting phase are central tenets of the mathematical theory of two-phase flow (Aziz and Settari, 1979; de Marsily, 1986).

There are some very sound reasons to suspect the general validity of two-phase flow theory when applied to gas flow in clay-rich formations (Harrington and Horseman, 1999). Calculations of the magnitude of the capillary displacement pressure for clays indicate that this quantity is often so large that the gas pressure can approach, or even exceed, the total stress acting on the clay formation (Clayton and Hay, 1992; Judd and Sim, 1998). Furthermore, there is a considerable body of evidence that gas moves along preferential pathways in clay-rich materials (Pusch *et al.*, 1985; Hovland and Judd, 1988; Volckaert *et al.*, 1995; Sim and Judd, 1995; Fioravante *et al.*, 1997; Ortiz *et al.*, 1997; Gallé, 1998; Harrington and Horseman, 1999). In the context of gas migration in shallow marine sediments, Judd and Sim (1998) go so far as to declare "Preliminary modelling suggests that, at shallow subsea depths, gas migration will be initiated by fracture failure rather than capillary migration for all sediment types except coarse sands". Clayton and Hay (1992) extend the debate to argillaceous rocks, suggesting that shaly caprocks at current depths of burial <500 m are always breached by fracturing, the fracture network providing the main route for leakage of liquid and gaseous hydrocarbons across the caprock. The specific problem of gas penetration of the original pores of argillaceous rocks is discussed by Tissot and Pelet (1971) and Hedberg (1974).

The maximum pressure of a free gas phase within a clay-rich formation is capped by the state of stress. The stress at any point in the subsurface can be quantified in terms of three mutually-perpendicular principal stresses (Amadei and Stephansson, 1997). If the gas pressure exceeds the minor principal stress, gas can usually migrate by exploiting existing weaknesses (i.e. paths of least resistance). In many cases, the incipient pathways of gas movement are present naturally as crack networks, fissures and faults. When subject to high gas pressure, these incipient flow pathways can dilate and become gas permeable (Horseman and Harrington, 1996). If the clay formation is extremely tight, then it is possible for the gas to make its own pathways by hydrofracturing.

Pockmark formation provides a very good example of gas flow along pressure-induced pathways in soft sediments (Hovland and Judd, 1988).

Pockmarks are crater-like depressions on the seabed. They are often circular or elliptical in shape and vary in size from some 10s to 100s of metres across. They occur in many offshore locations throughout the world and their presence is usually indicative of underlying hydrocarbon reservoirs. The widely-accepted explanation is that thermogenic gas (largely methane) leaks upwards from the reservoir following distinct migration pathways such as faults. When the gas encounters a layer of soft clay-rich sediment close to the seabed, it becomes trapped beneath this layer. It then accumulates until its pressure is sufficient to burst through the sediment at some critical overpressure. Gas release is reported to be a sudden event. The sediments can be lifted upwards into the seawater by the release of gas and then re-deposited in the region around the central depression (McQuillan *et al.*, 1979). The fabric of the seabed sediment is always disturbed by the passage of gas.

In shallow marine environments, gas migration may be triggered by disturbances such as earthquakes, storms or offshore construction operations (Sim and Judd, 1995). Gas migration pathways tend to follow the routes of least resistance (e.g. up the side of piles driven into the seabed).

Examination of gas-bearing seabed sediments reveals that the gas becomes trapped in cavities or bubbles with dimensions orders of magnitude larger than normal pores of the sediment (Wheeler, 1988; Sim and Judd, 1995). Such cavities are usually totally enclosed by water-saturated sediment. It is a common observation that gas moves upwards in buoyant gas bubbles in very wet muds and soft muddy sediments. Mud volcanoes provide one of the best examples of this behaviour (Hedberg, 1974). As the bubbles of volcanogenic gas move upward, there is a compensating downward flow of mud. The mud flows around each bubble from the region above the bubble to the region below. The rate of ascent of the gas bubbles must therefore be controlled by the rheological properties of the mud. Shear strength is probably a limiting factor. As water content decreases, the undrained shear strength of a clay increases (Schofield and Wroth, 1968). If the upward force due to gas buoyancy is insufficient to cause plastic or viscoplastic deformation of the mud, then the bubble is unable to rise.

Sim and Judd (1995) conducted a number of simple experiments in which air was injected into the base of a column of saturated sediment (sand, silt and clay) at a water content typical of seabed conditions. In all sediment types, gas pressure built up until the overpressure was sufficient to create very conspicuous gas migration pathways. Opening up of these pathways caused doming of the sediment–water interface, followed by a sudden and violent release of gas. In silt and clay, the migration pathways were sufficiently stable to act as a drain for continuing gas migration.

Harrington and Horseman (1999) describe a programme of gas injection experiments on initially saturated Boom Clay (Tertiary clay from Belgium) and on pre-compacted Wyoming bentonite. Both clays were isotropically consolidated at a fixed confining pressure while applying a back-pressure to the samples. According to these researchers, gas does not occupy, or flow through, the original pores of the clay matrix. In the absence of dilated pathways, water-saturated clays are described as being totally impermeable to a gaseous phase. The measured gas permeability of the clay is considered to be a dependent variable rather than a material property, since it depends on the number of pressure-induced pathways in the plane normal to the flow, together with the width and aperture distributions of these pathways.

In the controlled flow rate gas migration experiments reported in Harrington and Horseman (1999), gas pressure rose gradually to a well-defined peak, with no evidence of gas flow until just before the peak (Fig. 18.1). Gas pressure then fell away spontaneously, indicating that the clay had become gas permeable. The overall pressure-time response during gas injection was found to be similar to the hydrofracturing response of clay soils described by Murdoch (1993), suggesting that gas moves through an interconnected network of cracks formed by small-scale rupture of the clay fabric under high applied gas pressure. Very similar observations have been reported by Volckaert *et al.* (1995), Ortiz *et al.* (1997), Gallé (1998) and Gallé and Tanai (1998). Crack dilation and closure under varying internal gas pressure have been invoked to explain the nonlinearity of the flow law revealed by varying the gas flow rate (Harrington and Horseman, 1999).

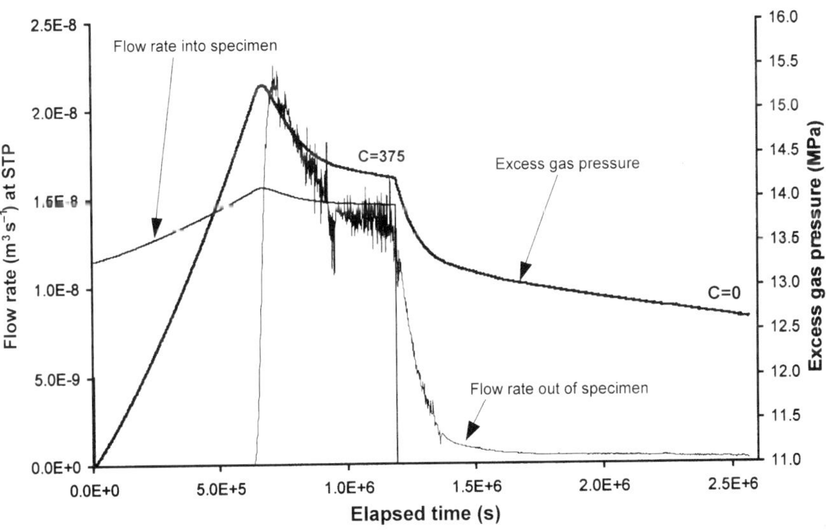


Fig. 18.1 Typical experimental history from a controlled flow rate gas injection experiment on Wyoming bentonite (from Horseman *et al.*, 1999). Excess gas pressure is the upstream gas pressure less the pressure of the water (i.e. back-pressure) at the downstream end of the specimen. Breakthrough is marked by a sudden outflow of gas from the specimen. The peak pressure response is suggestive of gas pathway propagation.

When gas injection was stopped in these experiments, the internal gas pressure of the flow pathways declined and gradually approached some finite limiting value. No gas flow was ever detected at excess gas pressures (relative to pore-water pressure) lower than this critical lower limit. Harrington and Horseman (1999) suggest that mechanical and surface tensional constraints on pathway stability make it impossible for gas to flow at excess pressures less than this lower threshold.

A number of researchers have noted a relationship between the swelling pressure of smectite-rich clays and the gas breakthrough pressure (Pusch *et al.*, 1985; Horseman *et al.*, 1997, 1999; Tanai *et al.*, 1997; Gallé, 1998). Horseman *et al.* (1999) conclude that the passage of gas into a pre-compacted and fully-hydrated bentonite (subject to a confining stress) is only possible if the gas pressure exceeds, by a small margin, the sum of the external equilibrium water pressure and the swelling pressure.

## 18.2 Gas entry and soil suction

Additional information on gas entry into clay-rich media can be obtained from the soil mechanics and soil physics literature. For soils, the gas pressure in equation 18.1 is assumed to remain constant with a value equal to mean atmospheric pressure, $p_0$. Pore-water pressure above the phreatic surface in a soil is less than atmospheric pressure and the quantity $(p_0 - p_w)$ is referred to as the soil suction. Suction is positive when the absolute pressure of the pore-water is less than atmospheric pressure. One possible interpretation of suction is that the pore-water is in a state of tension. Pore-water in tension is a thermodynamically metastable phase (Harrington and Horseman, 1999). When present in bulk it should undergo a phase change to form water vapour. It seems probable that the water in clay is unable to cavitate and form vapour bubbles because of the extreme narrowness of the interparticle spaces. Adsorption of water molecules by mineral surfaces may also be important (Graham, 1964; Newman, 1987; Schoonheydt, 1995).

Spontaneous air entry into a clay is believed to occur when the soil suction exceeds a critical limit known as the air-entry value (*AEV*). The gas entry criterion for a soil in suction can therefore be written:

$$(p_0 - p_w)_{entry} > AEV \tag{18.2}$$

The simplest way to increase the suction in a clay sample is to progressively remove the pore-water by drying under low humidity conditions. During the initial stages of the drying process, the volume of the clay sample decreases but the material remains fully saturated (Bronswijk, 1988; Konrad and Ayad, 1997). Most clays can sustain quite large suctions before they desaturate (Fleureau *et al.*, 1993; Abu-Heijleh and Znidarcic, 1995). The experimental results sketched in Fig. 18.2 suggest that spontaneous desaturation occurs at a water content which is somewhat larger than the shrinkage limit. Fox (1964)

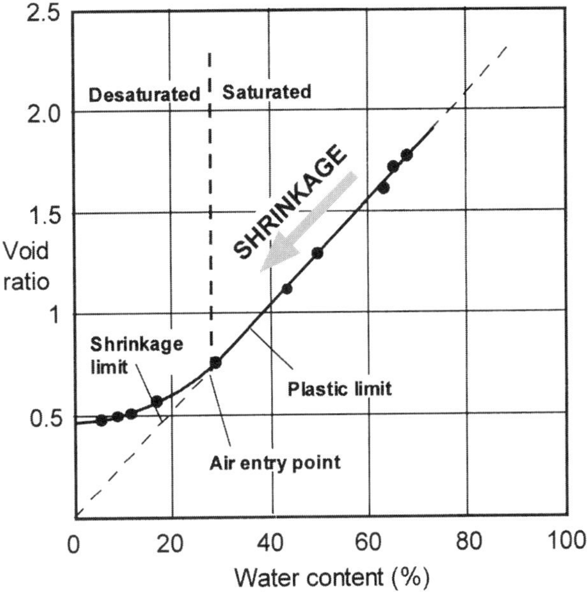


FIG. 18.2 Shrinkage path of an unconfined clay sample when progressively dried under low humidity conditions. The clay shrinks but remains fully-saturated up to a soil suction equivalent to the air-entry value (*AEV*). Air entry occurs at a water content which lies between the plastic limit and the shrinkage limit.

reports that air entered a desiccated sample of an expansive clay at a suction of 1000 kPa. Yule and Richie (1980) quote a value of 1500 kPa. Fleureau *et al.* (1993) give a correlation between the *AEV* and the liquid limit for a range of pure clays and clay soils. For soils with 60% or more particles smaller than 80 μm, the *AEV* can be predicted using the linear relationship

$$AEV = 32.4\ w_L - 466.7\ \text{(kPa)} \tag{18.3}$$

where $w_L$ is the liquid limit expressed as a percentage. The correlation coefficient, $r^2$, for this relationship was found to be 0.84.

The mathematical relationship between soil suction and water content is known as the water-retention function or moisture characteristic (Marshall and Holmes, 1979; Hillel, 1982). Figure 18.3 shows a suite of water-retention curves for clay soils of varying plasticity index (Black, 1962; Delage and Graham, 1995). Following soil mechanics practice, the water content is expressed in gravimetric terms. The plasticity index, *PI*, is simply the difference between the liquid and plastic limits. These curves are based on tests on remoulded clays and, in soil mechanics terms, are referred to as normally-consolidated (NC) suction curves. The locus of the *AEV* from equation 18.3 is sketched on this figure showing that spontaneous desaturation of a clay during a drying cycle occurs at a water content, $w_A$, which is substantially less than the plastic limit.

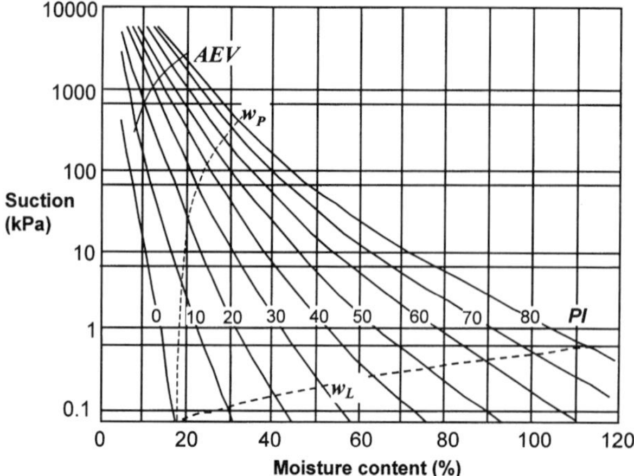


FIG. 18.3 Water-retention curves for clay soils of varying plasticity index showing the loci of the liquid and plastic limits (modified from Black, 1962). The locus of the air-entry value (*AEV*) is sketched on the basis of the empirical relationship of Fleureau *et al.* (1993). This locus separates the saturated domain (right) from the unsaturated domain (left).

We note that Russell and Mickle (1970) estimated that the soil suction at the liquid limit in clay-water systems is ~6 kPa which lies well above the locus of the liquid limit marked on Fig. 18.3.

Collectively, the Atterberg limits are indicative of mineralogy and particle sizes, and of the magnitude of the physicochemical (i.e. colloidal) interactions between the mineral grains of a soil (Warkentin, 1961; Muhunthan, 1991; Mitchell, 1993; Sridharan and Prakash, 1998). The liquid limit has been shown to correlate with the specific surface of the soil particles (Farrar and Coleman, 1967; Muhunthan, 1991). The plasticity index provides a reliable indicator of the drained compressibility of a normally-consolidated soil, where a higher plasticity index signifies higher compressibility (Schofield and Wroth, 1968; Wroth and Wood, 1978; Mitchell, 1993). Inspection of Fig. 18.3 shows that more compressible clays, with higher plasticity index, are likely to display higher values of *AEV*. This is an extremely important observation when we consider gas entry into very high plasticity clays such as bentonite (Harrington and Horseman, 1999).

To relate the *AEV* of a clay soil in suction to the excess gas pressure at gas entry of a clay in the positive pore pressure regime, we would normally invoke the principle of axis translation (Fredlund and Rahardjo, 1993; Delage and Graham, 1995). In effect, this states that gas entry into a saturated clay at some fixed dry density depends on the difference between the gas pressure, $p_g$, and the pore-water pressure, $p_w$, and is independent of the sign or the magnitude of water pressure. When the water content, $w$, of a clay is precisely the same as its water content at the air-entry point on the suction curve, $w_A$, the excess gas pressure (relative to pore-water pressure) necessary to produce gas entry should be given by:

$$(p_g - p_w)_{entry} > AEV \text{ (when } w = w_A \text{ only)} \qquad (18.4)$$

If the principle of axis translation can be demonstrated to be valid for clay-rich media, then we can use experimental data on gas entry during a drying cycle to help us understand gas migration processes occurring at depths which are well below the phreatic surface. Harrington and Horseman (1999) conclude that we should expect to find some degree of commonality between the mechanisms of gas entry during clay desiccation and those that operate in the realm of positive pore-water pressures.

## 18.3 Experiments

Although well suited to the measurement of gas transport properties, the controlled flow rate techniques of Harrington and Horseman (1999) provide very little information on the precise mechanism of gas entry and on the size and geometry of the gas flow paths. In order to address this problem, a programme of simple experiments was undertaken to study the entry and movement of gas in unconfined clay pastes prepared by slurrying clay powders with water. Specific objectives of the study were: (1) to investigate and quantify the relationships between gas entry pressure, water content, specific surface, and the Atterberg limits for a range of clay types; (2) to define the principal mechanisms of gas entry and flow by simple visual observations; and (3) to examine the effect of previous gas injection on gas entry pressure. It is evident that the fabric of a clay paste produced by slurrying a clay powder with water will differ considerably from that of a naturally-sedimented clay or that of an artificially-compacted mineral liner. In defence of this procedure, we would point to the considerable advances that have been made in soil mechanics by performing experiments on clays in which original fabric has been deliberately destroyed by mechanical remoulding procedures. Clay fabric is of secondary concern in the search for general principles.

The test materials represent a variety of 'pure' and 'natural' clay types selected to span a wide range of mineralogical and geotechnical properties. Some clays were obtained directly from suppliers, others were collected by sampling at the exposures. In order of their plasticity index, the clays are:

(1) Kaolinite: a commercial clay powder supplied by ECC International of St. Austell, Cornwall, UK, under the trade name Intrafill-B. This is a white clay of simple chemistry produced from the large deposits found in the southwest of England and used as a filler in paper manufacture. Typically the clay is of well ordered form, consisting of coarse hexagonal platelets (Highley, 1984). Particle sizes are reported to be: 55% <2 μm and 8% >10 μm.

(2) London Clay: a Tertiary (Eocene) clay taken from a road excavation (A34, new Newbury by-pass) at Reading Farm near Newbury in Oxfordshire;

described as a weathered, dark chocolate-brown, silty clay. Typically, the mineralogy of the clay to the west of the London Basin comprises illite (major), smectite and kaolinite, with quartz (major), feldspar and carbonates (Perrin, 1971).

(3) Ball Clay: a commercial clay supplied by ECC International from the 'A1 Seam' (Tertiary, Poole Formation, Oakdale Clay Member), at Arne Clay Pit near Wareham in Dorset. The clay is used in tile manufacturing. The typical composition of the clay seam is 37% kaolinite, 35% mica/illite and 26% quartz, together with some feldspar.

(4) Blauton (Blue Clay): a Tertiary (Eocene) clay from the Friedländer district of Germany supplied in powdered form by ORA Handels Gmbh & Co of Berlin. This clay is used in the construction of containment barriers for waste disposal sites. The material is dark blue-grey and is reported to contain 25–30% smectite, 20–35% illite, 10–15% kaolinite and ~25% quartz.

(5) Gault Clay: a Cretaceous (Albian) clay from the ECC International Munday's Hill quarry near Leighton Buzzard in Bedfordshire. This is a dark grey-green, silty and calcareous clay containing smectite (major), illite, kaolinite, quartz (major), calcite and mica, together with small amounts of gypsum. Particle sizes are reported to be: 48% $<2$ μm and 5% $>60$ μm.

(6) Bentonite: a commercial sodium bentonite from Wyoming supplied by Volclay Ltd., a subsidiary of the American Colloid Company, under the trade name Mx-80. This is a grey-green, high-swelling clay containing 88% montmorillonite, 4% quartz, 2% albite and minor amounts of calcite, pyrite and cristobalite (Herbert and Moog, 1999). This clay has many industrial applications, including high-integrity barriers for waste containment. Bentonite is formed by the chemical alteration of glassy volcanic ash (Odom, 1984).

Natural clay samples were oven-dried at 110°C for 24 h and then broken up and powdered using a hammer mill. Minimal milling was used in order to break down the soil aggregates and at the same time to avoid comminution of the primary particles. Weighed amounts of distilled water were added to the powders to produce pastes with water contents as close as possible to the liquid and plastic limits, together with a number of intermediate water contents lying in the plastic range. The hydrated pastes were thoroughly blended, placed in tightly-sealed plastic bags and stored in a cold room for 24 h to equilibrate. A 250 ml polythene beaker was then filled with the clay paste up to the level of the graduation mark. Pastes close to their plastic limit were carefully packed into the beaker using a spatula and tamped down with a plastic rod. Pastes near the liquid limit were poured into the beaker and agitated. Every effort was made to avoid the entrapment of air bubbles.

Liquid and plastic limits were measured according to BS 1377-Part 2 (1990) and are presented in Table 18.1. Total specific surfaces were measured using the ethylene glycol monoethyl ether (EGME) method. Powdered samples were equilibrated under vacuum in a desiccator containing anhydrous phosphorous pentoxide. Four sub-samples of each clay were analysed. The sub-samples

TABLE 18.1 Atterberg limits and total specific surfaces of the clay samples. Specific surfaces were measured using the ethylene glycol monoethyl ether (EGME) method.

| | Liquid limit (%) | Plastic limit (%) | Plasticity index (%) | Specific surface ($m^2$ $g^{-1}$) |
|---|---|---|---|---|
| Kaolinite | 60 | 35 | 25 | 29 |
| London Clay | 56 | 24 | 32 | 138 |
| Ball Clay | 72 | 33 | 39 | 101 |
| Blauton | 72 | 29 | 43 | 176 |
| Gault Clay | 91 | 45 | 46 | 260 |
| Bentonite | 332 | 38 | 294 | 750 |

were saturated with EGME and placed in a second desiccator containing dry calcium chloride. After waiting for 1.5 h, the desiccator was evacuated and left overnight. The samples were then weighed to determine the amount of adsorbed EGME. Paterson Court blue bentonite was used as the clay standard. Total surface areas were calculated and corrections applied for the slight variations between runs using the standard.

### *18.3.1 Gas injection*

Figure 18.4 is a schematic of the controlled flow rate gas injection apparatus. Air was used as the permeant. The clay paste was totally unconfined (i.e. at zero total stress) during testing. Volumetric flow rate was controlled by an ISCO-500, Series-D, syringe pump with a digital controller. A separate pressure transducer was used to record gas pressure. The apparatus was calibrated and leak-tested before use.

Gas injection into the pastes was through a hollow 16-gauge stainless-steel surgical needle, known in the medical profession as a blunt cannula, and manufactured by MDTech of Gainesville in Florida. The inside diameter of the cannula is 1.34 mm and the tip is externally chamfered at an angle of 15°. A removable stylet needle (steel rod with sharp faceted point) fits down the centre and projects a distance of 4 mm from the tip. The cannula has a moulded nylon Luer-lock fitting enabling a gas-tight connection to be made with flexible tubing.

With the stylet needle screwed in place, the cannula was inserted into the clay paste via a small hole in the bottom of the beaker. The tip of the needle was positioned 5 cm below the surface of the paste, the stylet needle removed, and the gap around the hole sealed. This procedure gives an injection pocket in the clay which comprises a cylinder ~2 mm long and 1.34 mm in diameter, terminated by a region which approximates to a 2 mm long cone.

Precisely 30 ml of water was drawn into the cylinder of the pump to cover the piston and seal so as to prevent gas leakage. An additional 50 ml of air

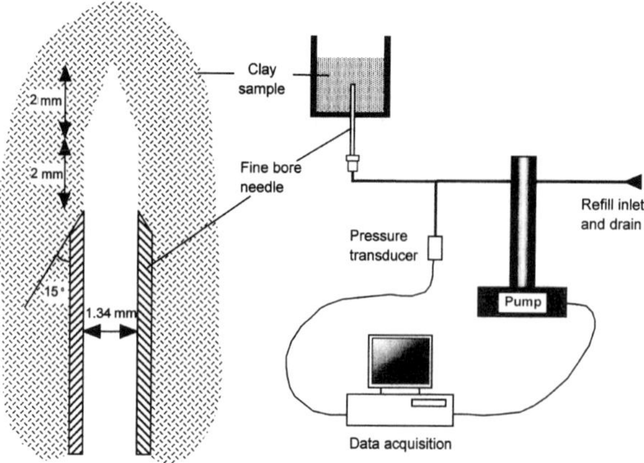


FIG. 18.4 Schematic of the controlled flow rate gas injection apparatus.

(at atmospheric pressure) was then drawn into the cylinder. The pump was set to deliver a constant air flow of 1 ml $min^{-1}$ in all experiments. Tests were conducted in a temperature-controlled laboratory at 20°C. The duration of each test ranged from 3 to 55 min, depending on breakthrough pressure. The relative humidity of the laboratory atmosphere was recorded.

Pressure and volume data were logged at 5 s intervals using a microcomputer running a simple data acquisition programme written in QBASIC. Raw data were transferred to an Excel 5.0 spreadsheet for processing and plotting.

Between five and eight individual gas injection tests were performed at each water content, enabling the means and standard deviations of the gas entry pressures to be calculated. Notes and sketches were made on the gas migration process and its visible effects. A total of 143 individual tests were performed on the suite of clays. After gas testing, the water content of each paste was determined to check for evaporative losses.

## 18.4 Results

Up to the point of gas entry, the pressure-time response was found to be governed by the isothermal compression of a fixed mass of gas in the injection system. All tests therefore followed a common path up to the point of gas entry. After entry, gas movement within the clay pastes was usually extremely rapid. In many tests, breakthrough occurred within a fraction of a second of gas entry and these two events could not be distinguished in the plots.

For the purpose of this summary, entry pressures above 100 kPa are classed as 'high' and those below 100 kPa as 'low'. There were no significant evaporative losses of water from the pastes during gas testing and variations in the relative humidity of the laboratory air over the testing period had no obvious effect on gas entry pressures.

The means and standard deviations of the gas entry pressures for the suite of clays are given in Table 18.2. It can be seen that water content and clay type have systematic effects on the gas entry pressure of these initially water-saturated clay pastes. Given that the samples were unconfined and that the injection pocket was only 5 cm from the surface of the clay, some gas entry pressures were surprisingly large, particularly for the high plasticity clays in the region of the plastic limit. The highest individual gas entry pressure recorded in these experiments was 1810 kPa for Wyoming bentonite.

After gas injection, numerous gas-filled cracks and voids remained trapped within the clay paste. These were found to re-open during repeated gas injection at a pressure which is only a fraction of the original entry pressure of the gas-free paste.

TABLE 18.2 Mean gas entry pressures for the suite of unconfined and initially water-saturated clay pastes.

| | Description | Water content (%) | Mean gas entry pressure (kPa) | Number of tests | Standard deviation (kPa) |
|---|---|---|---|---|---|
| Kaolinite | Plastic Limit | 31.9 | 98.0 | 5 | 18.6 |
| | Intermediate point | 44.8 | 12.0 | 6 | 8.2 |
| | Liquid limit | 58.7 | 3.6 | 5 | 1.9 |
| London Clay | Plastic limit | 22.9 | 520.5 | 6 | 132.1 |
| | Extra point | 28.8 | 186.8 | 6 | 70.2 |
| | Intermediate point | 38.3 | 27.3 | 6 | 12.4 |
| | Liquid limit | 55.3 | 4.8 | 6 | 2.3 |
| Ball Clay | Plastic limit | 31.3 | 343.0 | 6 | 56.6 |
| | Extra point | 46.2 | 18.5 | 6 | 3.7 |
| | Intermediate point | 51.4 | 17.4 | 6 | 2.5 |
| | Liquid limit | 70.2 | 9.5 | 6 | 5.3 |
| Blauton | Plastic limit | 27.1 | 423.9 | 5 | 95.0 |
| | Extra point | 38.2 | 128.1 | 8 | 61.2 |
| | Intermediate point | 50.8 | 23.7 | 6 | 12.9 |
| | Liquid limit | 71.5 | 11.7 | 6 | 1.5 |
| Gault Clay | Plastic limit | 41.8 | 269.5 | 6 | 64.3 |
| | Extra point | 51.4 | 121.4 | 7 | 33.2 |
| | Intermediate point | 66.0 | 21.1 | 6 | 5.5 |
| | Liquid limit | 92.1 | 18.8 | 6 | 9.5 |
| Bentonite | Plastic limit | 39.4 | 1614.5 | 6 | 210.5 |
| | Extra point | 52.2 | 501.6 | 5 | 44.8 |
| | Extra point | 103.0 | 73.4 | 6 | 24.7 |
| | Intermediate point | 175.4 | 26.8 | 6 | 4.0 |
| | Liquid limit | 332.5 | 11.7 | 6 | 4.6 |

General observations are as follows:

(1) For each clay tested, a decrease in the gravimetric water content of the saturated paste results in an increase in the gas entry pressure. This trend is completely consistent throughout the data set.

(2) For each clay tested, a clay paste near its plastic limit has a much higher gas entry pressure than one near its liquid limit. The ratio of the entry pressure at plastic limit to that at the liquid limit varies from 14 to 138, with an average of ratio of 60.

(3) Comparing gas entry pressures at ~50% gravimetric water content, the clays are in the sequence: kaolinite $\leqslant$ London Clay $<$ Ball Clay $<$ Blauton $<$ Gault Clay $<$ Wyoming bentonite. This sequence corresponds with the order of increasing plasticity index. It is also close to the order of increasing liquid limit, in general agreement with the observations of Fleureau *et al.* (1993) on the relationship between the *AEV* and liquid limit for clays in suction (see equation 18.3).

(4) The liquid limit corresponds with mean gas entry pressures in the fairly narrow range of 3.6–18.8 kPa. It is important to note that these entry pressures straddle the value of 6 kPa for the soil suction at the liquid limit quoted by Russell and Mickle (1970).

(5) The plastic limit corresponds with mean gas entry pressures which are generally in the range 89.0–520.5 kPa. The mean entry pressure of Wyoming bentonite (1614.5 kPa) is exceptionally high and lies well above the range for the other clay types.

(6) Repeatability of the entry pressure measurement is generally better when the clay paste exhibits a high entry pressure (i.e. at a low water content). Standard deviations are generally in the range 13–20% of the mean value, although these are as large as 56% in some sets of experiments performed at water contents close to the liquid limit.

The mean gas entry pressures are plotted against gravimetric water content in Fig. 18.5. Trend lines were fitted using the standard regression routine in the Excel spreadsheet and are based on the power-law relationship

$$(p_{\mathrm{g}} - p_{\mathrm{O}})_{\mathrm{entry}} = \frac{\mathrm{A}}{w^{\mathrm{n}}} \qquad (18.5)$$

where $(p_{\mathrm{g}} - p_{\mathrm{O}})_{\mathrm{entry}}$ is the gauge gas pressure at entry (kPa), A and n are constants for each clay type and $w$ (%) is the gravimetric water content.

The results of the regression analysis are presented in Table 18.3. The power-law evidently provides a good fit to the experimental data for the low plasticity clay types (kaolinite and silty London Clay) and a moderate fit to data for all other clays. The exponent, n, of the power-law decreases fairly systematically with the plasticity index. The gradual steepening of the slope with decreasing water content for bentonite (see Table 18.2) is real, rather than an artefact of experimental errors, indicating that an alternative to the power-law may be required to adequately quantify gas entry over the full range of water contents for this clay type. To illustrate this difficulty, Table 18.3 gives

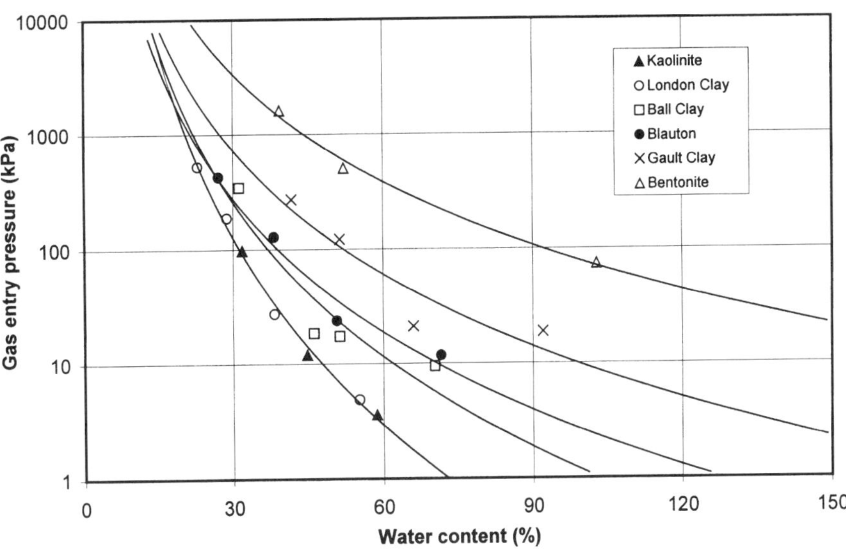


FIG. 18.5 Mean gas entry pressure plotted against gravimetric water content. Trend-lines are based on the power-law relationship of equation 18.5. For a fixed water content, gas entry pressure increases with increasing plasticity index of the clay paste.

two sets of fitting parameters for bentonite, the first for water contents <105% and the second (less satisfactory fit) for all available data.

It is interesting to note that parameter A is very nearly equal for the three smectite-rich clays (Blauton, Gault Clay and Wyoming bentonite, $w$<105%), suggesting that this parameter is largely governed by clay mineralogy and the presence of the high-swelling clay mineral.

TABLE 18.3 Power-law least-squares regression fits to the gas entry pressure against gravimetric water content relationship for each clay.

| | A (kPa) | n | $r^2$ |
|---|---|---|---|
| Kaolinite | $1.38 \times 10^{10}$ | 5.45 | 0.992 |
| London Clay | $1.39 \times 10^{10}$ | 5.44 | 0.994 |
| Ball Clay | $1.07 \times 10^{9}$ | 4.48 | 0.864 |
| Blauton | $1.45 \times 10^{8}$ | 3.87 | 0.969 |
| Gault Clay | $1.50 \times 10^{8}$ | 3.60 | 0.865 |
| Bentonite ($w$<105%) | $1.51 \times 10^{8}$ | 3.15 | 0.991 |
| Bentonite (all data) | $4.98 \times 10^{6}$ | 2.30 | 0.963 |

## 18.5 Mechanisms

Gas movement through these saturated clay pastes was entirely through pressure-induced pathways, including dilated fractures, flattened ellipsoidal cavities, bubbles and other interconnected voids. There was no evidence in any experiment that gas actually penetrated, or flowed through, the original intergranular porosity of the clay matrix.

Three basic modes of behaviour can be defined: (1) Mechanism BB – gas movement in buoyant bubbles and cavities with rapid and complete loss of gas pressure; (2) Mechanism TFR – Gas movement in tensile fractures with rapid and complete loss of gas pressure; and (3) Mechanism TFG – gas movement in tensile fractures characterized by slow crack propagation and gradual depressurization. Figure 18.6 shows a zonation of these mechanisms on a double logarithmic plot of gas entry pressure against water content. Individual test results are shown on this plot. Mechanism TFG would appear to be transitional between Mechanisms BB and TFR.

### *18.5.1 Mechanism BB*

This mode of behaviour was common to most clays at high to intermediate water contents and was observed in a total of 65 tests. Gas entry pressures were generally low (<100 kPa). Buoyancy forces had a strong effect on gas movement. Generally, a single bubble or cavity of gas moved upwards from the cannula tip, deforming the surrounding paste and discharging at surface. Cracking, pitting and doming of the surface were apparent in many of the tests

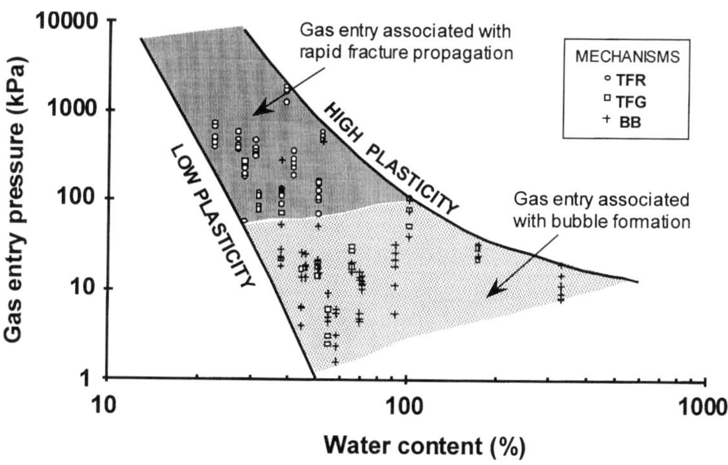


FIG. 18.6 Zonation of the gas entry mechanisms. Individual test results are shown. The high- and low-plasticity envelopes represent upper and lower bounds for the gas entry pressures of unconfined and initially water-saturated clay pastes.

(Fig. 18.7). Occasionally, small pockmarks were formed. Stretching of the clay during doming often produced networks of small extensile surface cracks. The region of cracking was from 1.5 to 5 cm in diameter, with a typical array of 10 to 30 individual cracks. Individual pits and pockmarks were from 2 to 5 mm in diameter. Very similar features were observed in the experimental study of Sim and Judd (1995).

### *18.5.2 Mechanism TFR*

This mode of behaviour occurred in all clay types at low water content and was observed in a total of 56 tests. Gas entry pressures were invariably high (>100 kPa). Gas entry was marked by an audible and occasionally violent release of pressure resulting in sub-horizontal tension fractures and very flattened ellipsoidal pockets (minor semi-axis generally sub-vertical) at the level of the injection point (Fig. 18.8). Buoyancy forces were far less important than in Mechanism BB. Rapid dilation of pathways by gas expansion often caused visible lifting and doming of the clay surface. Many of the pathways propagated dynamically to the walls of the beaker, a distance of ~30 mm. Fractures then moved upwards in the region of the clay/polythene interface and discharged at the surface. Tests were characterized by the very rapid and complete release of gas pressure. The apertures of the residual gas

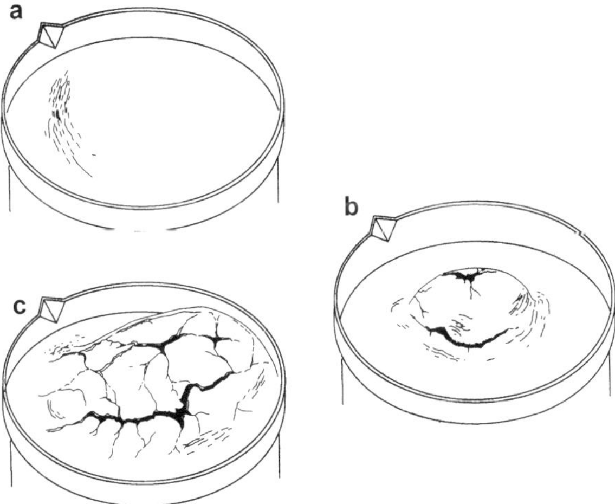


FIG. 18.7 Typical features of Mechanism BB. (*a*) A small pockmark in kaolinite paste (2 mm diam.), surrounded by an arc-shaped region of extensional fracturing ($w$ = 46%). (*b*) Typical doming of the free surface of Blauton paste ($w$ = 38%). The underlying gas-filled void is an inverted cone with its apex at the tip of the cannula. (*c*) Very extensive surface cracking in Wyoming bentonite paste ($w$ = 52%) produced by uplift and doming of the clay. The gas-filled void is a flattened ellipsoid with the minor semi-axis nearly vertical.

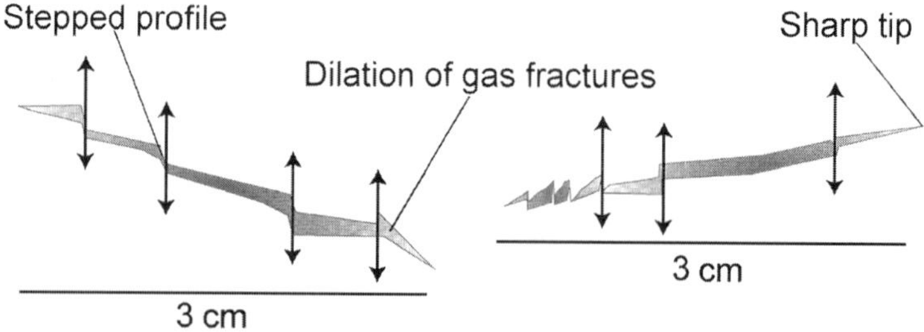


FIG. 18.8 Some typical features of Mechanism TFR. These tensile fractures propagate from the cannula tip, moving in a sub-horizontal direction until they reach the walls of the polythene beaker. The stepped profile is suggestive of localized shearing of the clay after crack formation. Crack dilation causes the free surface of the clay paste to move upwards. All gas in the clay is accommodated by the new void space generated by the opening of these cracks.

pathways were in the range 1–5 mm. Most cracks had sharp leading edges and some displayed the bifurcating tips characteristic of dynamic crack propagation. Surface extensile cracking of the type seen at higher water contents was rare.

### *18.5.3 Mechanism TFG*

In a variant of the above mechanism, the cracks were observed to slowly propagate at low residual gas pressures for periods of up to 10 min from the time of gas entry. Eventually these cracks became completely depressurized. Observed lightening and darkening of the clay in the region of slowly propagating crack tips were suggestive of very local changes in water content. This mode of behaviour occurred at intermediate to low water contents and was observed in a total of 22 tests.

## 18.6 Discussion

It is extremely surprising that an unconfined clay paste can withstand a locally-applied gas pressure as large as 1810 kPa before the gas actually moves into the paste. This observation is even more surprising when we consider that the gas injection pocket in these experiments was only 5 cm below the free surface of the clay. The most important question that arises from this study is: "What clay properties actually determine the measured gas entry pressures?"

The geometry of the gas source may be important. It seems probable that the sharp notch at the upper extremity of the injection pocket becomes rounded as soon as the stylet needle is withdrawn. For the purposes of analysis, we can therefore think of the source as a cylindrical hole.

Andersen *et al.* (1994) examine failure criteria relating to an internally-pressurized borehole in clay-rich seabed sediments. Four basic approaches to this problem are identified: (1) hydrofracture theory drawing on the theory of linear elasticity; (2) cavity expansion theory drawing on the theory of elastoplasticity; (3) an initial yielding criterion; and (4) the application of empirical formulae. The first two approaches will be outlined briefly.

According to hydrofracture theory, a tensile fracture will develop and propagate when the tensile effective stress acting in a direction tangential to the cylindrical hole is equal to the tensile strength, *T*, of the clay. Using linear elastic stress analysis, the breakdown pressure of the injection pocket is given by

$$(p_g - p_0)_{frac.} = T + 2\sigma_{eff} + p_w - p_0 \qquad (18.6)$$

where $\sigma_{eff}$ and $(p_w - p_0)$ are the isotropic effective stress and the gauge pore-water pressure acting in the clay outside the region of stress concentration and $(p_g - p_0)_{frac.}$ is the gauge pressure of gas at the point of breakdown. According to Terzaghi's theory of effective stress (Scott, 1965; Mitchell, 1993), the isotropic total stress, $\sigma$, is related to the effective stress by:

$$\sigma = \sigma_{eff} + p_w - p_0 \qquad (18.7)$$

Since the total stress was zero in our unconfined experiments, the effective stress must be equal to the soil suction for as long as the clay remains fully saturated, or:

$$\sigma_{eff} = p_0 - p_w \qquad (18.8)$$

Substituting equation 18.8 in 18.6 gives the breakdown pressure as:

$$(p_g - p_0)_{frac.} = T + p_0 - p_w \qquad (18.9)$$

Hydrofracture theory therefore predicts that a gas fracture will develop and propagate when the gauge pressure of the gas exceeds the sum of the tensile strength of the clay and the soil suction of the clay outside the region of stress concentration. Inspection of the NC suction curves of Fig. 18.3 suggests that each of our clay pastes was in suction (i.e. the gauge pore-water pressure was negative) at the onset of testing. The magnitude of the suction evidently depends on the water content of the paste and on its plasticity index. Suction increases with increasing plasticity index and decreasing water content. Our experiments show that the gas entry pressure of the clay pastes also increases with increasing plasticity index and decreasing water content. Even although the linear elastic analysis used in the derivation of equation 18.6 is questionable when applied to clays between the liquid and plastic limits, it appears that an interpretation based on simple hydrofracture theory provides a qualitative explanation of the overall trends in material behaviour revealed by our experiments.

Andersen *et al.* (1994) also discussed cavity expansion theory in the context of borehole breakdown in clayey sediments. The gas pressure needed to expand a cylindrical cavity in an initially-unstressed medium is given by

$$(p_g - p_0)_{exp.} = \alpha S_u = \left[1 + \ln\left(\frac{G}{S_u}\right)\right] S_u \qquad (18.10)$$

where G is the shear modulus in the elastic zone (outside the region of stress concentration) and $S_u$ is the undrained shear strength of the clay (Gibson and Anderson, 1961; Andersen *et al.*, 1994). A number of studies have shown that the shear strength of a clay slurry at the liquid limit is in the range 0.5 to 6 kPa (Skempton and Northey, 1953; Whyte, 1982; Wasti and Bezirci, 1986; Sridharan and Prakash, 1998). The shear strength at the plastic limit is generally found to be from 80 to 100 times larger than at the liquid limit. Thus the shear strength at the plastic limit is in the range 40 to 600 kPa. The measurements of Skempton and Northay (1953) give somewhat narrower limits of 80 to 140 kPa.

Skempton (1957) observed that the ratio of undrained shear strength to the effective stress is a constant for a normally-consolidated clay. Setting aside the subtle distinction between vertical effective stress and isotropic effective stress, this ratio can be estimated from plasticity index (%) using:

$$\frac{S_u}{\sigma_{eff}} = \beta = 0.11 + 0.37\frac{PI}{100} \tag{18.11}$$

Combining equations 18.8, 18.10 and 18.11 we obtain:

$$(p_g - p_0)_{exp.} = \alpha\beta(p_0 - p_w) \tag{18.12}$$

The gauge gas pressure necessary to expand the injection pocket therefore increases with the soil suction. We can think of suction as imposing some kind of compressive pre-stress on the clay around the injection pocket. Equation 18.11 indicates that expansion pressure is likely to be higher in a high plasticity clay than it is in a low plasticity clay. Both predictions are in qualitative agreement with the trend established from laboratory experiments.

Thus, the two alternative models of the gas entry process identify a number of material properties which can determine the magnitude of the entry pressure. Hydrofracture theory suggests that entry pressure is determined by the soil suction and by the small-scale tensile strength of the clay. Cavity expansion theory suggests that it is the suction-dependent undrained shear strength of the clay that is the primary determinant of the gas entry pressure. The simplest interpretation of these datasets would be to associate Mechanism TFR with hydrofracture theory and Mechanism BB with cavity expansion theory. The gradual transition from liquid-like behaviour at high water contents, through plastic behaviour at intermediate water contents, to solid-like (or brittle) behaviour at low water contents revealed by the experiments is entirely consistent with the general interpretation of the physical significance of the Atterberg limits (Sridharan and Prakash, 1998).

We make the final observation that the practitioners of soil mechanics make a basic subdivision of soils into cohesive and non-cohesive soil types on the basis of strength properties (Scott, 1965). Non-cohesive soils usually contain a very high proportion of sand- and/or silt-sized particles. They have very low liquid and plastic limits and their plasticity index is usually <20%. Soils with reasonably high clay contents are considered to be cohesive. Although cohesion is thought by many to be a surface energy effect, linked to capillarity and suction, the development of an adequate theoretical treatment of this phenomenon remains elusive. Mitchell (1993) even expresses doubt as to the existence of true cohesion in clay systems. Despite this controversy, we put forward the idea that the fundamental quantity that determines the gas entry pressure in clays is not the tensile strength of the clay fabric, nor is it the undrained shear strength of the fabric; entry pressure is governed by the basic cohesion of the colloidal particles when dispersed in water. In its broadest sense, cohesion can be interpreted as the resistance to dilation of the clay-water system when subject to undrained deformation. Gas entry is essentially a process of dilation, since new gas-filled void space is actually created within the clay-water system. Cohesion attains its maximum value in a very dense (i.e. concentrated) dispersion of ultra-small colloidal particles. Of the clay types examined in this study, the bentonite best fits this description. Cohesion is primarily a measure of the molecular bond strength of the dispersant (i.e. the interparticle water). Spontaneous air entry during a drying cycle occurs at a gravimetric water content somewhat larger than the shrinkage limit. Air entry under high suction can be interpreted as a local breakdown of cohesion. Atterberg, in his original studies of the soil index properties, used the phrase 'cohesion limit' to describe the quantity we now refer to as the shrinkage limit (Sridharan and Prakash, 1998). In our view, it is cohesion that distinguishes the gas transport properties of clays from those of coarse-grained sediments.

## 18.7 Conclusions

Gas movement in initially water-saturated, unconfined clay pastes was found to be entirely through pressure-induced pathways, including dilated cracks, flattened ellipsoidal cavities and bubbles. There was no evidence in any of the 143 tests performed in this study that gas actually penetrated, or flowed through, the original intergranular porosity of the clay matrix. In every case, gas made its own volume by pushing back the paste and lifting the free surface of the sample.

By examining entry mechanisms across the full range of water contents, it is possible to delineate three basic modes of behaviour. Mechanism BB, or gas movement in buoyant bubbles and cavities, is common to most clays at high to intermediate water contents and leads to modest gas entry pressures (<100 kPa). Mechanism TFR, or gas movement in tensile fractures with rapid and complete loss of gas pressure, occurs in all clays at low water contents and is associated with relatively high gas entry pressures (>100 kPa). Mechanism TFG is characterized by fractures which exhibit slow propagation and a gradual loss in internal gas pressure. It occurs at intermediate water contents and is probably transitional between mechanisms TFR and BB.

There is little doubt that Mechanism BB is the primary mechanism of vertical gas migration in wet, poorly-compacted, muddy sediments. A common characteristic of these sediments is their very low shear strength. The sediment can therefore deform plastically and flow around an ascending gas bubble. Mud volcanoes and seabed pockmarks provide two good examples of this mechanism operating in natural systems. Cavity expansion theory provides a simple criterion for the enlargement of gas-filled voids in soft mud. The suction-dependent undrained shear strength of the clay is the primary determinant of the critical gas pressure for cavity enlargement in an unconfined clay paste. Shear strength and cohesion are considered by many to be closely interrelated, although there is some debate in the literature on the actual existence of true cohesion is clay-water systems.

The ratio of the gas entry pressure at the plastic limit to that at the liquid limit varies from 14 to 138, with an average ratio of 60. The ratio of the undrained shear strength at the plastic limit to that at the liquid limit for a remoulded clay is reported to lie in the range 80 to 100. Cavity expansion theory provides a link between these two independent observations.

Gas entry and breakthrough pressures in the region of the plastic limit are surprisingly large, particularly for clay types with high plasticity index and large total specific surface. The highest individual entry pressure recorded in the study was 1810 kPa for Wyoming bentonite, confirming the findings of previous experimental studies which suggest that a high-swelling clay will display a commensurately large gas entry pressure (Horseman *et al.*, 1997, 1999; Harrington and Horseman, 1999). Gas entry and breakthrough at high pressures were often audible and were occasionally violent. The crack-like pathways formed at high pressure had sharp leading edges and some displayed the bifurcating tips characteristic of dynamic crack propagation. Hydrofracture theory suggests that gas enters an unconfined clay when the gauge gas pressure is larger than the sum of the small-scale tensile strength of the clay-water system and the soil suction outside the region of local stress concentration.

Gas entry pressure varies systematically with water content and plasticity index. Comparing entry pressures at ~50% water content, the clays are in the sequence: kaolinite $\leqslant$ silty London Clay < Ball Clay < Blauton < Gault Clay < Wyoming bentonite. The relationship between entry pressure and gravimetric water content can be quantified by a power-law function, although the fit to data for bentonite is not satisfactory at higher water contents. The exponent of the power law (see equation 18.5) decreases systematically with plasticity index. When at similar water contents, high plasticity clays will form better barriers to gas migration than low plasticity clays.

After gas injection, gas-filled voids and dilated cracks remain within the clay. These are re-opened during repeated injection at a pressure which is only a fraction of the entry pressure of the gas-free paste. The presence of residual gas-voids in a clay therefore causes a reduction in the critical pressure for gas entry and breakthrough.

Many features of gas movement in clay-rich media cannot be explained by invoking conventional two-phase flow theory, including the upward motion of large gas bubbles in wet clays and in mud volcanoes, the formation of pockmarks in seabed sediments, the development of gas-filled voids with dimensions many orders of magnitude larger than original pores, the visual observations of cracking and doming during gas injection, the lack of evidence for pore-water displacement, the conspicuous stress-sensitivity of the gas transport process, the relationship between gas entry pressure and swelling pressure in bentonite, the dynamics of the gas transport process, and, finally, the absence of any form of experimental confirmation that gas actually penetrates, and flows through, the original pores of a fully-hydrated and fully-saturated clay. Although specific observations on wet clays cannot be interpreted as general proof of difficulties with two-phase flow theory when applied to the broad class of argillaceous media (i.e. clay soils and mudrocks), the evidence is now sufficiently strong to prompt the development of alternative theories and models to explain the findings of a growing number of experimentalists and observational geoscientists.

## Acknowledgements

The experimental study was funded by a research grant from the EPSRC/NERC Waste Management and Pollution Programme. The authors wish to thank Mr Lee Jones of BGS Coastal and Engineering Geology Group for his help in obtaining the clay samples and Dr Vicky Hards of the BGS Mineralogy and Petrology Group for her measurements of total specific surface. The authors also wish to thank their research partners in this and previous projects on gas movement in clays for the stimulating and sometimes lively debate on this topic: University of Wales (UWCC, Cardiff); M.J. Carter Associates (Atherstone, UK); EnvirosQuantiSci (Henley on Thames, UK); AEA Technology plc (Harwell, UK); SCK-CEN (Mol, Belgium); ISMES SpA (Bergamo, Italy); and the University of Rome, La Sapienza (Italy). This paper is published with the permission of the Director of the British Geological Survey (NERC).

## References

Abu-Heijleh, A.N. and Znidarcic, D. (1995) Desiccation theory for soft cohesive clays. *J. Geotech. Eng.*, **121**, 493–501.

Amadei, B. and Stephansson, O. (1997) *Rock Stress and its Measurement*. Chapman & Hall, London.

Andersen, K.H., Rawlings, C.G., Lunne, T.A. and By, T.H. (1994) Estimation of hydraulic fracture pressure in clay. *Canad. Geotech. J.*, **31**, 817–28.

Aziz, K. and Settari, A. (1979) *Petroleum Reservoir Simulation*. Applied Science, London.

Black, W.P.M. (1962) A method of estimating the California Bearing Ratio of cohesive soils from plasticity data. *Géotechnique*, **12**, 271–82.

Bronswijk, J.J.B. (1988) Modelling of water balance, cracking and subsidence of clay soils. *J. Hydrol.*, **97**, 199–212.

Clayton, C.J. and Hay, S.J. (1992) Gas migration mechanisms from accumulation to surface. *Bull. Geol. Soc. Denmark*, **41**, 12–23.

Delage, P. and Graham, J. (1995) Mechanical behaviour of unsaturated soils: understanding the behaviour of unsaturated soils requires reliable conceptual models. Pp. 1223–56 in: *Unsaturated Soils. Proc. 1st Int. Conf., Paris* (E.E. Alonso and P. Delage, editors). Balkema, Rotterdam.

de Marsily, G. (1986) *Quantitative Hydrogeology for Engineers*. Academic Press, New York.

Farrar, D.M. and Coleman, J.D. (1967) The correlation of surface area with other properties of nineteen British clay soils. *J. Soil Sci.*, **18**, 118–24.

Fioravante, V., Airoldi, S. and Costantino, B.A. (1997) Physical soil model tests in a centrifuge. Pp. 177-94 in: Project on the Effects of Gas in Underground Storage Facilities for Radioactive Waste (PEGASUS Project). *Proceedings of a Progress Meeting held in Mol, Belgium*. EU Nuclear Science and Technology Series, Rept. EUR 18167 EN.

Fleureau, J.-M., Kheirbek-Saoud, S., Soemetro, R. and Taibi, S. (1993) Behaviour of clayey soils on drying-wetting paths. *Canad. Geotech. J.*, **30**, 287–96.

Fox, W.E. (1964) Study of bulk density and water in a swelling soil. *Soil Sci.*, **98**, 307–16.

Fredlund, D.G. and Rahardjo, H. (1993) *Soil Mechanics for Unsaturated Soils*. John Wiley & Sons, Ltd., New York.

Gallé, C. (1998) Migration des gaz et pression de rupture dans une argile compactée destinée à la barrier ouvragée d'un stockage profond. *Bull. Soc. Géol. France*, **169(5)**, 675–80.

Gallé, C. and Tanai, K. (1998) Evolution of the gas transport properties of backfill. Materials for waste disposal: $H_2$ migration experiments in compacted Fo-Ca clay. *Clays Clay Miner.*, **46**, 498–508.

Gibson, R.E. and Anderson, W.F. (1961) In situ measurement of soil properties with the pressuremeter. *Civil Engineering and Public Works Review*, **56**, 615–8.

Graham, J. (1964) Adsorbed water on clays. *Rev. Pure Appl. Chem.*, **14**, 81–90.

Gray, M.N., Kirkham, T.J., Wan, A.W.-L. and Graham, J. (1996) On the gas breakthrough resistance of engineered clay barrier materials proposed for use in nuclear fuel waste disposal. In: *International Conference on Deep Geological Disposal of Radioactive Waste, Winnipeg, Manitoba*. Canadian Nuclear Society.

Harrington, J.F. and Horseman, S.T. (1999) Gas transport properties of clays and mudrocks. Pp. 107–24 in: *Muds and Mudstones: Physical and Fluid Flow Properties* (A.C. Aplin, A.J. Fleet and J.H.S. Macquaker, editors). Geological Society, London, Spec. Publ. **158**, 107–24.

Hedberg, H.D. (1974) Relation of methane generation to undercompacted shales, shale diapirs and mud volcanoes. *Bull. Amer. Assoc. Petrol. Geol.*, **58**, 661–73.

Herbert, H.-J. and Moog, H.C. (1999) Cation exchange, interlayer spacing, and water content of Mx-80 bentonite in high molar saline solutions. *Eng. Geol.*, **54**, 55–65.

Highley, D.E. (1984) *China Clay. Mineral Dossier No. 26*. Mineral Resources Consultative Committee, HMSO, London.

Hillel, D. (1982) *Introduction to Soil Physics* (2nd edition). Academic Press, New York.

Horseman, S.T. and Harrington, J.F. (1996) *Evidence for thresholds, pathways and intermittent flow in argillaceous rocks*. Proceedings of the Joint EU-OECD/NEA International Workshop on Fluid Flow through Faults and Fractures in Argillaceous Rocks, Bern, Switzerland. Nuclear Energy Agency, Paris, ISBN 92-64-16021-3, 85-103 (available from the Stationery Office in the UK).

Horseman, S.T., Harrington, J.F. and Sellin, P. (1997) Gas migration in Mx80 buffer bentonite. Pp. 1003–10 in: *Scientific Basis for Nuclear Waste Management XX* (W.J. Gray, and I.R. Triay, editors). Symposia Proceedings **465**, Materials Research Society, Warrendale, Pennsylvania.

Horseman, S.T., Harrington, J.F. and Sellin, P. (1999) Gas migration in clay barriers. *Eng. Geol.*, **54**, 139–49.

Hovland, M. and Judd, A.G. (1988) *Seabed Pockmarks and Seepages: Impact on Geology, Biology and the Marine Environment*. Graham & Trotman, London.

Hume, H.B. (1999) *Gas breakthrough in compacted Avonlea bentonite*. MSc thesis, Univ.

Manitoba, Winnipeg, Manitoba.

Judd, A.G. and Sim, R.H. (1998) Shallow gas migration mechanisms in deep water sediments. Pp. 163–73 in: *Proc. S.U.T. Conference on Offshore Site Investigation and Foundation Behaviour.*

Konrad, J.-M. and Ayad, R. (1997) An idealized framework for the analysis of cohesive soils undergoing desiccation. *Canad. Geotech. J.*, **34**, 477–88.

Lineham, T.R. (1989) *A laboratory study of gas transport through clay samples.* UK Nirex Ltd., Harwell, Oxfordshire. Nirex Safety Studies Report No. NSS/R155.

Marshall, T.J. and Holmes, J.W. (1979) *Soil Physics.* Cambridge University Press, Cambridge.

McQuillan, R., Fannin, N.G.T. and Judd, A.G. (1979) IGS pockmark investigations, 1974–1978. Marine Geophysics Unit. British Geological Survey, Edinburgh, Technical Report No. 98.

Mitchell, J.K. (1993) *Fundamentals of Soil Behavior* (2nd Edition). John Wiley & Sons, Inc., New York.

Muhanthan, B. (1991) Liquid limit and the surface area of clays. *Géotechnique*, **41**, 135–8.

Murdoch, L.C. (1993) Hydraulic fracturing of soil during laboratory experiments. Part 1 – Methods and observations. *Géotechnique*, **43**, 255–65.

Newman, A.C.D. (1987) The interaction of water with clay mineral surfaces. Pp. 237–74 in: *Chemistry of Clays and Clay Minerals* (A.C.D. Newman, editor). Monograph **6**, Mineralogical Society, London.

Odom, I.E. (1984) Smectite clay minerals: properties and uses. *Phil. Trans. Royal Soc. Lond., Series A*, **311**, 391–409.

Ohsume, T. and Horibe, Y. (1984) Diffusivity of He and Ar in deep sea sediments. *Earth Planet. Sci. Lett.*, **70**, 61–8.

Ortiz, L., Volckaert, G., de Canniere, P., Put, M. Sen, M., Horseman, S.T., Harrington, J.F., Impey, M. and Einchcomb, S. (1997) *MEGAS – Modelling and experiments on gas migration in repository rocks: Final Rept. – Phase 2.* European Commission, Nuclear Science and Technology. EUR 17453 EN, Luxembourg.

Perrin, R.M.S. (1971) *The Clay Mineralogy of British Sediments.* Mineralogical Society, London.

Pusch, R., Ranhagen, L. and Nilsson, K. (1985) *Gas migration through MX-80 bentonite.* Nagra Technical Report NTB 85-36, Wettingen, Switzerland.

Rowe, R.K., Quigley, R.M. and Booker, J.R. (1995) *Clayey Barrier Systems for Waste Disposal Facilities.* E. & F.N. Spon, London.

Russell, E.R. and Mickle, J.L. (1970) Liquid limit values of soil moisture tension. *J. Soil Mechanics, Foundation Division, A.S.C.E.*, **96**, 967–87.

Schofield, A. and Wroth, P.W. (1968) *Critical State Soil Mechanics.* McGraw Hill, London.

Scott, R.F. (1965) *Principles of Soil Mechanics.* Addison-Wesley Publishing Company, Inc., Reading, Massachusetts.

Schoonheydt, R.A. (1995) Clay mineral surfaces. Pp. 303–27 in: *Mineral Surfaces* (D.J. Vaughan and R.A.D Pattrick, editors). Mineralogical Society Series, **5**. Chapman & Hall, London.

Sridharan, A. and Prakash, K. (1998) Characteristic water contents of a fine-grained soil-water system. *Géotechnique,* **48**, 337–46.

Sim, R.H. and Judd, A.G. (1995) Simple gas migration modelling. Pp. 35–8 in: *Proc. Workshop on Modelling Methane-Rich Sediments of Eckernförde Bay* (T.F. Wever, editor). FWR-Report 22, Forschungsanstalt der Bundeswehr fur Wasserschall- und Geophysik (FWR), Kiel.

Skempton, A.W. (1957) Discussion on the 'Planning and design of the new Hong Kong airport'. *Proc. Inst. Civil Engineers*, **7**, 306.

Skempton, A.W. and Northey, R.D. (1953) The sensitivity of clays. *Géotechnique,* **3**, 30–53.

Tanai, K., Kanno, T. and Gallé, C. (1997) Experimental study of gas permeabilities and breakthrough pressures in clays. Pp. 995–1002 in: *Scientific Basis for Nuclear Waste Management XX* (W.J. Gray and I.R. Triay, editors). Symposia Proceedings **465**, Materials Research Society, Warrendale, Pennsylvania.

Tissot, B. and Pelet, R. (1971) Nouvelles donées sur la mécanismes de genèse et de migration du pétrole: Simulation mathématique et application à la prospectation. Pp. 35–46 in: *Proc. 8th World Petroleum Congress, Moscow.*

Volckaert, G., Ortiz, L., de Canniere, P., Put, M., Horseman, S.T., Harrington, J.F., Fioravante, V. and Impey, M. (1995) *MEGAS – Modelling and experiments on gas migration in repository rocks: Final Rept.– Phase 1.* European Commission, Nuclear Science and Technology. EUR 16235 EN, Luxembourg.

Warkentin, B.P. (1961) Interpretation of the upper plastic limit of clays. *Nature,* **190**, 287–8.

Wasti, Y. and Bezirci, M.H. (1986) Determination of the consistency limits of soils by the fall cone test. *Canad. Geotech. J.*, **23**, 241–6.

Wheeler, S.J. (1988) A conceptual model for soils containing large gas bubbles. *Géotechnique*, **38**, 389–97.

Whyte, L.L. (1982) Soil plasticity and strength: a new approach using extrusion. *Ground Engineering*, **15**, 16–24.

Wroth, C.P. and Wood, D.M. (1978) The correlation of index properties with some basic engineering properties of soils. *Canad. Geotech. J.*, **15**, 121–32.

Yule, D.F. and Richie, J.T. (1980) Soil shrinkage relationships of Texas vertisols: Small cores. *Soil Sci. Soc. Amer. J.*, **44**, 1285–91.

CHAPTER NINETEEN

# Geosynthetic Clay Liners (GCLs) for municipal solid waste landfills

R. K. Rowe* and C. B. Lake†

*Department of Civil & Environmental Engineering, The University of Western Ontario, London, Ontario, Canada N6A 5B9*

**ABSTRACT**

This chapter provides a general overview of some of the current knowledge of geosynthetic clay liners (GCLs) in relation to use in municipal solid waste landfills. Engineering design issues for GCLs are discussed in the context of current research relating to hydraulic conductivity, clay-leachate compatibility, diffusion, contaminant transport through the entire barrier system, shear strength, and stability. Generally speaking, GCLs have considerable potential for use as part of a composite liner system in the base of municipal solid waste landfills provided issues such as those discussed in this chapter are addressed.

## 19.1 Introduction

The use of geosynthetic clay liners (GCLs) has become prevalent in engineering containment applications such as mining waste covers, lagoon and canal liners, and covers and liners for municipal solid waste landfills. Since these manufactured products are new relative to traditional compacted clay liner (CCL) systems, they have been the subject of a significant amount of research. The purpose of this chapter is to provide a concise review of recent GCL research and to highlight some factors to be considered when designing municipal solid waste landfill liner systems incorporating GCLs.

Geosynthetic clay liners are a manufactured hydraulic barrier comprising a thin layer of clay bonded to a layer or layers of geosynthetic. Bentonite, an altered volcanic ash composed predominately of smectite-group minerals such as montmorillonite with exchange sites primarily occupied by Na or Ca ions

---

* Current address: Department of Civil Engineering, Queen's University, Kingston, Ontario, Canada KT6 3N6. E-mail: kerry@civil.queensu.ca

† Current address: Jacques Whitford and Associates, Dartmouth, Nova Scotia, Canada. E-mail: clake@jacqueswhitford.com

Rowe, R.K. and Lake, C.B. (2000) Geosynthetic Clay Liners (GCLs) for municipal solid waste landfills. Pp. 395–406 in: *Environmental Mineralogy: Microbial Interactions, Anthropogenic Influences, Contaminated Land and Waste Management* (J.D. Cotter-Howells, L.S. Campbell, E. Valsami-Jones and M. Batchelder, editors). Mineralogical Society Series, **9**. Mineralogical Society, London. ISBN 0 903056 20 8.

(Grim and Güven, 1978), is used as the clay component. Typically, natural Na bentonite is used although, in some instances, 'Na-activated' Ca bentonites (i.e. where many of the exchangeable Ca cations have been replaced with Na ions to increase the swelling capability and decrease the hydraulic conductivity of the bentonite in the GCL) have been used.

Koerner (1998) summarizes the many different types of GCLs. The bentonite is either bonded to a very thin plastic membrane using adhesive, or enclosed between two geotextiles that are held together by adhesives, stitching, or needle-punching (see Fig. 19.1). The needle-punching process involves punching many small needles through the entire GCL, causing some fibres from the top needle-punched non-woven geotextile to extend through the bentonite and bottom geotextile, bonding the entire structure together (von Maubeuge and Heerten, 1994). The fibres that are punched through the bottom

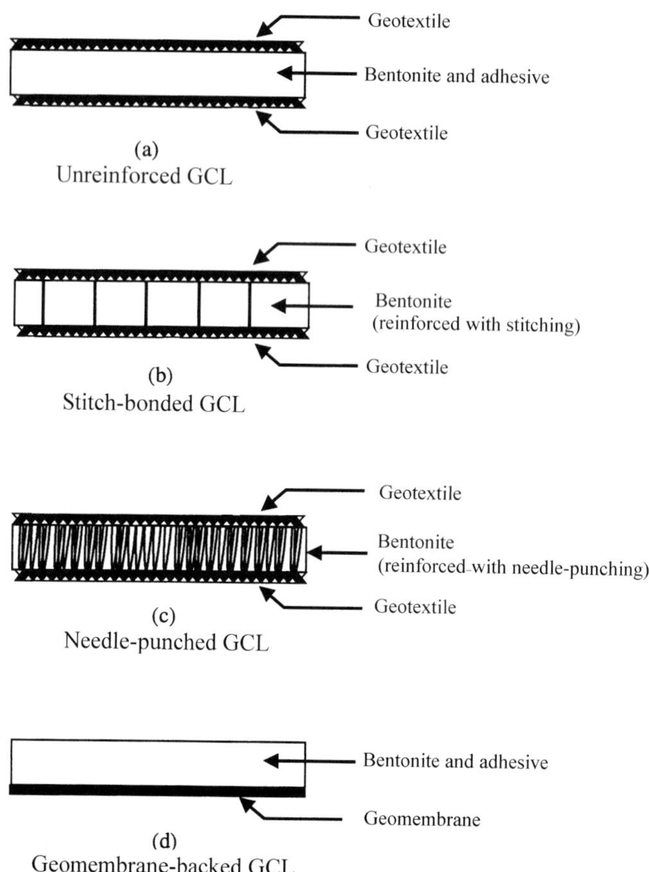


Fig. 19.1 Schematic of different types of GCLs (modified from Koerner, 1998).

geotextile rely on natural entanglement and friction to keep the GCL together or are heated (sometimes referred to as 'thermal locked', 'heat burnished', or 'thermally treated'), causing them to fuse together and/or attach to the bottom geotextile, potentially creating a stronger bond between the two geotextiles and bentonite.

Issues such as hydraulic conductivity, clay-leachate compatibility, diffusion, contaminant transport through the entire barrier system, shear strength, and stability must be considered in the engineering design phase. The manufacturing process may have an effect on the engineering characteristics of the GCL (Lake and Rowe, 2000*a*) and must be considered when comparing the engineering behaviour of different GCLs.

## 19.2 Hydraulic conductivity and leachate compatibility

The bentonite typically used in GCLs is usually primarily composed of Na montmorillonite and, when hydrated, the relatively immobile water in the diffuse double layer of the montmorillonite occupies a large portion of the clay volume (Mitchell, 1993; Rowe *et al.*, 1995*b*), resulting in a low GCL hydraulic conductivity (typical range of $5 \times 10^{-11}$ m/s to $7 \times 10^{-12}$ m/s; Rowe, 1998). However, pure montmorillonite also has a cation exchange capacity (CEC) ranging from 80 mEq/100 g to 150 mEq/100 g (Grim, 1962) and as a result, Na bentonite may experience exchange reactions when exposed to different cations found in municipal solid waste leachate. This may cause the double layers to contract, increasing the size of flow channels in the clay and hence increasing the hydraulic conductivity of the GCL. A discussion of factors affecting the size of the double layer is given by Van Olphen (1977) and Mitchell (1993).

A number of researchers (Shubert, 1987; Shan and Daniel, 1991; Ruhl and Daniel, 1997; Petrov *et al.*, 1997*a,b*; Petrov and Rowe, 1997) have examined the issue of GCL hydraulic conductivity and GCL compatibility with various permeants.

After reviewing existing hydraulic conductivity data and noticing significant scatter in the literature for different hydration conditions, different stress conditions, and different permeating fluids, Petrov *et al.* (1997*b*) and Petrov and Rowe (1997) demonstrated that there was a linear relationship between the hydraulic conductivity, k, and the final bulk GCL void ratio, $e_B$ (see Petrov *et al.*, 1997*b* for a definition). The different void ratios shown for each NaCl concentration in Fig. 19.2 were obtained by applying different stress levels to each sample either during hydration or after hydration under a low applied stress. Also plotted on Fig. 19.2 are results obtained after permeation with a synthetic municipal solid waste (MSW) leachate. Referring to Fig. 19.2, it can be seen that: (1) the hydraulic conductivity increases as the final bulk GCL void ratio increases; (2) as the NaCl concentration increases, the hydraulic conductivity increases for a particular void ratio; (3) the samples initially

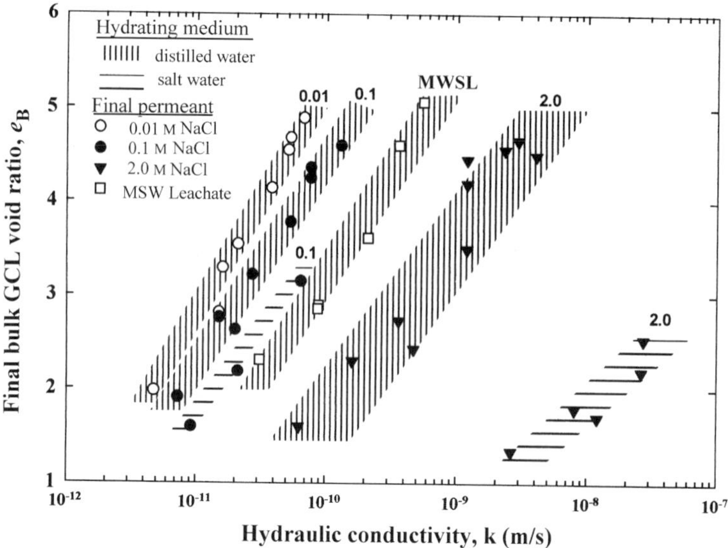


FIG. 19.2 Relationship between hydraulic conductivity, k, and final bulk GCL void ratio, $e_B$, for permeation of varying concentrations of NaCl and a MSW leachate (modified from Petrov and Rowe, 1997).

hydrated with water had lower hydraulic conductivities than those initially hydrated with the NaCl for a particular void ratio; and (4) relative to dilute 0.01 M NaCl concentrations, the hydraulic conductivity at a particular void ratio after leachate permeation increased by no more than an order of magnitude for the range of conditions examined.

The fact that interaction occurs between bentonite and leachate does not mean that they cannot be used. On the contrary, GCLs may be used provided that a hydraulic conductivity design value has been selected giving due consideration to this interaction and based on tests that have a stress history, hydrating conditions and permeating fluid which closely resemble those expected in the field. When performing these compatibility tests, it is important to monitor differences in permeant influent and effluent to ensure sufficient pore volumes of the permeant passed through the sample to reach chemical equilibrium. Results obtained prior to the sample reaching chemical equilibrium with the leachate (i.e. before the effluent concentration is essentially equal to the influent concentration) underestimate the effects of cation exchange and *c*-axis contraction (Rowe *et al.,* 1995*b*) on the hydraulic conductivity (see Rowe *et al.,* 1995*a,b*; Petrov *et al.,* 1997*a*).

Some GCLs are manufactured by bonding the clay to thin plastic membranes. These membranes may be manufactured from polymers similar to those used in geomembrane liners but are typically much thinner. In laboratory tests, those thin membranes may make the GCL essentially impermeable to fluids. However, because these plastic membranes are thin (typically <0.5 mm thick), one can not rely on the membrane being intact in the field. Thus, irrespective of whether or not a GCL has a membrane backing, a proper geomembrane (typically 1.5–2.5 mm thick, high density polyethylene) is still required to form a composite liner and only the bentonite component of GCLs with a membrane backing can be relied upon to provide hydraulic resistance in the field. Hence the bentonite component of these GCLs should be examined in the same way as other GCLs that do not have a plastic membrane backing.

## 19.3 Diffusion research

Diffusion is a chemical process by which contaminants can migrate from areas of higher concentration to areas of lower concentration even when there is no flow of water (e.g. see Rowe *et al.*, 1995*b*, for a detailed discussion). Recently, there has been a significant amount of research into the diffusive behaviour of municipal waste leachate contaminants through GCLs (Lo, 1992; Rowe *et al.*, 2000; Lake and Rowe, 2000*a*,*b*). Lo (1992) estimated diffusion coefficients for various contaminants by monitoring effluent from GCL hydraulic conductivity tests. Rowe *et al.* (1999) developed diffusion test cells that were used to establish inorganic GCL diffusion coefficients and to investigate the various factors that can affect contaminant diffusion through GCLs. It was shown that the effective porosity ($n_e$) of these ionic contaminants in the bentonite clay can be significantly lower than the total porosity ($n_t$) based on the moisture content of the sample when modelling diffusive migration through GCLs. However, it was also shown that for the single GCL examined (~10 mm thick), the ionic contaminants could be modelled, to sufficient accuracy for many practical applications, using a diffusion coefficient ($D_t$) that is deduced from laboratory data using the total porosity ($n_t$) of the bentonite in the GCL (see Rowe *et al.*, 2000 and Lake and Rowe, 2000*b*, for a more detailed explanation). Since single GCLs are generally used in field applications, it is expected that this behaviour will be exhibited when a single GCL is used as part of a composite liner. If layers of GCLs are stacked together, it may be necessary to establish both the effective porosity and its corresponding effective diffusion coefficient.

Figure 19.3 shows chloride diffusion coefficients ($D_t$) that were deduced using the total porosity of the bentonite and obtained for a GCL encased by two non-woven geotextiles. As is evident from this figure, there are a number of factors that affect the diffusion of contaminants through GCLs (Lake and Rowe, 2000*b*):

(1) The solid black circles of Fig. 19.3 represent diffusion coefficients obtained using a 0.08 M NaCl solution. The diffusion coefficient, $D_t$, obtained from these tests, increases linearly as $e_B$ increases for the range of void ratios examined. It was also shown that it is the final bulk GCL void ratio, $e_B$, that

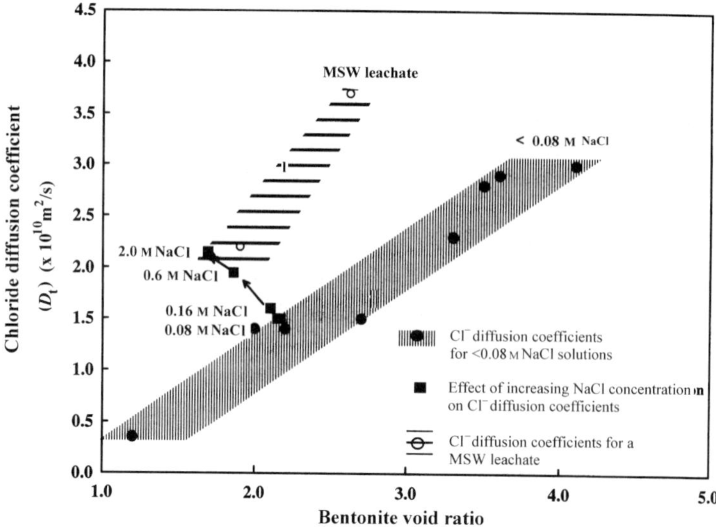


FIG. 19.3 $Cl^-$ diffusion coefficients ($D_t$) obtained using <0.08 M NaCl concentrations compared to $Cl^-$ diffusion coefficients obtained using increasing NaCl concentrations and MSW leachate.

controls diffusion coefficients for GCLs, regardless of the stress history of a GCL. For example, if a GCL was hydrated under a 10 kPa stress and subsequently consolidated under a 150 kPa stress, it is the void ratio at the end of the consolidation process that controls the diffusion coefficient. This void ratio at the end of the 150 kPa stress application is greater than would be present if the GCL was initially hydrated under the 150 kPa stress level. Thus, the engineering characteristics (such as hydraulic conductivity and diffusion coefficients) can be improved if the confining stress is applied to the GCL before hydration of the bentonite in the GCL, since this helps to minimize the void ratio of the GCL. Daniel *et al.* (1993) discuss the potential for the GCL to hydrate solely from the moisture of the underlying soil.

(2) The open circles in Fig. 19.3 are chloride diffusion coefficients obtained from tests with synthetic municipal solid waste leachate (Lake and Rowe, 2000*b*). The chloride diffusion coefficient obtained with the synthetic leachate examined is higher than that obtained with 0.08 M NaCl. In fact, the diffusion coefficient can be sensitive to concentration, as shown by the solid black squares that show the results of a diffusion test where the same GCL sample (under a stress of 150 kPa) was exposed to increasing NaCl concentrations. The diffusion coefficient of chloride is higher for the 2.0 M NaCl solution than for the 0.08 M NaCl solution. This illustrates the need to perform GCL diffusion testing with leachate representative of that expected in the field.

(3) Increasing the level of applied stress on the GCL prior to hydration and contact with leachate can: (*a*) lower the bulk void ratio of the sample after

hydration and hence lower the hydraulic conductivity and diffusion coefficient; and (*b*) lessen the negative interaction of MSW leachate in relation to increased diffusion coefficients. When comparing results for concentrations <0.08 M NaCl with those for the MSW leachate, it appears that as the stress level increases (i.e. a lower void ratio) the relative difference between the two sets of results appears to lessen. Both of these factors suggest that the higher the application of effective stress prior to contact with the leachate, the lower the diffusion coefficient, all other things being equal.

Both diffusion coefficients and hydraulic conductivities of a given GCL are required when designing a liner system involving a GCL. These parameters should be obtained under conditions as close as is practical to the expected field conditions.

## 19.4 GCL interaction with other components of the landfill system

### *19.4.1 Intimate contact of a GCL with a geomembrane in a composite liner*

Geosynthetic clay liners are often utilized in combination with geomembranes to form composite liners at the base of municipal solid waste landfills (e.g. Richardson, 1997*b*; Giroud *et al.*, 1997). Examples of some of these types of configurations are given in Fig. 19.4. The primary method of transport of ionic contaminants through composite liners is with the flow of leachate through defects (holes, tears, holes over wrinkles, etc.) in the geomembrane. Much has been written about leakage through these defects when a geomembrane is used in combination with a compacted clay liner for the composite liner system (Brown *et al.*,1987; Giroud and Bonaparte, 1989; Giroud *et al.*, 1992; Rowe, 1998). Equations have been developed empirically and analytically in an attempt to predict flow through these defects.

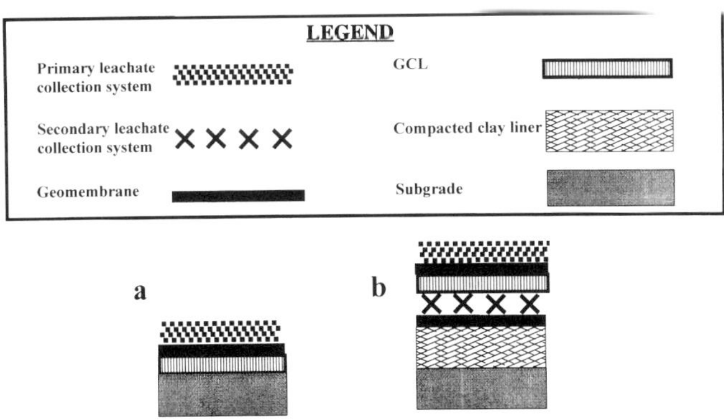


FIG. 19.4 Schematic of examples of GCLs utilized in (*a*) single and (*b*) double composite liner systems at the base of municipal solid waste landfills (simplifications have been made to aid clarity).

Wilson-Fahmy and Koerner (1995) looked specifically at geomembrane/GCL contact conditions and found that GCLs have the potential to limit leakage rates through a GM/GCL composite system to values less than would be expected for a GM/CCL system. The implications of this have been discussed in detail by Rowe (1998).

### *19.4.2 Contaminant transport*

The assessment of the suitability of a GCL liner system either alone or in combination with other geosynthetics or native soils at a proposed site should be related to the potential impact on the environment. This usually involves solving the advection-dispersion equation subject to the appropriate boundary conditions (Rowe and Booker, 1997). However, some regulatory agencies have standard designs for municipal solid waste landfills. Usually these standard designs call for a composite liner consisting of a geomembrane over a compacted clay liner (CCL) unless it can be demonstrated that another design (e.g. involving a GCL replacing a portion of, or, all of the CCL) is 'equivalent' to the standard design. 'Equivalency' comparisons of GCLs and CCLs vary in the literature (Koerner and Daniel, 1995; Rowe *et al.,* 1997*b*; Giroud *et al.,* 1997; Richardson, 1997*b*). Rowe (1998) outlines the various factors that should be included in assessing equivalency between compacted clay liners and geosynthetic clay liners in terms of contaminant migration. They are: (1) potential for clay/leachate interaction; (2) contact with the geomembrane; (3) advection/diffusion and sorption through the entire liner system, including any native soil separating the liner from the underlying groundwater; (4) service life of engineered components of the landfill including consideration of factors related to freeze-thaw prior to waste placement (Kraus *et al.,* 1997) and desiccation both before (Shan and Daniel, 1991) and after waste placement; (5) potential biodegradation of contaminants (Rowe *et al.,* 1997*a*); and (6) source concentration of the landfill and its decay characteristics.

Rowe (1998) compared two single composite liner systems: (1) a geomembrane and 0.6 m CCL over 3 m of native soil; and (2) a GM/GCL composite liner over 3.6 m of native soil and found that for all practical purposes, the two liner systems are equivalent when the factors mentioned above were considered (see Rowe, 1998 for details). The important conclusion to draw from this exercise is that many different factors must be considered when assessing equivalence of GCLs to CCLs and any comparative assessment must include equivalence in terms of minimising contaminant impact as one of the primary criteria.

Based on the present knowledge of GCLs, there is considerable potential for a GM/GCL composite liner to be used as part of a municipal solid waste landfill barrier system that provides good environmental protection.

### *19.4.3 Shear strength and slope stability*

The stability of landfill liner systems must be ensured before, during and after placement of the waste. Several failures involving geomembrane-lined landfills (Byrne *et al.*, 1992; Ouvry *et al.*, 1995) highlight the importance of this (see Rowe, 1998 for a review). The Na montmorillonite in GCLs has a very low shear strength (Mesri and Olsen, 1970) and hence the stability implications of the introduction of this potentially weak layer into a liner or cover system must be carefully considered. Not all GCLs have the same shear resistance and the manufacturing process and techniques such as stitching, needle-punching and thermal treatment of needle-punched GCLs can greatly increase the internal shear strength of the GCL.

Factors that need to be assessed when considering the stability of liner (or cover) systems involving a GCL include the potential for: (1) internal failure of the GCL (in the bentonite or at the interface between the bentonite and geosynthetics in the GCL); and (2) failure at one of the interfaces in contact with the GCL (GCL in contact with a geomembrane, another geosynthetic or soil). Evaluation of these two potential failure mechanisms requires geotechnical laboratory test data and stability analyses.

Generally, unreinforced GCLs have a low internal shear strength similar to that of montmorillonite and considerable caution is required when considering their use on slopes (Richardson, 1997*a*). Reinforced GCLs (stitched, needle-punched) have produced varying laboratory results depending on the type of GCL tested and the test method employed (Heerten *et al.*, 1995; Gilbert *et al.*, 1996; Stark and Eid, 1996; Fox *et al.*, 1998; Von Maubeuge and Eberle, 1998). The internal shear strength of one particular type of GCL can not be used for another GCL due to differences in the manufacturing processes. Some needle-punched GCLs have quite high peak strengths. However, if this value is to be used in design, care is required to either establish that large deformations cannot occur or will be dealt with by other mechanisms. It is believed that pullout of the reinforcement after peak shear stress results in residual values approaching residual values of unreinforced GCLs ($\tau_{residual}/\tau_{peak} \approx 0.05$ to 0.5: Fox *et al.*, 1998) at large displacements.

To prevent the residual or post peak GCL internal strength from being mobilized, liner systems involving GCLs in combination with other geosynthetics (geomembrane, geotextiles) are often designed such that interface strengths of one of the various components of the liner system is less than the GCL's peak internal shear strength but is greater than the residual strength of the GCL (Gilbert *et al.*, 1996). This is possible for low normal stresses, but as the normal stress increases, the failure plane may move internally to the GCL (Gilbert *et al.*, 1996).

It has been found that some low interface strengths in the laboratory result from bentonite extruding through woven geotextiles and non-woven geotextiles with a mass per unit area less than 220 g/m$^2$ of the GCL and being deposited on the geomembrane, reducing the frictional resistance along this interface (e.g. see Gilbert *et al.,* 1996). Techniques such as using a GCL with the non-woven geotextile with a mass per unit area equal to or greater than 250 g/m$^2$ in contact with textured geomembranes can help to reduce this problem.

Daniel *et al.* (1998) reported the findings from a field evaluation of the internal and interface shear strength behaviour of many different types of GCLs configured with other liner components such as geomembranes, geotextiles and native soils. All geosynthetic configurations on slopes inclined at 3(horizontal):1(vertical) were reported to be performing satisfactorily. Three slides have occurred on the steeper 2:1 slope. In one case, an unreinforced bentonite between two geomembranes became hydrated causing a slide to occur. The other two slides occurred 20 and 50 days after construction and were reported to have been caused by bentonite extruding from the GCL through a woven geotextile in contact with a textured geomembrane, thereby reducing the strength of the interface. Data from this investigation should also eventually provide valuable long-term data on the potential creep of the reinforcement of the GCLs and how it affects stability with time.

In summary, when designing GCL-lined slopes it is important to recognize the differences between different types of GCLs and consequently differences in interface and internal shear strengths. It is also important to consider the implications of low post-peak internal shear strength of the GCL and the need to prevent bentonite extrusion through the geotextiles in the GCL that decreases the frictional resistance of the interface between GCLs and other geosynthetics.

## 19.5 Conclusion

Geosynthetic clay liners have considerable potential for use either as an alternative to a conventional compacted clay liner or to augment the performance of a compacted clay liner as part of a composite liner system in the base of municipal solid waste landfills. Assessment of equivalency involves selecting hydraulic conductivity and diffusion parameters that reflect the likely field conditions, including the effect of potential interaction with leachate. Contaminant impact calculations should then be performed to demonstrate that the alternate liner system involving a GCL will provide the same (or better) environmental protection as a conventional design. As with the design of all liner systems, particular attention should be paid to ensuring adequate stability of the liner and landfill.

## References

Brown, K.W., Thomas, J.C., Lytton, R.L., Jayawickrama, P. and Bahrt, S. (1987) *Quantification of leakage rates through holes in landfill liners.* US EPA Report CR810940, Cincinnati.

Byrne, R.J., Kendall, J. and Brown, S. (1992) Cause and mechanism of failure of Kettlemans

Hills Landfill. Proceedings of ASCE Conference on Stability and Performance of Slopes and Embankments II, pp. 1–23.

Daniel, D.E., Shan, H.Y. and Anderson, J.D. (1993) Effects of partial wetting on the performance of the bentonite component of a geosynthetic clay liner. *Proc. Geosynthetics '93*, Vancouver, B.C., Canada: 1483–96. IFAI, St. Paul, MN, USA.

Daniel, D.E., Koerner, R.M., Bonaparte, R., Landreth, R.E., Carson, D.A. and Scranton, H.B. (1998) Slope stability of geosynthetic clay liner test plots. *J. Geotech. Geoenv. Eng.*, **124**, 628–37.

Fox, P.J., Rowland, M.G. and Scheithe, J.R. (1998) Internal shear strength of three geosynthetic clay liners. *J. Geotech. Geoenv. Eng.*, **124**, 933–44.

Gilbert, R.B., Fernandez, F. and Horsfield, D.W. (1996) Shear strength of reinforced geosynthetic clay liner. *J. Geotech. Geoenv. Eng.*, **122**, 259–66.

Giroud, J.P. and Bonaparte, R. (1989) Leakage through liners constructed with geomembranes – Part II. Composite liners. *Geotextiles Geomembranes*, **8**, 71–111.

Giroud, J.P., Badu-Tweneboah, K. and Bonaparte, R. (1992) Rate of leakage through a composite liner due to geomembrane defects. *Geotextiles Geomembranes*, **11**, 1–12.

Giroud, J.P., Badu-Tweneboah, K. and Soderman, K.L. (1997) Comparison of leachate flow through compacted clay and geosynthetic clay liners in landfill liner systems. *Geosynthetics Int.*, **4**, 391–431.

Grim, R.E. (1962) *Clay Mineralogy*. McGraw Hill, New York.

Grim, R.E. and Güven, N. (1978) Bentonites – geology, mineralogy, properties and uses. *Devel. Sedimentol.*, **24**.

Heerten, G., Saathoff, F., Scheu, C. and von Maubeuge, K.P. (1995) On the long-term shear behavior of geosynthetic clay liners (GCLs) in capping sealing systems. *Proc. Int. Symposium "Geosynthetic Clay Liners", Nurenberg*, 141–50.

Koerner, R.M. (1998) *Designing with Geosynthetics* 4th edition. Prentice-Hall, N.J.

Koerner, R.M. and Daniel, D.E. (1995) A suggested method for assessing the technical equivalency of GCLs to CCLs. *Proc. Int. Symposium "Geosynthetic Clay Liners", Nurenberg*, 175–81.

Kraus, J.F., Benson, C.H., Erickson, A.E. and Chamberlain, E.J. (1997) Freeze thaw cycling and hydraulic conductivity of bentonitic barriers. *J. Geotechnical Geoenv. Eng.*, **123**, 229–38.

Lake, C.B. and Rowe, R.K. (2000*a*) Swelling of needlepunched, thermally treated GCLs. *Geotextiles Geomembranes*, **18**, 77–101.

Lake, C.B. and Rowe, R.K. (2000*b*) Diffusion of sodium and chloride through geosynthetic clay liners. *Geotextiles Geomembranes*, **18**, 103–31.

Lo, M.C. (1992) *Development and evaluation of clay-liner materials for hazardous waste sites*, PhD thesis, Univ. Texas, Austin.

Mesri, G. and Olsen, R.E. (1970) Shear strength of montmorillonite. *Geotechnique*, **20**, 261–70.

Mitchell, J.K. (1993) *Fundamentals of Soil Behavior*, 2nd edition. John Wiley & Sons, New York.

Ouvry, J.F., Gisbert, T. and Closset, L. (1995) Back analysis of a slide in a waste storage centre. *Recontres '95*, 148–52.

Petrov, R.J. and Rowe, R.K. (1997) GCL – chemical compatibility by hydraulic conductivity testing and factors impacting its performance. *Canad. Geotech. J.*, **34**, 863–85.

Petrov, R.J., Rowe, R.K. and Quigley, R.M. (1997*a*) Comparison of laboratory measured GCL hydraulic conductivity based on three permeameter types. *Geotech. Testing J.*, **20**, 49–62.

Petrov, R.J., Rowe, R.K. and Quigley, R.M. (1997*b*) Selected factors influencing GCL hydraulic conductivity. *J. Geotech. Geoenv. Eng.*, **123**, 683–95.

Richardson, G.N. (1997*a*) GCL internal shear strength requirements. *Geotechnical Fabrics Report*, March, 20–5.

Richardson, G.N. (1997*b*) GCLs: Alternative subtitle D liner systems. *Geotech. Fabrics Report*, May, 36–42.

Rowe, R.K. (1998) Geosynthetics and the minimization of contaminant migration through barrier systems beneath solid waste. *Proc. 6th Int. Conf. on Geosynthetics, Atlanta*, **1**, 27–102.

Rowe, R.K. and Booker, J.R. (1997) *POLLUTE v.6.3 – 1-D pollutant migration through a non homogeneous soil* ©1983, 1990, 1994, 1997. Distributed by GAEA Environmental Engineering Ltd.

Rowe, R.K., Hrapovic, L. and Kosaric, N. (1995*a*) Diffusion of chloride and dichloromethane through an HDPE geomembrane, *Geosynthetics Int.*, **2**, 507–36.

Rowe, R.K., Quigley, R.M. and Booker, J.R. (1995*b*) *Clayey Barrier Systems for Waste Disposal Facilities*. E&FN Spon, London.

Rowe, R.K., Hrapovic, L., Kosaric, N. and Cullimore, D. (1997*a*) Anaerobic degradation of dichloromethane diffusing through clay, *J. Geotech. Geoenv. Eng.*, **123**, 1085–95.

Rowe, R.K., Lake, C.B., von Maubeuge, K. and Stewart, D. (1997*b*) Implications of diffusion of chloride through geosynthetic clay liners. *Proc. Geoenvironment '97, Melbourne*, 295–300.

Rowe, R.K., Lake, C.B. and Petrov, R.J. (2000) Apparatus and procedures for assessing inorganic diffusion coefficients through geosynthetic clay liners. *Geotech. Testing J.*, **23**, 206–14.

Ruhl, J.L. and Daniel, D.E. (1997) Geosynthetic clay liners permeated with chemical solutions and leachates. *J. Geotech. Geoenv. Eng.*, **123**, 369–81.

Schubert, W.R. (1987) Bentonite matting in composite lining systems. Pp. 784–96 in: *Geotechnical Practice for Waste Disposal '87* (R.D. Woods, editor). American Society of Chemical Engineers, New York.

Shan, H.S. and Daniel, D.E. (1991) Results of laboratory tests on geotextile/bentonite liner material. *Proc. Geosynthetics '91 Conf., Atlanta, USA*, 517–32.

Stark, T.D. and Eid, H.T. (1996) Shear behavior of reinforced geosynthetic clay liners. *Geosynthetics Int.*, **3**, 771–86.

Van Olphen, H. (1977) *An Introduction to Clay Colloid Chemistry*. Wiley Interscience Publishers, New York.

von Maubeuge, K.P. and Eberle, M.A. (1998) Can geosynthetic clay liners be used on slopes to achieve long-term stability? *3rd Int. Cong.: Environmental Geotechnics, Lisboa, Portugal*, 375–80.

von Maubeuge, K.P. and Heerten, G. (1994) Needle-punched geosynthetic clay liners (GCLs). *Proc. 8th GRI conference 'Geosynthetic resins, formulations and manufacturing', Philadelphia, PA, USA*, 199–207.

Wilson-Fahmy, R.F. and Koerner, R.M. (1995) Leakage rates through holes in geomembranes overlying geosynthetic clay liners. *Proc. Geosynthetics '95, Industrial Fabrics Assoc. Int.*, 655–68.

# INDEX


Abiotic weathering, and erosion of rock surfaces, 83
*Acarosporion sinopicae*, lichen colonizing spoil heaps at Mynydd Parys Cu-Pb-Zn mines, Wales, 171
Acid mine/rock drainage, 111
  in Mynydd Parys Cu-Pb-Zn mines, Anglesey, Wales, 173
  rates of sulphide oxidation in relation to, 118
Acid potential (AP) values in acid rock drainage (ARD), 143
Acid uranium leaching, 272
Acid, production by fungi, and implications for metal speciation, 57
Alkali-aggregate reactivity (AAR), 113
Alkali-silica reactivity (ASR), 113
AMD, see acid mine drainage
Analytical transmission electron microscopy (ATEM), in the identification of Pb phosphates, 293
Anglesite, from the Northern Pennine and Yorkshire Dales lead-zinc-fluorite-baryte orefield in northeast England, 212
Animals, *in vivo* uses of clinoptilolite to remove fission products from cattle and sheep, 347
Ankerite, from the Northern Pennine and Yorkshire Dales lead-zinc-fluorite-baryte orefield in northeast England, 212
Anthropogenic
  activities, decay effects associated with soluble salts on granite buildings, 181
  influences on mineral interactions, 109
Apatite, 293
Aphthitalite, chemical attack on building materials by, 183
Aqueous uranium chemistry, 250
ARD, see acid rock drainage
Arsenopyrite, as a source of ARD, 134
*Aspergillus*
  *niger*, fungus with the ability to dissolve metal phosphates, 297
  strains of, used to leach Ni, Co and Mn from low-grade laterite ores, 68
Atomic absorption spectrometry (AAS), of soluble salts from decaying granite, 185
Atomic force microscope (AFM), imaging bacteria using, 44
Auger spectroscopic analyses,
  for uranium complex characterization, 262
  in the investigation of Mn oxides, 209
  to determine a series of oxidation reactions, 132
'Azulejos', formation of trona efflorescences on panels of azulejos (tiles), 193

Bacteria
  biological leaching utilizes chemoautotrophic bacteria, 68
  catalysing oxidation reactions, 126
  cyanobacteria fulfilling the photosynthetic role in lichen, 79
  *Ferromicrobium acidophilus*, iron oxidizing bacteria, 170
  involved in acid genesis and metal solubilization in the waters and 'soils' at Mynydd Parys mine, Anglesey, Wales, 170
  iron oxidizing bacteria, 170
  *Leptospirillum ferrooxidans*, in the process of sulphide oxidation, 122
  mineral dissolution by, 27
  sulphide oxidation,
  *Sulfobacillus thermosulfidooxidans*, in the process of sulphide oxidation, 122
  *Thiobacillus ferrooxidans* (see entry for *T. ferrooxidans*)
  transport of uranium by, 263
  treatment of organic contaminants in soils, 292
  weathering of layer silicates in the rhizosphere of plants (root environment) as influenced by, 7

Bacterial
action, in ARD generation, 120
bacterially-induced mineral decomposition, 111
biology, 29
control on biologically-induced mineralization of magnetic iron minerals, 1
Ball Clay, gas entry experiments on, 378
Baryte, from the Northern Pennine and Yorkshire Dales lead-zinc-fluorite-baryte orefield in northeast England, 212
Bassanite, chemical attack on building materials by, 183
Bentonite
gas injection experiments on, 370
in geosynthetic clay liners (GCLs) for municipal solid waste landfills, 395
Biochemilithic zone, lichen-rock interactions in, 90
Biogeochemistry of metals, 64
Bioleaching
experiments, 17
of uraninite ores, 273
Biological process, and minerals, 1
Biology
of lichens, 79
of Mynydd Parys Cu-Pb-Zn mines, Anglesey, Wales, 170
Bioprecipitation of elements, 12
Bioremediation of solid wastes, 68
Blauton (Blue Clay), gas entry experiments on, 378
Brewsterite, see zeolite
Buildings
corrosion, role of organic acids in, 66
decay effects associated with soluble salts on buildings of granite, 181

Cadmium
in fluids associated with ARD, 119
solubility of cadmium oxalate, 61
strongly associated with Mn oxides, 210
Caesium, soil containing clinoptilolite taking up, 344
Calcite, from the Northern Pennine and Yorkshire Dales lead-zinc-fluorite-baryte orefield in northeast England, 212
Calcium oxalate, the most common metal oxalate in the terrestrial environment, 59
Carbonates
as environmental minerals, 202
in waste management, 315
to neutralize the acidity created by oxidizing iron sulphides, 152
Carbonic acid, as a weathering agent, 91
Cation exchange capacity (CEC) of zeolite, 324
Cation hydrolysis of metal cations, 230
Cd, see cadmium
Cerussite, from the Northern Pennine and Yorkshire Dales lead-zinc-fluorite-baryte orefield in northeast England, 212
Chabazite, see zeolite
Chalcopyrite
as a source of ARD, 131
from the Northern Pennine and Yorkshire Dales lead-zinc-fluorite-baryte orefield in northeast England, 211
rate of alteration of, 144
Chemical attack, by soluble salts on building materials, 182
Chemolithotrophic microorganisms, 7
Chemoorganotrophic microorganisms, 7
Chernobyl, Russia, amelioration of nuclear fall-out from, 343
Chromium, capacity of montmorillonite to sorb hydrolysed Cr, 236
Classification of uranium deposits, 269
Clay minerals (see also entries for bentonite, kaolinite, etc.)
as biochemical weathering products, 96
expanding layer silicates, intercalation of organic contaminants by, 227
sorption of uranyl ions on, 260
use in waste management, 314
Clay pastes, gas entry into unconfined clay pastes, 369
Clinoptilolite, see zeolite
Coffinite, present in some uranium deposits, 247
Colloids, surface-sorbed uranium transported by, 245
Compacted clay liners (CCLs), superseded by geosynthetic clay liners (GCLs) for municipal solid waste landfills, 395

Confocal laser scanning microscope, imaging bacteria using, 44
Contaminants, diffusion of, through geosynthetic clay liners (GCLs) of municipal solid waste, 399
Contaminated
  environments, minerals in, 201
  land, metal phosphates and remediation of, 291
Controlled pressure scanning electron microscopy (CP-SEM), of the lichen-rock surface interface, 81
Copper, solubility of copper oxalate, 61
*Coriolus versicolor*, fungus with the ability to dissolve metal phosphates, 297
Cr, see chromium
Cu, see copper
Cu-Pb-Zn mines, mineralogy of, from Mynydd Parys, Anglesey, Wales, 161
Cyanobacterial zone, organic-mineral interactions taking place in, 85
Cyclical transformations of salt, producing repeated crystallization and hydration pressures, 182

Decay effects associated with soluble salts on granite buildings, 181
Diffusion of contaminants through geosynthetic clay liners (GCLs) of municipal solid waste, 399
Dissolution of minerals, 77; by heterotrophic bacteria, 27; controlled by the mineral surface, 28

Earth, uranium distribution in, 245
Efflorescences, found on stone, mortar and tiles in decaying granite buildings, 185
Enargite, rate of alteration of, 144
Energy dispersive X-ray spectrometry (EDX), in the characterization of heavy metal-bearing Mn oxides in river channel sediments, 213
Energy generation by bacteria, 32
England, Tyne and Tees river sediments, heavy metal-bearing Mn oxides in, 211
Environment
  mineral-environment interactions, 109
  uranium behaviour in, 245
Environmental
  impact of ARD, 123
  minerals, e.g. carbonates, 201
  significance, of Mn oxides, in controlling the distribution of heavy metals in river channel sediment, 221
Epsomite, chemical attack on building materials by, 183
Erionite, see zeolite
*Euglena mutabilis*, from Mynydd Parys Cu-Pb-Zn mines, 171
Extended X-ray absorption fine structure spectroscopy (EXAFS)
  for uranium complex characterization, 262
  in the identification of Pb phosphates, 293

Faujasite, see zeolite
Ferrierite, see zeolite
*Ferromicrobium acidophilus*, iron oxidizing bacteria, 170
Ferrihydrite, adsorption of uranyl ions on, 261
Floodplain sediments, heavy metal-bearing Mn oxides in, 211
Fluorite, from the Northern Pennine and Yorkshire Dales lead-zinc-fluorite-baryte orefield in northeast England, 212
Fungi
  *Aspergillus niger*, with the ability to dissolve metal phosphates, 297
  *Coriolus versicolor*, with the ability to dissolve metal phosphates, 297
  effect on weathering rate of minerals, 78
  in surface waters at Mynnyd Parys mine, Anglesey, Wales, 172
  interaction between minerals and biological organisms, 2
  *Paxillus involutus*, weathering micas, 65
  *Penicillium bilaii*, with the ability to dissolve metal phosphates, 297
  *Pisolinthus tinctorius*, weathering micas, 65
  production of organic acids by, 57
  *Rhizopus arrhizus*, fungus with the ability to dissolve metal phosphates, 297

solubilization of metal-bearing minerals by, 57
strains of *Aspergillus*, used to leach Ni, Co and Mn from low-grade laterite ores, 68
strains of *Penicillium*, used to leach Ni, Co and Mn from low-grade laterite ores, 68
weathering of layer silicates in the rhizosphere of plants as influenced by, 7

Galena
as a source of ARD, 129
capable of reducing U(VI) under anoxic conditions, 266
from the Northern Pennine and Yorkshire Dales lead-zinc-fluorite-baryte orefield in northeast England, 211
rate of alteration of, 144
Gas injection experiments in clays, 369
Gaseous emissions, from nuclear facilities, treatment of, 352
Gault Clay, gas entry experiments on, 378
Geochemical engineering, waste management applications of minerals, 315
Geology, of Mynydd Parys Cu-Pb-Zn mines, Anglesey, Wales, 163
Geosynthetic clay liners (GCLs) for municipal solid waste landfills, 395
Gibbsite, adsorption of uranyl ions on, 245
Gismondine, see zeolite
Goethite
adsorption of uranyl ions on, 261
dissolution of, by Fe-reducing bacteria, 20
Granite buildings, decay effects associated with soluble salts on, 181
Groundwater, leaching uranium from volcanic rocks, 246
Gypsum, chemical attack on building materials by, 183

Halite, chemical attack on building materials by, 183
Harmotome, see zeolite
Hawaiian lavas, lichen-mediated weathering of, 97
Heavy metal-bearing Mn oxides in river channel sediments, 207
Hematite, adsorption of uranyl ions on, 261
Heterotrophs, role in mineral dissolution, 33
Heulandite, see zeolite
Hexahydrite, chemical attack on building materials by, 183
Hg, see mercury
Hydraulic conductivity of bentonite from geosynthetic clay liners (GCLs) for municipal solid waste, 397
Hydrochemistry, of Mynydd Parys Cu-Pb-Zn mines, Anglesey, Wales, 166
Hydroxypyromorphite, stable lead mineral, 293

Imaging bacteria, methods of, 42
Infrared (IR) spectroscopy, evidence for inner-sphere complexation of Cr, 236
Inorganic contaminants, intercalation of, by expanding layer silicates, 227
Intercalation of organic contaminants, 230, 239
Ion chromatography, of soluble salts from decaying granite, 186
Ion exchange properties of minerals used in waste management, 313
Ion exchangers, zeolites as, 322

Kaolinite (St. Austell, England), gas entry experiments on, 377

Landfill sites, geosynthetic clay liners (GCLs) for municipal solid waste, 395
Layer silicates, intercalation of organic contaminants by, 228
Layered crystal structure, in environmental mineralogy, 201
Lead
as a decay product of uranium, 248
from the Mynydd Parys mine, Anglesey, Wales, 161
in the fluid associated with ARD, 119
siderophores binding, 35
solubility of lead oxalate, 59
strongly associated with Mn oxide, 210
*Lecanora*
*Atra*, lichen producing Mg oxalate, 67

*epanora* lichen, from Mynydd Parys Cu-Pb-Zn mines, 171
*Lecidea* aff. *sarcogynoides*, lichen, rates of penetration into sandstone, 89
Lichen
*Acarosporion sinopicae*, colonizing spoil heaps at Mynydd Parys Cu-Pb-Zn mines, Wales, 171
copper oxalate growing in lichen on Cu-rich rocks, 64
*Euglena mutabilis*, from Mynydd Parys Cu-Pb-Zn mines, 171
*Lecanora atra*, producing Mg oxalate, 67
*Lecanora epanora*, from Mynydd Parys Cu-Pb-Zn mines, 171
*Lecidea* aff. *sarcogynoides*, lichen, rates of penetration into sandstone, 89
*Pertusaria corallina*, producing calcium oxalate, 67
*Porpidia albocaerulescens*, amphibole syenite substrate inhabited by, 84
*Psilolechia leprosa*, from Mynydd Parys Cu-Pb-Zn mines, 171
*Rhizocarpon furfurosum*, from Mynydd Parys Cu-Pb-Zn mines, 172
*Rhizocarpon geographicum*, BSEM image of thallus of (lichen), 80
*Rhizocarpon grande*, amphibole syenite substrate inhabited by lichen, 84
*Trapelia placodioides*, 82
uranium transported in groundwater by, 263
weathering of rocks by, 77, 84
Limestone biomineralization, role of fungal oxalate in, 67
London Clay, gas entry experiments on, 377

Malachite, from the Northern Pennine and Yorkshire Dales lead-zinc-fluorite-baryte orefield in northeast England, 212
Man, *in vivo* uses of mordenite to reduce the body burden of man fed with radiocaesium-contaminated reindeer meat, 347
Marcasite, as a source of ARD, 124
Mercury, in fluids associated with ARD, 119
Metabolic oxidation (direct and indirect), 120
Metal phosphates and remediation of contaminated land, 291
Metal-bearing minerals, solubilization of, by fungi, 57
Mica, weathering of, 13
Microbe
mineral-microbe interactions, 1, 7, 28
metal and anionic nutrition of, 69
Microbiology, of Mynydd Parys Cu-Pb-Zn mines, Anglesey, Wales, 171
Microorganisms, 1, 7
as agents of element cycling, 27
concentrating metals from dilute aqueous solutions, 263
dissolving metal phosphates, 297
impact of, on weathering rates, 78
in the solubilization of inorganic elements, 7
interaction between rock-forming minerals and, 2
oxalic acid produced by, 59
prime agents of element cycling, 27
which utilize sulphur species for energy, 112
Mineral
deposits formation, 7
interactions, anthropogenic influences on, 109
-microbe interactions, 1, 7, 28
-pollutant interactions, 201
Mineralogy
of Mynydd Parys Cu-Pb-Zn mines, Anglesey, Wales, 161
of the lichen-rock interface, 81
Mining
for uranium, 272
history of Mynydd Parys Cu-Pb-Zn mines, Anglesey, Wales, 163
Mirabilite, chemical attack on building materials by, 183
Mn oxides, heavy metal-bearing, 208
Molybdenum (Mo), strongly associated with Mn oxide, 210
Montmorillonite
expanding layer silicate, intercalation of organic contaminants by, 228
high capacity to sorb hydrolysed Cr, 236
in landfills, low shear strength, 402

Mordenite, see zeolite
Mynydd Parys Cu-Pb-Zn mines, Anglesey, Wales, Wales, 161

Natrolite, see zeolite
Neutralization potential (NP) values in acid rock drainage (ARD), 145
Niter, chemical attack on building materials by, 183
Nitratine, chemical attack on building materials by, 183
Non-ionic molecules as organic contaminants, 240
Non-polar molecules as organic contaminants, 240
Nuclear
  facilities, treatment of wastes from, 338
  reactors, uranium ions produced in, 247
  wastes, zeolites in the treatment of, 319

Ochre, oxidation of ferrous iron in solution, 119
Organic
  acid biosynthesis, 58
  contaminants, intercalation of, by expanding layer silicates, 227
  matter, transport of uranium complexed with, 263
  polymers, alteration of mineral surfaces by, 88
Oxalic acid, agent of biochemical weathering, 91
Oxyhydroxides
  adsorption of uranyl ions on, 261
  in waste management, 313

Particulate environments, soil and subsurface soils, 35
*Paxillus involutus*, fungus weathering micas, 65
Pb, see lead
Pearson correlation coefficients for Mn, 221
*Penicillium*
  *bilaii*, fungus with the ability to dissolve metal phosphates, 297
  strains of, used to leach Ni, Co and Mn from low-grade laterite ores, 68
*Pertusaria corallina*, lichen producing calcium oxalate, 67
Perovskite, resistance to alteration of, 316
Phillipsite, see zeolite
Phosphates, metal phosphates and remediation of contaminated land, 291
*Pisolinthus tinctorius*, weathering micas, 65
Pitchblende in high-$T$ hydrothermal vein deposits, 247
Plankton, transport of uranium by, 263
*Porpidia albocaerulescens*, amphibole syenite substrate inhabited by lichen, 84
Portugal, decay effects associated with soluble salts on granite buildings, 181
*Psilolechia leprosa*, from Mynydd Parys Cu-Pb-Zn mines, 171
Pyrite oxidation, at Mynydd Parys mines, Anglesey, Wales, 165
Pyrite
  as a source of ARD, 117, 125
  capable of reducing U(VI) under anoxic conditions, 266
  dissolution of, by *Thiobacillus ferrooxidans*, 45
  exposure of, 109
  from the Northern Pennine and Yorkshire Dales lead-zinc-fluorite-baryte orefield in northeast England, 212
  interactions between *T. ferrooxidans* and pyrite during bioleaching, 16
  mineral-microbe interactions, 1
  rate of alteration of, 144
Pyromorphite, 292
Pyrrhotite
  as a source of ARD, 128
  rate of alteration of, 144

Radiation exposure, 249
Radioactive waste, zeolites in the treatment of, 319
Radiochemical stability of the aluminosilicate framework of zeolites, 326
Rates of weathering, determined by rate of accumulation of calcium oxalate, 98
Redox conditions, dominating uranium transport in crustal fluids, 245
Remediation
  of contaminated land, 291
  processes, 201

*Rhizocarpon*
*furfurosum*, from Mynydd Parys Cu-Pb-Zn mines, 171
*geographicum*, BSEM image of thallus of (lichen), 80
*grande*, amphibole syenite substrate inhabited by lichen, 84
*Rhizopus arrhizus*, fungus with the ability to dissolve metal phosphates, 297
Rhizosphere
interaction phenomena between micro-organisms-roots and mineral soil constituents, 13
lichen-mineral interface as a model for, 79
oxalate concentrations in the, 65
River sediments, heavy-metal-bearing Mn oxides in, 209
Rock-forming minerals in ecosystems, 2

Salts, decay effects associated with soluble salts on granite buildings, 181
Scanning electron microscopy (SEM)
imaging bacteria using, 44
in the characterization of heavy metal-bearing Mn oxides in river channel sediments, 213
of soluble salts from decaying granite, 185
of the lichen-rock surface interface, 81
Scanning force microscopy (SFM), in the characterization of heavy metal-bearing Mn oxides in river channel sediments, 213
Scolecite, see zeolite
Sedimentary deposits of uranium, 270
Siderite, from the Northern Pennine and Yorkshire Dales lead-zinc-fluorite-baryte orefield in northeast England, 212
Silica gel, adsorption of uranyl ions on, 261
Sinks of uranium, 266
Sites of Special Scientific Interest (SSSI), Mynydd Parys Cu-Pb-Zn mines, Anglesey, Wales, 176
Slope stability of landfills after placement of waste, 402
Smithsonite, from the Northern Pennine and Yorkshire Dales lead-zinc-fluorite-baryte orefield in northeast England, 212
Soil containing clinoptilolite, uptake of Cs by, 344
Soils, remediation of metal-contaminated soils, 291
Solubility of metal phosphates, 294
Solubilization of metal-bearing minerals by fungi, 57
Sphalerite
as a source of ARD, 131
from the Northern Pennine and Yorkshire Dales lead-zinc-fluorite-baryte orefield in northeast England, 211
rate of alteration of, 144
Sr, see strontium
Static tests, used to assess the acid-generating potential of rocks, 142
Stilbite, see zeolite
Strontium
phosphate minerals and the retention of Sr, 276
soil containing clinoptilolite taking up, 344
solubility of strontium oxalate, 61
*Sulfobacillus thermosulfidooxidans*, bacteria, in the process of sulphide oxidation, 122
Sulphide
minerals, chemical reactivity of, 123
oxidation, acid rock drainage due to, 111; rates of, 117; secondary minerals forming as a result of, 152
resistance and persistence, rate of alteration of sulphide minerals in mine-waste settings, 144
Surface
mineralogy and geochemistry, of Mynydd Parys Cu-Pb-Zn mines, 165
pyrite surface oxidation, 121
Synroc, resistance to alteration of minerals incorporated in, 316

Thenardite, chemical attack on building materials by, 183
*Thiobacillus ferrooxidans*
dissolution of pyrite and others sulphides by, 45
formation of ARD promoted by the activity of, 120

from Mynydd Parys Cu-Pb-Zn mines, 172
in the formation of ferric sulphate and sulphuric acid, 273
interactions with sulphide minerals, 16
*Thiobacillus thiooxidans*
iron oxidizing bacteria, 170
oxidizing elemental sulphur and sulphide to sulphuric acid, 120
Thomsonite, see zeolite
Three Mile Island, Pennsylvania, USA, amelioration of nuclear fall-out from, 343
Time resolved laser-induced fluorescence, for uranium complex characterization, 262
Toxicology of uranium, 248
Transmission electron microscopy (TEM), imaging bacteria using, 44
*Trapelia placodioides*, crustose lichen, 82
Trona, chemical attack on building materials by, 183

U, see uranium
Unconformity deposits of uranium, 270
Uraninite in high-*T* hydrothermal vein deposits, 247
Uranium behaviour in natural environments, 245

Vermiculite, characterized by XRD, 14

Wairakite, see zeolite
Wales, Anglesey, mineralogy of Mynydd Parys Cu-Pb-Zn mines, 161
Waste
carbonate minerals in waste management, 315
clay minerals in waste management, 314
containment uses of natural zeolites, 348
management, how minerals are used, 313
Weathering
of rocks by lichens, 77
of silicate minerals, 1
rates of minerals, 146
West Valley, New York, USA, amelioration of nuclear fall-out from, 343
Wyoming bentonite, gas injection experiments on, 370

X-ray analysis, of thin sections of the lichen-rock surface interface, 81
X-ray diffraction (XRD)
in the characterization of heavy metal-bearing Mn oxides in river channel sediments, 213
in the identification of Pb phosphates, 293
of hydroxy-Cr polymers, 237
of minerals from plant roots, 13
of soluble salts from decaying granite, 185
patterns of lead phosphates, 291
used to characterize vermiculite, 14
X-ray photoelectron spectroscopy (XPS)
of surface reactions on pyrrhotite, 129
proof that pyrite is capable of reducing U(IV) under anoxic conditions, 266

Yucca Mountain, Nevada, USA, clinoptilolite in bedrock surrounding a repository for high-level nuclear waste, 351
Yugawaralite, see zeolite

Zeolites
brewsterite, 323
chabazite, 323
clinoptilolite, 327
erionite, 323
faujasite, 323
ferrierite, 323
gismondine, 323
harmotome, 323
heulandite, 323
in the treatment of nuclear waste and fall-out, 319
in waste management, 314
mordenite, 323
natrolite, 323
phillipsite, 323
scolecite, 323
stilbite, 323
thomsonite, 323
wairakite, 323
yugawaralite, 323
Zinc
concentration on basalt, caused by lichen, 97
strongly associated with Mn oxides, 210
Zirconolite, resistance to alteration of, 316
Zn, see zinc